THE ILLUSTRATED ENCYCLOPEDIA OF COUNTRY LIVING

THE ILLUSTRATED ENCYCLOPEDIA OF COUNTRY LIVING

Abigail R. Gehring

illustrations by
James Balkovek

Skyhorse Publishing

Skyhorse Publishing books may be purchased in bulk at special discounts for sales promotion, corporate gifts, fund-raising, or educational purposes. Special editions can also be created to specifications. For details, contact the Special Sales Department, Skyhorse Publishing, 307 West 36th Street, 11th Floor, New York, NY 10018 or info@skyhorsepublishing.com.

Skyhorse® and Skyhorse Publishing® are registered trademarks of Skyhorse Publishing, Inc.®, a Delaware corporation.

www.skyhorsepublishing.com

10 9 8 7 6 5

Library of Congress Cataloging-in-Publication Data

Gehring, Abigail R.
  The illustrated encyclopedia of country living / Abigail R. Gehring.
    p. cm.
  ISBN 978-1-61608-467-7 (pbk.)
  1. Agriculture—Encyclopedias. 2. Family farms—Encyclopedias. 3. Sustainable living—Encyclopedias. 4. Home economics, Rural—Encyclopedias. I. Title.
  S501.2.G445 2011
  640—dc23
                          2011016189

Printed in China

CONTENTS

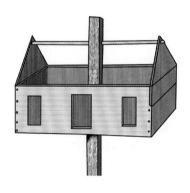

INTRODUCTION

To create an encyclopedia is a daunting task. I began this book with grandiose plans to include everything there is to know about living off the land—if anyone had ever planted it, raised it, built it, baked it, or crafted it, it would be in this book. And here it is! Okay, maybe not *quite*. The truth is, every time I flip through these pages I think of some other project, or some useful bit of information I would love to add. But I trust that I've compiled a resource that is comprehensive enough to teach even the most seasoned farmer a trick or two, and user-friendly enough to inspire the greenest green thumbs among my readers.

There was some lively debate leading up to publication regarding whether an encyclopedia must be in alphabetical order. Strictly speaking (according to the all-knowing Merriam Webster), it does not. Though Merriam states that encyclopedias are *usually* arranged alphabetically, the criteria for fitting a book into the category is simply "a work that contains information on all branches of knowledge or treats comprehensively a particular branch of knowledge . . ." When a colleague stopped by my desk, glanced at my nearly-finished work, and casually suggested, "Shouldn't it be in alphabetical order?" I got worried. My instinct was to keep the book arranged in a more chronological way—first you plan your garden, then you buy the seeds, then you till the soil, then plant, weed, and water, and finally you harvest.

My prior books have all been set up in this way, and to me it made perfect sense. This was my position when I began asking others for their opinions. "That's fine," was the general response "but you can't call it an 'encyclopedia.'" In the end, I hastily rearranged the entire book to begin with the letter "A." And I'm so glad I did.

A book of this size would be terribly difficult to navigate without the order imposed by the alphabet. As it stands now, you know that animals are at the start of the book, and that within that section goats will come before horses, and rabbits before sheep. In the Crafts section, baskets come well before candles, and soap making is easily found between sewing and spinning wool. You can thank my meddlesome friends and colleagues.

As you begin using this book, I hope you find exactly what you're looking for, whether it's instructions for companion planting techniques or just the chance to dream about a simpler way of living. When the book falls short of its aim to be all-inclusive, be sure to check out the extensive resource section at the end.

If you're like most homesteaders I know, you're eager to get started, so without further ado, I present *The Illustrated Encyclopedia of Country Living*. Here's to your adventures in living off the land!

—Abigail R. Gehring

PART ONE

ANIMALS

BEES

Beekeeping (also known as apiculture) is one of the oldest human industries. For thousands of years, honey has been considered a highly desirable food. Beekeeping is a science and can be a very profitable employment; it is also a wonderful hobby for many people in the United States. Keeping bees can be done almost anywhere—on a farm, in a rural or suburban area, and even, at times, in urban areas (even on rooftops!). Anywhere there are sufficient flowers from which to collect nectar, bees can thrive.

Apiculture relies heavily on the natural resources of a particular location and the knowledge of the beekeeper in order to be successful. Collecting and selling honey at your local farmers' market or just to family and friends can supply you with some extra cash if you are looking to make a profit from your apiary.

Why Raise Bees?

Bees are essential in the pollination and fertilization of many fruit and seed crops. If you have a garden with many flowers or fruit plants, having bees nearby will only help your garden flourish and grow year after year. Furthermore,

nothing is more satisfying than extracting your own honey for everyday use.

How to Avoid Getting Stung

Though it takes some skill, you can learn how to avoid being stung by the bees you keep. Here are some ways you can keep your bee stings to a minimum:

1. Keep gentle bees. Having bees that, by sheer nature, are not as aggressive will reduce the number of stings you are likely to receive. Carniolan bees are one of the gentlest species, and so are the Caucasian bees introduced from Russia.
2. Obtain a good "smoker" and use it whenever you'll be handling your bees. Pumping smoke of any kind into and around the beehive will render your bees less aggressive and less likely to sting you.
3. Purchase and wear a veil. This should be made out of black bobbinet and worn over your face. Also, rubber gloves help protect your hands from stings.

4. Use a "bee escape." This device is fitted into a slot made in a board the same size as the top of the hive. Slip the board into the hive before you open it to extract the honey, and it allows the worker bees to slip below it but not to return back up. So, by placing the "bee escape" into the hive the day before you want to gain access to the combs and honey, you will most likely trap all the bees under the board and leave you free to work with the honeycombs without fear of stings.

What Type of Hive Should I Build?

Most beekeepers would agree that the best hives have suspended, moveable frames where the bees make the honeycombs, which are easy to lift out. These frames, called Langstroth frames, are the most popular kind of frame used by apiculturists in the United States.

Whether you build your own beehive or purchase one, it should be built strongly and should contain accurate bee spaces and a close-fitting, rainproof roof. If you are looking to have honeycombs, you must have a hive that permits the insertion of up to eight combs.

Where Should the Hive Be Situated?

Hives and their stands should be placed in an enclosure where the bees will not be disturbed by other animals or humans and where it will be generally quiet. Hives should be placed on their own stands at least 3 feet from each other. Do not allow weeds to grow near the hives and keep the hives away from walls and fences. You, as the beekeeper, want to be able to easily access your hive without fear of obstacles.

Swarming

Swarming is simply the migration of honeybees to a new hive and is led by the queen bee. During swarming season (the warm summer days), a beekeeper must remain very alert.

If you see swarming above the hive, take great care and act calmly and quietly. You want to get the swarm into your hive, but this will be tricky. If they land on a nearby branch or in a basket, simply approach and then "pour" them into the hive. Keep in mind that bees will more likely inhabit a cool, shaded hive than one that is baking in the hot summer sun.

Sometimes it is beneficial to try to prevent swarming, such as if you already have completely full hives. Removing the new honey frequently from the hive before swarming begins will deter the bees from swarming. Shading the hives on warm days will also help keep the bees from swarming.

Bee Pastures

Bees will fly a great distance to gather food but you should try to contain them, as well as possible, to an area within 2 miles of the beehive. Make sure they have access to many honey-producing plants, which you can grow in your garden. Alfalfa, asparagus, buckwheat, chestnut, clover, catnip, mustard, raspberry, roses, and sunflowers are some of the best honey-producing plants and trees. Also make sure that your bees always have access to pure, clean water.

Preparing Your Bees for Winter

If you live in a colder region of the United States, keeping your bees alive throughout the winter months is difficult. If your queen bee happens to die in the fall, before a young queen can be reared, your whole colony will die throughout the winter. However, the queen's death can be avoided by taking simple precautions and giving careful attention to your hive come autumn.

Colonies are usually lost in the winter months due to insufficient winter food storages, faulty hive construction, lack of protection from the cold and dampness, not enough or too much ventilation, or too many older bees and not enough young ones.

If you live in a region that gets a few weeks of severe weather, you may want to move your colony indoors, or at least to an area that is protected from the outside elements.

knives specifically for this purpose). Then put the combs in a machine called a honey extractor to extract the honey. The honey extractor whips the honey out of the cells and allows you to replace the fairly undamaged comb into the hive to be repaired and refilled.

The extracted honey runs into open buckets or vats and is left, covered with a tea towel or larger cloth, to stand for a week. It should be in a warm, dry room where no ants can reach it. Skim the honey each day until it is perfectly clear. Then you can put it into cans, jars, or bottles for selling or for your own personal use.

Making Beeswax

Beeswax from the honeycomb can be used for making candles (see page 434), can be added to lotions or lip balm, and can even be used in baking. Rendering wax in boiling water is especially simple when you only have a small apiary.

Collect the combs, break them into chunks, roll them into balls if you like, and put them in a muslin bag. Put the bag with the beeswax into a large stockpot and bring the water to a slow boil, making sure the bag doesn't rest on the bottom of the pot and burn. The muslin will act as a strainer for the wax. Use clean, sterilized tongs to occasionally squeeze the bag. After the wax is boiled out of the bag, remove the pot from the heat and allow it to cool. Then, remove the wax from the top of the water and then re-melt it in another pot on very low heat, so it doesn't burn.

Pour the melted wax into molds lined with wax paper or plastic wrap and then cool it before using it to make other items or selling it at your local farmers' market.

Extra Beekeeping Tips

General Tips

1. Clip the old queen's wings and go through the hives every 10 days to destroy queen cells to prevent swarming.
2. Always act and move calmly and quietly when handling bees.
3. Keep the hives cool and shaded. Bees won't enter a hot hive.

But the essential components of having a colony survive through the winter season are to have a good queen; a fair ratio of healthy, young, and old bees; and a plentiful supply of food. The hive needs to retain a liberal supply of ripened honey and a thick syrup made from white cane sugar (you should feed this to your bees early enough so they have time to take the syrup and seal it over before winter).

To make this syrup, dissolve 3 pounds of granulated sugar in 1 quart of boiling water and add 1 pound of pure extracted honey to this. If you live in an extremely cold area, you may need up to 30 pounds of this syrup, depending on how many bees and hives you have. You can either use a top feeder or a frame feeder, which fits inside the hive in the place of a frame. Fill the frame with the syrup and place sticks or grass in it to keep the bees from drowning.

Extracting Honey

To obtain the extracted honey, you'll need to keep the honeycombs in one area of the hive or packed one above the other. Before removing the filled combs, you should allow the bees ample time to ripen and cap the honey. To uncap the comb cells, simply use a sharp knife (apiary suppliers sell

When Opening the Hive

1. Have a smoker ready to use if you desire.
2. Do not stand in front of the hive while the bees are entering and exiting.
3. Do not drop any tools into the hive while it's open.
4. Do not run if you become frightened.
5. If you are attacked, move away slowly and smoke the bees off yourself as you retreat.
6. Apply ammonia or a paste of baking soda and water immediately to any bee sting to relieve the pain. You can also scrape the area of the bee sting with your fingernail or the dull edge of a knife immediately after the sting.

When Feeding Your Bees

1. Keep a close watch over your bees during the entire season, to see if they are feeding well or not.
2. Feed the bees during the evening.
3. Make sure the bees have ample water near their hive, especially in the spring.

Making a Beehive

The most important parts of constructing a beehive are to make it simple and sturdy. Just a plain box with a few frames and a couple of other loose parts will make a successful beehive that will be easy to use and manipulate. It is crucial that your beehive be well adapted to the nature of bees and also the climate where you live. Framed hives usually suffice for the beginning beekeeper. Below is a diagram of a simple beehive that you can easily construct for your backyard beekeeping purposes.

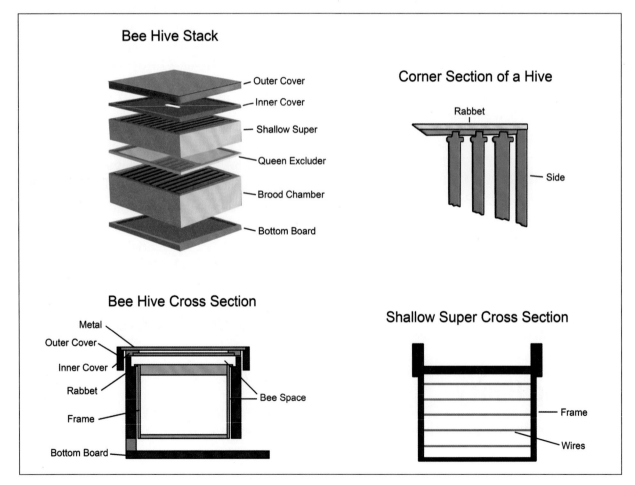

ILLUSTRATIONS BY TIMOTHY W. LAWRENCE

CHICKENS

Raising chickens in your yard will give you access to fresh eggs and meat, and since chickens are some of the easiest creatures to keep, even families in very urban areas are able to raise a few in a small backyard. Four or five chickens will supply your whole family with eggs on a regular basis.

Housing Your Chickens

You will need to have a structure for your chickens to live in—to protect them from predators and inclement weather, and to allow the hens a safe place to lay their eggs. See "Poultry Houses" on page 378 to see several types of structures you can make for housing your chickens and other poultry.

Placing your henhouse close to your home will make it more convenient to feed the chickens and to gather eggs. Es-

tablish the house and yard in dry soil and to stay away from areas in your yard that are frequently damp or moist, as this is the perfect breeding ground for poultry diseases. The henhouse should be well-ventilated, warm, protected from the cold and rain, have a few windows that allow the sunlight to shine in (especially if you live in a colder climate), and have a sound roof.

The perches in your henhouse should not be more than 2 ½ feet above the floor, and you should place a smooth platform under the perches to catch the droppings so they can easily be cleaned. Nesting boxes should be kept in a darker part of the house and should have ample space around them.

A simple movable chicken coop can be constructed out of two-by-fours and two wheels. The floor of the coop should have open slats so that the manure will fall onto the ground and fertilize the soil. An even simpler method is to

construct a pen that sits directly on the ground, making sure that it has a roof to offer the chickens suitable shade. The pen can be moved once the area is well fertilized.

Selecting the Right Breed of Chicken

Take the time to select chickens that are well suited for your needs. If you want chickens solely for their eggs, look for chickens that are good egg-layers. Mediterranean poultry are good for first-time chicken owners as they are easy to care for and only need the proper food in order to lay many eggs. If you are looking to slaughter and eat your chickens, you will want to have heavy-bodied fowl (Asiatic poultry) in order to get the most meat from them. If you are looking to have chickens that lay a good amount of eggs and that can also be used for meat, invest in the Wyandottes or Plymouth Rock breeds. These chickens are not incredibly bulky but they are good sources of both eggs and meat.

Wyandottes have seven distinct breeds: Silver, White, Buff, Golden, and Black are the most common. These breeds are hardy and they are very popular in the United States. They are compactly built and lay excellent dark brown eggs. They are good sitters and their meat is perfect for broiling or roasting.

Plymouth Rock chickens have three distinct breeds: Barred, White, and Buff. They are the most popular breeds in the United States and are hardy birds that grow to a medium size. These chickens are good for laying eggs, roost well, and also provide good meat.

Black Wyandotte

White Wyandotte

Buff Wyandotte

Silver Wyandotte

Golden Wyandotte

White Rock

Silver Rock

Buff Rock

Feeding Your Chickens

Chickens, like most creatures, need a balanced diet of protein, carbohydrates, vitamins, fats, minerals, and water. Chickens with plenty of access to grassy areas will find most of what they need on their own. However, if you don't have the space to allow your chickens to roam free, commercial chicken feed is readily available in the form of mash, crumbles, pellets, or scratch. Or you can make your own feed out of a combination of grains, seeds, meat scraps or protein-rich legumes, and a gritty substance such as bone meal, limestone, oyster shell, or granite (to aid digestion, especially in winter). The correct ratio of food for a warm, secure chicken should be 1 part protein to 4 parts carbohydrates. Do not rely too heavily on corn as it can be too fattening for hens; combine corn with wheat or oats for the carbohydrate portion of the feed. Clover and other green foods are also beneficial to feed your chickens.

How much food your chickens need will depend on breed, age, the season, and how much room they have to exercise. Often it's easiest and best for the chickens to leave feed

available at all times in several locations within the chickens' range. This will ensure that even the lowest chickens in the pecking order get the feed they need.

Hatching Chicks

To hatch a chick, an egg must be incubated for a sufficient amount of time with the proper heat, moisture, and position. The period for incubation varies based on the species of chicken. The average incubation period is around 21 days for most common breeds.

If you are only housing a few chickens in your backyard, natural incubation is the easiest method. Natural incubation is dependent upon the instinct of the mother hen and the breed of hen. Plymouth Rocks and Wyandottes are good hens to raise chicks. It is important to separate the setting hen from the other chickens while she is nesting and to also keep the hen clean and free from lice. The nest should also be kept clean and the hens should be fed grain food, grit, and clean, fresh water.

When considering hatching chicks, make sure your hens are healthy, have plenty of exercise, and are fed a balanced

Chicken Feed

4 parts corn (or more in cold months)

3 parts oat groats

2 parts wheat

2 parts alfalfa meal or chopped hay

1 part meat scraps, fish meal, or soybean meal

2 to 3 parts dried split peas, lentils, or soybean meal

2 to 3 parts bone meal, crushed oyster shell, granite grit, or limestone

½ part cod-liver oil

You may also wish to add sunflower seeds, hulled barley, millet, kamut, amaranth seeds, quinoa, sesame seeds, flax seeds, or kelp granules. If you find that your eggs are thin-shelled, try adding more calcium to the feed (in the form of limestone or oystershell). Store feed in a covered bucket, barrel, or other container that will not allow rodents to get into it. A plastic or galvanized bucket is good, as it will also keep mold-causing moisture out of the feed.

Chickens don't need a lot of space, but they will be healthiest if they have a green area where they can forage.

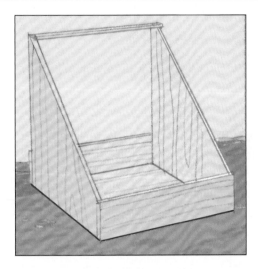

You can use scrap wood to construct a simple nesting box.

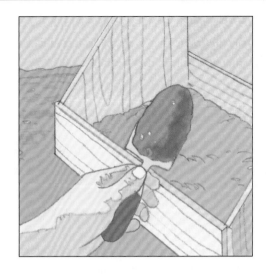

Add damp soil to the bottom of your nesting box.

Cover soil with a layer of hay or straw.

Place the nesting box in a quiet and secluded place away from the other chickens. If space permits, you can construct a smaller shed in which to house your nesting hen. A hen can generally sit on anywhere between 9 and 15 eggs. The hen should only be allowed to leave the nest to feed, drink water, and take a dust bath. When the hen does leave her box, check the eggs and dispose of any damaged ones. An older hen will generally be more careful and apt to roost than a younger female.

Once the chicks are hatched, they will need to stay warm and clean, have lots of exercise, and have access to food regularly. Make sure the feed is ground finely enough that the chicks can easily eat and digest it. They should also have clean, fresh water.

diet. They need materials on which to scratch and should not be infested with lice and other parasites. Free range chickens, which eat primarily natural foods and get lots of exercise, lay more fertile eggs than do tightly confined hens. The eggs selected for hatching should not be more than 12 days old and they should be clean.

You'll need to construct a nesting box for the roosting hen and the incubated eggs. The box should be roomy and deep enough to retain the nesting material. Treat the box with a disinfectant before use to keep out lice, mice, and other creatures that could infect the hen or the eggs. Make the nest of damp soil a few inches deep placed in the bottom of the box, and then lay sweet hay or clean straw on top of that.

Bacteria Associated with Chicken Meat
- Salmonella—This is primarily found in the intestinal tract of poultry and can be found in raw meat and eggs.
- *Campylobacter jejuni*—This is one of the most common causes of diarrheal illness in humans, and is spread by improper handling of raw chicken meat and not cooking the meat thoroughly.
- *Listeria monocytogenes*—This causes illness in humans and can be destroyed by keeping the meat refrigerated and by cooking it thoroughly.

Storing Eggs

Eggs are among the most nutritious foods on earth and can be part of a healthy diet. Hens typically lay eggs every 25 hours, so you can be sure to have a fresh supply on a daily basis, in many cases. But eggs, like any other animal byproduct, need to be handled safely and carefully to avoid rotting and spreading disease. Here are a few tips on how to best preserve your farm-fresh eggs:

1. Make sure your eggs come from hens that have not been running with male roosters. Infertile eggs last longer than those that have been fertilized.
2. Keep the fresh eggs together.
3. Rinse eggs thoroughly before storing.
4. Make sure not to crack the shells, as this will taint the taste and make the egg rot much quicker.
5. Place your eggs directly in the refrigerator where they will keep for several weeks.

COWS

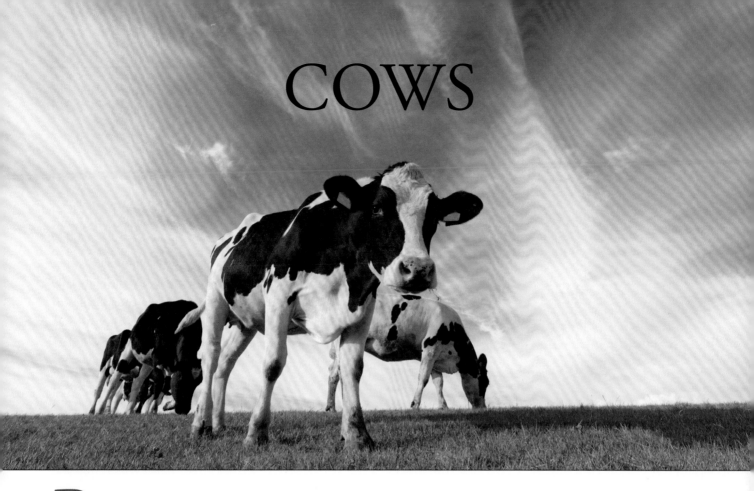

Raising dairy cows is difficult work. It takes time, energy, resources, and dedication. There are many monthly expenses for feeds, medicines, vaccinations, and labor. However, when managed properly, a small dairy farmer can reap huge benefits, like extra cash and the pleasure of having fresh milk available daily.

Breeds

There are thousands of different breeds of cows, but what follows are the three most popular breeds of dairy cows.

The Holstein cow has roots tracing back to European migrant tribes almost two thousand years ago. Today, the breed is widely popular in the United States for their exceptional milk production. They are large animals, typically marked with spots of jet black and pure white.

The Ayrshire breed takes its name from the county of Ayr in Scotland. Throughout the early 19th century, Scottish breeders carefully crossbred strains of cattle to develop a cow well suited to the climate of Ayr and with a large flat udder best suited for the production of Scottish butter and cheese. The uneven terrain and the erratic climate of

their native land explain their ability to adapt to all types of surroundings and conditions. Ayrshire cows are not only strong and resilient, but their trim, well rounded outline, and red and predominantly white color has made them eas-

Holstein

Ayshire

ily recognized as one of the most beautiful of the dairy cattle breeds.

The Jersey breed is one of the oldest breeds, originating from Jersey of the Channel Islands. Jersey cows are known for their ring of fine hair around the nostrils and their milk rich in butterfat. Averaging to a total body weight of around

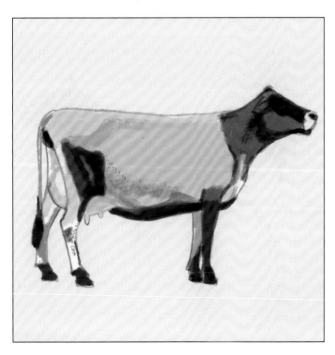

Jersey

900 pounds, the Jersey cow produces the most pounds of milk per pound of body weight of all other breeds.

Housing

There are many factors to consider when choosing housing for your cattle, including budget, preference, breed, and circumstance.

Free stall barns provide a clean, dry, comfortable resting area and easy access to food and water. If designed properly, the cows are not restrained and are free to enter, lie down, rise, and leave the barn whenever they desire. They are usually built with concrete walkways and raised stalls with steel dividing bars. The floor of the stalls may be covered with various materials, ideally a sanitary inorganic material such as sand.

A flat barn is another popular alternative, which requires tie-chains or stanchions to keep the cows in their stalls. However, it creates a need for cows to be routinely released into an open area for exercise. It is also very important that the stalls are designed to fit the physical characteristics of the cows. For example, the characteristically shorter Jersey cows should not be housed in a stall designed for much larger Holsteins.

A compost-bedded pack barn, generally known as a compost dairy barn, allows cows to move freely, promising increased cow comfort. Though it requires exhaustive pack and ventilation management, it can notably reduce manure storage costs.

Grooming

Cows with sore feet and legs can often lead to losses from milk production, diminished breeding efficiency, and lameness. Hoof trimming is essential to help prevent these outcomes, though it is often very labor intensive, allowing it to be easily neglected. Hoof trimming should be supervised or taught by a veterinarian until you get the hang of it.

A simple electric clipper will keep your cows well-groomed and clean. Mechanical cow brushes are another option. These brushes can be installed in a free-stall dairy barn, allowing cows to groom themselves using a rotating brush that activates when rubbed against.

Feeding and Watering

In the summer months, cows can receive most of their nutrition from grazing, assuming there is plenty of pasture. You may need to rotate areas of pasture so that the grass has an opportunity to grow back before the cows are let loose in that area again. Grazing pastures should include higher protein grasses, such as alfalfa, clover, or lespedeza. During the winter, cows should be fed hay. Plan to offer the cows two to three pounds of high-quality hay per 100 pounds of body weight per day. This should provide adequate nutrition for the cows to produce 10 quarts of milk per day, during peak production months. To increase production, supplement feed with ground corn, oats, barley, and wheat bran. Proper mixes are available from feed stores. Allowing cows access to a salt block will also help to increase milk production.

Water availability and quantity is crucial to health and productivity. Water intake varies, however it is important that cows are given the opportunity to consume a large amount of clean water several times a day. Generally, cows consume 30 to 50 percent of their daily water intake within an hour of milking. Water quality can also be an issue. Some of the most common water quality problems affecting livestock are high concentrations of minerals and bacterial contamination. One to two quarts of water from the source should be sent to be tested by a laboratory recommended by your veterinarian.

If you intend to run an organic dairy, cows must receive feed that was grown without the use of pesticides, commercial fertilizers, or genetically-modified ingredients along with other restrictions.

Breeding

You may want to keep one healthy bull for breeding. Check the bull for STDs, scrotum circumference, and sperm count before breeding season begins. The best cows for breeding have large pelvises and are in general good health. An alternative is to use the artificial insemination (A.I.) method. There are many advantages to A.I., including the prevention of spreading infectious genital diseases, the early detection of infertile bulls, elimination of the danger of handling unruly bulls, and the availably of bulls of high genetic material. The disadvantage is that implementing a thorough breeding program is difficult and requires a large investment of time and resources. In order to successfully execute an A.I. program, you may need a veterinarian's assistance in determining when your cows are in heat. Cows only remain fertile for 12 hours after the onset of heat, and outside factors such as temperature, sore feet, or tie-stall or stanchion housing can drastically hinder heat detection.

Calf Rearing

The baby calf will be born approximately 280 days after insemination. Keep an eye on the cow once labor begins, but try not to disturb the mother. If labor is unusually long (more than a few hours), call a veterinarian to help. It is also crucial that the newborns begin to suckle soon after birth to receive ample colostrum, the mother's first milk, after giving birth. Colostrum is high in fat and protein with antibodies that help strengthen the immune system, though it is not suitable for human consumption. You'll need to continue milking the mother cow, even though the calf will be nursing, as she's likely to produce much more milk than the calf can consume. After the first 4 days or so, the milk is fine for people to drink.

If you want to maximize the milk you get from the mother, you can separate the calf after it's nursed for at least 4 or 5 days. Teach the calf to drink from a bucket by gently pulling its head toward the pail. A calf should consume about 1 quart of milk for every 20 pounds of body weight. A calf starter can be used to help ensure proper ruminal development. You can find many types of starters on the market, each meeting the nutritional requirements for calves. Calves are usually weaned from milk at about 4 or 5 months, by gradually introducing them to hay and grain and giving them opportunity to graze. Allowing a calf to nurse from the mother longer will likely raise a larger cow, but will mean less milk for you.

Calf vaccination is also very important. You should consult your veterinarian to design a vaccination program that best fits your calves' needs.

Cows need to be milked every 12 hours. It's best to do it early in the morning, before the cow goes to pasture, and at the end of the day. If you skip a milking or are late, the cow will be in pain. Before milking, clean your bucket with warm, soapy water and wash your hands while you're at it. Wash her udders and teats with warm water and a soft cloth. Let her air dry for one minute, and then grasp the teat with your thumb and forefinger near the top of the udder. A gentle upward pressure will help the teat fill with milk. Then gently slide your hand down the teat, closing your fingers around it. The first two or three squirts should just be aimed at the ground, as they contain bacteria. The rest (of course), goes into the bucket. Continue until the udder is completely empty. To strain the milk, filter it through a kitchen strainer lined with several layers of cheesecloth. Be sure to chill the milk to 40 degrees within an hour and keep it in the coldest part of your fridge.

Common Diseases

Pinkeye and foot rot are two of the most prevalent conditions affecting all breeds of cattle of all ages year-round. Though both diseases are non-fatal, they should be taken seriously and treated by a qualified veterinarian.

Wooden tongue occurs worldwide, generally appearing in areas where there is a copper deficiency or the cattle graze on land with rough grass or weeds. It affects the tongue, causing it to become hard and swollen so that eating is painful for the animal. Surgical intervention is often required.

Brucellosis or bangs is the most common cause of abortion in cattle. The milk produced by an infected cow can also contain the bacteria, posing a threat to the health of humans. It is advisable to vaccinate your calves, to prevent exorbitant costs in the long run, should your herd contract the disease.

Mastitis will make itself known by the milk, which will develop a lumpy or flaky consistency. It can occur from forcing your cow to dry up too quickly. For your cow to have a calf every year, you have to dry her up after 10 months of milking. To do this, milk her a little less every day, and give her a little less grain or other supplements. Milk her once a day for several days, without emptying the udder completely, then once every few days, and then once a week. Her milk production will slow as you milk her less and less, but never leave her full of milk so long that her udders become painful.

If you suspect your cow has mastitis, feel her udders for swelling or hardened lumps, and look to see if they are red. The milk may become yellow or brown. Never drink the milk from a cow with mastitis, nor feed it to any animal, as it contains the infection. Get the cow antibiotics as quickly as possible.

DOGS

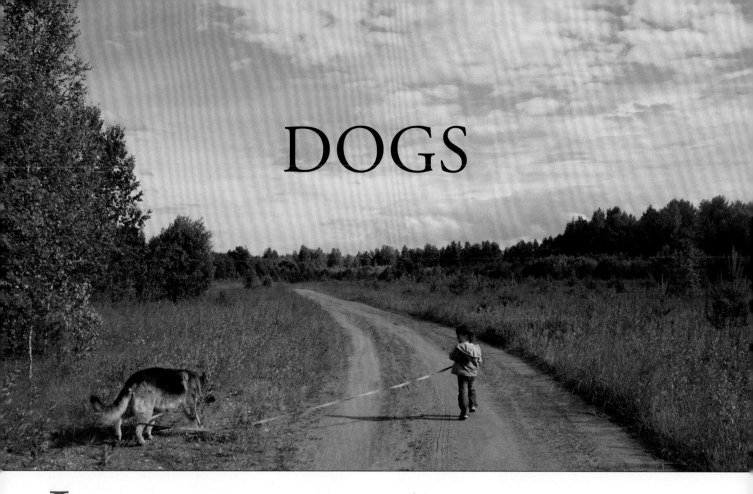

Loveable, loyal, and eager to please, dogs are a welcome addition to any family. The first few months with your new dog, regardless of its age—an adopted middle-aged dog may be as poorly housebroken as a puppy—can be a trying time for your patience and your carpet. Housetraining an animal is difficult, whether you're a recent college graduate looking to share an apartment with a furry friend or whether your history with animals is long and varied.

Start a relationship with your vet within the first 48 hours of bringing your dog home. Doing this can ensure that your dog doesn't have any physical conditions that may prevent it from being trained properly. You should also have the dog food ready; a high quality brand of dry (kibble) food is recommended. Keep your dog on a consistent diet by refraining from offering table scraps or changing the food brand. Not only do available table scraps encourage later bad habits, but scraps, wavering food brands, and wet food itself can give your dog diarrhea or loose stool resulting in accidents that are even more difficult to clean up. If you ever do need to change your dog's diet for any reason, do so gradually, over a span of 4-7 days.

Housetraining

The three things you need to remember about housetraining are:

Constant supervision is absolutely necessary.

Concentrate on *prevention* rather than punishment.

Be patient!

If you can keep these in mind, housetraining will not only be quicker, but it will be a better experience for both you and your dog.

As soon as you bring your dog home, take it outside. Let it get acquainted with the yard, the exercise area, the route you may walk most often. Wait until the dog "goes" before taking it back inside, and once that happens, make sure you praise your dog. The next few days you may be offering a ridiculous amount of praise, but you should understand that dogs want to please. If you can make them understand what you want from them, what makes you happy, they'll be more receptive through praise than punishment.

However, be realistic; your dog will most likely have an accident in the house, which will undoubtedly frustrate you.

You can't punish the dog after the damage is done, though. Dogs live moment-to-moment; they assume that whatever they're doing at the exact second they're punished is what made you upset. Catching your dog in the act of piddling on the floor and showing your disapproval will make more sense to the dog, but this is dependent on constant supervision. Watch for signs of your dog needing to relieve itself, like turning in circles, sniffing the floor, or abruptly running out of sight. Older dogs may be able to control their bowels and bladder much better, but puppies aren't able to control the same muscles until they're about 12 weeks old. Puppies need to pee around six times a day, so by keeping close watch, you can avoid those six accidents.

Once you catch your dog in the act, your tone of voice is plenty to get the point across that you're not pleased. You can say whatever you want, as long as you keep it consistent every time. Promptly bring your dog outside to the appropriate piddling place and wait for it to "go" again. If and when it does, bestow praise on the dog.

It's smart to confine your dog to a certain part of the house (perhaps the rooms without carpets) or to a crate until it is housebroken. This is especially good if you work all day. Dogs should not be stuck inside their crates all day, however; that dog is likely to become neurotic, unhappy, destructive, and noisy.

Your dog's crate should be big enough for the puppy's bed, but no larger. This is because dogs dislike the idea of soiling their bed, so using a crate that is essentially their bed, teaches them to control the urge to "go." You should keep the puppy in the crate when you aren't with him, but do not use it as punishment. It should be a safe space for your dog. Also, remember that if you keep your dog in a crate, that time must be balanced with a sufficient amount of exercise and companionship.

Define a bathroom schedule for your dog: it should be the first thing you do in the morning, the last thing you do every night, and what you do after every meal and play session. When you go for a walk, try not to return home (or inside) with your dog until it has gone to the bathroom. Pay attention to your dog's behavior and where it goes to the bathroom and share the information with any other person engaging in the housebreaking process. Using a word, like "outside," and visiting the same spot helps cement the idea that there is a certain place and time to "go."

These frequent trips outside will continue until you reach the five month mark. Once there, you can start to trust that your dog understands where to go and is able to "hold it" until you are there to make it possible.

If you've brought home a small, toy breed that will be inside most of the time, you can opt to paper train it. Start with some sort of dog pen or confined space and place your dog's food and water at one end with layered newspaper at the opposite. Not only do dogs dislike soiling their bed, but they feel similarly about relieving themselves near their sustenance. Once your dog has eaten, lead it towards the newspaper, wait until it "goes," and make sure to praise him. Afterwards, replace the newspaper with new sheets, but leave the bottom layer. This leaves their smell behind and reminds them of where to "go" next time. Make sure to remain consistent with the placement of the newspaper so as to not confuse your dog.

Because there will undoubtedly be accidents, neutralize the odor—dogs like to return to places they have "gone" before—with a pet odor neutralizer.

With patience, a keen eye, and praise and prevention in favor of punishment, housetraining your dog will become a simpler, shorter process.

Grooming

Make sure that you keep your dog's fur clean by brushing the coat a few times a week (more frequently for long-haired dogs), or whenever it looks dirty. Every time you brush your dog check it for any injury or illness; notice the eyes, ears, teeth, and nose. How frequently you should bathe your dog depends on what type of hair it has, but generally a good bath every few weeks will keep your pup clean and healthy. Keeping nails trimmed is essential—a dog's nails are constantly growing and usually do not wear away on their own. Untrimmed nails can lead to infection or injury. You can purchase dog nail clippers at any pet supply store. Be careful when clipping your dog's nails—they still have living tissue in the upper part of the nails, and clipping them too short can hurt your dog. If you feel unable to trim your dog's nails yourself, a vet or groomer can do it for you.

Dogs love to play outdoors, which means they'll be regularly exposure to bugs and parasites. Be sure to keep your dog protected from fleas, ticks, and other parasites. Unwanted pests can bite and spread disease to your dog, and be carried into your house. There are many different flea/pest control medications available; just ask your vet which one is best for your dog.

Socializing

Dogs are social creatures and need continual stimulation and interaction to be happy. You can "socialize" your dog by taking it on walks, having it encounter other dogs and people, and playing with your dog. Encourage your dog to explore its environment. Socialization is especially important for puppies, as they are still learning about their environment, and socialization will lead to a better behaved and well-adjusted dog. A bored or unhappy dog can be a destructive dog.

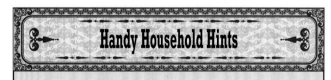

Handy Household Hints

Natural Flea Repellants

Coconut Oil: Add a tablespoon or two to your dog's food.

Apple Cider Vinegar: Put a few tablespoons in your dog's food or water, or rinse your dog in diluted vinegar.

Mix essential oils of eucalyptus, citronella, tea tree oil, geranium, and/or lavender and rub into your dog's collar or a bandana.

DUCKS

ucks tend to be somewhat more difficult than chicks to raise, but they do provide wonderful eggs and meat. Ducks tend to have pleasanter personalities than chickens and are often prolific layers. The eggs taste similar to chicken eggs, but are usually larger and have a slightly richer flavor. Ducks are happiest and healthiest when they have access to a pool or pond to paddle around in and when they have several other ducks to keep them company.

Breeds of Ducks

There are six common breeds of ducks: White Pekin, White Aylesbury, Colored Rouen, Black Cayuga, Colored Muscovy, and White Muscovy. Each breed is unique and has its own advantages and disadvantages.

1. White Pekin—The most popular breed of duck, these are also the easiest to raise. These ducks are hardy and do well in close confinement. They are timid and must be handled carefully. Their large frame gives them lots of meat and they are also prolific layers.

White Pekins were originally bred from the Mallard in China and came to the United States in 1873.

2. White Aylesbury—This breed is similar to the Pekin but the plumage is much whiter and they are a bit heavier than the former. They are not as popular in the United States as the White Pekin duck.

According to Mrs. Beeton in her *Book of Household Management*, published in 1861, "[Aylesbury ducks'] snowy plumage and comfortable comportment make it a credit to the poultry-yard, while its broad and deep breast, and its ample back, convey the assurance that your satisfaction will not cease at its death."

3. Colored Rouens—These darkly plumed ducks are also quite popular and fatten easily for meat purposes.

Colored Rouen

4. Black Cayuga and Muscovy breeds—These are American breeds that are easily raised but are not as productive as the White Pekin.

Black Cayuga

Housing Ducks

You don't need a lot of space in which to raise ducks—nor do you need water to raise them successfully, though they will be happier if you can provide at least a small pool of water for them to bathe and paddle around in. Housing for ducks is relatively simple. The houses do not have to be as warm or dry as for chickens but the ducks cannot be confined for as long periods as chickens can. They need more exercise out of doors in order to be healthy and to produce more eggs. A house that is protected from dampness or excess rain water and that has straw or hay covering the floor is adequate for ducks. If you want to keep your ducks somewhat confined, a small fence about 2 ½ feet high will do the trick. Ducks don't require nesting boxes, as they lay their eggs on the floor of the house or in the yard around the house.

Feeding and Watering Ducks

Ducks require plenty of fresh water to drink, as they have to drink regularly while eating. Ducks eat both vegetable and animal foods. If allowed to roam free and to find their own food stuff, ducks will eat grasses, small fish, and water insects (if streams or ponds are provided).

Ducks need their food to be soft and mushy in order for them to digest it. Ducklings should be fed equal parts corn meal, wheat bran, and flour for the first week of life. Then, for the next 50 days or so, the ducklings should be fed the above mixture in addition to a little grit or sand and some green foods (green rye, oats, clover) all mixed together. After this time, ducks should be fed on a mixture of two parts cornmeal, one part wheat bran, one part flour, some coarse sand, and green foods.

Hatching Ducklings

The natural process of incubation (hatching ducklings underneath a hen) is the preferred method of hatching ducklings. It is important to take good care of the setting hen. Feed her whole corn mixed with green food, grit, and fresh water. Placing the feed and water just in front of the nest for the first few days will encourage the hen to eat and drink without leaving the nest. Hens will typically lay their eggs on the ground, in straw or hay that is provided for them. Make sure to clean the houses and pens often so the laying ducks have clean areas in which to incubate their eggs.

Caring for Ducklings

Young ducklings are very susceptible to atmospheric changes. They must be kept warm and free from getting chilled. The ducklings are most vulnerable during the first three weeks of life; after that time, they are more likely to thrive to adulthood. Construct brooders for the young ducklings and keep them very warm by hanging strips of cloth over the door cracks. After three weeks in the warm brooder, move the ducklings to a cold brooder as they can now withstand fluctuating temperatures.

Common Diseases

On a whole, ducks are not as prone to the typical poultry diseases, and many of the diseases they do contract can be prevented by making sure the ducks have a clean environment in which to live (by cleaning out their houses, providing fresh drinking water, and so on).

Two common diseases found in ducks are botulism and maggots. Botulism causes the duck's neck to go limp, making it difficult or even impossible for the duck to swallow. Maggots infest the ducks if they do not have any clean water in which to bathe, and are typically contracted in the hot summer months. Both of these diseases (as well as worms and mites) can be cured with the proper care, medications, and veterinary assistance.

A Muscovy duck with her ducklings.

FISH

Starting a Fish Farm

Raising your own fish ensures that they are healthy and safe to eat and you can sell or trade your fresh fish to neighbors for other things that you need. You can also use your fish to stock your local pond or lake, increasing the fish population in your area.

Before starting your fish farm, check your state and local laws—some states require a license to run a fish farm. Also, pick the particular fish you want to raise and research everything about it; you will need to know the species' specific life cycle, habitat, and dietary needs in order to properly care for your fish.

You'll need a pond to raise your fish in. If you do not have one naturally, you will have to build one. Dig a hole in the ground to the desired width and depth. Check the dirt at the bottom; if it is too porous the water will seep into the ground. If this is the case use tamped down clay or thick plastic to cover the bottom. If you want your pond to be specifically for fish farming, put topsoil at the bottom of the pond for plants. If you are going to release fingerlings, place some tree stumps or dead bushes in the water for the smaller fish to hide in.

Be sure to choose freshwater fish that are adapted to life in non-flowing waters. It is preferable to raise vegetarian fish, as they will eat any plant matter, while carnivorous fish require expensive protein-rich fish food.

Make sure that you keep records of everything; how many fish/fingerlings you put in the pond, how many fish you take out, and the basic conditions the pond is in on a consistent basis. This will help you decide when it is time to restock or go fishing!

Carp—Carp is a very popular fish farming choice. They are hardy, and a lot of fish can be got for little effort. They will eat any sort of rotting vegetation in the water, and you can supplement their diet with oatmeal, barley, or vegetable waste. Keep the carp in your larger pond—they will breed naturally. Choose the grown ones you would like to eat or sell and isolate them in a smaller, shallow "stew pond." Keep in mind the stew ponds need to be deep enough to stay ice-free in the winter.

Tilapia—Since Tilapia are a tropical fish, they will require a heated tank kept at around 80 degrees Fahrenheit. They do not require running water, although it is important to keep an eye on the tank's filters and air pumps. They are a very hardy species, and do not often get sick. They can produce tilapia fry almost every week year round. This combination of continuous production and high survival rate allows the tilapia farmer to have a constant supply of fingerlings to replace those that get big enough to eat.

One way to get started is to purchase a "hen mother" and several other tilapia. Alternatively, you could buy some fingerlings from a hatchery. Keep a few "breeding fish" in a separate tank.

Tilapia eat algae or any vegetation you put in the water. Tilapia fry, however, need some protein so live zooplankton, brine shrimp, or protein flakes are necessary until they grow. You can move the fingerlings to the main tank when they are about one inch long.

Aquaponics

Aquaponics is a sustainable food production structure that incorporates traditional aquaculture, like raising aquatic animals such as fish, prawns, or crayfish in tanks, with hydroponics, the method of cultivating plants in water. The fish and plants have a cooperative relationship and live in a harmonious, symbiotic environment.

Fish produce waste and effluents that accumulate in the water, and since too much waste can be toxic, the hydroponic system filters it out and utilizes it as vital nutrients for the plants. After this is done, the cleansed water re-circulates and returns to the animals.

Regardless of the size of an aquaponic system, the technology used is the same, and it is usually relegated into several subsystems. There is a rearing tank used for raising and feeding fish, and there is a unit that removes the solids: uneaten food, detached biofilms, fine particles, etc. In addition, there is a biofilter, an area where nitrification bacteria can grow and convert the ammonia into nitrates that plants can use. Nitrification is one of the most important processes because high concentrations of ammonia can kill the fish and aren't easily absorbed by plants. The biofilter contains *Nitrosomonas*, bacteria that convert ammonia into nitrites, and then *Nitrobacter*, bacteria convert the nitrites into nitrates.

The hydroponics subsystem is the portion where plants are grown by absorbing the excess nutrients; their roots are immersed in the nutrient-rich, effluent water. This may be done in various ways: Styrofoam rafts floating in an aquaculture basin in troughs, closed loop aquaponics (utilizing solid media like gravel or clay beads held in a container that is then flooded by water from the aquaculture), flood-and-drain aquaponics (solid media in a container that is alternately flooded and drained utilizing different types of siphon drains), towers that trickle the water from the top, nutrient film technique channels, etc. Finally, there is also the sump which is the lowest point in the system and where the water flows to and from once it is cleaned.

An aquaponics system relies on the relationship between the plants and animals; a stable aquatic environment must be maintained with minimal fluctuations in the nutrient and oxygen levels. Occasionally, water must be added when water is lost, whether because the absorption and transpiration lead to water loss or evaporation of the surface water.

The three main components of the system are water, food given to aquatic animals, and the electricity that pumps water between the aquaculture and the hydroponics. With those three things, you can run a successful aquaponic system that can help you have an even more successful fish farm.

Set Up an Aquaponics System

Thanks to aquaponics master Marek Broadstock for the use of these directions for setting up your own aquaponics system!

For a very basic system on a small scale, you'll need:

250 L pond bin
150 L grow tub
1000 L/ph pond pump capable of pushing water to a height of 1 meter
Double outlet air pump, 10 m extension cable
Large garden hose—bird wire
Plant choice (for example, tomato plants from seed packets)
Grow medium (for example, 50 kg of fine gravel)
Digital timer/surge protector; this is used to run the pump and to circulate the water to the plants
fFsh (for example, 6 Comet goldfish)
Bricks and roof tiles

Directions

Dig a hole in which the fish tub will go. Ensure that the tub has some shade to avoid evaporation and is also below ground level to maintain a good, steady water temperature.

Run a 10-meter long outdoor/waterproof extension cable from a power outlet (whether in a shed, garage, house, etc.) to the fish area, and connect the pump and the air pump to the power outlet. Place all leads and sensitive electrical parts in a waterproofed box. Bury it in a small, easily accessible hole.

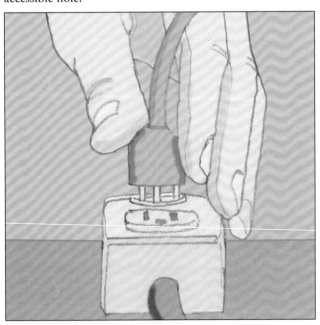

Drill or make some holes in the bottom of the grow tub so that the water can drain out and back into the fish tub. Next, set up a place for the grow tub to sit, preferably as close as possible to the fish tub so the pump won't have to work as hard to get the water into the grow tub from the fish tub. For example, you can set your grow tub on a higher level than the fish tub, letting gravity control the water flow back into the fish tub with the help of a basic roof tile.

Connect a hose from the 1000 L pond pump to the grow tub, and set it up with the digital timer to control when you want the pond pump to turn on and off.

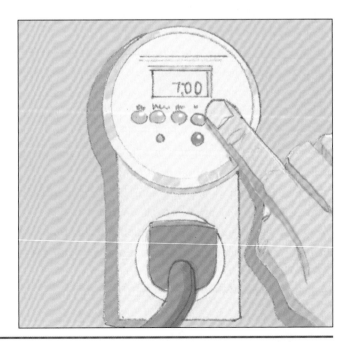

Fill the fish tub with non chlorinated water and fish. Have a few trial runs to see if the water flows from the fish tub to the grow tub and back out correctly. Monitor how fast the water escapes as this will slow down after you add your grow medium (for example, fine gravel).

Wash your grow medium and add it to the grow tub (make sure that you're happy with the location because it gets extremely heavy and impossible to move). From here, you can simply drop seeds into the grow tub and cover lightly with the medium and wait for germination to occur.

When you first put the seeds in, you will have doubts that the plants will even come up, but be patient. You will see plenty of little seedlings start to spring up. Otherwise, you can always add plants transplanted from a soil garden. Germination is always faster in hotter weather.

Maintenance

Fish feed: Some try to keep their systems totally organic by feeding the fish organic feed such as duckweed and worms. This method is perfectly fine and, in fact, truer to the idea of aquaponics, but you may not be able to breed worms, grow duckweed, or buy expensive feed. Cheap feed from supermarkets will work just as well as the organic kinds.

Pump cycle: This is determined by how fast your grow tub fills with water. Ideally, you want the water to fill up to about 3-5 cm. below the surface of the grow medium. Play around with the time setting, until you find a nice spot.

Plant selection and care: Most plants will work very well in an aquaponics type system, especially ones like tomatoes, spinach, and most green vegetables. The choice of plants to grow depends on the size of your system and how much experimentation you're willing to try as to which plants/vegetables will grow best. Like any plants, even aquaponically grown plants can show deficiencies in some vitamins and minerals. It's wise to add an organic, fish-friendly fertilizer, like a seaweed fertilizer, to your system every 2 weeks or so as you would a dirt garden to fix any signs of deficiencies.

If you'd like to increase the size of your aquaponics system, it's actually an easy process. Abide to the general rule: For a 100 L fish tub, you can have a 200 L grow tub worth of plants.

GOATS

Goats provide us with milk and wool and thrive in arid, semitropical, and mountainous environments. In the more temperate regions of the world, goats are raised as supplementary animals, providing milk and cheese for families and acting as natural weed killers.

Breeds of Goats

There are many different types of goats. Some breeds are quite small (weighing roughly 20 pounds) and some are very large (weighing up to 250 pounds). Depending on the breed, goats may have horns that are corkscrew in shape, though many domestic goats are dehorned early on to lessen any potential injuries to humans or other goats. The hair of goats can also differ—various breeds have short hair, long hair, curly hair, silky hair, or coarse hair. Goats come in a variety of colors (solid black, white, brown, or spotted).

Feeding Goats

Goats can sustain themselves on bushes, trees, shrubs, woody plants, weeds, briars, and herbs. Pasture is the lowest cost feed available for goats, and allowing goats to graze in the summer months is a wonderful and economic way to keep goats, even if your yard is quite small. Goats thrive best when eating alfalfa or a mixture of clover and timothy. If you have a lawn and a few goats, you don't need a lawn mower if you plant these types of plants for your goats to eat. The one drawback to this is that your goats (depending on how many you own) may quickly deplete these natural resources, which can cause weed growth and erosion. Supplementing pasture feed with other food stuff, such as greenchop, root crops, and wet brewery grains will ensure that your yard does not become overgrazed and that your goats remain well-fed and healthy. It is also beneficial to supply your goats with unlimited access to hay while they are grazing. Make sure that your goats have easy access to shaded areas and fresh water, and offer a salt and mineral mix on occasion.

Dry forage is another good source of feed for your goats. It is relatively inexpensive to grow or buy and consists of good quality legume hay (alfalfa or clover). Legume hay is high in protein and has many essential minerals beneficial to your goats. To make sure your forages are highly nutritious, be sure that there are many leaves that provide protein and minerals and that the forage had an early cutting date, which

Six Major U.S. Goat Breeds

Alpine—Originally from Switzerland, these goats may have horns, are short haired, and are usually white and black in color. They are also good producers of milk.

Anglo-Nubian—A cross between native English goats and Indian and Nubian breeds, these goats have droopy ears, spiral horns, and short hair. They are quite tall and do best in warmer climates. They do not produce as much milk, though it is much higher in fat than other goats'. They are the most popular breed of goat in the United States.

LaMancha—A cross between Spanish Murciana and Swiss and Nubian breeds, these goats are extremely adaptable, have straight noses, short hair, may have horns, and do not have external ears. They are not as good milk producers as the Saanen and Toggenburg breeds, and their milk fat content is much higher.

Pygmy—Originally from Africa and the Caribbean, these dwarfed goats thrive in hotter climates. For their size, they are relatively good producers of milk.

Saanen—Originally from Switzerland, these goats are completely white, have short hair, and sometimes have horns. Goats of this breed are wonderful milk producers.

Toggenburg—Originally from Switzerland, these goats are brown with white facial, ear, and leg stripes; have straight noses; may have horns; and have short hair. This breed is very popular in the United States. These goats are good milk producers in the summer and winter seasons and survive well in both temperate and tropical climates.

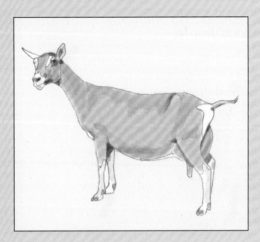

will allow for easier digestion of the nutrients. If your forage is green in color, it most likely contains more vitamin A, which is good for promoting goat health.

Goat Milk

Goat milk is a wonderful substitute for those who are unable to tolerate cow's milk, or for the elderly, babies, and those suffering from stomach ulcers. Milk from goats is also high in vitamin A and niacin but does not have the same amount of vitamins B6, B12, and C as cow's milk.

Lactating goats do need to be fed the best quality legume hay or green forage possible, as well as grain. Give the grain to the doe at a rate that equals ½ pound grain for every pound of milk she produces.

Common Diseases Affecting Goats

Goats tend to get more internal parasites than other herd animals. Some goats develop infectious arthritis, pneu-

monia, coccidiosis, scabies, liver fluke disease, and mastitis. It is advisable that you establish a relationship with a good veterinarian who specializes in small farm animals to periodically check your goats for various diseases.

Milking a Goat

Milking a goat takes some practice and patience, especially when you first begin. However, once you establish a routine and rhythm to the milking, the whole process should run relatively smoothly. The main thing to remember is to keep calm and never pull on the teat, as this will hurt the goat and she might upset the milk bucket. The goat will pick up on any anxiousness or nervousness on your part and it could affect how cooperative she is during the milking.

Supplies

- A grain bucket and grain for feeding the goat while milking is taking place
- Milking stand
- Metal bucket to collect the milk

- A stool to sit on (optional)
- A warm sterilized wipe or cloth that has been boiled in water
- Teat dip solution (2 tbsp bleach, 1 quart water, one drop normal dish detergent mixed together)

Directions

1. Ready your milking stand by filling the grain bucket with enough grain to last throughout the entire milking. Then retrieve the goat, separating her from any other goats to avoid distractions and unsuccessful milking. Place the goat's head through the head hold of the milking stand so she can eat the grain and then close the lever so she cannot remove her head.

2. With the warm, sterilized wipe or cloth, clean the udder and teats to remove any dirt, manure, or bacteria that may be present. Then, place the metal bucket on the stand below the udder.

3. Wrap your thumb and forefinger around the base of one teat. This will help trap the milk in the teat so it can be squirted out. Then, starting with your middle finger, squeeze the three remaining fingers in one single, smooth motion to squirt the milk into the bucket. Be sure to keep a tight grip on the base of the teat so the milk stays there until extracted. Remember: the first squirt of milk from either teat should not be put into the bucket as it may contain dirt or bacteria that you don't want contaminating the milk.

4. Release the grip on the teat and allow it to refill with milk. While this is happening, you can repeat this process on the other teat and can alternate between teats to speed up the milking process.

6. When the teats begin to look empty (they will be somewhat flat in appearance), massage the udder just a little bit to see if any more milk remains. If so, squeeze it out in the same manner as above until you cannot extract much more.

7. Remove the milk bucket from the stand and then, with your teat dip mixture in a disposable cup, dip each teat into the solution and allow to air dry. This will keep bacteria and infection from going into the teat and udder.

8. Remove the goat from the milk stand and return her to the pen.

Making Cheese from Goat Milk

Most varieties of cheese that can be made from cow's milk can also be successfully made using goats' milk. Goats' milk cheese can easily be made at home. In order to make the cheese, however, at least one gallon of goat milk should be available. Make sure that all of your equipment is washed and sterilized (using heat is fine) before using it.

Cottage Cheese

1. Collect surplus milk that is free of strong odors. Cool it to around 40ºF and keep it at that temperature until it is used.

2. Skim off any cream. Use the skim milk for cheese and the cream for cheese dressing.

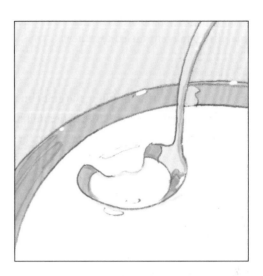

3. If you wish to pasteurize your milk (which will allow it hold better as a cheese) collect all the milk to be processed into a flat bottomed, straight-sided pan and heat to 145ºF on low heat. Hold it at this temperature for about 30 minutes and then cool to around 80ºF. Use a dairy thermometer to measure the milk's temperature. Then, inoculate the cheese milk with a desirable lactic acid fermenting bacterial culture (you can use commercial buttermilk for the initial source). Add about 7 ounces to 1 gallon of cheese milk, stir well, and let it sit undisturbed for about 10 to 16 hours, until a firm curd is formed.

4. When the curd is firm enough, cut the curd into uniform cubes no larger than ½ inch using a knife or spatula.

5. Allow the curd to sit undisturbed for a couple of minutes and then warm it slowly, stirring carefully, at a temperature no greater than 135°F. The curd should eventually become firm and free from whey.

6. When the curd is firm, remove from the heat and stop stirring. Siphon off the excess whey from the top of the pot. The curd should settle to the bottom of the container. If the curd is floating, bacteria that produces gas has been released and a new batch must be made.

7. Replace the whey with cold water, washing the curd and then draining the water. Wash again with ice-cold water from the refrigerator to chill the curd. This will keep the flavor fresh.

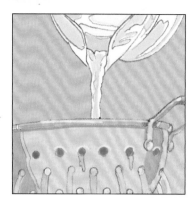

8. Using a colander, drain the excess water from the curd. Now your curd is complete.

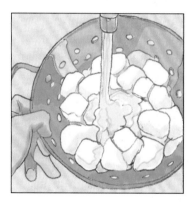

9. In order to make the curd into a cottage cheese consistency, separate the curd as much as possible and mix with a milk or cream mixture containing salt to taste.

Domiati Cheese

This type of cheese is made throughout the Mediterranean region. It is eaten fresh or aged 2 to 3 months before consumption.

1. Cool a gallon of fresh, quality milk to around 105°F, adding 8 ounces of salt to the milk. Stir the salt until it is completely dissolved.

2. Pasteurize the milk as described in step 3 of the cottage cheese recipe.

3. This type of cheese is coagulated by adding a protease enzyme (rennet). This enzyme may be purchased at a local drug store, health food store, or a cheese maker in your area. Dissolve the concentrate in water, add it to the cheese milk, and stir for a few minutes. Use 1 milliliter of diluted rennet liquid in 40 milliliters of water for every 2 ½ gallons of cheese milk.

4. Set the milk at around 105°F. When the enzyme is completely dispersed in the cheese milk, allow the mix to sit undisturbed until it forms a firm curd.

5. When the desired firmness is reached, cut the curd into very small cubes. Allow for some whey separation. After 10 to 20 minutes, remove and reserve about ⅓ the volume of salted whey.

6. Put the curd and remaining whey into cloth-lined molds (the best are rectangular stainless steel containers with perforated sides and bottom) with a cover. The molds should be between 7 and 10 inches in height. Fill the molds with the curd, fold the cloth over the top, allow the whey to drain, and discard the whey.

7. Once the curd is firm enough, apply added weight for 10 to 18 hours until it is as moist as you want.

8. Once the pressing is complete and the cheese is formed into a block, remove the molds, and cut the blocks into 4-inch-thick pieces. Place the pieces in plastic containers with airtight seals. Fill the containers with reserved salted whey from step 5, covering the cheese by about an inch.

9. Place these containers at a temperature between 60 and 65°F to cure for 1 to 4 months.

Feta Cheese

This type of cheese is very popular to make from goats' milk. The same process is used as the Domiati cheese except that salt is not added to the milk before coagulation. Feta cheese is aged in a brine solution after the cubes have been salted in a brine solution for at least 24 hours.

Angora Goats

Angora goats may be the most efficient fiber producers in the world. The hair of these goats is made into mohair, a long, lustrous hair that is woven into fine garments. Angora goats are native to Turkey and were imported to the United States in the mid-1800s. Now, the United States is one of the two biggest produces of mohair on Earth.

Angora goats are typically relaxed and docile. They are delicate creatures, easily strained by their year-round fleeces. Angora goats need extra attention and are more high-maintenance than other breeds of goat. While these goats can adapt to many temperate climates, they do particularly well in the arid environment of the southwestern states.

Angora goats can be sheared twice yearly, before breeding and before birthing. The hair of the goat will grow about ¾ inch per month and it should be sheared once it reaches 4 to 6 inches in length. During the shearing process, the goat is usually lying down on a clean floor with its legs tied. When the fleece is gathered (it should be sheared in one full piece), it should be bundled into a burlap bag and should be free of contaminants. Mark your name on the bag and make sure there is only one bag per fleece. For more thorough rules and regulations about selling mohair through the government's direct-payment program, contact the USDA Agricultural Stabilization and Conservation Service online or in one of their many offices.

Shearing can be accomplished with the use of a special goat comb, which leaves ¼ inch of stubble on the goat. It is important to keep the fleeces clean and to avoid injuring the animal. The shearing seasons are in the spring and fall. After a goat has been sheared, it will be more sensitive to changes in the weather for up to six weeks. Make sure you have proper warming huts for these goats in the winter and adequate shelter from rain and inclement weather.

A horse can be a chauffeur (whether you're in a saddle or in a carriage trailing behind), a farmer's best friend when the fields need to be plowed, a spotlight-loving performer, or a wonderful pet and companion. The versatility of horses is what makes them such appealing animals. Yet with their great size, comes even greater responsibility. Horses have very specific needs: their living quarters, their diet, their exercise regimen, and their grooming, to name a few. They are complicated animals, but if taken care of the correct way, all the reaped rewards will absolutely be worth the work.

Breeds of Horses

There are hundreds of horse breeds, from the rare Abtenauer bred in a secluded valley just south of Salzburg, Austria, to the elegant Zweibrücken hailing from Germany. While it is fascinating to really explore all the differences in each breed, it is more practical to break the breeds up into three main categories based on their body type. The first two are "light horses" and "draft horses"; light horses are used for undemanding work and for their speed, whereas draft horses are able to complete more arduous tasks. You are likely to see light horses galloping around a racetrack or in a dressage show, while draft horses and their carriages are a popular method of transportation for princesses or, more likely, tourists in New York's Central Park. The final type of horse is a pony, sizing in at a mere 14.2 hands or under. Ponies have always been especially popular amongst children.

1. Light horses—An example of a light horse is the thoroughbred. This breed's native home is England, where their human counterparts have fostered racing for over a thousand years. They have slender necks, deep, wide chests with long, slanting shoulders, and hind quarters that are high and muscular. Their legs are long and they end in rounded, well-shaped feet. Their lithe, angular bodies favor speed and are in complete opposition to a draft horse.

2. Draft horses—A Clydesdale is a well known draft breed. Bred in Scotland and later imported to America, this horse is famous for its intimidating size and its gigantic, hairy feet. Clydesdales are tall, and long limbed; they

Abtenauer

Thoroughbred

Clydesdale

Zweibrucken

Shetland

have medium-sized necks joining with slanting shoulders that meet strong legs, heavily fringed below the knee. This hair was often criticized by horsemen as being a fault of the breed; it was too hard to keep clean and free of disease, but many are fans of the defining characteristic.

3. Ponies—A Shetland pony is one of the more popular breeds. Shetlands resemble small draft horses, with their foretops, manes, and tails heavy and long. Being tinier than other horses, a pony is a great option for a child learning how to ride—they are exceedingly intelligent and can be readily trained.

Decide what you want to do with your horse before you select a breed; if you want your horse to be ridden, who will be the primary rider? Do you plan on leisurely trail rides or do you want your horse to be a racer? Once you figure these things out you can narrow down your choices. For many people, especially if they are new to having horses, the breed is not necessarily the deciding factor. Most care more about a horse's personality and how well trained or how green the horse is. If selecting a horse to begin breeding, then the breed would obviously have to be taken into deep consideration.

Housing

Horses should have some kind of shelter; all horses need to be protected from inclement weather. Housing can be indoor or outdoor, or even a combination of the two. It is decided by the use of the horse on a day-to-day basis. Indoor housing is ideal for horses that are being ridden or used every day in the winter time. The size of the stall should be around 12x12, which is the industry standard, but if you are housing a draft horse then the dimensions should be increased to 16x16. Horses should be provided with bedding for their stall (different types of straw or shavings); they need a dry, soft bed. Their bedding should also be kept fresh and clean, which means cleaning the stall daily and removing soiled bedding and manure. If an owner neglects the horse's stall, it lessens the comfort and promotes disease. Outdoor housing has a myriad of advantages, for example lower construction costs and less labor where cleaning the building is concerned. The building should be a three-sided structure, like a run-in shed.

Grooming

Unlike other farm animals, horses must be groomed. Without grooming, a horse is susceptible to discomfort and diseases; this includes taking care of a horse's hooves because feet are the horse's most complicated structures, and the most integral to look after. A horse's hooves should be cleaned (with a pick) every day, before and after riding or being sent out to do work. The feet are the most liable to injury from the effects of hard work and mismanagement, and

subsequently, there is no body part that more requires care in both health and disease. Prevention, in the foot's case, is much better than cure. It is also a good idea to get the horse shoes, which prevent the hoof from wearing down—a hoof wears down faster than it grows back. A stable horse should be thoroughly groomed each day, before and after the horse's work; it is necessary to scrub away the "scurf" (the small shreds of epidermis that are continually exfoliated from the skin) that obstructs the pores. Doing so, admits free perspiration and promotes circulation to the extremities. Cleaning and rubbing the skin is obviously important, but it is also imperative for the legs and feet because it preserves soundness.

Feeding and Watering

With horses, it is unnecessary to limit the amount of water; it should be left up to the horse's discretion. He will take only as much as he wants. Food, however, is a different story. Horses can be overfed with hay (horses will eat it just for the amusement) and with their oats or grain. They should be fed according to however much they work—a work horse would be fed a more substantial amount than an idle horse, for example. If a horse is continually overfed, its appetite will eventually increase and, of course, the horse will become overweight. Horses should be fed regularly, at a scheduled time, as they anticipate the hour they will be fed and will become nervous it is too long delayed. As for what to feed horses—oats have been proven especially great for them; they favor speed and endurance more than any other food. If oats are too expensive, grains like barley, wheat, rye, and bran can be substituted. Horses are sometimes even known to eat corn. It should also be noted that if one is changing a horse's feed, it needs to be gradual, otherwise, there could be digestive problems.

Care and Keeping of Horse Tack

The "tack" of a horse is a general term meaning the equipment needed to ride and/or work a horse. This includes the saddle, saddle blanket/pad, reins, bridle and bit, and stirrups. This equipment, especially any tack directly in contact with the horse such as the saddle, should be cleaned after every use.

Before cleaning your saddle, unfasten any buckles and remove any fittings. Using a damp towel, remove any dust, dirt, mud, hair, or other debris.

To clean the leather saddle; get basic saddle soap, a sponge, and water. Lather up the soap and scrub the saddle, using the sponge or a small soft bristled brush. Rinse the saddle with water and lather it again–if the suds are grey it means the saddle is still dirty. It is important that the pores of the leather are clean and do not become clogged with dirt.

Use the cleaning and disassembly of your tack to check for any weaknesses or tears. Fix any small problem immediately before it becomes a bigger problem.

Occasionally it will be necessary to condition the saddle with oil. It is usually time to apply the oil if the saddle is "thirsty" and soaking up the washing water. After the saddle is cleaned, but not all the way dry, apply the oil, making sure not to apply too much. For a new saddle three coats of oil is necessary, but for an older saddle one heavier followed by one light one is enough. Buff the saddle with a soft cloth after oiling to wipe off excess. Riders use everything from expensive, special leather saddle oil to plain cooking oil.

Clean any other leather tack, including the bridle and reins in the same way as the saddle. Wipe off the bit and any other metal parts immediately after use.

Once a week the bit and bridle should be entirely taken apart and cleaned thoroughly.

Any nylon tack should also be cleaned. Simply wash the nylon with soapy water and allow it to air dry completely. No special cleaning or oil is required.

Brush off the saddle blanket/pad after every use. This is the gear that touches the horse the most, and any dirt, thorns, or grass will irritate the horse's skin.

Make sure to store tack properly between uses. Do not leave any leather in damp places, such as barns or basements; mold and mildew will develop and ruin the saddle in as little as one month. Be sure to keep your tack out of the elements; snow and rain will dampen the leather while direct sunlight will cause the leather to dry out and crack.

Breeding

For horse owners, the idea of breeding from their own mare has much appeal. The prospect of producing a foal with qualities similar to its mother, or even better, has many attractions.

Before any breeding decisions are made, it is best to have knowledge about normal breeding behavior, what should happen at foaling, and how a newborn foal should behave and develop, is essential. For this reason, it is best for a novice horse breeder to seek professional help with mating and foaling from a stud.

Common Diseases

One of the most common, and most dangerous, diseases a horse can contract is colic. Colic is a broad term that covers any acute gastrointestinal problem: colic could be anything from a stomachache or cramp from changing food too fast to impaction of waste in a horse's intestines. Colic can be fatal; it is the leading cause of premature death among domesticated horses, and therefore if an owner suspects his or her horse has colic, it is absolutely necessary to call a vet. Some symptoms of colic are restlessness, lying down, kicking with the hind feet upward and toward the belly, jerky swishing of the tail, groaning, frequent position changes, and stretching as if to urinate, but with greater intensity, the movements become violent: the horse may throw himself down, roll, assume unnatural positions (for example, sitting on his haunches), and grunt loudly. With colic, the pain is not constant, so during the periods of peace, the horse may act completely normal. However, during the periods of pain, the horse will be sweating profusely.

As stated earlier, it is imperative that the owner takes care of their horses' sensitive hooves. There are many diseases that result from neglected feet. One common disease is Laminitis, where the horse's digital laminae (attached to the hoof wall and coffin bone) become inflamed. It will eventually become impossible for the horse to walk without pain. It is impossible to discern that a horse has laminitis without radiographs (so if an owner suspects it, they should consult a vet); but the owner should be wary of laminitis if the horse has any of the following symptoms: sweating, increased vital signs, and a tendency to favor the afflicted foot.

Founder is not a disease, exactly, but a complication of many horse diseases. Founder occurs when a disease (like laminitis) goes untreated for a very long time. What will happen is that the coffin bone will sink through the frog of the hoof making moving, and even standing, impossible without extreme pain and discomfort. A horse that founders and refuses to stand could very likely end up with colic—and that means the horse is in a lot of danger of dying. Any horses that founder will need constant attention and possibly even hospitalization in an equine clinic.

Choosing a Team of Work Horses

Choosing your team of work horses, the number, the size, and appearance, is dependent upon what you plan to do with them. If you want them to pull a carriage and become mascots for beer, then their appearance suddenly becomes high priority and their pulling capacity less so. However, if you need a team of work horses to help your farming operation, the size of the field and the nature of the work will dictate the horses you'll need.

A general rule to follow is 25 acres of land per draft horse; it is always wise to have an extra horse in case of injury or to lighten the load. And speaking of the load, a horse can exert 10 percent of its body weight in a horizontal pull steadily—meaning a 1,000 lb horse could exert a 100 lb pull while a 2000 lb horse could do twice that. It is good to know that for a brief time, a horse can exert a pull of half its own weight. The breed is much less important than the build, health, training, and manners of the horse.

When looking for a team, you must pay close attention to how well-matched they are. Notice whether they go the same (walk the same pace, work well together) and whether they are close in size. Short, sturdy draft horses will be easier to harness on a daily basis, but bigger ones may be able to handle larger loads.

It's best to procure horses privately from someone you know to be trustworthy. Buying from an auction is very risky; "broke" can mean a wide range of levels of experience and training, so make sure you have all the facts and speak to the owner, if you can, if you are intent on buying from an auction. After finding a horse in which you're interested, check its feet to be sure they're sound. If the feet pass inspection, check for clean, straight limbs. Next, look at the breadth of the brisket and the width of the barrel back through the hips. Carefully inspect the hocks for any sign of defects as this joint takes the most strain in a heavy pull. As you do this, notice the horse's reactions and personalities. Having a horse that enjoys kicking is dangerous for someone who will work so closely with it.

The best thing to do when choosing a team of work horses is to drive them yourself, pulling or doing something similar to what you'll do at home. Seeing them in action is the only way to know how much training they've had, how they react to one another and the work they are doing, and how they react to you as a teamster. If the owner objects to something you ask him to show you, make sure to ask why. If a horse has flaws or defects, think first about how you would handle it, monetarily and otherwise, before you decide that horse is or isn't the one for you.

LLAMAS

Llamas often make excellent pets and are a great source of wooly fiber (their wool can be spun into yarn). Llamas are being kept more and more by people in the United States as companion animals, sources of fiber, pack and light plow animals, therapy animals for the elderly, "guards" for other backyard animals, and good educational tools for children. Llamas have an even temperament and are very intelligent. Their intelligence and gentle nature make them easy to train, and their hardiness allows them to thrive well in both cold and warmer climates (although they can have heat stress in extremely hot and humid parts of the country).

Before you decide to purchase a llama or two for your yard, check your state requirements regarding livestock. In some places your property must also be zoned for livestock.

Llamas come in many different colors and sizes. The average adult llama is between 5 ½ and 6 feet tall and weighs between 250 and 450 pounds. Llamas, being herd animals, like the company of other llamas, so it is advisable that you raise a pair to keep each other company. If you only want to care for one llama, then it would be best to also have a sheep, goat, or other animal that can be penned with the llama for camaraderie. Although llamas can be led well on a harness and lead, never tie one up as it could potential break its own neck trying to break free.

Llamas tend to make their own communal dung heap in a particular part of their pen. This is quite convenient for

cleanup and allows you to collect the manure, compost it, and use it as a fertilizer for your garden.

Feeding Llamas

Llamas can subsist fairly well on grass, hay (an adult male will eat about one bale per week), shrubs, and trees, much like sheep and goats. If they are not receiving enough nutrients, they may be fed a mixture of rolled corn, oats, and barley, especially during the winter season when grazing is not necessarily available. Make sure not to overfeed your llamas, though, or they will become overweight and constipated. You can occasionally give cornstalks to your llamas as an added source of fiber, and you may add mineral supplements to the feed mixture or hay if you want. Salt blocks are also acceptable to have in your llama pen, and a constant supply of fresh water is necessary. Nursing female llamas should receive a grain mixture until the cria (baby) is weaned.

Be sure to keep feed and hay off the ground. This will help ward off parasites that establish themselves in the feed and are then ingested by the llamas.

Housing Your Llamas

Llamas may be sheltered in a small stable or even a converted garage. There should be enough room to store feed and hay, and the shelter should be able to be closed off during wet, windy, and cold weather. Llamas prefer light, open spaces in which to live, so make sure your shed or shelter has large doors and/or big windows. The feeders for the hay and grain mixture should be raised above the ground. Adding a place where a llama can be safely restrained for toenail clippings and vet checkups will help facilitate these processes but is not absolutely necessary.

The llamas should be able to enter and exit the shelter easily and it is a good idea to build a fence or pen around the shelter so they do not wander off. A fence about four feet tall should be enough to keep your llamas safe and enclosed. If you happen to have both a male and female llama, it is necessary to have separate enclosures for them to stave off unwanted pregnancies.

Toenail Trimming

Llamas need their toenails to be trimmed so they do not twist and fold under the toe, making it difficult for the llama to move around. Laying gravel in the area where your llamas frequently walk will help to keep the toenails naturally trimmed, but if you need to cut them, be careful not to cut too deeply or you may cause the tip of the toe to bleed and this could lead to an infection in the toe. Use shears designed for this purpose to cut the nails. Use one hand to hold the llama's "ankle" just above where the foot bends. Hold the clippers in your other hand, cutting away from the foot toward the tip of the nail. The nail's are easiest to clip in the early morning or after a rain, since the wetness of the ground will soften them.

Shearing

You'll need to groom and shear your llama, especially during hot weather. Brushing the llama's coat to remove dirt and keep it from matting will not only make your llamas look clean and healthy but it will improve the quality of their coats. If you want to save the fibers for spinning into yarn, it is best to brush, comb, and use a hair dryer to remove any dust and debris from the llama's coat before you begin shearing.

Shearing is not necessarily difficult, but if you are a first-time llama owner, you should ask another llama farmer to teach you how to properly shear your llama. In order to shear your llama, you can purchase battery-operated shears to remove the fibers for sale or use. Different llamas will respond in different ways to shearing. Try holding the llama with a halter and lead in a smaller area to begin the shearing process. Do not completely remove the llama from any other llamas you have, though, as their presence will help calm the llama you are shearing. It is best to have another person with you to aid in the shearing (to hold the llama, give it treats, and offer any other help). When shearing a llama, don't shear all the way down to the skin. Allowing a thin coating of hair to cover the llama's body will help protect it from the sun and from being scratched when it rolls in the dirt.

Start by shearing a flat top the length of the llama's back. Next, taking the shears in one hand, move them in a downward position to remove the coat. Shear a strip the length of the neck from the chin to the front legs about 3 inches wide to help cool the llama. Shearing can take a long time, so it may be necessary for both you and the llama to take a break. Take the llama for a quiet walk and allow it to go to the bathroom so it will not become antsy during the rest of the shearing process.

Collect the sheared fibers in a container and make sure you are working on a clean floor so you can collect any excess fibers and use them for spinning. Do not store the fiber in a plastic bag, as moisture can easily accumulate, ruining the fiber and making it unusable for spinning.

Caring for the Cria

Baby llamas, called cria, require some additional care in their first few days of life. It is important for the cria to receive the colostrum milk from their mothers, but you may need to aid in this process. Approach the mother llama and pull gently on each teat to remove the waxy plugs covering the milk holes. Sometimes, you may need to guide the cria into position under its mother in order for it start nursing.

Weigh the cria often (at least for the first month) to see that it's gaining weight and growing strong and healthy. A bathroom scale, hanging scale, or larger grain scale can be used for this.

If the cria seems to need extra nourishment, goat or cow milk can be substituted during times when the mother llama cannot produce enough milk for the cria. Feed this additional milk to the cria in small doses, several times a day, from a milking bottle.

Diseases

Llamas are prone to getting worms and should be checked often to make sure they do not have any of these parasites. There is special worming paste that can be mixed in with their food to prevent worms from infecting them. You should also establish a relationship with a good veterinarian who knows about caring for llamas and can determine if there are any other vaccinations necessary in order to keep your llamas healthy. Other diseases and pests that can affect llamas are tuberculosis, tetanus, ticks, mites, and lice.

Using Llama Fibers

Llama fiber is unique from other animal fibers, such as sheep's wool. It does not contain any lanolin (an oil found in sheep's wool); thus, it is hypoallergenic and not as greasy. How often you can shear your llama will depend on the variety of llama, its health, and environmental conditions. Typically, though, every year llamas grow a fleece that is 4 to 6 inches long and that weighs between 3 and 7 pounds. Llama fiber can be used like any other animal fiber or wool, making it the perfect substitute for all of your fabric and spinning needs.

Llama fiber is made up of two parts: the undercoat (which provides warmth for the llama) and the guard hair (which protects the llama from rain and snow). The undercoat is the most desirable part to use due to its soft, downy texture, while the coarser guard hair is usually discarded.

Gathering llama hair is easy. To harvest the fiber, you must shear the llama. However, the steps involved in shearing when you are gathering the fiber are slightly different than when you are simply shearing to keep the llama cooler in the summer months. To shear a llama for fiber collection:

1. Clean the llama by blowing and brushing until the coat is free from dirt and debris.

2. Wash the llama. Be sure to rinse out all of the soap from the hair and let the llama air-dry.

3. You can use scissors or commercial clippers to shear the llama. Start at the top of the back, behind the head and neck and work backwards. If using clippers, sheer with long sweeping motions, not short jerky ones. If using scissors, always point them downward. Leave about an inch of wool on the llama for protection against the sun and insect bites. You can sheer just the area around the back and belly (in front of the hind legs and behind the front legs) if your main purpose is to offer the llama relief from the heat. Or you can sheer the entire llama—from just below the head, down to the tail—to get the most wool. Once the shearing is complete, skirt the fleece by removing any little pieces or belly hair from the shorn fleece.

The fiber can be hand-processed or sent to a mill (though sending the fibers to a mill is much more expensive and is not necessary if you have only one or two llamas). Processing the fiber by hand is definitely more cost-effective but you will initially need to invest in some equipment (such as a spinning wheel, drop spindle, or felting needle).

To process the fiber by hand:

1. Pick out any remaining debris and unwanted (coarse) fibers.

2. Card the fiber. This helps to separate the fiber and will make spinning much easier. To card the fiber, put a bit of fiber on one end of the cards (standard wool cards do the trick nicely) and gently brush it until it separates. This will produce a rolag (log) of fiber.

3. Once the fiber is carded, you can use it in a few different ways:
   a. Wet felting: To wet felt, lay the fiber out in a design between two pieces of material and soak it in hot, soapy water. Then, agitate the fiber by rubbing or rolling it. This will cause it to stick together. Rinse the fiber in cold water. When it dries, you will have produced a strong piece of felt that can be used in many crafting projects.
   b. Needle felting: For this type of manipulation, you will need a felting needle (available at your local arts and crafts or fabric store). Lay out a piece of any material you want over a pillow or Styrofoam piece. Place the fiber on top of the material in any design of your choosing. Push the needle through the fiber and the bottom material and then gently draw it back out. Continue this process until the fiber stays on the material of its own accord. This is a great way to make table runners or hanging cloths using your llama fiber.
   c. Spinning: Spinning is a great way to turn your llama fiber into yarn. Spinning can be accomplished by using either a spinning wheel or drop spindle, and a piece of fiber that is either in a batt, rolag, or roving. A spinning wheel, while larger and more expensive, will easily help you to turn the fiber into yarn. A drop spindle is convenient because it is smaller and easier to transport, and if you have time and patience, it will do just as good a job as the spinning wheel. To make yarn, twist two or more pieces of spun wool together.
   d. Other uses: carded wool can also be used to weave, knit, or crochet.

If you become very comfortable using llama fiber to make clothing or other craft items, you may want to try to sell these crafts (or your llama fiber directly) to consumers. Fiber crafts may be particularly successful if sold at local craft markets or even at farmers' markets alongside your garden produce.

PIGS

igs can be farm-raised on a commercial scale for profit, in smaller herds to provide fresh, homegrown meat for your family or to be shown and judged at county fairs or livestock shows. Characterized by their stout bodies, short legs, snouts, hooves, and thick, bristle-coated skin, pigs are omnivorous, garbage-disposing mammals that, on a small farm, can be difficult to turn a profit on but yield great opportunities for fair showmanship and quality food on your dinner table.

along desirable traits, such as durability, leanness, and quality of meat, while others are known for their reproductive and maternal qualities. The breed you choose to raise will depend on whether you are raising your pigs for show, for profit, or to put food on your family's table.

Breeds

Pigs of different breeds have different functionalities—some are known for their terminal sire (the ability to produce offspring intended for slaughter rather than for further breeding) and have a greater potential to pass

Pig Terminology

pig, hog, or swine	Refers to the species as a whole or any member thereof.
shoat or piglet (or "pig" when species is referred to as "hog")	Any unweaned or immature young pigs.
sucker	A pig between birth and weaning.
runt	An unusually small and weak piglet. Often one per litter.
boar or hog	A male pig of breeding age.
barrow	A male pig castrated before reaching puberty.
stag	A male pig castrated later in life.
gilt	A young female not yet mated (farrowed) or has birthed fewer than two litters.
sow	An active breeding female pig.

Eight Majors U.S. Pig Breeds

1. **Yorkshire**—Originally from England, this large white breed of hog has a long frame, comparable to the Landrace. They are known for their quality meat and mothering ability and are likely the most widely distributed breed of pig in the world. Farmers will also find that the Yorkshire breed generally adapts well to confinement.

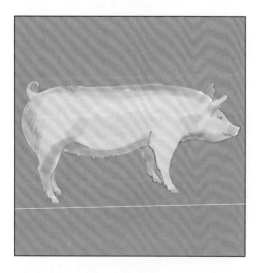

2. **Landrace**—This white-haired hog is a descendent from Denmark and is known for producing large litters, supplying milk, and exhibiting good maternal qualities. The breed is long-bodied and short-legged with a nearly flat arch to its back. Its long, floppy ears are droopy and can cover its eyes.

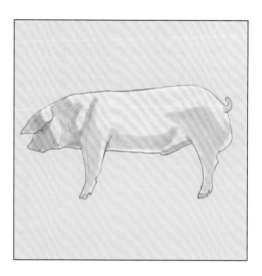

3. **Chester**—Like the Landrace, this popular white hog is known for its mothering abilities and large litter size. Originating from cross breeding in Pennsylvania, Chester hogs are medium-sized and solid white in color.

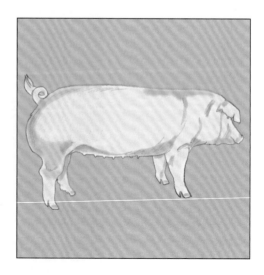

4. **Berkshire**—Originally from England, the black and white Berkshire hog has perky ears and a short, dished snout. This medium-sized breed is known for its siring ability and quality meat.

5. **Poland China**—Known for often reaching the maximum weight at any age bracket, this black and white breed is of the meaty variety.

6. **Hampshire**—A likely descendent of an Old English breed, the Hampshire is one of America's oldest original breeds. Characterized by a white belt circling the front of their black bodies, this breed is known for its hardiness and high-quality meat.

7. **Spot**—Known for producing pigs with high growth rates, this black and white spotted hog gains weight quickly while maintaining a favorable feed efficiency. Part of the Spot's ancestry can be traced back to the Poland China breed.

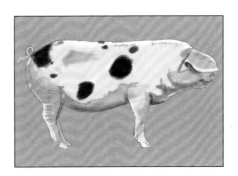

Housing Pigs

Keeping your pigs happy and healthy and preventing them from wandering off requires two primary structures: a shelter and a sturdy fence. A shelter is necessary to protect your pigs from inclement weather and to provide them with plenty of shade, as their skin is prone to sunburns. Shelters can be relatively simple three-sided, roofed structures with slanted, concrete flooring to allow you to spray away waste with ease. To help keep your pigs comfortable, provide them with enough straw in their shelter and an area to make a wallow—a muddy hole they can lie in to stay cool.

Because pigs will use their snouts to dig and pry their way through barriers, keeping these escape artists fenced in can post a challenge. "Hog wire," or woven fence wire, at least 40 inches high is commonly used for perimeter fencing. You can line the top and especially the bottom of your fence with a strand of barbed or electric wire to discourage your pigs from tunneling their way through. If you use electric wiring, you may have a difficult time driving your herd through the gate. Covering the gate with non-electric panels using woven wire, metal, or wood can make coaxing your pigs from the pasture an easier task.

Feeding Your Pigs

Pigs are of the omnivorous variety, and there isn't much they won't eat. Swine will consume anything from table and

---

Tips for Selecting Breeder Sows

- Look for well-developed udders on a gilt (a minimum of six pairs of teats, properly spaced and functional)
- Do not choose those with inverted teats which do not secrete milk, and do not choose sows that are otherwise unable to produce milk
- Opt for longer-bodied sows (extra space promotes udder development)
- Look for a uniform width from the front to the rear
- Check for good development in the ham, loin, and shoulder regions to better assure good breeding
- Choose the biggest animals within a litter
- Choose female breeders from litters of eight or more good-sized piglets that have high rates of survivability
- Choose hardy pigs from herds raised in well-sanitized environments and avoid breeding any pigs with physical abnormalities

---

garden scraps to insects and worms to grass, flowers, and trees. Although your pigs won't turn their snouts up to garbage, a cost-effective approach to assuring good health and a steady growth rate for your pigs is to supply farm grains (mixed at home or purchased commercially), such as oats, wheat, barley, soybeans, and corn. Corn and soybean meal are a good source of energy that fits well into a pig's low fiber, high protein diet requirements. For best results, you should include protein supplements and vitamins to farm grain diets.

As pigs grow, their dietary needs change, which is why feeding stages are often classified as starter, grower, and finisher. Your newly weaned piglets make up the starting group, pigs 50-125 lbs are growing, and those between 125 and the 270 lb market weight are finishing pigs.

As your pigs grow, they will consume more feed and should transition to a less dense, reduced-protein diet. You should let your pigs self-feed during every stage. In other words, allow them to consume the maximum amount they will take in a single feeding. Letting your pigs self-feed once or twice a day allows them to grow and gain weight quickly.

Another essential part of feeding is to make sure you provide a constant supply of fresh, clean water. Your options range from automatic watering systems to water barrels. Your pigs can actually go longer without feed than they can without water, so it's important to keep them hydrated.

Diseases

You can prevent the most common pig diseases from affecting your herd by asking your veterinarian about the right vaccination program. Common diseases include E. Coli—a bacteria typically caused by contaminated fecal matter in the living environment that causes piglets to experience diarrhea. You should vaccinate your female pigs for E. Coli before they begin farrowing.

Another common pig disease is Erysipelas, which is caused by bacteria that pigs secrete through their saliva or waste products. Heart infections or chronic arthritis are possible ailments the bacteria causes in pigs that can lead to death. You should inoculate pregnant females and newly-bought feeder pigs to defend against this prevalent disease.

Other diseases to watch out for are Atrophic Rhinitis, characterized by inflammation of a pig's nasal tissues; Leptospirosis, an easily spreadable bacteria-borne disease; and Porcine Parvovirus, an intestinal virus that can spread without showing symptoms. Consult your veterinarian to discuss vaccinating against these and other fast-spreading diseases that may affect your herd.

RABBITS

Rabbits are very social and docile animals, and easy to maintain. They like to play, but because of their skittish nature, are not necessarily the best pets for young children. Larger rabbits, bred for eating, often make good pets because of their more relaxed personalities. Rabbits are easier to raise than chickens and can provide you with beautiful fur and lean meat. In fact, rabbits will take up less space and use less money than chickens.

1. **Californian:** 6-10 lbs. Short fur. Relaxed personality. Choice for eating.

Breeds

There are over forty breeds of domestic rabbits. Below are ten of the most commonly owned varieties, along with their traits and popular uses.

2. **Dutch:** 3-5 lbs. Short fur. Relaxed personality. Choice pet. Good for young children.

3. **Flemish Giant:** 9+ lbs. Medium-length fur. Calm personality. Choice for eating.

4. **Holland Lop:** 3-5 lbs. Medium-length fur. Curious personality. Choice pet. One of the lop-eared rabbits, its ears flop down next to its face. A similar popular breed is the American Fuzzy Lop.

5. **Jersey Wooly:** 2-4 lbs. Long fur. Relaxed personality. Choice pet.

6. **Mini Lop:** 4-7 lbs. Medium-length fur. Relaxed personality. Choice pet. Lop-eared. Some reports of higher biting tendencies.

7. **Mini Rex:** 3-5 lbs. Very short, velvety fur. Curious personality. Choice pet. Tend to have sharp toenails.

8. **Netherland Dwarf:** 2-4 lbs. Medium-length fur. Excitable personality. Choice pet.

9. **New Zealand:** 9+ lbs. Short fur. Curious personality. Choice for eating. Variable reputation for biting.

10. **Satin:** 9+ lbs. Medium-length fur. Relaxed personality. Fur is finer and denser than other furs.

Housing

Rabbits should be kept in clean, dry, spacious homes. You will need a hutch, similar to a henhouse, to house your rabbits. It is important to provide your rabbits with lots of air. The best hutch will have a wide, over-hanging roof and is elevated about six inches off the ground. This way, your rabbits will not only have shade, but their homes will be prevented from getting damp.

Food

A rabbit's diet should be made up of three things: a small portion pellets (provided they are high in fiber), a continual source of hay, and vegetables. Rabbits love vegetables that are dark and leafy or root vegetables. Avoid feeding them beans or rhubarb, and limit the amount of spinach they eat. If you want to give rabbits a treat, try a small piece of fruit, such as a banana or apple. Remember that all of their food needs to be fresh (pellets should not be more than six weeks old), and like all other animals, be careful not to overfeed them. Also, to keep them from dehydrating, provide them with plenty of clean water every day.

Note: If you have a pregnant doe, allow her to eat a little more than usual.

Breeding

When you want to breed rabbits, put a male and female together in the morning or evening. After they have mated, you may separate them again. A female's gestation period is approximately a month in length, and litters range from 6-10 babies. Baby rabbits' eyes will not open until two weeks after birth. Their mother will nurse them for a month, and for at least the first week, you must not touch any of the litter; you can alter their smell and the mother may stop feeding them. At two months, babies should be weaned from their mother, and at four months, or approximately 4.5 lbs, they are old enough to sell, eat, or continue breeding. Larger rabbit varieties may take 6 to 12 months to sexually mature.

Health Concerns

The main issues that may arise in your rabbits' health are digestive problems and bacterial infections. Monitor your rabbit's droppings carefully. Diarrhea in rabbits can be fatal. Some diarrhea is easy to identify, but also be on the lookout for droppings that are misshapen, softer in consistency, a lack of droppings altogether, and loud tummy growling. Diarrhea requires antibiotics from your veterinarian. In bacterial infections, your rabbit may have a runny nose or eyes, a high

temperature, or a rattling or coughing respiratory noise. This also requires medical attention and an antibiotic specific to the type of infection.

Hairballs are another issue you may encounter and also require some attention. Every three months, rabbits shed their hair, and these sheds will alter between light and heavy. Since rabbits will attempt to groom themselves as cats do, but cannot vomit hair as cats can, you must groom them additionally, to prevent too much hair ingestion. Brush and comb them when their shedding begins, and provide them with ample fresh hay and opportunity for exercise. The fiber in the hay will help the hair to pass through their digestive tracts, and the exercise will keep their metabolisms active.

If your rabbit has badly misaligned teeth, they may interfere with his or her ability to eat and will need to be trimmed by the veterinarian.

Never give your rabbit amoxicillin or use cedar or pine shavings in their hutches. Penicillin-based drugs carry high risks for rabbits, and the shavings emit a carbon that can cause respiratory or liver damage to small animals like rabbits.

SHEEP

Sheep were possibly the first domesticated animals, and are now found all over the world on farms and smaller plots of land. Almost all the breeds of sheep that are found in the United States have been brought here from Great Britain. Raising sheep is relatively easy, as they only need pasture to eat, shelter from bad weather, and protection from predators. Sheep's wool can be used to make yarn or other articles of clothing and their milk can be made into various types of cheeses and yogurt, though this is not normally done in the United States.

Sheep are naturally shy creatures and are extremely docile. If they are treated well, they will learn to be affectionate with their owner. If a sheep is comfortable with its owner, it will be much easier to manage and to corral into its pen if it's allowed to graze freely. Start with only one or two sheep; they are not difficult to manage but do require a lot of attention.

Breeds of Sheep

There are many different breeds of sheep—some are used exclusively for their meat and others for their wool. Six quality wool-producing breeds are as follows:

1. **Cotswold**—This breed is very docile and hardy and thrives well in pastures. It produces around 14 lbs of fleece per year, making it a very profitable breed for anyone wanting to sell wool.

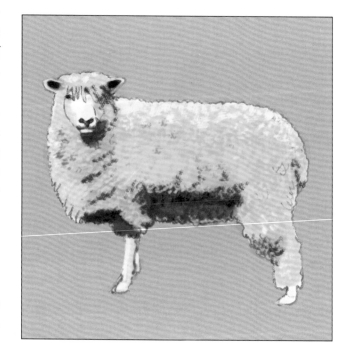

2. **Leicester**—This is a hardy, docile breed of sheep that is a very good grazer. This breed has 6-inch-long, coarse wool that is desirable for knitting. It is a very popular breed in the United States.

3. **Merino**—Introduced to the United States in the early twentieth century, this small- to medium-sized sheep has lots of rolls and folds of fine white wool and produces a fleece anywhere between 10 and 20 lbs. It is considered a fine-wool specialist, and though its fleece appears dark in color, the wool is actually white or buff. It is a wonderful foraging sheep, is hardy, and has a gentle disposition, but is not a very good milk producer.

4. **Oxford Down**—A more recent breed, these dark-faced sheep have hardy constitutions and good fleece.

5. **Shropshire**—This breed has longer, more open, and coarser fleece than other breeds. It is quite popular in the United States, especially in areas that are more moist and damp, as they seem to better in these climates than other breeds of sheep.

6. **Southdown**—One of the oldest breeds of sheep, they are popular for their good quality wool and are deemed the standard of excellence for many sheep owners. Docile, hardy, and good grazing on pastures, their coarse and light-colored wool is used to make flannel.

Housing Sheep

Sheep do not require much shelter—only a small shed that is open on one side (preferably to the south so it can stay warmer in the winter months) and is roughly 6 to 8 feet high. The shelter should be ventilated well to reduce any unpleasant smells and to keep the sheep cool in the summer. Feeding racks or mangers should be placed inside of the shed to hold the feed for the sheep. If you live in a colder region of the country, building a sturdier, warmer shed for the sheep to live in during the winter is recommended.

Straw should be used for the sheep's bedding and should be changed daily to make sure the sheep do not become ill from an unclean shelter. Especially for the winter months, a dry pen should be erected for the sheep to exercise in. The fences should be strong enough to keep out predators that may enter your yard and to keep the sheep from escaping.

What Do Sheep Eat?

Sheep generally eat grass and are wonderful grazers. They utilize rough and scanty pasturage better than other grazing animals and, due to this, they can actually be quite beneficial in cleaning up a yard that is overgrown with undesirable herbage. Allowing sheep to graze in your yard or in a small pasture field will provide them with sufficient food in the summer months. Sheep also eat a variety of weeds, briars, and shrubs. Fresh water should always be available for the sheep every time of year.

During the winter months especially, when grass is scarce, sheep should be fed on hay (alfalfa, legume, or clover hay) and small quantities of grain. Corn is also a good winter food for the sheep (it can also be mixed with wheat bran), and straw, salt, and roots can also be occasionally added to their diet. Good food during the winter season will help the sheep grow a healthier and thicker wool coat.

Shearing Sheep

Sheep are generally sheared in the spring or early summer before the weather gets too warm. To do your own shearing, invest in a quality hand shearer and a scale on which to weigh the fleece. An experienced shearer should be able to take the entire wool off in one piece.

You may want to wash the wool a few days to a week before shearing the sheep. To do so, corral the sheep into a pen on a warm spring day (make sure there isn't a cold breeze blowing and that there is a lot of sunshine so the sheep does not become chilled). Douse the sheep in warm water, scrub the wool, and rinse. Repeat this a few times until most of the dirt and debris is out of the wool. Diffuse some natural oil throughout the wool to make it softer and ready for shearing.

The sheep should be completely dry before shearing and you should choose a warm—but not overly hot—day. If you are a beginner at shearing sheep, try to find an experienced sheep owner to show you how to properly hold and shear a sheep. This way, you won't cause undue harm to the sheep's skin and will get the best fleece possible. When you are hand-shearing a sheep, remember to keep the skin pulled taut on the part where you are shearing to decrease the potential of cutting the skin.

Once the wool is sheared, tag it and roll it up by itself, and then bind it with twine. Be sure not to fold it or bind it too tightly. Separate and remove any dirty or soiled parts of the fleece before binding, as these parts will not be able to be carded and used.

Carding and Spinning Wool

To make the sheared wool into yarn you will need only a few tools: a spinning wheel or drop spindle and wool-cards. Wool-cards are rectangular pieces of thin board that have many wire teeth attached to them (they look like coarse brushes that are sometimes used for dogs' hair). To begin, you must clean the wool fleece of any debris, feltings, or other imperfections before carding it; otherwise your yarn will not spin correctly. Also wash it to remove any additional sand or dirt embedded in the wool and then allow it to dry completely. Then, all you need is to gather your supplies and follow these simple instructions:

Carding Wool

1. Grease the wool with rape oil or olive oil, just enough to work into the fibers.

2. Take one wool-card in your left hand, rest it on your knee, gather a tuft of wool from the fleece, and place it onto the wool-card so it is caught between the wired teeth of the card.

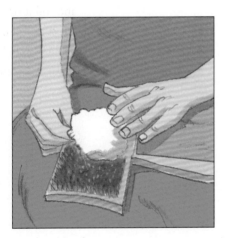

Note: Carded wool can also be used for felting, in which case no spinning is needed. To felt a small blanket, place large amounts of carded wool on either side of a burlap sack. Using felting needles, weave the wool into the burlap until it is tightly held by the jute or hemp fabrics of the burlap.

3. Take the second wool-card in your right hand and bring it gently across the other card several times, making a brushing movement toward your body.

4. When the fibers are all brushed in the same direction and the wool is soft and fluffy to the touch, remove the wool by rolling it into a small fleecy ball (roughly a foot or more in length and only 2 inches in width) and put it in a bag until it is used for spinning.

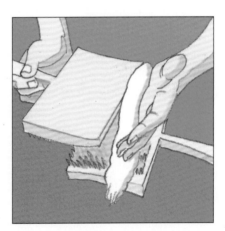

Spinning Wool

1. Take one long roll of carded wool and wind the fibers around the spindle.
2. Move the wheel gently and hold the spindle to allow the wool to "draw," or start to pull together into a single thread.

3. Keep moving the wheel and allow the yarn to wind around the spindle or a separate spool, if you have a more complex spinning wheel.

4. Keep adding rolls of carded wool to the spindle until you have the desired amount of yarn.

Note: If you are unable to obtain a spinning wheel of any kind, you can spin your carded wool by hand, although this will not produce the same tightness in your yarn as regular spinning. All you need to do is take the carded wool, hold it with one hand, and pull and twist the fibers into one, continuous piece. Winding the end of the yarn around a stick, spindle, or spool and securing it in place at the end will help keep your fibers tight and your yarn twisted. See page 507-508 for more on spinning yarn.

If you want your yarn to be different colors, try dying it with natural berry juices or with special wool dyes found in arts and crafts stores.

Milking Sheep

Sheep's milk is not typically used in the United States for drinking, making cheese, or other familiar dairy products. Sheep do not typically produce milk year-round, as cows do, so milk will only be produced if you bred your sheep and had a lamb produced. If you do have a sheep that has given birth and the lamb has been sold or taken away, it is important to know how to milk her so her udders do not become caked. Some ewes will still have an abundance of milk even after their lambs have been weaned and this excess milk should be removed to keep the ewe healthy and her udder free from infection.

To milk a ewe, bring her rear up to a fence so she cannot step backwards and, placing two knees against her shoulders to prevent her from moving forward, reach under with both hands and squeeze the milk into a bucket. When the udder is still soft but the ewe has been partly milked out, set her loose and then milk her again a few days later. If there is still milk to be had, wait another three days and then milk her again. By milking the ewes in this manner, you can prevent their udders from becoming infected and the milk from spoiling.

Diseases

The main diseases to which sheep are susceptible are foot rot and scabs. These are contagious and both require proper treatment. Sheep may also acquire stomach worms if they eat hay that has gotten too damp or has been lying on the floor of their shelter. As always, it is best to establish relationship with a veterinarian who is familiar with caring for sheep and have your flock regularly checked for any parasites or diseases that may arise.

TURKEYS

Turkeys are generally raised for their meat (especially for holiday roasts) though their eggs can also be eaten. Turkeys are incredibly easy to manage and raise as they primarily subsist on bugs, grasshoppers, and wasted grain that they find while wandering around the yard. They are, in a sense, self-sustaining foragers.

If you are looking to raise a turkey for Thanksgiving dinner, it is best to hatch the turkey chick in early spring, so that by November, it will be about 14 to 20 pounds.

Breeds of Turkeys

The largest breeds of turkeys found in the United States are the Bronze and Narragansett. Other breeds, though not as popular, include the White Holland, Black turkey, Slate turkey, and Bourbon Red.

Bronze breeds are most likely a cross between a wild North American turkey and domestic turkey, and they have beautiful rich plumage. This is the most common type of turkey to raise, as it is the largest, is very hardy, and is the most profitable. The White Holland and Bourbon Red, however, are said to be the most "domesticated" in their habits and are easier to keep in a smaller roaming area.

Bronze turkey

Black turkey

Bourbon Red turkey

Narragansett turkey

Slate turkey

White turkey

Housing Turkeys

Turkeys flourish when they can roost in the open. They thrive in the shelter of trees, though this can become problematic as they are more vulnerable to predators than if they are confined in a house. If you do build a house for them, it should be airy, roomy, and very clean.

It is important to allow turkeys freedom to roam; if you live in a more suburban or neighborhood area, raising turkeys may not be the best option for you, as your turkeys may wander into a neighboring yard, upsetting your neighbors. Turkeys need lots of exercise to be healthy and vigorous. When turkeys are confined for large periods of time, it is more

Slaughtering Poultry

If you are raising your own poultry, you may decide that you'd like to use them for consumption as well. Slaughtering your own poultry enables you to know exactly what is in the meat you and your family are consuming, and to ensure that the poultry is kept humanely before being slaughtered. Here are some guidelines for slaughtering poultry:

1. To prepare a fowl for slaughter, make sure the bird is secured well so it is unable to move (either hanging down from a pole or laid on a block that is used for chopping wood).

2. Killing the fowl can be done in two ways: one way is to hang the bird upside down and to cut the jugular vein with a sharp knife. It is a good idea to have a funnel or vessel available to collect the draining blood so it does not make a mess and can be disposed of easily. The other option is to place the bird's head on a chopping block and then, in one clean movement, chop its head off at the middle of the neck. Then, hang the bird upside down and let the blood drain as described above.

3. Once the bird has been thoroughly drained of blood, you can begin to pluck it. Have a pot of hot water (around 140°F) ready, into which to dip the bird. Holding the bird by the feet, dip it into the pot of hot water and leave it for about 45 seconds—you do not want the bird to begin to cook! Then, remove the bird from the pot and begin plucking immediately. The feathers should come off fairly easily, but this process takes time, so be patient. Discard the feathers.

4. Once the bird has been completely rid of feathers, slip back the skin from the neck and cut the neck off close to the base of the body. Then, remove the crop, trachea, and esophagus from the bird by loosening them and pulling them out through the hole created by chopping off the neck. Cut off the vent to release the main entrails (being careful not to puncture the intestines or bacteria could be released into the meat) and make a horizontal slit about an inch above it so you can insert two fingers. Remove the entrails, liver (carefully cutting off the gallbladder), gizzard, and heart from the bird and set the last three aside if you want to eat them later or make them into stuffing. If you are going to save the heart, slip off the membrane enclosing it and cut off the veins and arteries. Make sure to clean out the gizzard as well if you will be using it later.

5. Wash the bird thoroughly, inside and out, and wipe it dry.

6. Cut off the feet below the joints and then carefully pull out the tendons from the drumsticks.

7. Once the carcass is thoroughly dry and clean, store it in the refrigerator if it will be used that same day or the next. If you want to save the bird for later use, place it in a moisture-proof bag and set it in the freezer (along with any innards that you may have saved).

8. Make sure you clean and disinfect any surface you were working on to avoid the spread of bacteria and other diseases.

difficult to regulate their feeding (turkeys are natural foragers and thrive best on natural foods), and they are more likely to contract disease than if they are allowed to range freely.

What Do Turkeys Eat?

Turkeys gain most of their sustenance from foraging, either in lawns or in pastures. They typically eat green vegetation, berries, weed seeds, waste grain, nuts, and various kinds of acorns. In the summer months, turkeys especially like to get grasshoppers. Due to their love of eating insects that can damage crops and gardens, turkeys are quite useful in keeping your growing produce free from harmful insects and parasites.

Turkeys may be fed grain (similar to a mixture given to chickens) if they are going to be slaughtered, in order to make them larger.

Hatching Turkey Chicks

Turkey hens lay eggs from the middle of March to the first of April. If you are looking to hatch and raise turkey chicks, it is vital to watch the hen closely for when she lays the eggs, and then gather them and keep the eggs warm until the weather is more stable. Turkey hens generally aim to hide their nests from predators. It is best, for the hen's sake, to provide her with a coop of some sort, which she can freely enter and leave. Or, if no coop is available, encourage the hen to lay her eggs in a nest close to your house (putting a large barrel on its side and heaping up brush near the house may entice the hen to nest there). This way, you can keep an eye on the eggs and hatchlings.

Hens are well adapted to hatch all of the eggs that they lay. It takes 27 to 29 days for turkey eggs to hatch. While the hens are incubating the eggs, they should be given adequate food and water, placed close to their nest. Wheat and corn are the best food during the laying and incubation period.

Raising the Poults

Turkey chicks, also known as "poults," can be difficult to raise and require lots of care and attention for their first few weeks of life. In this sense, a turkey raiser must be "on call" to come to the aid of the hen and her poults at any time during the day for the first month or so. Many times, the hens can raise the poults quite well, but it is important that they receive enough food and warmth in the early weeks to allow them to grow healthy and strong. The poults should stay dry, as they become chilled easily. If you are able, encouraging the poults and their mother into a coop until the poults are stronger will aid their growth to adulthood.

Poults should be fed soft and easily digestible foods. Stale bread, dipped in milk and then dried until it crumbles, is an excellent source of food for the young turkeys.

Diseases

Turkeys are hardy birds but they are susceptible to a few debilitating or fatal diseases. It is a fact that the mortality rate among young turkeys, even if they are given all the care and exercise and food needed, is relatively high (usually due to environmental and predatory factors).

The most common disease in turkeys is blackhead. Blackhead typically infects young turkeys between 6 weeks and 4 months old. This disease will turn the head darker colored or even black and the bird will become very weak, will stop eating, and will have an insatiable thirst. Blackhead is usually fatal.

Another disease that turkeys occasionally contract is roup. Roup generally occurs when a turkey has been exposed to extreme dampness or cold drafts for long periods of time. Roup causes the turkey's head to swell around the eyes and is highly contagious to other turkeys. Nutritional roup is caused by a vitamin A deficiency, which can be alleviated by adding vitamin A to the turkeys' drinking water. It is best to consult a veterinarian if your turkey seems to have this disease.

PART TWO

BAKING, PRESERVING, AND MORE

BAKING: BREAD, CAKES, AND COOKIES

Breads

Bread has been a dining staple for thousands of years. The art of breadmaking has evolved over time, but the basic principles remain unchanged. Bread is made from flour of wheat or other grains, with the addition of water, salt, and a fermenting ingredient (such as yeast or another leavening agent). After you've baked a few loaves, you'll start to get a feel for what the dough should look and feel like. Then you can start experimenting with different flours, or additions of fruits, nuts, seeds, herbs, and more.

Quick Breads

Muffins, banana bread, zucchini bread, and many other sweet breads are often leavened with agents other than yeast, such as baking soda or baking powder. These breads are easy to make and require far less preparation time than yeast breads. They're also very versatile; once you master the basic recipe you can add almost any fruit, nut, or flavoring to make a uniquely delicious treat.

BASIC QUICK BREAD RECIPE

This basic recipe will make 2 loaves or 12 large muffins. Fold in 1 to 2 cups of mashed fruit, whole berries, nuts, or chocolate chips before pouring the batter into the pans.

> 3 ½ cups flour (use at least 2 cups of a gluten-rich flour)
> 2 tsp baking powder
> 1 tsp baking soda
> 1 tsp salt
> 1 to 2 tsp spices or herbs, if desired
> 1 ¼ cups sugar
> ¾ cup butter, oil, or fruit puree
> 3 eggs
> ¾ cup milk

1) In a large mixing bowl combine all dry ingredients except sugar.
2) In a separate bowl, beat together sugar and butter, oil, or fruit puree. Add eggs and beat until light and fluffy.
3) Add butter and sugar mixture and milk alternately to the dry ingredients, stirring just until combined. Fold in additional fruit, nuts, or flavors of your choice.
4) For bread, pour into a greased bread pan and bake at 350°F for 1 hour. For muffins, fill muffin cups ⅔ full and bake at 350°F for 20 to 25 minutes.

CINNAMON BREAD

2 eggs

½ cup butter

1 cup sugar

½ cup milk

1 ¼ cups flour

2 ½ tsp baking powder

1 tsp cinnamon

1 tsp butter, melted

2 Tbs sugar and 2 Tbs cinnamon, mixed together

1) Beat together the eggs, butter, and sugar until fluffy.

2) In a separate bowl, combine the dry ingredients. Add the dry mixture and the milk to the butter mixture and mix until combined.

3) Bake in a greased bread pan at 300°F for almost an hour. When done pour melted butter over top and sprinkle with cinnamon and sugar mixture.

ONE-HOUR BROWN BREAD

1 cup cornmeal

1 cup white flour

½ tsp salt

1 tsp baking soda

1 cup water, boiling

1 egg

½ cup molasses

½ cup sugar

1) Combine cornmeal, flour, and salt.

2) Add the baking soda to boiling water and stir. Add to dry ingredients.

3) Beat together egg, molasses, and sugar and add to dry ingredients. Mix until combined. Pour batter into an empty coffee can with a cover (or cover with foil).

4) Place a cake rack in the bottom of a dutch oven or large pot. Place the covered can on the rack and pour boiling water into the pot until it reaches half way up the cans. Cover the pot, turn the unit on very low, and steam for one hour.

Cinnamon Bread

CRANBERRY COFFEE CAKE

2 Tbs butter

¼ cup firmly packed brown sugar

1 cup cooked or canned cranberry sauce

¼ cup pecans, chopped

1 Tbs grated orange rind

1 ½ cups sifted flour

2 tsp double acting baking powder

¼ cup sugar

⅓ cup shortening

1 beaten egg

½ cup milk

1) Melt butter in 9-inch ring mold. Spread brown sugar over bottom of pan.

2) Combine cranberry sauce, pecans, and orange rind. Spread over brown sugar in bottom of pan.

3) Sift together flour, baking powder, and sugar.

4) Cut in shortening until dough resembles coarse meal. Combine egg with milk. Add all at once, mixing only to dampen flour. Turn into pan.

5) Bake at 400°F for 25 to 30 minutes. Cool 5 minutes and invert onto plate. Serve warm.

DATE-ORANGE BREAD

2 Tbs butter or margarine melted

¾ cup orange juice

2 Tbs grated orange rind

½ cup finely cut dates

1 cup sugar

1 egg, slightly beaten

½ cup coarsely chopped pecan

2 cups sifted all-purpose flour

½ tsp baking soda

1 tsp baking powder

½ tsp salt

1) Combine first 7 ingredients.

2) Mix and sift remaining ingredients; stir in. Mix well, but quickly, being careful not to overbeat.

3) Turn into greased loaf pan. Bake in moderate oven, 350°F, for 50 minutes or until done. Remove from pan and let cool right side up, on a wire rack.

PINEAPPLE NUT BREAD

2 ¼ cups sifted flour

¾ cup sugar

1 ½ tsp salt

3 tsp baking powder

½ tsp baking soda

1 cup prepared bran cereal

¾ cup chopped walnuts

1 ½ cups crushed pineapple, undrained

1 egg beaten

3 Tbs shortening, melted

1) Sift flour, sugar, salt, baking powder, and soda together.

2) Mix together remaining ingredients and combine with dry mixture.

3) Bake in greased loaf pan at 350°F for 1 ¼ hrs. This bread keeps moist a week or ten days, and slices best when a day or more old.

DATE MUFFINS

1 ¾ cups sifted enriched flour

2 Tbs sugar

2 ¼ baking powder

¾ tsp salt

½ to ¾ cup coarsely cut pitted dates

1 well-beaten egg

¾ cup milk

⅓ cup melted shortening or salad oil

1) Sift dry ingredients into mixing bowl and stir in dates. Make a well in center.

2) Combine egg, milk, and salad oil; add all at once to dry ingredients. Stir quickly only till dry ingredients are moistened.

3) Drop batter by tablespoons into greased muffin pans. Fill ⅔ full. Bake in hot oven, 400°F, for about 25 minutes. Makes 1 dozen.

BREAD-BRAN MUFFINS

2 cups bran

1 cup dried, ground breadcrumbs

½ cup bread flour

3 ½ Tbs sugar

1 tsp salt

1 egg

2 cups milk

4 tsp baking powder

Mix together dry ingredients, beat egg, add milk, or half and half of milk and water, and stir into first mixture. Bake about twenty-five minutes in well-oiled gem pans. This will make eighteen muffins.

CARAMEL BISCUITS

2 cups bread flour

4 tsp baking powder

1 tsp salt

1 Tbs lard

1 Tbs butter

⅓ cup milk

⅓ cup water

1 cup light brown sugar

½ cup butter

Nutmeg

Mix and sift the flour, baking powder, and salt twice. Work in the butter and lard with the tips of the fingers until it is thoroughly blended. Add the milk and water and mix to a soft dough, using a knife. (A trifle more liquid may be needed.) Toss on a floured board, roll lightly to one-fourth inch thickness. Cream the brown sugar and butter together till it is smooth, then spread lightly over the dough. Roll up like a jelly roll, fasten end by moistening with milk or water, and cut in pieces three-fourths inch thick. Sprinkle just a little nutmeg over each slice and bake in a hot oven fifteen minutes. Serve hot.

BERRY LOAF

⅓ cup shortening

⅔ cup brown sugar

⅓ cup sour milk

1 egg

1 ½ cups pastry flower

1 tsp baking powder

½ tsp baking soda

½ tsp cinnamon

½ tsp salt

½ tsp nutmeg

1 cup cooked berries, drained

Cream together the shortening and the brown sugar; add the milk, egg well beaten, and all the dry ingredients sifted together. Then add the berries. Mix thoroughly together and bake in a well-greased loaf-cake pan.

COCOA MUFFINS

2 Tbs shortening

2 Tbs cocoa

3 tsp baking powder

1 cup sugar

½ cup water

1 ½ cup pastry flour

2 eggs

½ tsp vanilla

¼ tsp salt

Cream shortening and sugar, add egg yolks. Combine flour, cocoa, and baking powder; add alternatively with the water. Add vanilla and fold in stiffly beaten whites of eggs. Bake in muffin pans in quick oven.

CRUMPETS

3 cups tepid milk

1 tsp salt

½ compressed yeast-cake

2 Tbs melted butter

4 cups bread flour

¼ tsp baking soda

Soften the yeast in a little warm milk, add to remaining milk with the salt, and stir in about four cups of flour or enough to make a muffin-batter; let stand overnight, and in the morning add the melted butter and the soda dissolved in a tablespoon of hot water. Beat thoroughly, place in well-oiled muffin tins, filling them half full, let raise about 25 minutes, and bake in a hot oven.

DELICIOUS QUICK TEA-ROLLS

    1 compressed yeast-cake
    ½ cup scalded milk
    ½ cup tepid water
    1 Tbs sugar
    3 cups bread flour
    2 Tbs lard or butter
    1 tsp salt

Add shortening to milk, and let cool till lukewarm. Soften yeast and dissolve sugar in the water, combine mixtures, beat in half of the flour, whipping till very smooth, t hen add the balance together with the salt. Knead thoroughly; roll one-fourth inch thick, brush lightly with melted butter, shape with a two-inch biscuit-cutter, crease and fold over in a pocket-shape. Set to raise in a warm place for about two hours, and bake fifteen minutes in a moderate oven. This makes two dozen small rolls.

ENGLISH MUFFINS

    1 ½ cups milk
    1 tsp sugar
    1 tsp salt
    3 cups bread flour
    ½ yeast-cake
    1 egg
    ⅛ tsp baking soda

Heat milk till tepid; add sugar and salt and the yeast-cake softened in a little warm water. Beat in flour to make a soft batter, from two or three cups according to the brand of flour, and let rise till light, about three hours. Stir in the egg well beaten and the baking soda, beat thoroughly, and cook in muffin-rings on a griddle. This must be done slowly, about twelve minutes.

GRANDMOTHER'S MUFFINS

    1 cup cornmeal
    1 cup bread flour
    ½ cup light brown or maple sugar
    2 cups cold water
    1 Tbs shortening

    ½ tsp baking soda
    1 Tbs hot water
    1 tsp salt

Mix the cornmeal, flour, sugar, salt, and cold water together. Let stand overnight. In the morning add the shortening, melted, and the baking soda dissolved in the hot water. Beat very thoroughly, drop into hot, well-greased gem-pans and bake in a moderate oven about twenty-five minutes.

MAPLE ROLLS

    2 cups bread flour
    4 tsp baking powder
    1 tsp salt
    ¾ cup milk
    Melted butter

Scraped or grated maple-sugar

2 Tbs lard

Make a baking powder biscuit-dough of the flour, baking powder, salt, lard, and milk. Roll in oblong shape one-half inch thick, brush with the melted butter and spread with the maple-sugar. Dampen the outer edges with a little cold water and roll up firmly. Cut in crosswise slices about one-half inch thick, place in a well-oiled baking-pan, cut side down, and bake in a hot oven twenty to twenty-five minutes.

APPLE CORN BREAD

⅞ cup cornmeal

½ cup bread flour

1 tsp salt

1 Tbs molasses

¾ cup buttermilk

½ tsp baking powder

1 Tbs melted shortening

3 medium-sized apples

½ tsp baking soda

Mix all the dry ingredients thoroughly together, add the buttermilk, molasses, and shortening, and mix well. Pour into a shallow greased tin, and place the apples, peeled and cut in eighths, over the top. Bake in a hot oven three-quarters of an hour. When done, dust with powdered sugar.

PEANUT BUTTER BREAD

½ cup peanut butter

½ cup sugar

1 egg

3 ½ cups bread flour

3 tsp baking powder

1 cup milk

½ tsp salt

Cream the peanut butter and sugar. Add the egg, well beaten. Mix and sift the dry ingredients and add alternatively with the milk. Beat the entire mixture well and place in a well-greased bread tin. Bake in a 350° F. oven for 50 minutes.

RICE SPOON BREAD

2 cups boiled rice

3 eggs

1 cup cornmeal

1 quart sweet milk

1 Tbs melted margarine

4 tsp baking powder

3 tsp salt

Beat the eggs lightly, and add the rice, milk and margarine. Sift the dry ingredients together, and add to the first mixture. Pour into a hot, greased baking pan and bake in a 350° F. oven for 45 minutes.

SPOON BREAD

1 pint milk

½ cup cornmeal

½ tsp baking powder

1 tsp salt

3 eggs

Heat the milk nearly to boiling. Stir in cornmeal gradually and cook until the consistency of mush. Add the baking powder, salt, and egg yolks beaten until light. Fold in the egg whites beaten stiff. Pour into a greased baking dish and bake for 30 minutes at 350° F. Serve at once with plenty of butter, from the dish in which it was baked.

OLIVE BREAD

1 ½ cups bread flour

1 ½ cups Graham flour

2 tsp baking powder

½ cup molasses

½ tsp baking soda

1 cup ripe olives

1 ¾ cups milk

1 ½ tsp salt

Mix thoroughly the flour, salt, and baking powder. Add soda to the molasses, and combine with milk. Mix all together and beat well. Lastly add the ripe olives, stoned and cut into pieces not too fine, turn into a well-greased loaf-pan, bake in a moderate oven for one hour.

SWEET MILK SCONES

½ tsp salt
1 rounding tsp baking powder
2 cups flour
1 Tbs butter
1 cup sweet milk

Add salt and baking powder to flour; sift once or twice, and rub in butter. Stir in sweet milk. The dough must be sufficiently soft to drop from the spoon. Dip a tablespoon in boiling water, then take up a tablespoonful of dough and drop into a greased pan, keeping the scones sufficiently far apart not to touch in the baking. Bake quickly for twenty minutes.

Sour milk may be used in the place of sweet milk, substituting half a teaspoonful of baking soda for the baking powder.

Scotch scones are made from sour milk and flour, the dough being sufficiently thick to roll and cut. These are baked on a griddle, turning them several times during the baking.

PINWHEEL BISCUITS (Fifteen biscuits)

2 cups flour
4 tsp baking powder
3 Tbs lard
½ tsp salt
¾ cup milk
⅓ cup stoned raisins
2 Tbs sugar
2 Tbs melted butter

Sift together the flour, baking powder, and salt, work in the lard with a knife, add gradually the milk, mixing with the knife to a soft dough. Toss on the floured board, roll one inch thick, spread with butter, and sprinkle with the sugar and cinnamon, which have been well mixed. Press in the raisins. Roll up the mixture evenly as you would a jelly roll. Cut off slices, an inch thick—flatten a little and place in a tin pan. Bake in a hot oven for fifteen minutes. (These are similar to the cinnamon rolls made from yeast sponge.)

TWIN MOUNTAIN MUFFINS

2 cups flour
4 tsp baking powder
¼ tsp salt
1 egg
1 cup milk
1 Tbs melted butter
¼ cup sugar

Mix and sift together the flour, baking powder, salt and sugar. Beat the egg, add the milk; add these liquid ingredients to the dry ones. Beat two minutes. Add the melted butter. Fill well bettered muffin pans one-half full. Bake in a moderate oven twenty minutes.

PEANUT BREAD (Twelve slices)

2 cups flour
4 tsp baking powder
½ tsp salt
4 Tbs sugar
1 egg
½ cup chopped peanuts
¾ cup milk

Mix thoroughly the flour, baking powder, salt, sugar, and peanuts. Add the egg and milk. Stir vigorously two minutes. Place in a well-buttered bread pan, and bake 35 minutes in a moderate oven.

DATE MUFFINS (Ten muffins)

¼ cup sugar
¼ cup dates cut fine
1 egg
¼ tsp salt
¾ cup milk
1 ¾ cups flour
4 tsp baking powder
2 Tbs butter (melted)

Mix the sugar, dates, baking powder, flour and salt. Add milk in which one egg has been beaten. Beat two minutes. Add butter, melted. Fill well-buttered muffin pans half full of the mixture, and place in the oven. Bake 20 minutes. Serve hot or cold.

CORN BREAD (Three portions)

½ cup cornmeal
⅔ cup all-purpose flour
3 Tbs sugar
2 tsp baking powder
½ tsp salt
1 egg yolk
⅔ cup milk
1 Tbs melted butter

Mix the cornmeal, flour, sugar, baking powder and salt thoroughly. Add the egg yolk and milk, and beat two minutes. Add the melted butter. Mix well. Pour into a well buttered square or round cake pan. Bake in a moderate oven 20 minutes.

NUT BREAD (Twenty-four sandwiches)

1 ½ cups graham flour
2 cups white flour
4 tsp baking powder
1 cup confectioner's sugar
2 tsp salt
1 ½ cups milk
⅔ cup chopped nuts, dates, or raisins

Sift together all the dry ingredients, add the nuts and fruit. Add the milk. Stir well, and pour into two well-buttered loaf pans. Allow to stand and rise for twenty minutes. Bake three-fourths of an hour in a moderate oven. Use bread twenty-four hours old for the sandwiches. "C" sugar is light brown sugar and gives food a delicious flavor.

LIGHT ROLLS

2 Tbs sugar
¼ tsp salt
½ cup scalded milk
½ yeast cake
¾ cup flour
2 Tbs melted butter

1 egg, well-beaten
2 Tbs lukewarm water
Flour

Add the sugar and salt to the scalded milk and when lukewarm, add the yeast dissolved in the lukewarm water, and three-fourths of a cup of flour. Cover and set in a warm place to rise. Then add the melted butter, the well-beaten egg, and enough flour to knead. Let rise in a warm place. Roll to one-half an inch in thickness and shape with a biscuit cutter. Butter the top of each. Fold over, place in a buttered pan, close together. Let rise again for 45 minutes and then bake in a quick oven for 20 minutes.

IRISH BREAD

1 cup milk
1 egg
1 tsp sugar
1 tsp salt
⅓ cup butter
¼ cake yeast

Heat milk and butter, add sugar and salt; next yeast and a little flour. Then the beaten egg and flour until it is stiffer than muffins. Let raise and put in greased pan. Bake half an hour.

OLD-FASHIONED RUSK

1 cup sugar
1 egg
1 Tbs butter
2 ½ cups flour
1 tsp cinnamon
½ tsp cloves
½ tsp nutmeg
½ tsp salt
1 small cup stoned raisins
1 cup sour milk
1 tsp baking soda

Mix all ingredients together. Bake in slow oven.

Cornbread

SQUASH MUFFINS

1 cup cooked and sifted squash
1 egg
½ cup sugar
1 cup milk
A little salt
1 ½ cups flour
2 tsp baking powder

Sift the baking powder in the flour. Combine ingredients. Mix well and bake in gem pans in a hot oven.

ALLEGHANY SPRINGS BROWN BREAD

½ tsp baking soda
½ cup hot molasses
2 cups sweet milk
A little salt
4 cups graham flour
2 tsp baking powder

Stir baking soda into hot molasses; add sweet milk, salt, and graham flour, baking powder; cover with a bread tin and bake one hour in a moderate oven.

Yeast Bread

Once you've made a loaf of homemade yeast bread, you'll never want to go back to buying packaged bread from the grocery store. Homemade bread tastes and smells heavenly and the baking process itself can be very rewarding. Store homemade bread in a paper or resealable plastic bag and eat within a day or two for best results. Bread that begins to get stale can be cubed and made into stuffing or croutons.

Before you start baking, it's helpful to understand the various components that make up bread.

Wheat

Wheat is the most common flour used in bread making, as it contains gluten in the right proportion to make bread rise. Gluten, the protein of wheat, is a gray, tough, elastic substance, insoluble in water. It holds the gas developed in bread dough by fermentation, which otherwise would escape. Though there are many ways to make gluten-free bread, flour that naturally contains gluten will rise more easily than gluten-free grains. In general, combining smaller amounts of other flours (rye, corn, oat, etc.) with a larger proportion of wheat flour will yield the best results.

A grain of wheat consists of (1) an outer covering, or husk, which is always removed before milling; (2) bran, a hard shell that contains minerals and is high in fiber; (3) the germ, which contains the fat and protein content and is the part that can be planted and cultivated to grow more wheat; and (4) the endosperm, which is the wheat plant's own food source and is mostly starch and protein. Whole wheat contains all of these components except for the husk. White flour is only the endosperm.

FLOUR	DESCRIPTION
All-purpose	A blend of high- and low-gluten wheat. Slightly less protein than bread flour. Best for cookies and cakes.
Amaranth	Gluten-free. Made from seeds of amaranth plants. Very high in fiber and iron.
Arrowroot	Gluten-free. Made from the ground-up root. Clear when cooked, which makes it perfect for thickening soups or sauces.
Barley	Ground barley grain. Very low in gluten. Use as a thickener in soups or stews or mixed with other flours in baked goods.
Bran	Made from the hard outer layer of wheat berries. Very high in protein, fiber, vitamins, and minerals.
Bread	Made from hard, high-protein wheat with small amounts of malted barley flour and vitamin C or potassium bromate. It has a high gluten content, which helps bread to rise. Excellent for bread, but not as good for use in cookies or cakes.
Chickpea	Gluten-free. Made from ground chickpeas. Used frequently in Indian, Middle Eastern, and some French Provençal cooking.

FLOUR	DESCRIPTION
Buckwheat	Gluten free. Highly nutritious with a slightly nutty flavor.
Oat	Gluten-free, though many people with gluten allergies or sensitivities are also adversely affected by oats. Made from ground oats. High in fiber.
Quinoa	Gluten-free. Made from ground quinoa, a grain native to the Andes in South America. Slightly yellow or ivory-colored with a mild nutty flavor. Very high in protein.
Rye	Milled from rye berries and rye grass. High in fiber and low in gluten. Light rye has had more of the bran removed through the milling process than dark rye. Slightly sour flavor.
Semolina	Finely ground endosperm of durum wheat. Very high in gluten. Often used in pasta.
Soy	Gluten-free. Made from ground soybeans. High in protein and fiber.
Spelt	Similar to wheat, but with a higher protein and nutrient content. Contains gluten but is often easier to digest than wheat. Slightly nutty flavor.
Tapioca	Gluten-free. Made from the cassava plant. Starchy and slightly sweet. Generally used for thickening soups or puddings, but can also be used along with other flours in baked goods.
Teff	Gluten-free. Higher protein content than wheat and full of fiber, iron, calcium, and thiamin.
Whole wheat	Includes the bran, germ, and endosperm of the wheat berry. Far more nutritious than white flour, but has a shorter shelf life.

The Junior Homesteader

The Grain Game

Play a "brain game" to learn about grain-based foods. Taste different foods made from grains and learn where grains are grown in the United States.

Materials Needed

- Food samples
- Paper cups, napkins, or paper towels
- Container of water

Goals

- To identify different foods made from grains.
- To expand the variety of foods eaten by tasting different kinds of foods made from grains.
- To identify where different grains are grown in the United States.

Key Concepts

- There are a wide variety of foods made from grains.
- Eating foods made from different grains adds a variety of tastes to meals.

Preparation

Purchase foods that are made from the grains that will be discussed in this lesson. Examples of foods that might be purchased are: corn tortilla, rye bread, pumpernickel bread, oatmeal muffins or oatmeal cookies, and rice cakes. Additional breads you might consider including, if they are available, are:

scones—a British sweet biscuit

chapatis—a flat bread eaten in India and in East Africa

pita bread—a flat bread also known as "pocket bread"

lavash—a paper-thin Russian bread used for wrapping food

matzoh—a flat, cracker-like bread

corn bread—a bread made from cornmeal

Cut foods into bite-sized pieces.

Background

A grain is a single seed of a cereal grass. Some of the cereal grains grown in the United States are wheat, corn, rye, rice, barley, and oats. More foods are made with wheat than any other cereal grain.

Each grain tastes differently and adds delicious taste, nutrition, and variety to meals. Grain-based foods provide complex carbohydrates, which are an important source of energy for the body.

Grains also provide vitamins that help keep the body strong and healthy such as B vitamins, minerals such as iron, and dietary fiber, which keeps the digestive systems healthy. Grain products belong in the Breads, Cereals, Rice, and Pasta Group of the Food Guide Pyramid. The following are examples of grain-based foods categorized by their main grain ingredient.

Wheat

white bread	wheat bread
noodles	spaghetti
biscuit	fry bread
flour tortilla	wonton wrapper
cracker	waffle
graham cracker	scone
pita bread	matzoh
pancake	crepe
cream-of-wheat cereal	wheat flakes
popover	couscous
tabbouleh	cake

Corn

corn bread	corn tortilla
popcorn	hominy
grits	corn flakes
cornmeal mush	hushpuppy

Oats

oatmeal	oatmeal cookie
oatmeal muffin	ready-to-eat oat cereal
granola	muesli

Rye

rye bread	rye flatbread
pumpernickel bread	rye crackers

Rice

wild rice	white rice
basmati rice	texmati rice
jasmine rice	brown rice
Spanish rice	ready-to-eat rice cereal
risotto	sticky rice
rice noodles	rice cake
rice pudding	rice cereal (infant)
rice balls	popped wild rice
cream-of-rice cereal	

Setup and Introduction

1. Tell children that they will be playing a game about grains.

2. Explain that before they begin the Grain Game they will first need to understand that a grain is a seed from a cereal grass. Some cereal grains grown in the United States for food are wheat, corn, rice, oat, and rye (barley and millet are also cereal grains but are used less often in the United States).

3. Divide the group into two teams. The group leader should moderate the game. First, explain to children how the game is played:

- The group leader will call out a type of grain (e.g., wheat, corn, rice, oats, rye).
- One team will begin the game by calling out a food made from that grain.
- The other team will respond by calling out a different food made from that grain.
- Each time a correct food is called out, that team gets 1 point. When a team calls out an incorrect food, that team will not get a point but the point will go to the other team.
- The teams will alternate calling out a different food until no more can be named.
- When no more can be named, the group leader calls out the name of a new grain and the game continues.
- Play rounds of this game until wheat, corn, rice, oats, and rye are covered.

Game Closure

Suggested discussion questions following the game:
What foods made from grains do you eat now?
What new foods made from grains did you learn about today?
Which of the foods named today would you want to taste?

Tasting Activity

1. Tell the kids that now they are going to taste some foods made from different grains.
2. Have everyone wash their hands.
3. Have a volunteer distribute cups of water and paper plates or napkins. The water is for sipping between food samples.
4. Children should take a sample from each food plate, taste the food, and try to figure out which grain it's made with. Have volunteers name each food and its grain ingredient.
5. Suggested discussion questions:
Which foods did you like best? Why?
Which were new to you?

Suggested discussion starters:
Name the states where grains are grown.
Where are most grains grown? Why?
How do the grains grown in the different states get to other parts of the country to be made into foods?

Milling Your Own Grains

You can grind grains into flour at home using a mortar and pestle, a coffee or spice mill, manual or electric food grinders, a blender, or a food processor. Grains with a shell (quinoa, wheat berries, etc.) should be rinsed and dried before milling to remove the layer of resin from the outer shell that can impart a bitter taste to your flour. Rinse the grains thoroughly in a colander or mesh strainer, then spread them on a paper or cloth towel to absorb the extra moisture. Transfer to a baking sheet and allow to air dry completely (to speed this process you can put them in a very low oven for a few minutes). When the grains are dry, they're ready to be ground.

TIP

1 package active dry yeast = about 2 ¼ teaspoons = ¼ ounce

4-ounce jar active dry yeast = 14 tablespoons

1 (6-ounce) cube or cake of compressed yeast (also know as fresh yeast) = 1 package of active dry yeast

Multiply the amount of instant yeast by 3 for the equivalent amount of fresh yeast.

Multiply the amount of active dry yeast by 2.5 for the equivalent amount of fresh yeast.

Multiply the amount of instant yeast by 1.25 for the equivalent of active dry yeast.

Yeast

Yeast is a microscopic fungus that consists of spores, or germs. These spores grow by budding and division, multiply very rapidly under favorable conditions, and produce fermentation. Fermentation is the process by which, under influence of air, warmth, and a fermenting ingredient, sugar (or dextrose, starch converted into sugar) is changed into alcohol and carbon dioxide.

Dry yeast is most commonly used for baking. Most grocery stores sell regular active dry and instant yeast. Instant yeast is more finely ground and thus absorbs the moisture faster, speeding up the leavening process and making the bread rise more rapidly.

Active dry yeast should be proofed before using. Mix one packet of active dry yeast with ¼ cup warm water and 1 teaspoon sugar. Stir until yeast dissolves. Allow it to sit for 5 minutes, or until it becomes foamy.

Gluten-Free Bread

Making good gluten-free bread isn't always easy, but there are several things you can do to improve your chances of success:

- Choose flours that are high in protein, such as sorghum, amaranth, millet, teff, oatmeal, and buckwheat

- Use all room temperature ingredients. Yeast thrives in warm environments.

- Add a couple teaspoons of xantham gum to your dry ingredients.

- Add eggs and dry milk powder to your bread. These will add texture and help the bread to rise.

- Crush a vitamin C tablet and add it to your dry ingredients. The acidity will help the yeast do its job.

- Substitiute carbonated water or gluten-free beer for other liquids in the recipe.

- If you're following a traditional bread recipe, add extra liquid (water, carbonated water, milk, fruit juice, or olive oil) to get a soft and sticky consistency. The batter should be a little too sticky to knead. For this reason, bread machines are great for making gluten-free bread.

Making Bread

Making bread is a fairly simple process, though it does require a chunk of time. Keep in mind, though, that you can be doing other things while the bread is rising or baking. The process is fairly straightforward and only varies slightly by kind of bread.

① Mix together the flour, sugar, salt, and any other dry ingredients. Form a well in the center and add the dissolved yeast and any other wet ingredients. Mix all the ingredients together.

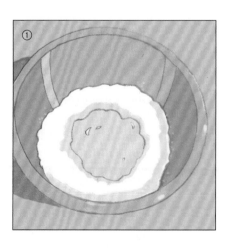

② Gather the dough into a ball and place it on a lightly floured surface. Flour your hands to keep the dough from sticking to your fingers. Knead the dough by folding it toward you and then pushing it away with the palms of your hands. Continue kneading for five to ten minutes, or until the dough is soft and elastic.

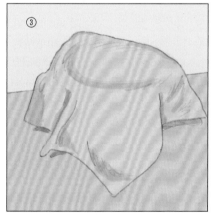

③ Place the dough in a lightly greased pan, cover with a dish towel, and allow to rise in a warm place until it doubles in size.

④ Punch the dough down to expel the air and place it in a greased and lightly floured baking pan. Cover and let rise a second time until it doubles in size.

⑤   Bake the bread in a preheated oven according the recipe. Bread is done when it is golden brown and sounds hollow when you tap the top.

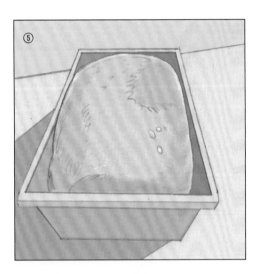

⑥   Remove bread from the pan by loosening the sides with a knife or spatula and tipping the pan upside down onto a wire rack.

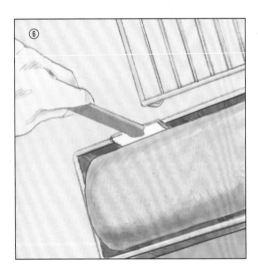

Biscuits

Any bread recipe can be made into biscuits instead of one large loaf. To shape bread dough into biscuits, pull or cut off pieces, making them all as close to uniform in size as possible. Flour palms of hands slightly and shape each piece individually. Using the thumb and first two fingers of one hand, and holding it in the palm of the other hand, move the dough round and round, folding the dough towards the center. When smooth, turn it over and roll between palms of

hands. Place in greased pans near together, brushed between with a little melted butter, which will allow biscuits to separate after baking.

Multigrain Bread

¼ cup yellow cornmeal

¼ cup packed brown sugar

1 tsp salt

2 Tbs vegetable oil

1 cup boiling water

1 package active dry yeast

¼ cup warm (105 to 115°F) water

⅓ cup whole wheat flour

¼ cup rye flour

2 ¼–2 ¾ cups all-purpose flour

1. Mix cornmeal, brown sugar, salt, and oil with boiling water; cool to lukewarm (105 to 115°F).

2. Dissolve yeast in ¼ cup warm water; stir into cornmeal mixture. Add whole wheat and rye flours and mix well. Stir in enough all-purpose flour to make dough stiff enough to knead.

3. Turn dough onto lightly floured surface. Knead until smooth and elastic, about 5 to 10 minutes.

4. Place dough in lightly oiled bowl, turning to oil top. Cover with clean towel; let rise in warm place until double, about 1 hour.

5. Punch dough down; turn onto clean surface. Cover with clean towel; let rest 10 minutes. Shape dough and place in greased 9 x 5 inch pan. Cover with clean towel; let rise until almost double, about 1 hour.

6. Preheat oven to 375°F. Bake 35 to 45 minutes or until bread sounds hollow when tapped. Cover with aluminum foil during baking if bread is browning too quickly. Remove bread from pan and cool on wire rack.

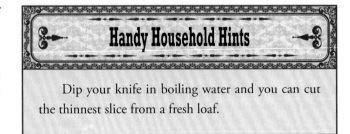

Handy Household Hints

Dip your knife in boiling water and you can cut the thinnest slice from a fresh loaf.

The Junior Homesteader

Bread in a Bag

Materials needed:

- A heavy-duty zipper-lock freezer bag (1 gallon size)
- Measuring cup
- Measuring spoons
- Cookie sheet
- Pastry towel or cloth
- 13-inch x 9-inch baking pan
- 8½-inch x 4½-inch glass loaf pan

Ingredients:

2 cups all-purpose flour, divided
1 package rapid rise yeast
3 Tbs sugar
3 Tbs nonfat dry milk
1 teaspoon salt
1 cup hot water (125ºF)
3 Tbs vegetable oil
1 cup whole-wheat flour

1. Combine 1 cup all-purpose flour, yeast, sugar, dry milk, and salt in a freezer bag. Squeeze upper part of the bag to force out air and then seal the bag.
2. Shake and work the bag with fingers to blend the ingredients.
3. Add hot water and oil to the dry ingredients in the bag. Reseal the bag and mix by working with fingers.
4. Add whole-wheat flour. Reseal the bag and mix ingredients thoroughly.
5. Gradually add remaining cup of all-purpose flour to the bag. Reseal and work with fingers until the dough becomes stiff and pulls away from sides of the bag.
6. Take dough out of the bag, and place on floured surface.
7. Knead dough 2 to 4 minutes, until smooth and elastic.
8. Cover dough with a moist cloth or pastry towel; let dough stand for 10 minutes.
9. Roll dough to 12-inch x 7-inch rectangle. Roll up from narrow end. Pinch edges and ends to seal.
10. Place dough in a greased glass loaf pan; cover with a moist cloth or pastry towel.
11. Place baking pan on the counter; half fill with boiling water.
12. Place cookie sheet over the baking pan and place loaf pan on top of the cookie sheet; let dough rise 20 minutes or until dough doubles in size.
13. Preheat oven, 375ºF, while dough is rising (about 15 minutes).
14. Place loaf pan in oven and bake at 375ºF for 25 minutes or until baked through.

Oatmeal Bread

1 cup rolled oats
1 tsp salt
1 ½ cups boiling water
1 package active dry yeast
¼ cup warm water (105 to 115ºF)
¼ cup light molasses
1 ½ Tbs vegetable oil
2 cups whole wheat flour
2–2 ½ cups all-purpose flour

1. Combine rolled oats and salt in a large mixing bowl. Stir in boiling water; cool to lukewarm (105 to 115ºF).
2. Dissolve yeast in ¼ cup warm water in small bowl.
3. Add yeast water, molasses, and oil to cooled oatmeal mixture. Stir in whole wheat flour and 1 cup all-purpose flour. Add additional all-purpose flour to make a dough stiff enough to knead.
4. Knead dough on lightly floured surface until smooth and elastic, about 5 minutes.
5. Place dough in lightly oiled bowl, turning to oil top. Cover with clean towel; let rise in warm place until double, about 1 hour.

6. Punch dough down; turn onto clean surface. Shape dough and place in greased 9 x 5 inch pan. Cover with clean towel; let rise in a warm place until almost double, about 1 hour.

7. Preheat oven to 375ºF. Bake 50 minutes or until bread sounds hollow when tapped. Cover with aluminum foil during baking if bread is browning too quickly. Remove bread from pan and cool on wire rack.

Raised Buns (Brioche)

½ cup milk

⅓ cup butter

¼ cup sugar

¾ tsp salt

1 package yeast (active, dry or compressed)

¼ cup lukewarm water

3 eggs, well beaten

2 ½ cups enriched flour

¼ tsp lemon rind, or ⅛ tsp crushed cardamom seeds

1. Save 2 tablespoons egg to brush brioches before baking. Scald milk; stir in butter, sugar, and salt; cool to lukewarm. Sprinkle or crumble yeast into water in large bowl; stir to dissolve. Add lukewarm milk mixture and beaten eggs; mix well.

2. Sift flour; add 1 ½ cups of it to mixture and beat by hand 8 minutes or with electric mixer at medium speed 3 minutes. Add remaining sifted flour and lemon rind; beat to smooth, heavy batter. Cover with towel; let rise in warm place 2 hours or until doubled. Stir down; cover tightly; chill at least 5 hours or overnight.

3. Stir down again. Mixture is soft now. Grease hands slightly, place dough on lightly floured board, and knead a few times. With sharp knife, cut off small pieces of dough. Roll with hands to about ½ inch in diameter, and about 10 inches long. Coil loosely in circle, winding around toward center. Top with small ball of dough. Cover with a towel and let rise ½ hr or until doubled. Brush with egg yolk beaten with 1 teaspoon water. If desired, sprinkle with coarse sugar. Bake in hot oven 400°F for 12 to 15 minutes. For an extra touch, top with thin frosting while still warm. Makes 12 to 18 buns.

SWISS ROLLS

1 cup sweet milk

1 tsp sugar

2 Tbs butter

½ yeast cake

4 cups flour (measured after being sifted twice)

1 tsp salt

Scald the sweet milk to which has been added sugar and butter; when milk is warm add yeast, flour, salt; put in milk. Put it in a moderately warm place to rise; when raised, place in refrigerator for at least three hours (or over night). When ready to use turn the dough on a well floured board and roll to half an inch thickness, cut any shape, rise slowly until very light, bake in a quick oven twenty minutes. Just before they are done brush over the top with a little melted butter.

POTATO TEA BISCUIT

¾ cup hot sifted potato

¾ cup butter

1 tsp salt

1 tsp sugar

1 cup scalded milk (cooled)

1 compressed yeast cake

1 egg white, slightly beaten

Mix salt and sugar well together. Combine ingredients. Knead with flour to a smooth dough. Let it rise, and knead

down once. Then after second rising shape as desired. When very light bake 15 minutes.

POTATO BREAD

4 or 5 good-sized mealy potatoes
3 lbs. flour
1 cake compressed yeast
2 Tbs salt
¼ lb. unsalted butter
Milk, boiled and cooled, enough to make a stiff dough

Boil potatoes with skins on two days before making bread. Set a dough of flour, yeast, potatoes, salt, butter, and milk. Cover and keep warm until risen enough. Butter pans, make into loaves, let rise again and bake in hot oven about 45 minutes.

WATER BREAD

1 cake of yeast
1 cup lukewarm water
4 cups flour
½ tsp salt

Soften a cake of yeast in half a cup of lukewarm water, then stir into it enough flour to make a very stiff dough (nearly two cups). Knead thoroughly, shaping into a ball. Make two cuts on the top about one-fourth an inch deep, then place the paste in a small saucepan of tepid water, the cut side up. In a few minutes it will begin to swell and float on the top of the water. When quite light remove with a skimmer to a bowl containing half a cup of lukewarm water and salt. Stir in enough flour to make a dough stiff enough to knead,—nearly two cups,—and let stand in a temperature of about 68 degrees Fahrenheit until light. Then shape into a loaf, and, when again light, bake.

TEA ROLLS

1 pint sweet milk
1 Tbs butter
1 Tbs lard
2 Tbs white sugar
1 tsp salt
½ cup yeast

Scald together the milk, butter, lard, and sugar, let cool and add salt; then make a soft batter, stir in yeasts; let rise over night, knead in the morning, let rise again. When light, roll out tin, cut with a large biscuit cutter, butter the top and fold over, let rise again and bake a light brown.

Substitutions

Spices	
Allspice	Cinnamon, cassia, dash of nutmeg, mace; or cloves
Aniseed	Fennel seed or a few drops anise extract
Cardamom	Ginger
Chili Powder	Dash bottled hot pepper sauce plus a combination of oregano and cumin
Cinnamon	Nutmeg or allspice (use only ¼ of the amount)
Cloves	Allspice, cinnamon, or nutmeg
Cumin	Chili powder
Ginger	Allspice, cinnamon, mace, or nutmeg
Mace	Allspice, cinnamon, ginger, or nutmeg
Nutmeg	Cinnamon; ginger; or mace
Saffron	Dash turmeric (for color)

Leavens	
Baking Powder (1 tsp)	• ⅝ tsp cream of tartar plus ¼ tsp baking soda
	• 2 parts cream of tartar plus 1 part baking soda plus 1 part cornstarch
	• Add ¼ tsp baking soda to dry ingredients and ½ C. buttermilk or yogurt or sour milk to wet ingredients. Decrease another liquid in the recipe by ½ C.
	• 1 tsp baker's ammonia
Baking Soda	Potassium BiCarbonate
Baker's Ammonia (1 tsp)	• 1 tsp baking powder
	• 1 tsp baking powder plus 1 tsp baking soda

Herbs	
Basil	Oregano
	Thyme
	Tarragon
	Equal parts parsley and celery leaves
	Cilantro
	Mint
Bay Leaf	Indian bay leaves
	Boldo leaves
	Juniper berries
Chervil	Parsley plus tarragon
	Fennel leaves plus parsley
	Parsley plus dill
	Tarragon
	Chives
	Dill Weed
Chives	Green onion tops
Cilantro	Italian parsley (add some mint or lemon juice or a dash of ground coriander, if you want)
	Equal parts parsley and mint
	Parsley plus dash of lemon juice
	Dill
	Basil
	Parsley plus ground coriander
Curly Parsley	Italian parsley
	Chervil
	Celery tops
	Cilantro
Dill Weed	Tarragon

	Fennel leaves
Mint	Fresh parsley plus pinch of dried mint
	Basil
Oregano	Marjoram (2 parts of oregano or 2 parts of marjoram)
	Thyme
	Basil
	Summer savory
Parsley	Chervil
	Celery tops
	Cilantro
Rosemary	Sage
	Savory
	Thyme
Sage	Poultry seasoning
	Rosemary
	Thyme
Summer Savory	Thyme
	Thyme plus dash of sage or mint
Sweet Basil	Pesto
	Oregano
	Thyme
	Tarragon
	Summer savory
	Equal parts parsley and celery leaves
	Cilantro
	Mint
Tarragon	Dill
	Basil
	Marjoram
	Fennel seed
Thyme	Omit from recipe
	Italian seasoning
	Poultry seasoning
	Herbes de Provence
	Savory
	Oregano
Winter Savory	Summer savory
	Thyme
	Thyme plus dash of sage or mint

Extracts	
Almond Extract	Vanilla extract
	Almond liqueur (use 4-8 times as much)
Brandy Extract	Brandy (1 Tbsp brandy extract=5 Tbsp brandy)
	Vanilla extract
	Rum extract
Cherry Flavoring	Juice from a jar of maraschino cherries plus vanilla extract
Cinnamon Extract	Cinnamon oil (⅛ tsp oil per tsp of extract)
Cinnamon Oil	Cinnamon extract (2 units of extract per unit of oil)
Imitation Vanilla Extract	Vanilla extract
	Vanilla powder
Lemon Extract	Lemon zest (1 tsp extract = 2 tsp zest)
	Oil of lemon (⅛ tsp of oil per tsp of extract)
	Orange extract
	Vanilla extract
	Lemon-flavored liqueur (1 or 2 Tbsp liqueur per tsp extract)
Almond Oil	Almond extract (4 units of extract to 1 unit of oil)
Oil of Lemon	Lemon extract (4 units of extract to 1 unit of oil)
Oil of Orange	Orange extract (4 units of extract to 1 unit of oil)
Orange Extract	Orange juice plus minced orange zest (reduce another liquid in recipe)
	Rum extract
	Vanilla extract
	Orange liqueur (1 tsp orange extract is 1 Tbsp orange liqueur)
Peppermint Extract	Peppermint oil (⅛ tsp of oil per 1 unit of extract)
	Crème de menthe
	Peppermint schnapps (1 or 2 Tbsp schnapps for each tsp extract)
	Vanilla extract (use more)
Peppermint Oil	Wintergreen oil
	Peppermint extract (4 units of extract to 1 unit of oil)
Rose Essence	Rose syrup
	Rose water (1 part rose essence is 4-8 parts rose water)
	Saffron
Rose Syrup	Rose essence
	Rose water
Rose Water	Orange flower water
	Rose syrup (few drops)
	Rose essence (few drops)
	Almond extract (use less)

	Vanilla extract (use less)
Rum Extract	Rum (1 tsp rum extract = 3 Tbsp rum)
	Orange extract (use less)
Vanilla Extract	Vanilla powder (half as much)
	Imitation vanilla extract (less potent, so use more)
	Vanilla-flavored liqeur
	Rum (1 tsp extract = 1 Tbsp rum)
	Almond extract (use less)
	Peppermint extract (use ⅛ as much)
Vanilla Powder	Vanilla extract (twice as much as powder)
	Imitation vanilla extract (twice as much as powder)

Weights and Measures

Abbreviations.

t. stands for teaspoon.

T. stands for tablespoon.

c. stands for cup.

pt. stands for pint.

qt. stands for quart.

m. stands for minute.

hr. stands for hour.

Approximate Measure of One Pound.

2 c. milk

2 c. butter

2 c. chopped meat

2 2-3 c. powdered sugar .

3 1-2 c. confectioners sugar

4 c. patent flour

4 c. whole wheat flour

4 1-2 c. Graham flour

2 5-6 c. granulated sugar

2 2-3 c. oatmeal

6 c. rolled oats

4 1-3 c. rye meal

1 7-8 c. rice

2 1-3 c. dry beans

4 1-3 c. coffee

8 large eggs

9 medium eggs

10 small eggs

Proportions Thickening Agents.

1 T flour will thicken 1 c of liquid for soup.

2 T flour will thicken 1 c of drippings or liquid for gravies.

5 T of browned flour will thicken 1 c of liquid for gravy.

Thickening power of corn-starch is twice that of flour.

4 T of corn-starch will stiffen 1 pt. of liquid for pudding.

2 good sized eggs to 1 pt. of milk for custard.

1 egg to 1 c milk for soft custard or baked cup custard.

3 eggs to 1 pt. milk for large mold custard.

1 T gellatin to 1 pt. liquid for jelly cooled on ice.

Weights and Measures.

3 t.=l T.

16 T.=l c.

2 c.=l pt.

2 pt.=l qt.

4 qt.=l gal.

1 c.=½ lb. or 8 ounces.

One quart wheat equals one pound and two ounces.

Four large tablespoonfuls equals one-half gill.

Eight large tablespoons equals one gill.

Sixteen large tablespoons equal one-half pint.

A common-sized wine glass holds half a gill.

A common-sized tumbler holds half a pint.

Four ordinary teacups liquid equals one quart.

Two and one-half teaspoons make one tablespoon.

Four tablespoons makes one wine glass.

Two wine glasses makes one gill.

Two gills makes one teacup.

Two teacups makes one pint.

Four teaspoons salt makes one ounce.

One and one-half tablespoons sugar makes one ounce.

Two tablespoons flour makes one ounce.

Two cups or one pint of sugar makes one pound.

One scant quart wheat flour makes one pound.

Ten ordinary size eggs makes one pound.

Piece of butter size of egg makes one and one-half ounces.

Two cups butter makes one pound.

Four teaspoonfuls equal one tablespoon, liquid.

Four tablespoonfuls equal one wine glass.

Two wine glassfuls equal one gill or half cup.

Two gills equal one coffee cup or sixteen tablespoons.

Two coffee cups equals one pint.

Two pints equal one quart.

Four quarts equal one gallon.

Two tablespoons equal one ounce, liquid.

One tablespoon of salt equals one ounce.

Sixteen ounces equal one pound, or one pint, liquid.

Four coffee cups sifted flour equals one pound.

One quart unsifted flour equals one pound.

Eight or ten ordinary-sized eggs equal one pound.

One pint sugar equals on pound (white granulated).

Two coffee-cups powdered sugar equal one pound.

One coffee-cup cold butter, pressed down, equals one-half pound.

One tablespoon soft butter, well rounded, one ounce.

An ordinary tumberful equals one coffee-cup or one-half pint.

Twenty-five drops thin liquid will fill a teaspoon.

One pint chopped meat, packed, equals one pound.

One pound of brown sugar, or one pound of white sugar, powdered or loaf, equals one quart.

One pound soft butter equals one quart.

One quart Indian meal equals one pound and two ounces.

A tablespoonful is measured level.

A cupful is all the cup will hold leveled with a knife.

One teaspoonful of soda to one pint of sour milk.

One teaspoonful of soda to one cup of molasses.

Three heaping teaspoonfuls of baking powder to one quart of flour.

Half a cupful of yeast, or quarter of compressed cake, to one pint of liquid.

One teaspoonful of salt to two quarts of flour.

One teaspoonful of salt to one quart of soup.

One scant cupful of liquid to two full cupfuls of flour for muffins.

One cup of water to each pound of meatbone for soup stocks.

One saltspoonful of white pepper to each quart of soup stock.

One teaspoonful of extract to one quart of custard.

One tablespoonful of extract to one quart of cream or custard, for freezing.

One teaspoonful of extract to one plain loaf cake.

A pinch of salt or spice is a saltspoonful.

A few grains is less than a saltspoonful.

To blend seasonings, sift them thoroughly together before adding them to mixture.

Four peppercorns, four cloves, one teaspoonful minced herbs, and one tablespoonful each of chopped vegetables to each quart of water for soup stock.

Cookies

If you can master a good cookie recipe, you can make your way into almost anyone's heart. Who doesn't melt just a little when biting into a warm chocolate chip cookie, the outside just slightly browned, the inside soft and gooey? Is there anything that makes one feel more loved than opening a package and discovering a batch of homemade cookies? Cookies are simple and almost as fun to make as they are to eat.

BROWN SUGAR COOKIES

2 cups brown sugar

1 cup softened margarine

3 eggs

5 cups pastry flour

½ tsp salt

1 tsp baking soda

¼ cup milk

1 tsp vanilla extract

Cream together the margarine and sugar. Beat the eggs well and add to the mixture. Add milk, vanilla, and half the flour sifted with the salt and soda. Add more flour, enough to make dough which may be rolled. Cut in any desired shapes and bake about ten minutes in a 400°F. oven. Drizzle with icing and sprinkle with coarse sugar if desired.

CHOCOLATE ALMOND COOKIES

3 cups brown sugar

4 eggs

2 tsp cinnamon

½ tsp ground cloves

2 squares melted chocolate

½ lb. chopped blanched almonds

1 Tbs melted butter

2 tsp baking powder

1 cup whole-wheat flour

Beat eggs well, add sugar and melted butter. Sift together one cup of whole-wheat flour with cinnamon, cloves, and baking powder. Add to the mixture. Then add the chocolate and the almonds and enough additional flour to make a dough stiff enough to shape in the hands. Roll in small balls. Place in greased pans not too close together with an almond on each cookie. Bake at 400°F.

AUNT SUG'S NUT COOKIES

1 ½ cups sugar

1 cup butter

3 eggs

1 Tbs cinnamon

½ tsp salt

1 tsp baking soda, dissolved in 4 Tbs hot water

3 cups flour

1 cup raisins

1 cup currants

1 cup walnuts or almonds, broken up

1 tsp vanilla extract

To well beaten eggs add creamed butter and sugar, cinnamon, soda, water, flour, fruits, which must be dredged with flour, and nuts. Lastly add vanilla and drop by spoonful into greased pans. Bake in moderate oven.

SOUR CREAM COOKIES (Three dozen)

1 cup sugar

½ cup butter

2 eggs

½ cup sour cream or sour milk

½ tsp baking soda

½ tsp salt

2 tsp grated nutmeg

2 cups flour, or as little as possible

Cream the butter, add the sugar. Cream again. Add the eggs well beaten, sour cream, one cup flour, soda, salt and nutmeg mixed and sifted together. Add the rest of the flour. Roll out to one-third of an inch thickness, cut any desired shape, and bake in a moderately hot oven for 15 minutes. Sugar mixed with a little flour may be sifted over the dough before cutting. Raisins may also be pressed into the top of each cookie.

CHOCOLATE COOKIES (Three dozen)

1 cup sugar

⅓ cup butter

1 egg

¼ cup milk

2 cups flour

½ tsp cinnamon

½ tsp salt

3 tsp baking powder

1 square chocolate

1 tsp vanilla

Cream the butter, add the sugar and cream well. Add alternately the sifted flour, salt, baking powder and egg beaten in milk. Add the chocolate and vanilla. Turn out on a floured board and roll a small portion at a time to one-fourth of an inch in thickness. Cut with a floured cookie cutter. Place on a buttered, floured pan and bake in a moderate oven until slightly brown. (About 10 minutes).

Handy Household Hints

Vanilla Essence or Extract

This is an expensive article when of fine quality, and you may prepare it yourself either with brandy or alcohol. With brandy, the flavor is superior. Cut into very small shreds three vanilla beans, put them in a bottle with a pint of brandy and cork the bottle tightly. Shake it occasionally and it will be ready for use after three months. You may shorten the process to three weeks by using alcohol at 95 percent. Chop three vanilla beans and pound them in a mortar. Cover them with a little powdered sugar and put them in a pint bottle, adding a tablespoonful of water. Let it stand twelve hours, then pour over it a half pint of alcohol or spirits of wine. Cork tightly, shake it every day, and it will be ready for use in three weeks.

DROP COOKIES (Twenty-four cookies)

⅓ cups butter

1 cup sugar

1 egg

½ cup sour milk

½ tsp baking soda

¼ tsp salt

1 tsp vanilla extract

¼ cup chopped raisins

2 ½ cups flour

½ tsp baking powder

Cream the butter, add the sugar, then the whole egg. Mix well. Add the sour milk and the vanilla. Mix the baking powder, soda and flour well, add the raisins and add to the first mixture. Beat well. Drop from a spoon onto a buttered and floured pan, leaving three inches between the cookies. Bake 15 minutes in a moderate oven.

NUTS COOKIES (Three dozen cookies)

⅓ cup butter and lard mixed

⅔ cup "C" sugar

1 egg

4 Tbs milk

2 cups flour

2 tsp baking powder

⅓ cups chopped nuts (preferable walnuts)

1 tsp powdered cinnamon

¼ tsp powdered cloves

¼ tsp mace

¼ tsp nutmeg

Cream the butter, add the sugar and mix well. Add the egg and milk and then the flour, nuts, cinnamon, cloves, mace, nutmeg and baking powder. Place the dough on a floured board. Roll it out one-fourth of an inch thick and cut with a cookie cutter. Place on a well-buttered and floured baking sheet. Bake 12 minutes in a moderate oven.

PEANUT COOKIES

2 Tbs butter

2 Tbs milk

1 tsp baking powder

¼ cup sugar

1 egg

¼ tsp salt

½ cup flour

½ cup chopped peanuts

½ tsp vanilla extract

Mix ingredients together. Drop from a teaspoon on unbuttered sheets one inch apart. Bake 12 to 15 minutes in a slow oven.

SUGAR COOKIES

2 cups sugar

1 cup butter

2 eggs, whites and yolks beaten separately

Flour to make stiff dough

1 tsp baking powder

1 tsp salt

1 tsp lemon or desired flavoring

Cinnamon

Chopped nuts

Cream butter, beat sugar well into it. Add well beaten yolks of the eggs, then the whites, then sift in the flour through which the baking powder has been sifted. Add salt and flavoring last of all. Roll out very flat, cut out with round or heart-shaped cutter, sprinkle with sugar, cinnamon and chopped nuts and place in floured pans. Cook in moderate oven.

OATMEAL COOKIES

2 Tbs butter

1 cup sugar

2 eggs

2 cups Quaker Oats

1 tsp vanilla

2 tsp baking powder

Beat eggs separately. Combine all ingredients. Drop by teaspoonfuls on sheet iron pans. Bake in moderate oven. Allow them to stand several minutes before removing from the pans.

Oatmeal Cookies

BUTTERSCOTCH COOKIES

2 cups medium brown sugar

1 cup melted butter

1 Tbs baking soda

1 Tbs vinegar

1 Tbs cream of tartar

2 eggs

Vanilla extract

Flour to make stiff

Beat eggs well. Combine all ingredients. Mix into rolls and put into refrigerator over night. In the morning slice off into the shape of cookies and bake in hot oven.

MOLASSES COOKIES

1 cup brown sugar

⅔ cup molasses

½ cup shortening

½ cup hot water

1 egg

1 tsp ginger

2 tsp soda

Salt

Flour to make stiff

Combine ingredients. Roll thin and bake in quick oven.

DATE BARS

1 cup nuts

1 cup dates

2 eggs

3 Tbs flour

1 tsp baking powder

1 tsp vanilla

¾ cup powdered sugar

2 Tbs cream

Combine nuts, dates, flour, baking powder, and vanilla. Beat eggs separately and add. Add powdered sugar and cream. Bake on buttered tin.

PECAN PRALINES

2 cups brown sugar

¼ cup boiling water

1 cup pecans

2 Tbs butter

Boil sugar and water, add butter and nuts. Boil 5 minutes, remove and beat. Then drop by spoonfuls on buttered pan and bake.

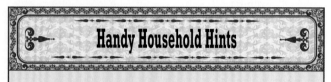

Handy Household Hints

Never stir sugar and butter together in a tin basin or with an iron spoon—a wooden spoon is better than any other kind.

CHOCOLATE JUMBLES

2 cups sugar

3 eggs

1 cup melted shortening

2 squares melted chocolate

1 tsp baking soda dissolved in

2 Tbs warm water

4 cups flour

Roll. Cut out with cake-cutter and frost the top when the cookies are cool, using boiled frosting.

Very good results may be obtained by using one-half of the recipe and taking two eggs, saving the white of the one for the frosting.

It is easy to vary the amount of chocolate, if one likes it a little stronger or with less.

Another very delicious cookie is made with a filling of suitable fruits. It is unusual and extremely satisfying.

CHOCOLATE COOKIES

2 eggs

1 cup sugar or syrup

2 squares of chocolate

¾ cup flour

¼ tsp salt

Mix with eggs, sugar or syrup, and chocolate, shaved into small pieces, and melted over hot water. Add flour, mixed with salt. Mix all well, take up by spoonfuls and place on greased and floured tin, flatten out rather thin, and bake in a moderate oven.

WALNUT MACAROONS

¼ lb. walnuts
¼ lb. sugar
2 egg whites

Pound fine walnuts with sugar. Mix with the egg whites beaten stiff. Shape with a spoon or pastry tube and bag on unbuttered paper, and bake in a slow oven until firm.

Cakes

There's really no better way to mark a special celebration than with a homemade cake. If you're not a regular cake baker, you'll be amazed that making one from scratch is hardly more difficult than using a pre-packaged mix, and the results are so worth the tiny bit of extra effort. Here are a range of old favorites that will do you proud for birthdays, holidays, or any day!

APPLE JOHNNY-CAKE

½ cup yellow cornmeal
½ cup white cornmeal
1 Tbs sugar
½ tsp salt
1 tsp shortening
¾ cup boiling water
3 Tbs milk
1 cup chopped apple (optional)

Mix the yellow and white cornmeal, sugar, salt, and shortening together. Scald with the boiling water, until the mixture is a little thicker than will spread; then add the milk and chopped apple. Spread on well-greased tins to a thickness of one-fourth inch, crease in squares with the back of a knife, and bake in a hot oven until light brown. Split, and eat buttered, with coffee. The apple may be omitted, if desired. Either variety makes delicious cream or milk toast.

BUCKWHEAT CAKES

1 quart milk
2 tsp salt
¾ tsp soda
1 yeast cake
¼ cup lukewarm water
1 cup buttermilk
5 cups buckwheat flour

Scald milk, and cool to lukewarm; add yeast-cake softened in the water, salt, and flour to make a rather thick batter. Beat well and let rise overnight. In the morning add the soda dissolved in the buttermilk. Bake on a hot, greased griddle, and serve with butter and syrup.

CREAM BLUEBERRY GINGERBREAD

1 cup blueberries
1 cup molasses
1 cup sour cream
2 cups pastry flour
½ tsp salt
1 tsp ginger
2 tsp baking soda

Mix together the molasses and cream, but reserve one-fourth cup of the flour to mix with the blueberries. Add the dry ingredients to the molasses mixture. Fold in the blueberries, and place in the oven as quickly as possible. Bake gently for 30 minutes. The gingerbread should be about one and one-half inches thick.

EGGLESS NUGGET CAKE

2 cups light-brown sugar
½ cup shortening
1 cup sour milk
2 cups pastry flour
½ tsp salt
½ tsp baking soda
1 tsp baking powder
1 cup raisins or chopped dates

1 cup walnut or pecan meats

2 squares chocolate

Cream the shortening and the sugar. Sift the salt, baking powder and soda with the flour. Mix the flour and milk with shortening and sugar, and add the nuts, raisins and lastly chocolate, which has been melted over hot water. Bake 35 minutes in a moderate oven. Frost with mocha frosting.

LIGHT FRUIT CAKE

2 cups butter

2 cups sugar

4 cups pastry flour

8 eggs

Brandy

Rosewater

1 lb. candied pineapple

1 lb. candied cherries

1 lb. blanched almonds

½ pound citron

1 coconut, grated

Shred the cherries, cut the pineapple and citron in bits, and chop the almonds. With the fruit mix three tablespoons of brandy, and with the almonds two tablespoons of rosewater, and allow all to stand overnight. Wash the butter in rosewater, cream it with the sugar, add the egg yolks well beaten, then the coconut, flour, one cup of brandy, and the egg whites (whipped stiff), putting them in alternately. Finally stir in the fruit and nuts. Bake very slowly from four to five hours. One-half pound of desiccated coconut, chopped fine and allowed to stand in two tablespoons of rosewater or milk, may be substituted for the fresh coconut. The amounts given will make three medium-sized cakes or one large one, which will keep indefinitely. Cider or white grape juice may be used in place of the brandy.

LUNCHEON CAKE

1 cup brown sugar

½ cup shortening

1 tsp cinnamon

1 tsp cloves

1 cup raisins

1 cup strained stewed tomatoes

2 cups pastry flour

2 tsp baking soda

Cream sugar and shortening together, add spice and raisins; then to the stewed tomato add the soda; beat well and stir all together rapidly with the sifted flour. Bake in a loaf pan. A moist and excellent cake, very wholesome for children's school luncheons.

ORANGE CHOCOLATE CAKE

¼ lb. bitter chocolate

½ cup shortening

1 ½ cups sugar

1 ½ cups milk

2 cups pastry flour

2 eggs

2 tsp vanilla

1 tsp baking soda

¾ tsp salt

3 Tbs hot water

Orange icing

Cut up the chocolate and put in a saucepan with the milk and one-half cup of the sugar, letting it heat to boiling point, stirring occasionally, and boiling five minutes. Cool and add the vanilla. In the meantime cream the shortening and the remaining sugar thoroughly together, add the eggs beaten and stir vigorously. Add the hot water. Alternate the flour sifted with soda and salt, and the chocolate mixture, stirring them in thoroughly. Bake in two layers in a moderate oven (375°F.) for 30 minutes. Put together and cover the top with orange icing, or put together and cover the top with orange icing, or put together with strawberry cinnamon filling just before serving, and cover the top with whipped and sweetened cream.

Orange Icing:

2 cups confectioners' sugar

1 tsp melted butter

Juice ½ orange

Beat sugar and butter into the orange juice and when stiff enough to hold its shape spread on the cake.

Light Fruit Cake

HONEY AND NUT GINGERBREAD

1 cup honey

⅓ cup butter or margarine

1 cup cold water

2 cups whole wheat flour

1 cup chopped nuts

2 eggs

2 tsp baking powder

1 tsp ginger

1 tsp cinnamon

¼ tsp baking soda

½ tsp salt

Cream the butter and honey together. Add the eggs well-beaten. Mix and sift the dry ingredients together, and add alternatively with the water. Add the chopped nuts last. Bake in a 350°F. oven for 45 minutes, or until done.

MOTHER'S GINGERBREAD

1 cup brown sugar

½ cup margarine

½ cup cream

2 eggs

1 cup molasses

2 ½ cups pastry flour

½ tsp baking soda

1 Tbs ginger

½ tsp salt

½ cup milk

Cream the margarine and sugar together. Beat the eggs well. Add the cream and eggs, then the molasses. Sift the dry ingredients together and add them to the other mixture, alternating with the milk. Pour into a small dripping-pan, well-greased and floured. Bake in a 350°F. oven

MAPLE LAYER CAKE

3 eggs

1 cup soft maple sugar

1 cup pastry flour, sifted

½ tsp salt

1 tsp baking-powder

1 cup cream

¼ cup grated maple sugar

Beat the egg yolks until light and fluffy. Add the maple sugar and the flour with the salt and baking powder. In a separate bowl, beat the egg whites until stiff. Fold the egg whites into the mixture. Mix quickly and bake for 30 minutes in two greased and floured layer-cake pans at 320°F. Put together with the cream, whipped and sweetened with the grated maple sugar. Spread cream on top or ice, if desired with maple fondant, or merely sprinkle with powdered sugar.

AUNT REBECCA'S OLD-FASHIONED SPONGE CAKE

1 ⅓ cups powdered sugar

1 cup bread flour

½ lemon, juice and rind

5 eggs

Pinch of salt

Separate eggs, beat yolks with sugar, till light and creamy, at least ten minutes. Add lemon juice and rind. Beat whites stiff, fold into mixture, and then sift in the flour and salt slowly, stirring gently. Bake 50 minutes in a slow oven.

COCOA APPLE-SAUCE CAKE

1 cup sugar

½ cup sour cream

1 cup hot, sour apple sauce

1 ¼ tsp baking soda

1 Tbs cocoa

1 tsp cinnamon

½ tsp cloves

2 cups bread flour

1 cup raisins

Mix together cocoa, spices, flour, and raisins. In a separate bowl, put the sugar, cream, and hot apple sauce into which the baking soda has been stirred. Beat in the flour mixture. Bake in a well-lined loaf for 45 minutes in a moderate oven. Frost with sour cream icing.

WHITE LAYER CAKE

1 ½ cup granulated sugar

½ cup butter

1 cup milk

3 cups pastry flour
4 tsp baking powder
1 tsp vanilla
6 egg whites

Cream the butter, add sugar gradually, and cream till very light and fluffy. Sift flour and baking powder together three times, add this alternately with the milk to the first mixture until all is used. Then fold in the egg whites beaten stiff and the vanilla. Bake in three small or two large, well-oiled layer-cake pans for 40 minutes in a moderate oven, 375°F. for the first 30 minutes, then at a lower heat. Put together with thick white icing.

BUTTER CAKES

3 cups bread flour
¾ tsp baking soda
½ tsp baking powder
1 tsp salt
1 ½ cups sour or buttermilk

Sift together the flour, soda, baking powder, and salt. Then mix lightly with enough sour or buttermilk to moisten. Turn out on a floured board and knead very gently. Roll thin and cut into large rounds. Bake on a well-greased griddle, turning frequently to insure even browning. They will require at least 8 minutes on the griddle. Tear them apart and drop a piece of butter in each cake. Wrap them in a napkin and serve piping hot. These are delicious when served hot and buttery with a saucer of berries at afternoon tea.

DEVIL'S FOOD CAKE

2 eggs
2 cups brown sugar
1 cup butter
1 cup buttermilk
3 cups flour
½ cake of melted chocolate
1 tsp vanilla
1 ½ Tbs cinnamon
1 tsp cloves
1 tsp allspice
1 tsp baking soda dissolved in ½ cup of boiling water

Cream butter and sugar and add to well beaten egg, next add milk, melted chocolate, flour beaten in lightly, vanilla and spices, and lastly the boiling water and baking soda. Bake in layer tins and moderate oven.

Filling:

2 cups white sugar
1 cup sweet milk
½ cake of chocolate
1 egg yolk
⅓ cup butter
1 tsp vanilla

Put all the ingredients except vanilla on to cook. Cook until thick, then beat until creamy, add vanilla and spread on layers.

MAGNOLIA SPONGE CAKE

3 eggs
1 ½ cups sugar
1 ½ cups flour
1 ½ tsp baking powder
½ cup boiling water

Beat eggs separately and then together. Mix ingredients together in the order given and cook in biscuit pan in moderate oven.

MARTHA WASHINGTON CAKE

4 eggs
2 cups sugar
1 cup butter
1 cup sweet milk
3 cups flour
1 tsp baking powder
½ lb. raisins
1 tsp cinnamon
1 tsp cloves
1 tsp allspice

Cream butter and sugar and add to well beaten eggs. Next add the milk, flour, baking powder, raisins and spices. Bake in layers in moderate oven.

Filling:

- 2 cups sugar
- 2 lemons, juice and grated rind
- 2 cups grated coconut
- 1 cup boiling water
- 1 Tbs cornstarch, dissolved in a little cold water

When water begins to boil, add the cornstarch mixture. Cook until it spins a thread, then beat until creamy and spread between layers.

HICKORY-NUT CAKE

- ½ cup butter
- 1 ½ cup sugar
- ¾ cup water
- 2 cups flour
- 3 tsp baking powder
- 1 cup hickory-nut kernels, chopped
- 4 egg whites

Beat butter to a cream; add sugar, beating all the while. Sift flour with baking powder. Add first a little of the water to the butter and sugar, then more water and flour until both are used; beat thoroughly for 3 minutes. Beat the egg whites to a stiff froth, add one-half to the mixture, then add the nuts, and last the remaining egg whites. Bake in square or round pans in a moderate oven for 45 minutes. Ice with soft icing and decorate with halves of nuts.

DUTCH APPLE CAKE

- 2 eggs
- 1 ½ cups milk
- 1 Tbs butter, melted
- ½ tsp salt
- 2 cups flour
- 3 tsps baking powder

Separate two eggs; add to the yolks milk, butter, and salt; mix and add flour, sifted with baking powder; beat quickly, fold in the well-beaten egg whites, and turn in a greased cake-pan. Cover the top thickly with apples that have been pared, cored, and quartered, putting the rounding sides up, and dust over half a cupful or sugar. Bake at 375°F for half an hour, or until the apples are tender.

Peaches, huckleberries, blackberries, or elderberries may be substituted for the apples, or use a combination.

SPICED CAKE (Sixteen pieces)

- ⅓ cup butter
- 1 cup sugar
- 2 egg yolks
- ⅔ cup sour milk
- 1 ½ tsp cinnamon
- ¼ tsp ground cloves
- ¼ tsp mace
- 1 tsp baking soda
- 2 cups flour
- 1 egg white
- 1 tsp vanilla

Cream the butter, add the sugar and egg yolks. Mix well. Mix and sift all dry ingredients. Sift and add alternately with sour milk. Add vanilla and stiffly beaten egg white. Bake in a loaf cake pan, prepared with waxed paper, in a moderate oven for 25 minutes. Cover with "C" sugar icing.

"C" sugar icing (Sixteen pieces):

- 1 cup "C" (confectioner's) sugar
- ⅓ cup water
- ⅛ tsp cream of tartar
- 1 egg white
- ½ tsp vanilla

Mix the sugar, water and cream of tartar. Cook until the syrup clicks when a little is dropped in the cold water. Do not stir while cooking. Have the mixture boil evenly but not too fast. Pour gently over the beaten white of the egg. Stir and beat briskly until creamy. Add vanilla. Place on the cake. If too hard, add a tablespoon of water.

Dutch Apple Cake

HOT WATER SPONGE CAKE (Eight portions)

- 2 egg yolks
- 1 cup sugar
- 1 cup boiling water
- 1 Tbs lemon juice
- 1 tsp grated rind lemon
- 2 egg whites
- 1 cup flour
- 1 tsp baking powder
- ¼ tsp salt

Beat the yolks until thick and lemon colored, add the sugar gradually and beat for two minutes. Add the flour, sifted with the baking powder, and salt. Add the boiling water, lemon juice, and grated rind. Beat with an egg-beater, two minutes. Fold in egg whites. Bake 35 minutes in a moderate oven in an unbuttered pan. Do not cut sponge cake, except through the crust, then break apart.

GINGER DROP-CAKES (Fifteen cakes)

- 1 cup molasses
- ½ cup boiling water
- 2 ¼ cup flour
- 1 tsp soda
- 2 tsp ginger
- ½ tsp salt
- ½ cup chopped raisins
- 4 Tbs melted butter

Put the molasses in a bowl, add the bowling water and the dry ingredients, sifted. Then add the raisins and the melted butter. Beat well for 2 minutes. Pour into buttered muffin pans, filling the pans ½ full. Bake 20 minutes in a moderate oven.

BURNT SUGAR CAKE (Sixteen pieces)

- ½ cup butter
- 1 ½ cup sugar
- 2 eggs
- ¼ tsp salt
- 2 ½ cups flour
- 4 tsp baking powder
- 1 cup bowling water
- 1 tsp vanilla

Caramelize ⅔ of a cup of sugar. When the sugar is melted and reaches the light brown or the "caramel" stage, add the water. Cook until the sugar is thoroughly dissolved in the water. Allow it to cool. Cream the butter, add the rest of the uncooked sugar, and then add the egg-yolks. Mix well. Add the salt, flour, baking-powder, vanilla and the cooled liquid. Beat 2 minutes and add the egg whites stiffly beaten. Pour into 2 pans prepared with buttered paper. Bake 25 minutes in a moderate oven. Ice with confectioner's icing.

Confectioner's Icing (Sixteen portions):

- 2 Tbs cream or milk
- ½ tsp vanilla
- 1 Tbs caramelized syrup or maple syrup
- 1 ½ cup powdered sugar

Mix the cream, vanilla and syrup. Add the sugar (sifted) until the right consistency to spread. Spread carefully between the layers and on the top. Set aside to cool, and to allow the icing to "set." (More sugar may be needed in making the icing.)

POP-OVERS (Eight)

- 1 cup flour
- 1 cup milk
- ½ tsp salt
- 1 egg, beaten well

Add the milk slowly to the flour and salt, stirring constantly until a smooth paste is formed. Beat and add the remainder of the milk, and the egg. Beat vigorously for three minutes. Fill very hot gem pans ¾ full. Bake 30 minutes in a hot oven. They are done when they have "popped" at least twice their size, and when they slip easily out of the pan. Iron pans are the best.

GRAHAM CRACKER CAKE (Twelve pieces)

- ⅓ cup butter
- ⅔ cup sugar
- 2 egg yolks

1 cup milk

⅔ lb. graham crackers rolled fine

3 tsp baking powder

2 egg whites, beaten

½ tsp ground cinnamon

½ tsp vanilla

Cream the butter, add the sugar and heat. Add all the dry ingredients mixed together alternately with the milk. Beat 2 minutes. Add the vanilla and the egg whites, stiffly beaten. Bake in square tin pans for 25 minutes in a moderate oven.

White Icing:

¼ cup sugar

¼ cup water

Sifted powdered sugar

½ tsp vanilla

Boil the sugar and the water 5 minutes without stirring. Remove from the fire. Add the flavoring, and sufficient sifted powdered sugar to spread evenly on the cake.

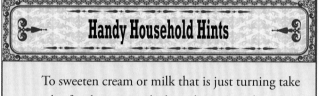

Handy Household Hints

To sweeten cream or milk that is just turning take a pinch of soda, wet it a little and stir it in the cream or milk.

SMALL CAKES (Fourteen cakes)

1 ¼ cups sugar

⅓ cup butter

2 cups flour

4 tsp baking powder

⅛ tsp salt

⅔ cup milk

1 tsp vanilla

½ tsp lemon extract

Cream the butter, add the sugar slowly and continue creaming. Mix and sift the flour, baking powder and salt and add these and the milk, vanilla and lemon extracts to the butter and sugar. Mix well and beat 2 minutes. Beat the egg whites till very stiff and fold tehse very carefull into the cake mixture. When thoroughly mixed, fill the cake pans (which have been prepared with wax paper) ⅔ of an inch deep with the mixture.

Bake 24 minutes in a moderate oven, allow to stand 5 minutes, then slip a knife around the edges and remove the cake carefully from the pan. Turn over, remove the paper and allow the cake to cool. Ice on the bottom side. When ready for serving, cut in 2-inch squares.

Bettina Icing:

1 egg white

1 Tbs cream

2 cups powdered sugar

1 tsp vanilla

½ tsp lemon extract

Beat the egg whites, add part of the sugar. Add the cream, vanilla and lemon extracts. Keep beating. Add the rest of the sugar gradually. (A little more sugar may be needed.) Beat the icing till very fluffy and until it will spread without running off the cake. Spread each layer.

SOUR CREAM BISCUITS (Four portions)

2 cups flour

½ tsp salt

3 tsp baking powder

3 Tbs fat

¼ tsp baking soda

⅔ cup sour milk

Mix the flour, salt and baking powder. Cut in the fat with a knife. Add the soda to the milk, and when the effervescing ceases, add slowly to the dry ingredients. (All the milk may not be needed.) When a soft dough is formed, toss onto a floured board. Pat into shape, cut with a biscuit cutter, and place side by side on a tin pan or baking sheet. Bake 15 minutes in a moderately hot oven.

FANCY CAKES (Eighteen cakes)

½ cup butter

1 cup sugar

8 egg yolks

½ cup milk

1 ¾ cup flour

2 tsp baking powder

2 tsp lemon extract

Cream the butter, add the sugar and mix well. Beat the egg yolks until very thick, and add to the first mixture. Mix and sift together the flour and baking powder and add the milk alternately with the flour mixture, beating well. Eat 2 minutes after mixing. Add the extract. Pour to the thickness of 1 inch into flat pans lined with buttered paper. Bake 12 minutes in a moderate oven. Remove from the fire and when cool, cut into shapes with fancy animal cutters. The individual cakes may be iced if desired.

WASHINGTON PIE (Six portions)

1 ⅓ cup sugar

3 eggs

½ cup water

½ tsp lemon extract

2 cups flour

2 tsp baking powder

Beat the egg yolks 5 minutes, add the sugar and beat 3 minutes. Add the water, lemon extract, flour and baking powder. Mix thoroughly. Fold in the beaten egg whites very carefully. Bake 25 minutes in two round shallow pans in a moderate oven. When cool, put the following filling between the layers. Sprinkle the top with powdered sugar.

Cream Filling for Washington Pie:

⅔ cup sugar

⅓ cup flour

½ tsp salt

1 ½ cup milk

1 egg yolk

½ tsp vanilla extract

½ tsp lemon extract

Mix thoroughly the sugar, salt and flour. Gradually add the milk, stirring constantly. Pour into the top of a double boiler, and cook until very thick. Add the egg yolk, vanilla and lemon extract, and cook 2 minutes. Beat until creamy and cool. Spread on the cake. Serve Washington pie with whipped cream if desired.

BURNT SUGAR CAKE (Sixteen pieces)

½ cup butter

1 ½ cup sugar

2 eggs

¼ tsp salt

2 ½ cup flour

4 tsp baking powder

1 cup boiling water

1 tsp vanilla

Caramelize ⅔ of a cup of sugar. When the sugar is melted and reaches the light brown or the "caramel" stage, add the water. Cook until the sugar is thoroughly dissolved in the water. Allow it to cool. Cream the butter, add the rest of the uncooked sugar, and then add the egg yolks. Mix well. Add the salt, flour, baking powder, vanilla and the cooled liquid. Beat 2 minutes and add the egg whites stiffly beaten. Pour into two pans prepared with buttered paper. Bake 25 minutes in a moderate oven. Ice with confectioner's icing.

Confectioner's Icing (Sixteen portions):

2 Tbs cream or milk

½ tsp vanilla

1 Tbs caramelized syrup or maple syrup

1 ½ cups powdered sugar

Mix the cream, vanilla and syrup. Add the sugar (sifted) until the right consistency to spread. Spread carefully between the layers and on the top. Set aside to cool, and to allow the icing to "set." (More sugar may be needed in making the icing.)

Sour cream biscuits are delicious served like scones with fresh whipped cream and jam.

CHOCOLATE NOUGAT CAKE

4 Tbs butter

⅔ cup sugar

2 squares chocolate

2 Tbs sugar

2 Tbs water

1 egg

½ cup milk

⅓ cup flour

2 tsp baking powder

½ tsp baking soda

½ tsp vanilla

Cook the two tablespoons of sugar, water and chocolate together for one minute, stirring constantly. Cream the butter, add the sugar, the whole egg and the flour, baking powder and soda sifted together. Add the vanilla. Beat two minutes. Pour into two square layer-cake pans prepared with waxed paper. Bake twenty-two minutes in a moderate oven. Chocolate cakes burn easily and they should be carefully watched while baking. Ice with White Mountain Cream Icing.

White Mountain Cream Icing for Cakes:

1 cup granulated sugar

⅛ tsp cream tartar

¼ cup water

1 egg white

½ tsp vanilla

Boil the sugar, water and cream of tartar together without stirring. Remove from fire as soon as the syrup hairs when dropped from a spoon. Pour very slowly onto the stiffly beaten egg whites. Beat vigorously with sweeping strokes until cool. If icing gets too hard to spread, add a little warm water and keep beating. Add extract and spread on cakes. Decorate with tiny pink candies.

"LIGHTNING" TEA CAKES (Twelve cakes)

1 ½ cup flour

¾ cup granulated or powdered sugar

2 tsp baking powder

⅓ tsp salt

3 Tbs butter (melted)

1 egg

½ cup milk

½ tsp vanilla

Sift and mix together the flour, sugar, baking powder and salt. Make a "well" in the center of the mixture and pour in the melted butter, egg, milk and vanilla. Stir all together and beat vigorously for two minutes. Fill well buttered muffin pans half full of the mixture and bake fifteen minutes in a moderate oven.

(From Southern Recipes Tested By Myself by Laura Thornton Knowles)

SWISS ROLLS

1 cup sweet milk

1 tsp sugar

2 Tbs butter

½ yeast cake

4 cups flour (measured after being sifted twice)

1 tsp salt

Scald the sweet milk to which has been added sugar and butter; when milk is warm add yeast, flour, salt; put in milk. Put it in a moderately warm place to rise; when raised, place in refrigerator for at least three hours (or over night). When ready to use turn the dough on a well floured board and roll to half an inch thickness, cut any shape, rise slowly until very light, bake in a quick oven twenty minutes. Just before they are done brush over the top with a little melted butter.

STRAWBERRY SHORTCAKE

4 cups flour

2 tsp baking powder

1 Tbs butter

1 Tbs lard

1 cup milk

1 Tbs sugar

Strawberries

Sift flour and baking powder together, then mix in butter and lard, add sugar and milk. Bake in a cake pan, or like biscuits on a cookie sheet, and split while hot, butter and put in strawberries, which you have previously prepared with sugar. Serve with fresh whipped cream.

Strawberry Shortcake

ALMOND CAKE

1 ½ cups flour
1 cup sugar
2 eggs
2 cups butter
¾ cup milk
1 ½ tsp baking powder
½ lb. shelled almonds

Brown almonds slightly in oven, grind quite fine and mix with a little pulverized sugar, set aside until cake is done. Cream, butter and sugar, beat yolks and add, then the flour (sifted with baking powder in it) and milk alternately, beat whites stiff and fold them in. Bake in square pans. When cold, cut in finger lengths and butter each piece of cake on all sides, then roll in the mixture of almonds and sugar.

LEMON SPONGE

1 ¼ oz. gelatin
4 egg whites
3 lemons
2 cups sugar
1 cup cold water
2 cups boiling water

Soak the gelatin in cold water and pour boiling water on it, then add the lemon juice, and sugar. Whip the whites of eggs to a stiff froth, whip gelatin, etc., till frothy; when cool add the egg whites and beat all well together, dip your mold into hot, then into cold water and put your sponge in; set quickly on ice.

SNIPPY DOODLE CAKE

1 cup granulated sugar
1 cup flour
½ cup milk
2 eggs, beaten light
1 Tbs butter
1 Tbs cinnamon
1 ½ tsp baking powder

Cream butter and sugar, add eggs, beaten all together and light; then the flour and milk, stirring briskly. Mix cinnamon and baking powder together with the flour. Bake in a sheet and sprinkle granulated sugar on top when nearly done.

(From *Recipes, Old and New* by Women of the First Unitarian Church)

SUNSHINE CAKE

7 egg whites
5 egg yolks
1 ¼ cups sugar
1 cup flour
1 pinch salt
⅓ tsp cream of tartar
Flavoring

Sift sugar and flour five times, measure and set aside. Separate eggs, beating yolks to stiffen froth; whip whites to a foam. Add cream of tartar and whip stiff. Add sugar to whites, then beat. Next add yolks and beat, then flour and flavoring and fold lightly through. Moderate oven 40 to 50 minutes in tube tin which is not greased.

COFFEE CAKE

1 cup shortening
1 cup molasses
1 cup sugar
1 cup chopped raisins
1 cup strong coffee
1 egg
1 tsp baking soda
1 tsp cinnamon
1 tsp lemon extract
2 ½ cups flour

Combine ingredients. Bake in two loaves. Never fails, and keeps well.

MOCK ANGEL FOOD

1 cup sugar
1 cup flour

2 tsp baking powder (heaping)

Pinch of salt

1 cup sweet milk, hot but not boiling

2 eggs

Combine sugar, flour, baking powder, and salt. Sift 6 times. Add milk. Add eggs, whites beaten stiff. Do not grease tin; bake in slow oven. When taken from oven turn bottom side up on 2 cups to give it air. Frost with boiled frosting.

SCOTCH CAKE

½ lb. bread flour

¼ lb. brown sugar

½ lb. butter

Work together bread flour, brown sugar, and butter. Chill, roll out one-third an inch thick, and cut in triangles or other shape. Bake in a slow oven.

SOUR CREAM NUT LAYER CAKE

2 eggs

1 cup granulated sugar

½ cup rich sour cream

2 cups flour measured before sifting

½ tsp baking soda

1 tsp baking powder

Pinch of salt

Beat the eggs till whites and yolks are well blended, and add sugar. Dissolve the soda in cream, stirring it then into the eggs and sugar. Sift into the mixture the flour, baking powder, and salt, and beat well together. Bake in three layer cake tins.

Filling:

1 cup of pecan or walnuts crushed with rolling-pin on bread-board

1 small egg

¾ cup confectioner's sugar

½ cup sour cream

A few drops of vanilla

Beat the egg well, white and yolk together: add the sugar and nuts, and last of all the cream and vanilla, stirring it then only enough to mix all together. Spread between the layers and over the top of cake when cold.

CIDER CAKE

3 cups flour

2 cups sugar

1 cup butter

3 eggs

½ tsp salt

1 tsp baking soda

1 cup cider

1 tsp cinnamon and allspice, mixed

1 cup raisins and currants, optional

Mix flour, sugar, salt, and spices together. Work in the butter until no lumps remain. Add the eggs, well beaten, and the cider in which the soda has been dissolved. The dough should be fairly stiff. Bake in a moderate oven. Cover with a brown sugar frosting.

HONEY CAKES

1 cup sugar

1 ¾ cups honey

1 tsp cloves

½ tsp ginger

1 tsp cinnamon

½ tsp salt

½ tsp nutmeg

¼ tsp pepper

1 tsp anise

1 ½ cups rye flour

1 ½ cups wheat flour

Sift together spices, salt, and flour. Put honey and sugar in a pan and let the mixture boil up. Then pour it on the flour mixture and stir until a thick dough is formed. If necessary, add more honey or flour until the paste is stiff enough to roll. Roll into small balls and bake in a moderate oven. When cool, dip each ball separately in a thin white frosting.

BEER

Home brewing has grown in popularity to the point where the equipment and ingredients are fairly easy to find. The Internet is the easiest place to find the tools you'll need.

What You'll Need

1. Brewpot: a huge, stainless steel (or other enamel-coated metal) pot of at least sixteen quart capacity. This will be used to boil all the ingredients, otherwise known as "wort."

2. Primary fermenter: where the wort goes after it's been boiled. It's where beer begins to ferment. The fermenter must have a capacity of seven gallons and an airtight lid that can accommodate an airlock and rubber stopper. Look for one made of food-grade plastic.

3. Airlock and stopper: the airlock allows carbon dioxide to escape without allowing any outside air in and fits into a rubber stopper with a hole drilled in it. The stopper goes on top of the primary fermenter; they are sized by number, so make sure to match the size of the hole with the stopper that fits it.

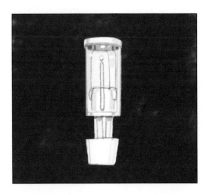

4. Plastic hose: a five-foot length of hose made out of food-grade plastic is ideal to transfer beer. Make sure to keep it clean and clear of kinks or leaks.

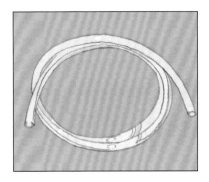

5. Bottling bucket: a one gallon, food-grade plastic bucket with a spigot at the bottom. It must be at least as big as the primary fermenter.

6. Beer bottles. After primary fermentation, you place beer in bottles for the second stage of fermentation and then finally, storage. You need enough bottles to hold all the beer you'll make (a five gallon batch is 640oz). Use solid dark glass bottles (to keep the light out) with smooth tops (not the ones that accommodate twist-off caps) that will accept a cap from a bottle capper.

7. Bottle brush: thin, curvy brush to clean the beer bottle.

8. Stick-on thermometer.

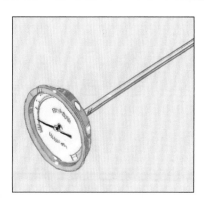

9. Household items: a small bowl, saucepan, rubber spatula, over mitts, a big mixing spoon (stainless steel or plastic).

Choose a recipe and buy ingredients, or buy a "beer kit," which includes a can of hopped malt concentrate and a packet of yeast. For your first time brewing, a beer kit may be the best choice since it would eliminate the possibility of error and allow you to get used to the procedure. Purchase other fermentables (more fermentables mean more alcohol) like dry malt extract, rice syrup, brewers' sugar, liquid malt extract, Belgian candi sugar, or demera sugar or any combination of the above. You need at least two pounds of fermentables, but no more than three.

Clean and sanitize all the equipment you'll use. Sanitizing is the use of heat, chlorine, or iodine mixed with water, but if your dishwasher has a "heat dry" cycle, cut down the preparation time by using it.

1. Bring 2 quarts of water to 160-180 degrees F: steaming, but not boiling. Remove from heat.

2. Add your beer kit (or recipe) and additional fermentables according to the directions. Each fermentable adds its own unique flavor.

3. Stir aggressively to dissolve everything in the pot. Put the lid on the pot and let it sit for 10-15 minutes on the lowest heat setting.

4. Add the contest of the pot to 4 gallons of cold water, which should already be waiting in the primary fermenter. Mix well, at least a minute or two, to add oxygen to the wort prior to adding the yeast. When the side of your fermenter is cool to the touch, it's safe to add the yeast.

5. Ferment as close to the recommended temperature range as possible. It will begin to ferment within the first day

and continue to do so for 3-5 days. You can tell because of the air bubbles. When there no bubbles, or a pause of two minutes in between bubbles, the beer is ready to be bottled.

6. Make sure you have enough bottles ready and cleaned and sanitized! Also have pure dextrose on hand to make the priming solution, which is what helps the yeast already in the beer to carbonate. Take the saucepan and put 2 or 3 cups of water in it. Dissolve ¾ cups of dextrose in the water. Then, bring it to a boil over medium heat, cover it, and set aside to cool for 15-20 minutes.

7. After this is done, place the bottling bucket on the floor. Place the primary fermenter on a chair, table, or counter directly above the bottling bucket; do not shake the beer up inside the fermenter as the sediment should stay at the bottom. Attach the plastic hose to the spigot on the fermenter and put the other end of the hose in the bottom of the bottling bucket. Pour priming solution into bottling bucket and open the spigot on the fermenter, allowing the beer to flow in and mix with the solution. Don't worry about saving the last of the beer in the fermenter since it contains sediment.

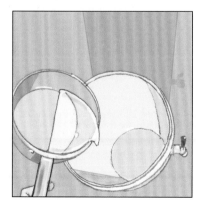

8. Move the fermenter and put the bottling bucket where it was. Hook the hose to its spigot. Line up the bottles on the floor underneath and stick the hose into one, all the way into the bottle. Open up the spigot and fill the bottle until the beer gets to the top, leaving about an inch of airspace. Quickly yank the hose out and stick it in the next bottle.

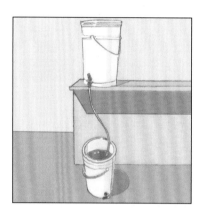

9. Once all the beer is drained out of the bucket, put caps on the bottles. Every second your beer is exposed to the air is bad. Find a cool, dark place to put the bottles while the second round of fermentation takes place. Do not put it in a refrigerator. Leave the beer to ferment for a minimum of two weeks before you drink it. Once the cloudiness caused by the yeast has dissipated (after the two weeks), you can put it in the fridge. If it hasn't cleared, leave it to sit for longer. When you finally drink your beer, it's best to pour it in the glass rather than drinking straight from the bottle to avoid the sediment and leftover yeast.

BUTCHERING

By raising your own meat, you can ensure that the animals are kept and slaughtered humanely and can avoid the hormones and other health hazards found in most commercially-raised meats. In addition, you'll develop a better understanding and appreciation for the meat you're eating, which will likely lead to less waste. The best way to learn how to slaughter an animal is to assist someone else several times, but you may want an overview of the process before even attempting that.

Choosing Animals to Slaughter

A healthy animal yields good meat. If the animal is sick in any way, there is always some danger that disease may be transmitted to the person who eats the meat. Do not butcher animals that are losing weight—they may be suffering from some health problem, and a reasonable amount of fat gives juiciness and flavor to the meat. In general, the more humanely an animal is raised, the better quality the meat will be. Animals that are raised with proper food, housing, space

to graze, and attention to their health will yield flavorful and healthy meat.

Hogs or cows may be slaughtered any time after 8 weeks of age, but the most profitable time is between 8 and 12 months. The meat from young hogs or cows lacks flavor and is watery, and that from older animals may be tough.

Equipment:

* .22 to .30 caliber rifle
* Tractor with a hydraulic lift or a block and tackle chain hoist
* Singletree with hooks on both ends
* 6-inch boning knife and 8-inch butcher knife
* Meat saw
* Large pans or buckets
* Plenty of clean, cold water

Killing and Dressing

Lead the animal to the area where you want to do the slaughtering. Choose a grassy area that is clean and dry, preferably near your hoist. Walk at an easy pace and try to keep the animal calm. Shoot the animal in the center of the forehead, slightly above the eyes—imagine an "X" across the forehead and aim at the center of it. Immediately after the animal falls, stand behind it and slit the neck lengthwise from the jaw down to the breastbone—this is called "sticking." From there, point the knife toward the rear of the animal and cut toward the backbone to sever the arteries just below the point of the breastbone. Do all this as quickly as safely possible—if the heart is still beating, the blood will drain more quickly. Pumping a rear leg back and forth will help expedite the draining. Roll the animal to its back and use a knife to cut through the flat joints in the front knees, removing the front feet and shanks. You'll need a saw to remove the hind legs from the knees down, being careful not to cut the tendons.

"Hoisting" the animal is the process of hanging it upside down to allow the remaining blood to drain out. Insert the singletree hooks into the hind leg tendons. Attach the singletree chain to your forklift and raise the animal until it's head is suspended 1 ½ to 2 feet above the ground.

At this point, the process differs for pigs and cows.

Gutting a Cow

Cut around the circumference of the neck and twist the head with a solid jerk to break the spine. Use a saw to finish cutting off the head. Remove the tongue and save it. If using the head meat for sausage, save this as well. The cow will be skinned after it is gutted.

If the animal was male, start by cutting around genitalia, and then slit up the midline. Cut the hide first and then around the anus, freeing the end in order to tie off. Free the genitalia and then cut down the rest of the belly line. Cut through the "aitch" bone (the H-shaped bone between the hind legs). Place a large wash-tub under the carcass to catch the innards. Make sure to cut all the way down the front and then begin to pull the innards out. They are large and very heavy, and so will fall into the bucket by themselves once released. If desired, find and save the heart and liver. The liver is large and easy to see-make sure you remove the gall-bladder and the part of the liver connected to it. The heart will be hard to find- it is encased in a thick layer of fat and tissue.

Cut through the breastbone all the way down to the throat in order to open up the rest of the carcass. Remove any remaining innards.

Skinning a Cow

From the center cut of the carcass you created to gut the cow, cut along the inside of the legs. Using a sharp knife, cut the skin away from the flesh. The skinning of a cow is done after the gutting, unlike other animals, because the innards should not be left in the cow for too long; it may spoil the meat.

Quartering and Aging the Beef

First halve the carcass by cutting all the way down and through the backbone with a saw. This is a tough job and better if two people can trade off. To quarter the carcass look into the cavity at the belly cut and cut between the third and fourth rib from the rear. After separating the ribs, but before slicing all the way through, prepare to hang the front quarters by making a slit between the fourth and fifth rib, and tying a rope through the slit. Have a second person hold the front quarters off the ground when the cut is made to fully separate them, and then to hang the quarters.

Move the quarters to where you are going to hang them for aging. Aging beef should hang in an animal and pest free environment that is kept at around 40 degrees Fahrenheit the entire time. An increase in temperature will reduce aging time, but may also lead to bacterial growth. A carcass with only a thin layer of fat covering the meat needs to be aged only 3-5 days, while a carcass with more fat covering the meat can be aged for 5-7 days.

Cutting Up a Cow

The simplest way to approach the daunting task of cutting up a cow carcass is to do one quarter a day. This will also allow the meat to adequately freeze- too much fresh meat in the freezer at once will slow freezing time and may lead to the meat going bad, as well as defrosting everything else in the freezer.

How to make cuts of meat- cut all the way through the meat around the bone, then cut through the bone. The thickness of steaks is really a matter of personal preference–anything between ½ to ¾ inch is fine. A thinner cut will yield more steaks. Any mistakes can be made into hamburger!

Forequarters- Cut off the foreleg at the shoulder joint. This is a two person job; one person should cut from under

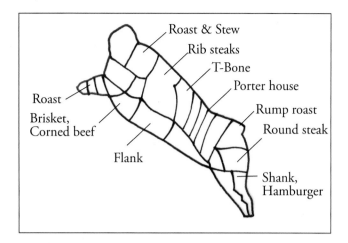

the leg while a second pulls the leg upward. Pull the leg out from under the shoulder joint. The top third of the leg can become a "blade roast", while the middle third can be one or more regular roasts. Directly behind the blade roast is a piece that can become a brisket, corned beef, or pastrami. All of the fat does not need to be trimmed from the roasts–it adds flavor! The bottom third of the leg can become soup bones with some meat attached, or you can slice off and grind that meat into hamburger.

Saw about eight inches away from the backbone to separate the ribs from the spine. You can then cut between the ribs to create rib steaks of any size you like.

The meat above the ribs, the shoulder and the neck can be cut up into roasts and stew meat. Once all your chosen cuts have been removed any remaining and scraps can be made into hamburger.

Below the rib area are the spare ribs and a lot of extra meat, an area called the plate. This area is a bit more difficult to butcher. Cut between each rib and separate them. Then cut through each individual rib, creating either halves or thirds to create short ribs. A few cuts from the outside part of the plate can become skirt steak. Most of this area can become hamburger, as the meat is tougher and fatty.

Hindquarters- Cut the leg from the body just below the hip bone. Starting from the top of the leg, cut round steaks about ½ to ¾ inch thick. These steaks may be cut in half again for ease of cooking and freezing. Create round steaks for all of the meat down to the knee joint. After the knee joint the leg meat can be cut into roasts of any size desired. The rest of the lower leg that is too small for roasts can become either soup bones or hamburger.

The back of the carcass can become a rump roast. From the rump moving forward cut your different types of steak. The area closest to where the ribs were is known as the short loin—that area can be cut into t-bone and porterhouse steaks. The sirloin area behind that can be cut into steaks as well; the area closest to the backbone will yield top sirloin, while just below that will become tri-tip and bottom sirloin.

Cut away the abdominal muscle flap found below the area that becomes the short loin and sirloin (also called the flank) and make a few cuts of steak. This steak is tougher, and best when marinated. Any leftovers of that cut can be made into hamburger.

Scalding a Pig

Once a pig has been killed, but before gutting, the pig needs to be "scalded," a process of placing the carcass in boiling water in order to remove the hair. When the pig is being killed a scalding vat should be prepped and waiting nearby.

The pig carcass will need to be hoisted with a homemade pulley type system. You will need two ropes of about 30 feet in length. This is a two person job. Loop both ropes around the carcass by laying it over the ropes, creating a U-shape with both ropes, and bring the ends of the rope over and around the pig and through the inside of the U-shape.

Once the ropes are ready put the pig in the vat, slowly so as not to splash. Roll the carcass gently in the water by pulling on one end of the two ropes, then the other. Turn the pig over and keep rolling it. The pig's hair will begin "slipping" around the ropes; it will look like the hair has been worn off around the ropes. Once "slipping" begins to occur it means that the pig is ready to be scraped. To test the pig's readiness scrape the bit that is sticking out of the water; if the hair comes off easily it is ready. With all four ends of the rope on one side haul the pig out of the water. The scalding process should take anywhere between 3 and 6 minutes.

There are special hog scrappers available, or you can use several knives. Knives should be held at both ends and pulled across the pig's skin, being careful not to slice into it. Knives will dull quickly while scraping tough pig hair. Scrappers will not cut the skin. Be sure to keep plenty of hot water ready to rinse the skin as you scrape.

It is best to scrape the feet and legs first, especially if you are going to use the feet for pickling. Pull out the nails by putting a hook through the top of the nail and pulling. Grip a leg with both hands and twist to pull the hair off. Repeat this process for all legs.

Using a scraper or knives, work around the eyes, snout and jaw, scraping hair off. The head has a lot of parts to work around and will therefore take time. You do not have to remove the feet or ears at this time.

Skinning the Pig

You do not have to skin the pig if you scrape, and some people prefer to leave the skin on, as it adds flavor to bacon and ham. However, skinning the carcass is a lot less time intensive than scraping, so if you wish to skin the carcass you simply follow a similar method to skinning a cow.

Lay the pig flat on a clean surface turned on its back and held there with blocks. Cut around the rear legs. Cut (only though the hide) along the legs, over the hocks, and to the center of the hams. Remove the skin from around each leg, ending just below the hock. Open the hide down the center of the pig, from the throat cut to the pubic area, cutting around each side of that, and on to the anus. Be careful not to puncture the intestines. Remove the skin from the hams, being careful to remove as little fat as possible.

Continue cutting along the side of the carcass, from the hams up to the breast. It is helpful for someone else to grab the loose skin at the other end to smooth it out and make cutting easier. Skin all the way down the sides of the carcass, excluding the front legs. At this point it will be necessary to hoist the carcass up with a singletree and hang it to complete the skinning process.

Skin around the hams. Remove skin from around the anus and cut off the tail. Pull the hide down over the hips; it can be pulled off by this point. Cut along the forelegs and skin those along with the neck. Cut along the shoulders, and about halfway up the back of the carcass. Pull the skin out and down as far as possible, using a knife if it gets stuck. Skin the head and face if you are saving that part, removing the skin all the way to the snout. Saw off the front feet right below the knee.

Gutting a Pig

If you skinned the carcass, it has already been hung up for gutting. If you scalded the pig instead hang it using a singletree and then proceed with gutting.

For a Male Pig- Cut around the penis and up to and around the anus. There will be a cord running from the penis- tie that off. Also tie off the end of the rectum. Be careful not to cut into the intestine.

For a Female Pig- Starting around the anus, cut all around it and tie of the end of the rectum. From here both male and female are gutted in the same way.

Cut down the middle of the belly all the way to the jaw. Split the breastbone in half and cut between the hind legs. Finish cutting into the belly by putting your hand inside and between your knife and the intestines. It is very important that you do not cut the intestines. Have a large bucket or washbasin underneath the carcass, and then bring the organs out. Cut the diaphragm and release the remaining organs. If you want to save the heart and liver, they need to be separated from the rest of the innards at this point. Trim off the top of the heart and set aside. Cut the gallbladder away from the liver, cutting into the liver (avoid slicing the gallbladder in any way). When you pull all the guts out everything will be removed all the way up to the tongue.

Cut off the head regardless of whether you are going to use it or not; it needs to be removed in order to split the pig carcass in half. To do this cut all the way through the meat to the bone just behind the ears. Grabbing the head firmly and twisting it around until it snaps off is usually all that is necessary to remove it.

Finish halving the carcass by sawing through the middle of the backbone, all the way from one end to the other. Leave the carcass hanging in order to cool the meat. Pork does not have to be aged, but it should be cooled down and left for a day (a few days in cold weather is alright too).

Dividing the Meat
Shoulder

Cut off the front foot about 1 inch below the knee. The shoulder cut is made through the third rib at the breastbone and across the fourth. Remove the ribs from the shoulders, also the piece backbone which is attached. Cut close to the ribs in removing them so as to leave as much meat on the shoulder as possible. These are shoulder or neck ribs and make an excellent dish when fried or baked. If only a small quantity of cured meat is desired, the top of the shoulder

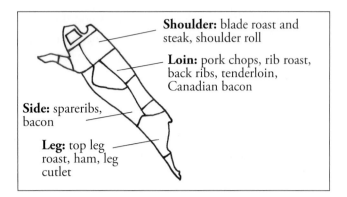

Shoulder: blade roast and steak, shoulder roll

Loin: pork chops, rib roast, back ribs, tenderloin, Canadian bacon

Side: spareribs, bacon

Leg: top leg roast, ham, leg cutlet

may be cut off about one-third the distance from the top and parallel to it. The fat of the shoulder top may be used for lard and the lean meat for steak or roasts. It should be trimmed smoothly. In case the shoulders are very large divide them crosswise into two parts (see fig. 6). This enables the cure mixture to penetrate more easily and therefore lessens the danger of souring. The fat trimmings should be used for lard and the lean trimmings for sausages.

The ham is removed from the middling by cutting just at the rise in the backbone and at a right angle to the shank, as shown in figure 5.

The loin and fatback are cut off in one piece, parallel with the back just below the tenderloin muscle on the rear part of the middling. Remove the fat on the top of the loin, but do not cut into the loin meat. The lean meat is excellent for canning or it may be used for chops or roasts and the fatback for lard. The remainder should then be trimmed for middling or bacon. Remove the ribs cutting as close to them as possible. If it is a very large side, it may be cut into two pieces. Trim all sides and edges as smoothly as possible (fig. 9).

Ham

Cut off the foot 1 inch below the hock joint. All rough and hanging pieces of meat should be trimmed from the ham. It should then be trimmed smoothly, exposing as little lean meat as possible, because the curing hardens it (fig. 8). All lean trimmings should be saved for sausage and fat trimmings for lard. The other half of the carcass should be cut up in similar manner.

Loins

Separate the loin from the belly by sawing thru the ribs, starting at the point of greatest curvature of the fourth rib.

Skin the fat from the loin, leaving the lean muscle barely covered with the fat. The loins can be boned out and used for sausage if a large amount of that is desired, or if the weather will not permit holding them as fresh meat they can be given a middle cure as boneless loins. The loin is best adapted for the pork (loin) roast or for pork chops. The latter are cut in such a way as to have the rib end in each alternate piece or chop.

Meat trimmings and fat trimmings

After the carcass has been cut up and the pieces are trimmed and shaped properly for the curing process, there are many pieces of lean meat, fat meat, and fat which can be used for making sausage and lard. The fat should be separated from the lean and used for lard. The meat should be cut into convenient-sized pieces to pass through the grinder.

Rendering lard

The leaf fat makes lard of the best quality. The back strip of the side also makes good lard, as do the trimmings of the ham, shoulder, and neck. Intestinal or gut fat makes an inferior grade and is best rendered by itself. This should be thoroughly washed and left in cold water for several hours before rendering, thus partially eliminating the offensive odor. Leaf fat, back strips, and fat trimmings may be rendered together. If the gut is included, the lard takes on a very offensive odor.

First, remove all skin and lean meat from the fat trimmings. To do this cut the fat into strips about 1 ½ inches wide, then place the strip on the table, skin down, and cut the fat from the skin. When a piece of skin large enough to grasp is freed from the fat, take it in the left hand and with the knife held in the right hand inserted between the fat and skin, pull the skin. If the knife is slanted downward slightly, this will easily remove the fat from the skin. The strips of fat should then be cut into pieces 1 or 1 ½ inches square, making them about equal in size.

Pour into the kettle about a quart of water, then fill it nearly full with fat cuttings. The fat will then heat and bring out the grease without burning. Render the lard over a moderate fire. At the beginning the temperature should be about 160° F, and it should be increased to 240°F. When the cracklings begin to brown, reduce the temperature to 200°F or a little more, but not to exceed 212° F, in order to prevent scorching. Frequent stirring is necessary to prevent burning. When the cracklings are thoroughly browned, and light enough to float, the kettle should be removed from the fire. Press the lard from the cracklings. When the lard is removed from the fire allow it to cool a little. Strain it through a muslin cloth into the containers. To aid cooling, stir it, which also tends to whiten it and make it smooth.

Lard which is to be kept for a considerable time should be placed in air-tight containers and stored in the cellar or other convenient place away from the light, in order to avoid rancidity. Fruit jars make excellent containers for lard, because they can be completely sealed. Glazed earthenware containers, such as crocks and jars, may be also be used. All containers should be sterilized before filling, and if covers are placed on the crocks or jars, they also should be sterilized before use. Lard stored in air-tight containers away from the light has been found to keep in perfect condition for a number of years.

When removing lard from a container for use, take it off evenly from the surface exposed. Do not dig down into the lard and take out a scoopful, as that leaves a thin coating around the sides of the container, which will become rancid very quickly through the action of the air.

Curing Pork

The first essential in curing pork is to make sure that the carcass is thoroughly cooled, but meat should never be allowed to freeze either before or during the period of curing.

The proper time to begin curing is when the meat is cool and still fresh, or about 24 to 36 hours after killing.

Selecting Quality Meat

As a general rule, the best meat is that which is moderately fat. Lean meat tends to be tough and tasteless. Very fat meat may be good, but is not economical. The butcher should be asked to cut off the superfluous suet before weighing it.

1. *Beef.* The flesh should feel tender, have a fine grain, and a clear red color. The fat should be moderate in quantity, and lie in streaks through the lean. Its color should be white or *very light* yellow. Ox beef is the best, heifer very good if well fed, cow and bull, decidedly inferior.

2. *Mutton.* The flesh, like that of beef, should be a good red color, perhaps a shade darker. It should be fine-grained, and well mixed with fat, which ought to be pure white and firm. The mutton of the black-faced breed of sheep is the best, and may be known by the shortness of the shank; the best age is about five years, though it is seldom to be had so old. Whether mutton is superior to either ram or ewe, and may be distinguished by having a prominent lump of fat on the broadest part of the inside of the leg. The flesh of the ram has a very dark color and is of a coarse texture; that of the ewe is pale, and the fat yellow and spongy.

3. *Veal.* Its color should be white, with a tinge of pink; it ought to be rather fat, and feel firm to the touch. The flesh should have a fine delicate texture. The leg-bone should be small; the kidney small and well covered with fat. The proper age is about two or three months; when killed too young it is soft, flabby, and dark colored. The bull-calf makes the best veal, though the cow-calf is preferred for many dishes on account of the udder.

4. *Lamb.* This should be light-colored and fat, and have a delicate appearance. The kidneys should be small and imbedded in fat, the quarters short and thick, and the knuckle stiff. When fresh, the vein in the fore quarter will have a bluish tint. If the vein look green or yellow, it is a certain sign of staleness, which may also be detected by smelling the kidneys.

5. *Pork.* Both the flesh and the fat must be white, firm, smooth, and dry. When young and fresh, the lean ought to break when pinched with the fingers, and the skin, which should be thing, yield to the nails. The breed having short legs, thick neck, and small head, is the best. Six months is the right age for killing, when the leg should not weigh more than 6 or 7 lb.

BUTTER

Making butter the old-fashioned way is incredibly simple and very gratifying. It's a great project to do with kids, too. All you need are a jar, a marble, some fresh cream, and about 20 minutes.

1. Start with about twice as much heavy whipping cream as you'll want butter. Pour it into the jar, drop in the marble, close the lid tightly, and start shaking.

2. Check the consistency of the cream every three to four minutes. The liquid will turn into whipped cream, and then eventually you'll see little clumps of butter forming in the jar. Keep shaking for another few minutes and then begin to strain out the liquid into another jar. This is buttermilk, which is great for use in making pancakes, waffles, biscuits, and muffins.

3. The butter is now ready, but it will store better if you wash and work it. Add ½ cup of ice cold water and continue to shake for two or three minutes. Strain out the water and repeat. When the strained water is clear, mash the butter to extract the last of the water, and strain.

4. Scoop the butter into a ramekin, mold, or wax paper.

If desired, add salt or chopped fresh herbs to your butter just before storing or serving. Butter can also be made in a food processor or blender to speed up the processing time.

CANNING

Canning began in France, at the turn of the nineteenth century, when Napoleon Bonaparte was desperate for a way to keep his troops well fed while on the march. In 1800 he decided to hold a contest, offering 12,000 francs to anyone who could devise a suitable method of food preservation. Nicolas François Appert, a French confectioner, rose to the challenge, considering that if wine could be preserved in bottles, perhaps food could be as well. He experimented until he was able to prove that heating food to boiling after it had been sealed in airtight glass bottles prevented the food from deteriorating. Interestingly, this all took place about 100 years before Louis Pasteur found that heat could destroy bacteria. Nearly ten years after the contest began, Napoleon personally presented Nicolas with the cash reward.

Canning practices have evolved over the last two centuries, but the principles remain the same. In fact, the way we can foods today is basically the same way our grandparents and great grandparents preserved their harvests for the winter months.

On the next few pages you will find descriptions of proper canning methods, with details on how canning works and why it is both safe and economical. Much of the information here is from the USDA, which has done extensive research on home canning and preserving. If you are new to home canning, read this section carefully as it will help to ensure success with the recipes that follow.

Whether you are a seasoned home canner or this is your first foray into food preservation, it is important to follow directions carefully. With some recipes it is okay to experiment with varied proportions or added ingredients, and with others it is important to stick to what's written. In many instances it is noted whether or not creative liberty is a good idea for a particular recipe, but if you are not sure, play it safe—otherwise you may end up with a jam that is too runny, a vegetable that is mushy, or a product that is spoiled. Take time to read the directions and prepare your foods and equipment adequately and you will find that home canning is safe, economical, tremendously satisfying, and a great deal of fun!

The Benefits of Canning

Canning is fun, economical, and a good way to preserve your precious produce. As more and more farmers' markets make their way into urban centers, city dwellers are also discovering how rewarding it is to make seasonal treats last all year round. Besides the value of your labor, canning home-grown or locally grown food may save you half the cost of buying commercially canned food. Freezing food may be simpler, but most people have limited freezer space, whereas cans of food can be stored almost anywhere. And what makes a nicer, more thoughtful gift than a jar of homemade jam, tailored to match the recipient's favorite fruits and flavors?

The nutritional value of home canning is an added benefit. Many vegetables begin to lose their vitamins as soon as they are harvested. Nearly half the vitamins may be lost within a few days unless the fresh produce is kept cool or preserved. Within one to two weeks, even refrigerated produce loses half or more of certain vitamins. The heating process during canning destroys from one-third to one-half of vitamins A and C, thiamin, and riboflavin. Once canned, foods may lose from 5 percent to 20 percent of these sensitive vitamins each year. The amounts of other vitamins, however, are only slightly lower in canned compared with fresh food. If vegetables are handled properly and canned promptly after harvest, they can be more nutritious than fresh produce sold in local stores.

The advantages of home canning are lost when you start with poor quality foods, when jars fail to seal properly, when food spoils, and when flavors, texture, color, and nutrients deteriorate during prolonged storage. The tips that follow explain many of these problems and recommend ways to minimize them.

How Canning Preserves Foods

The high percentage of water in most fresh foods makes them very perishable. They spoil or lose their quality for several reasons:

- Growth of undesirable microorganisms—bacteria, molds, and yeasts
- Activity of food enzymes
- Reactions with oxygen
- Moisture loss

Microorganisms live and multiply quickly on the surfaces of fresh food and on the inside of bruised, insect-damaged, and diseased food. Oxygen and enzymes are present throughout fresh food tissues.

Proper canning practices include:

- Carefully selecting and washing fresh food
- Peeling some fresh foods
- Hot packing many foods
- Adding acids (lemon juice, citric acid, or vinegar) to some foods
- Using acceptable jars and self-sealing lids
- Processing jars in a boiling-water or pressure canner for the correct amount of time

Collectively, these practices remove oxygen; destroy enzymes; prevent the growth of undesirable bacteria, yeasts, and molds; and help form a high vacuum in jars. High vacu-

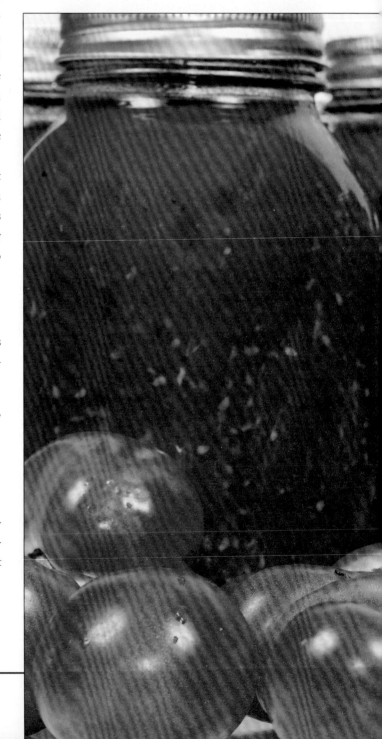

ums form tight seals, which keep liquid in and air and microorganisms out.

Canning Glossary

Acid foods—Foods that contain enough acid to result in a pH of 4.6 or lower. Includes most tomatoes; fermented and pickled vegetables; relishes; jams, jellies, and marmalades; and all fruits except figs. Acid foods may be processed in boiling water.

TIP

A large stockpot with a lid can be used in place of a boiling-water canner for high-acid foods like tomatoes, pickles, apples, peaches, and jams. Simply place a rack inside the pot so that the jars do not rest directly on the bottom of the pot.

Ascorbic acid—The chemical name for vitamin C; commonly used to prevent browning of peeled, light-colored fruits and vegetables.

Blancher—A 6- to 8-quart lidded pot designed with a fitted, perforated basket to hold food in boiling water or with a fitted rack to steam foods. Useful for loosening skins on fruits to be peeled or for heating foods to be hot packed.

Boiling-water canner—A large, standard-sized, lidded kettle with jar rack designed for heat-processing seven quarts or eight to nine pints in boiling water.

Botulism—An illness caused by eating a toxin produced by growth of *Clostridium botulinum* bacteria in moist, low-acid food containing less than 2 percent oxygen and stored between 40°F and 120°F. Proper heat processing destroys this bacterium in canned food. Freezer temperatures inhibit its growth in frozen food. Low moisture controls its growth in dried food. High oxygen controls its growth in fresh foods.

Canning—A method of preserving food that employs heat processing in airtight, vacuum-sealed containers so that food can be safely stored at normal home temperatures.

Canning salt—Also called pickling salt. It is regular table salt without the anti-caking or iodine additives.

Citric acid—A form of acid that can be added to canned foods. It increases the acidity of low-acid foods and may improve their flavor.

Cold pack—Canning procedure in which jars are filled with raw food. "Raw pack" is the preferred term for describing this practice. "Cold pack" is often used incorrectly to refer to foods that are open-kettle canned or jars that are heat-processed in boiling water.

Enzymes—Proteins in food that accelerate many flavor, color, texture, and nutritional changes, especially when food is cut, sliced, crushed, bruised, or exposed to air. Proper blanching or hot-packing practices destroy enzymes and improve food quality.

Exhausting—Removing air from within and around food and from jars and canners. Exhausting or venting of pressure canners is necessary to prevent botulism in low-acid canned foods.

Headspace—The unfilled space above food or liquid in jars that allows for food expansion as jars are heated and for forming vacuums as jars cool.

Heat processing—Treatment of jars with sufficient heat to enable storing food at normal home temperatures.

Hermetic seal—An absolutely airtight container seal that prevents reentry of air or microorganisms into packaged foods.

Hot pack—Heating of raw food in boiling water or steam and filling it hot into jars.

Low-acid foods—Foods that contain very little acid and have a pH above 4.6. The acidity in these foods is insufficient to prevent the growth of botulism bacteria. Vegetables, some varieties of tomatoes, figs, all meats, fish, seafood, and some dairy products are low-acid foods. To control all risks of botulism, jars of these foods must be either heat processed in a pressure canner or acidified to a pH of 4.6 or lower before being processed in boiling water.

Microorganisms—Independent organisms of microscopic size, including bacteria, yeast, and mold. In a suitable environment, they grow rapidly and may divide or reproduce every 10 to 30 minutes. Therefore, they reach high populations very quickly. Microorganisms are sometimes intentionally added to ferment foods, make antibiotics, and for other reasons. Undesirable microorganisms cause disease and food spoilage.

Mold—A fungus-type microorganism whose growth on food is usually visible and colorful. Molds may grow on many foods, including acid foods like jams and jellies and canned fruits. Recommended heat processing and sealing practices prevent their growth on these foods.

Mycotoxins—Toxins produced by the growth of some molds on foods.

Open-kettle canning—A non-recommended canning method. Food is heat-processed in a covered kettle, filled while hot into sterile jars, and then sealed. Foods canned this way have low vacuums or too much air, which permits rapid loss of quality in foods. Also, these foods often spoil because they become recontaminated while the jars are being filled.

Pasteurization—Heating food to temperatures high enough to destroy disease-causing microorganisms.

pH—A measure of acidity or alkalinity. Values range from 0 to 14. A food is neutral when its pH is 7.0. Lower values are increasingly more acidic; higher values are increasingly more alkaline.

PSIG—Pounds per square inch of pressure as measured by a gauge.

Pressure canner—A specifically designed metal kettle with a lockable lid used for heat processing low-acid food. These canners have jar racks, one or more safety devices, systems for exhausting air, and a way to measure or control pressure. Canners with 20- to 21-quart capacity are common. The minimum size of canner that should be used has a 16-quart capacity and can hold seven one-quart jars. Use of pressure saucepans with a capacity of less than 16 quarts is not recommended.

Raw pack—The practice of filling jars with raw, unheated food. Acceptable for canning low-acid foods, but allows more rapid quality losses in acid foods that are heat-processed in boiling water. Also called "cold pack."

Style of pack—Form of canned food, such as whole, sliced, piece, juice, or sauce. The term may also be used to specify whether food is filled raw or hot into jars.

Vacuum—A state of negative pressure that reflects how thoroughly air is removed from within a jar of processed food; the higher the vacuum, the less air left in the jar.

Proper Canning Practices

Growth of the bacterium *Clostridium botulinum* in canned food may cause botulism—a deadly form of food poisoning. These bacteria exist either as spores or as vegetative cells. The spores, which are comparable to plant seeds, can survive harmlessly in soil and water for many years. When ideal conditions exist for growth, the spores produce vegetative cells, which multiply rapidly and may produce a deadly toxin within three to four days in an environment consisting of:

- A moist, low-acid food
- A temperature between 40°F and 120°F, and
- Less than 2 percent oxygen.

Botulinum spores are on most fresh food surfaces. Because they grow only in the absence of air, they are harmless on fresh foods. Most bacteria, yeasts, and molds are difficult to remove from food surfaces. Washing fresh food reduces their numbers only slightly. Peeling root crops, underground stem crops, and tomatoes reduces their numbers greatly. Blanching also helps, but the vital controls are the method

of canning and use of the recommended research-based processing times. These processing times ensure destruction of the largest expected number of heat-resistant microorganisms in home-canned foods.

Properly sterilized canned food will be free of spoilage if lids seal and jars are stored below 95°F. Storing jars at 50 to 70°F enhances retention of quality.

Food Acidity and Processing Methods

Whether food should be processed in a pressure canner or boiling-water canner to control botulism bacteria depends on the acidity in the food. Acidity may be natural, as in most fruits, or added, as in pickled food. Low-acid canned foods contain too little acidity to prevent the growth of these bacteria. Other foods may contain enough acidity to block their growth or to destroy them rapidly when heated. The term "pH" is a measure of acidity: the lower its value, the more acidic the food. The acidity level in foods can be increased by adding lemon juice, citric acid, or vinegar.

Low-acid foods have pH values higher than 4.6. They include red meats, seafood, poultry, milk, and all fresh vegetables except for most tomatoes. Most products that are mixtures of low-acid and acid foods also have pH values above 4.6 unless their ingredients include enough lemon juice, citric acid, or vinegar to make them acid foods. Acid foods have a pH of 4.6 or lower. They include fruits, pickles, sauerkraut, jams, jellies, marmalade, and fruit butters.

Although tomatoes usually are considered an acid food, some are now known to have pH values slightly above 4.6. Figs also have pH values slightly above 4.6. Therefore, if they are to be canned as acid foods, these products must be acidified to a pH of 4.6 or lower with lemon juice or citric acid. Properly acidified tomatoes and figs are acid foods and can be safely processed in a boiling-water canner.

Botulinum spores are very hard to destroy at boiling-water temperatures; the higher the canner temperature, the more easily they are destroyed. Therefore, all low-acid foods should be sterilized at temperatures of 240 to 250°F, attainable with pressure canners operated at 10 to 15 PSIG. (PSIG means pounds per square inch of pressure as measured by a gauge.) At these temperatures, the time needed to destroy bacteria in low-acid canned foods ranges from 20 to 100 minutes. The exact time depends on the kind of food being

canned, the way it is packed into jars, and the size of jars. The time needed to safely process low-acid foods in boiling water ranges from seven to 11 hours; the time needed to process acid foods in boiling water varies from five to 85 minutes.

Know Your Altitude

It is important to know your approximate elevation or altitude above sea level in order to determine a safe processing time for canned foods. Since the boiling temperature of liquid is lower at higher elevations, it is critical that additional time be given for the safe processing of foods at altitudes above sea level.

What Not to Do

Open-kettle canning and the processing of freshly filled jars in conventional ovens, microwave ovens, and dishwashers are not recommended because these practices do not prevent all risks of spoilage. Steam canners are not recommended because processing times for use with current models have not been adequately researched. Because steam canners may not heat foods in the same manner as boiling-water canners, their use with boiling-water processing times may result in spoilage. So-called canning powders are useless as preservatives and do not replace the need for proper heat processing.

It is not recommended that pressures in excess of 15 PSIG be applied when using new pressure-canning equipment.

Ensuring High-Quality Canned Foods

Examine food carefully for freshness and wholesomeness. Discard diseased and moldy food. Trim small diseased lesions or spots from food.

Can fruits and vegetables picked from your garden or purchased from nearby producers when the products are at their peak of quality—within six to 12 hours after harvest for most vegetables. However, apricots, nectarines, peaches, pears, and plums should be ripened one or more days between harvest and canning. If you must delay the canning of other fresh produce, keep it in a shady, cool place.

Fresh, home-slaughtered red meats and poultry should be chilled and canned without delay. Do not can meat from sickly or diseased animals. Put fish and seafood on ice after harvest, eviscerate immediately, and can them within two days.

Maintaining Color and Flavor in Canned Food

To maintain good natural color and flavor in stored canned food, you must:

- Remove oxygen from food tissues and jars
- Quickly destroy the food enzymes, and
- Obtain high jar vacuums and airtight jar seals.

Follow these guidelines to ensure that your canned foods retain optimal colors and flavors during processing and storage:

- Use only high-quality foods that are at the proper maturity and are free of diseases and bruises
- Use the hot-pack method, especially with acid foods to be processed in boiling water
- Don't unnecessarily expose prepared foods to air; can them as soon as possible
- While preparing a canner load of jars, keep peeled, halved, quartered, sliced or diced apples, apricots, nectarines, peaches, and pears in a solution of 3 grams (3000 milligrams) ascorbic acid to 1 gallon of cold water. This procedure is also useful in maintaining the natural color of mushrooms and potatoes and for preventing stem-end discoloration in cherries and grapes. You can get ascorbic acid in several forms:

Pure powdered form—Seasonally available among canning supplies in supermarkets. One level teaspoon of pure powder weighs about 3 grams. Use 1 teaspoon per gallon of water as a treatment solution.

Vitamin C tablets—Economical and available year-round in many stores. Buy 500-milligram tablets; crush and dissolve six tablets per gallon of water as a treatment solution.

Commercially prepared mixes of ascorbic and citric acid—Seasonally available among canning supplies in supermarkets. Sometimes citric acid powder is sold in supermarkets, but it is less effective in controlling discoloration. If you choose to use these products, follow the manufacturer's directions.

- Fill hot foods into jars and adjust headspace as specified in recipes
- Tighten screw bands securely, but if you are especially strong, not as tightly as possible
- Process and cool jars
- Store the jars in a relatively cool, dark place, preferably between 50 and 70°F
- Can no more food than you will use within a year.

Advantages of Hot Packing

Many fresh foods contain from 10 percent to more than 30 percent air. The length of time that food will last at premium quality depends on how much air is removed from the food before jars are sealed. The more air that is removed, the higher the quality of the canned product.

Raw packing is the practice of filling jars tightly with freshly prepared but unheated food. Such foods, especially fruit, will float in the jars. The entrapped air in and around the food may cause discoloration within two to three months of storage. Raw-packing is more suitable for vegetables processed in a pressure canner.

Hot packing is the practice of heating freshly prepared food to boiling, simmering it three to five minutes, and promptly filling jars loosely with the boiled food. Hot packing is the best way to remove air and is the preferred pack style for foods processed in a boiling-water canner. At first, the color of hot-packed foods may appear no better than that of raw-packed foods, but within a short storage period both color and flavor of hot-packed foods will be superior.

Whether food has been hot packed or raw packed, the juice, syrup, or water to be added to the foods should be heated to boiling before it is added to the jars. This practice helps to remove air from food tissues, shrinks food, helps keep the food from floating in the jars, increases vacuum in sealed jars, and improves shelf life. Preshrinking food allows you to add more food to each jar.

Controlling Headspace

The unfilled space above the food in a jar and below its lid is termed headspace. It is best to leave a ¼-inch headspace for jams and jellies, ½-inch for fruits and tomatoes to be pro-cessed in boiling water, and from 1 to 1 ¼ inches in low-acid foods to be processed in a pressure canner.

This space is needed for expansion of food as jars are processed and for forming vacuums in cooled jars. The extent of expansion is determined by the air content in the food and by the processing temperature. Air expands greatly when heated to high temperatures—the higher the temperature, the greater the expansion. Foods expand less than air when heated.

Jars and Lids

Food may be canned in glass jars or metal containers. Metal containers can be used only once. They require special sealing equipment and are much more costly than jars.

Mason-type jars designed for home canning are ideal for preserving food by pressure or boiling-water canning. Regular and wide-mouthed threaded mason jars with self-sealing lids are the best choices. They are available in half-pint, pint, 1 ½-pint, and quart sizes. The standard jar mouth opening is about 2 ⅜ inches. Wide-mouthed jars have openings of about 3 inches, making them more easily filled and emptied. Regular-mouth decorative jelly jars are available in eight-ounce and 12-ounce sizes.

With careful use and handling, mason jars may be reused many times, requiring only new lids each time. When lids are used properly, jar seals and vacuums are excellent.

Look for scratches in the glass because while they may appear inconsequential, they could cause breakages and cracking while being processed in a canner. Mayonnaise-type jars are also notorious for jar breakage when being used for foods to be processed in a pressure canner and therefore aren't recommended. Other commercial jars are also not recommended if they cannot be sealed with two-piece canning lids.

Handy Household Hints

How to open a jar of fruit or vegetables that has stuck: Place the jar in a deep saucepan half full of cold water, bring it to a boil and allow to boil a few minutes. The jar will then open easily.

Jar Cleaning

Before reuse, wash empty jars in hot water with detergent and rinse well by hand, or wash in a dishwasher. Rinse thoroughly, as detergent residue may cause unnatural flavors and colors. Scale or hard-water films on jars are easily removed by soaking jars several hours in a solution containing 1 cup of vinegar (5 percent acid) per gallon of water. Jars should be kept hot until they are ready to be filled with food. Submerge the jars in a pot of simmering water (like a boiling water canner or a large stockpot) that can hold enough water to cover them and keep them simmering until it's time to fill the jars. Alternatively, a dishwasher could be used for the preheating process if the jars are washed and dried on a regular cycle.

Sterilization of Empty Jars

Use sterile jars for all jams, jellies, and pickled products processed less than 10 minutes. To sterilize empty jars, put them right side up on the rack in a boiling-water canner. Fill the canner and jars with hot (not boiling) water to 1 inch above the tops of the jars. Boil 10 minutes. Remove and drain hot sterilized jars one at a time. Save the hot water for processing filled jars. Fill jars with food, add lids, and tighten screw bands.

Empty jars used for vegetables, meats, and fruits to be processed in a pressure canner need not be sterilized beforehand. It is also unnecessary to sterilize jars for fruits, tomatoes, and pickled or fermented foods that will be processed 10 minutes or longer in a boiling-water canner.

Lid Selection, Preparation, and Use

The common self-sealing lid consists of a flat metal lid held in place by a metal screw band during processing. The flat lid is crimped around its bottom edge to form a trough, which is filled with a colored gasket material. When jars are processed, the lid gasket softens and flows slightly to cover the jar-sealing surface, yet allows air to escape from the jar. The gasket then forms an airtight seal as the jar cools. Gaskets in unused lids work well for at least five years from date of manufacture. The gasket material in older unused lids may fail to seal on jars.

It is best to buy only the quantity of lids you will use in a year. To ensure a good seal, carefully follow the manufacturer's directions in preparing lids for use. Examine all metal lids carefully. Do not use old, dented, or deformed lids or lids with gaps or other defects in the sealing gasket.

After filling jars with food, release air bubbles by inserting a flat plastic (not metal) spatula between the food and the jar. Slowly turn the jar and move the spatula up and down to allow air bubbles to escape. Adjust the headspace and then clean the jar rim (sealing surface) with a dampened paper towel. Place the lid, gasket down, onto the cleaned jar-sealing surface. Uncleaned jar-sealing surfaces may cause seal failures.

Then fit the metal screw band over the flat lid. Follow the manufacturer's guidelines enclosed with or on the box for tightening the jar lids properly.

- If screw bands are too tight, air cannot vent during processing, and food will discolor during storage. Overtightening also may cause lids to buckle and jars to break, especially with raw-packed, pressure-processed food.
- If screw bands are too loose, liquid may escape from jars during processing, seals may fail, and the food will need to be reprocessed.

Do not retighten lids after processing jars. As jars cool, the contents in the jar contract, pulling the self-sealing lid firmly against the jar to form a high vacuum. Screw bands are not needed on stored jars. They can be removed easily after jars are cooled. When removed, washed, dried, and stored in a dry area, screw bands may be used many times. If left on stored jars, they become difficult to remove, often rust, and may not work properly again.

Selecting the Correct Processing Time

When food is canned in boiling water, more processing time is needed for most raw-packed foods and for quart jars than is needed for hot-packed foods and pint jars.

To destroy microorganisms in acid foods processed in a boiling-water canner, you must:

- Process jars for the correct number of minutes in boiling water;
- Cool the jars at room temperature.

To destroy microorganisms in low-acid foods processed with a pressure canner, you must:

- Process the jars for the correct number of minutes at 240°F (10 PSIG) or 250°F (15 PSIG);
- Allow canner to cool at room temperature until it is completely depressurized.

The food may spoil if you fail to use the proper processing times, fail to vent steam from canners properly, process at lower pressure than specified, process for fewer minutes than specified, or cool the canner with water.

Processing times for half-pint and pint jars are the same, as are times for 1 ½-pint and quart jars. For some products, you have a choice of processing at 5, 10, or 15 PSIG. In these cases, choose the canner pressure (PSIG) you wish to use and match it with your pack style (raw or hot) and jar size to find the correct processing time.

Recommended Canners

There are two main types of canners for heat-processing home-canned food: boiling-water canners and pressure canners. Most are designed to hold seven one-quart jars or eight to nine one-pint jars. Small pressure canners hold four one-quart jars; some large pressure canners hold eighteen one-pint jars in two layers but hold only seven quart jars. Pressure saucepans with smaller volume capacities are not recommended for use in canning. Treat small pressure canners the same as standard larger canners; they should be vented using the typical venting procedures.

Low-acid foods must be processed in a pressure canner to be free of botulism risks. Although pressure canners also may be used for processing acid foods, boiling-water canners are recommended because they are faster. A pressure canner would require from 55 to 100 minutes to can a load of jars;

the total time for canning most acid foods in boiling water varies from 25 to 60 minutes.

A boiling-water canner loaded with filled jars requires about 20 to 30 minutes of heating before its water begins to boil. A loaded pressure canner requires about 12 to 15 minutes of heating before it begins to vent, another 10 minutes to vent the canner, another five minutes to pressurize the canner, another eight to 10 minutes to process the acid food, and, finally, another 20 to 60 minutes to cool the canner before removing jars.

Boiling-Water Canners

These canners are made of aluminum or porcelain-covered steel. They have removable perforated racks and fitted lids. The canner must be deep enough so that at least 1 inch of briskly boiling water will cover the tops of jars during processing. Some boiling-water canners do not have flat bottoms. A flat bottom must be used on an electric range. Either a flat or ridged bottom can be used on a gas burner. To ensure uniform processing of all jars with an electric range, the canner should be no more than 4 inches wider in diameter than the element on which it is heated.

Using a Boiling-Water Canner

Follow these steps for successful boiling-water canning:

1. Fill the canner halfway with water.

2. Preheat water to 140°F for raw-packed foods and to 180°F for hot-packed foods.

3. Load filled jars, fitted with lids, into the canner rack and use the handles to lower the rack into the water; or fill the canner, one jar at a time, with a jar lifter.

4. Add more boiling water, if needed, so the water level is at least 1 inch above jar tops.

5. Turn heat to its highest position until water boils vigorously.

6. Set a timer for the minutes required for processing the food.

7. Cover with the canner lid and lower the heat setting to maintain a gentle boil throughout the processing time.

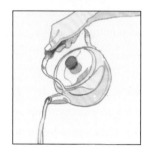

Pressure Canners

Pressure canners for use in the home have been extensively redesigned in recent years. Models made before the 1970s were heavy-walled kettles with clamp-on lids. They were fitted with a dial gauge, a vent port in the form of a petcock or counterweight, and a safety fuse. Modern pressure canners are lightweight, thin-walled kettles; most have turn-on lids. They have a jar rack, gasket, dial or weighted gauge, an automatic vent or cover lock, a vent port (steam vent) that is closed with a counterweight or weighted gauge, and a safety fuse.

Pressure does not destroy microorganisms, but high temperatures applied for a certain period of time do. The success of destroying all microorganisms capable of growing in canned food is based on the temperature obtained in pure steam, free of air, at sea level. At sea level, a canner operated at a gauge pressure of 10 pounds provides an internal temperature of 240°F.

Air trapped in a canner lowers the inside temperature and results in under-processing. The highest volume of air

8. Add more boiling water, if needed, to keep the water level above the jars.

9. When jars have been boiled for the recommended time, turn off the heat and remove the canner lid using a jar lifter, remove the jars and place them on a towel, leaving at least 1 inch of space between the jars during cooling.

trapped in a canner occurs in processing raw-packed foods in dial-gauge canners. These canners do not vent air during processing. To be safe, all types of pressure canners must be vented 10 minutes before they are pressurized.

To vent a canner, leave the vent port uncovered on newer models or manually open petcocks on some older models. Heating the filled canner with its lid locked into place boils water and generates steam that escapes through the petcock or vent port. When steam first escapes, set a timer for 10 minutes. After venting 10 minutes, close the petcock or place the counterweight or weighted gauge over the vent port to pressurize the canner.

Weighted-gauge models exhaust tiny amounts of air and steam each time their gauge rocks or jiggles during processing. The sound of the weight rocking or jiggling indicates that the canner is maintaining the recommended pressure and needs no further attention until the load has been processed for the set time. Weighted-gauge canners cannot correct precisely for higher altitudes, and at altitudes above 1,000 feet must be operated at a pressure of 15.

Check dial gauges for accuracy before use each year and replace if they read high by more than 1 pound at 5, 10, or 15 pounds of pressure. Low readings cause over-processing and may indicate that the accuracy of the gauge is unpredictable. If a gauge is consistently low, you may adjust the processing pressure. For example, if the directions call for 12 pounds of pressure and your dial gauge has tested 1 pound low, you can safely process at 11 pounds of pressure. If the gauge is more than 2 pounds low, it is unpredictable, and it is best to replace it. Gauges may be checked at most USDA county extension offices, which are located in every state across the country. To find one near you, visit www.csrees. usda.gov.

Handle gaskets of canner lids carefully and clean them according to the manufacturer's directions. Nicked or dried gaskets will allow steam leaks during pressurization of canners. Gaskets of older canners may need to be lightly coated with vegetable oil once per year, but newer models are pre-lubricated. Check your canner's instructions.

Lid safety fuses are thin metal inserts or rubber plugs designed to relieve excessive pressure from the canner. Do not pick at or scratch fuses while cleaning lids. Use only canners that have Underwriter's Laboratory (UL) approval to ensure their safety.

Replacement gauges and other parts for canners are often available at stores offering canner equipment or from canner manufacturers. To order parts, list canner model number and describe the parts needed.

Using a Pressure Canner

Follow these steps for successful pressure canning:

1. Put 2 to 3 inches of hot water in the canner. Place filled jars on the rack, using a jar lifter. Fasten canner lid securely.

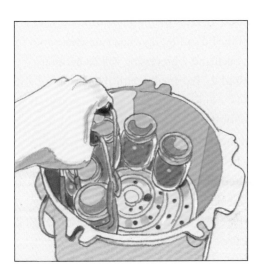

2. Open petcock or leave weight off vent port. Heat at the highest setting until steam flows from the petcock or vent port.

3. Maintain high heat setting, exhaust steam 10 minutes, and then place weight on vent port or close petcock. The canner will pressurize during the next three to five minutes.

4. Start timing the process when the pressure reading on the dial gauge indicates that the recommended pressure has been reached or when the weighted gauge begins to jiggle or rock.

5. Regulate heat under the canner to maintain a steady pressure at or slightly above the correct gauge pressure. Quick and large pressure variations during processing may cause unnecessary liquid losses from jars. Weighted gauges on Mirro canners should jiggle about two or three times per minute. On Presto canners, they should rock slowly throughout the process.

When processing time is completed, turn off the heat, remove the canner from heat if possible, and let the canner depressurize. Do not force-cool the canner. If you cool it with cold running water in a sink or open the vent port before the canner depressurizes by itself, liquid will spurt from jars, causing low liquid levels and jar seal failures. Force-cooling also may warp the canner lid of older model canners, causing steam leaks.

Depressurization of older models should be timed. Standard size heavy-walled canners require about 30 minutes when loaded with pints and 45 minutes with quarts. Newer thin-walled canners cool more rapidly and are equipped with vent locks. These canners are depressurized when their vent lock piston drops to a normal position.

1. After the vent port or petcock has been open for two minutes, unfasten the lid and carefully remove it. Lift

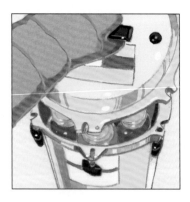

the lid away from you so that the steam does not burn your face.

2. Remove jars with a lifter, and place on towel or cooling rack, if desired.

Cooling Jars

Cool the jars at room temperature for 12 to 24 hours. Jars may be cooled on racks or towels to minimize heat damage to counters. The food level and liquid volume of raw-packed jars will be noticeably lower after cooling because air is exhausted during processing and food shrinks. If a jar loses excessive liquid during processing, do not open it to add more liquid. As long as the seal is good, the product is still usable.

Testing Jar Seals

After cooling jars for 12 to 24 hours, remove the screw bands and test seals with one of the following methods:

Method 1: Press the middle of the lid with a finger or thumb. If the lid springs up when you release your finger, the lid is unsealed and reprocessing will be necessary.

Method 2: Tap the lid with the bottom of a teaspoon. If it makes a dull sound, the lid is not sealed. If food is in contact with the underside of the lid, it will also cause a dull sound. If the jar lid is sealed correctly, it will make a ringing, high-pitched sound.

Method 3: Hold the jar at eye level and look across the lid. The lid should be concave (curved down slightly in the center). If center of the lid is either flat or bulging, it may not be sealed.

Reprocessing Unsealed Jars

If a jar fails to seal, remove the lid and check the jar-sealing surface for tiny nicks. If necessary, change the jar, add a new, properly prepared lid, and reprocess within 24 hours using the same processing time.

Another option is to adjust headspace in unsealed jars to 1 ½ inches and freeze jars and contents instead of reprocessing. However, make sure jars have straight sides. Freezing may crack jars with "shoulders."

Foods in single unsealed jars could be stored in the refrigerator and consumed within several days.

Storing Canned Foods

If lids are tightly vacuum-sealed on cooled jars, remove screw bands, wash the lid and jar to remove food residue, then rinse and dry jars. Label and date the jars and store them in a clean, cool, dark, dry place. Do not store jars at temperatures above 95°F or near hot pipes, a range, a furnace, in an un-insulated attic, or in direct sunlight. Under these conditions, food will lose quality in a few weeks or months and may spoil. Dampness may corrode metal lids, break seals, and allow recontamination and spoilage.

Accidental freezing of canned foods will not cause spoilage unless jars become unsealed and re-contaminated. However, freezing and thawing may soften food. If jars must be stored where they may freeze, wrap them in newspapers, place them in heavy cartons, and cover them with more newspapers and blankets.

Identifying and Handling Spoiled Canned Food

Growth of spoilage bacteria and yeast produces gas, which pressurizes the food, swells lids, and breaks jar seals. As each stored jar is selected for use, examine its lid for tightness and vacuum. Lids with concave centers have good seals.

Next, while holding the jar upright at eye level, rotate the jar and examine its outside surface for streaks of dried food originating at the top of the jar. Look at the contents for rising air bubbles and unnatural color.

While opening the jar, smell for unnatural odors and look for spurting liquid and cotton-like mold growth (white, blue, black, or green) on the top food surface and underside of lid. Do not taste food from a stored jar you discover to have an unsealed lid or that otherwise shows signs of spoilage.

All suspect containers of spoiled low-acid foods should be treated as having produced botulinum toxin and should be handled carefully as follows:

- If the suspect glass jars are unsealed, open, or leaking, they should be detoxified before disposal.
- If the suspect glass jars are sealed, remove lids and detoxify the entire jar, contents, and lids.

Detoxification Process

Wear rubber or heavy plastic gloves when handling the suspect foods or cleaning contaminated work areas and equipment. The botulinum toxin can be fatal through ingestion or through entering the skin.

Carefully place the suspect containers and lids on their sides in an eight-quart-volume or larger stockpot, pan, or boiling-water canner. Wash your hands thoroughly. Carefully add water to the pot. The water should completely cover the containers with a minimum of 1 inch of water above the containers. Avoid splashing the water. Place a lid on the pot and heat the water to boiling. Boil 30 minutes to ensure detoxifying the food and all container components. Cool and discard lids and food in the trash or bury in the soil.

Thoroughly clean all counters, containers, and equipment including can opener, clothing, and hands that may have come in contact with the food or the containers. Discard any sponges or washcloths that were used in the clean-up. Place them in a plastic bag and discard in the trash.

Canned Foods for Special Diets

The cost of commercially canned special diet food often prompts interest in preparing these products at home. Some low-sugar and low-salt foods may be easily and safely canned at home. However, it may take some experimentation to create a product with the desired color, flavor, and texture. Start with a small batch and then make appropriate adjustments before producing large quantities.

Fruit

There's nothing quite like opening a jar of home-preserved strawberries in the middle of a winter snowstorm. It takes you right back to the warm early-summer sunshine, the smell of the strawberry patch's damp earth, and the feel of the firm berries as you snipped them from the vines. Best of all, you get to indulge in the sweet, summery flavor even as the snow swirls outside the windows.

Preserving fruit is simple, safe, and it allows you to enjoy the fruits of your summer's labor all year round. On the next pages you will find reference charts for processing various fruits and fruit products in a dial-gauge pressure canner or a weighted-gauge pressure canner. The same information is also included with each recipe's directions. In some cases a boiling-water canner will serve better; for these instances, directions for its use are offered instead.

Adding syrup to canned fruit helps to retain its flavor, color, and shape, although it does not prevent spoilage. To maintain the most natural flavor, use the Very Light Syrup listed in the table found on page 148. Many fruits that are typically packed in heavy syrup are just as good—and a lot better for you—when packed in lighter syrups. However, if you're preserving fruit that's on the sour side, like cherries or tart apples, you might want to splurge on one of the sweeter versions.

Process Times for Fruits and Fruit Products in a Dial-Gauge Pressure Canner*

Type of Fruit	Style of Pack	Jar Size	Process Time	Canner Pressure (PSI) at Altitudes of:			
				0–2,000 ft	2,001–4,000 ft	4,001–6,000 ft	6,001–8,000 ft
Applesauce	Hot	Pints	8 minutes	6 lbs	7 lbs	8 lbs	9 lbs
	Hot	Quarts	10 minutes	6 lbs	7 lbs	8 lbs	9 lbs
Apples, sliced	Hot	Pints or Quarts	8 minutes	6 lbs	7 lbs	8 lbs	9 lbs
Berries, whole	Hot	Pints or Quarts	8 minutes	6 lbs	7 lbs	8 lbs	9 lbs
	Raw	Pints	8 minutes	6 lbs	7 lbs	8 lbs	9 lbs
	Raw	Quarts	10 minutes	6 lbs	7 lbs	8 lbs	9 lbs
Cherries, sour or sweet	Hot	Pints	8 minutes	6 lbs	7 lbs	8 lbs	9 lbs
	Hot	Quarts	10 minutes	6 lbs	7 lbs	8 lbs	9 lbs
	Raw	Pints or Quarts	10 minutes	6 lbs	7 lbs	8 lbs	9 lbs
Fruit purées	Hot	Pints or Quarts	8 minutes	6 lbs	7 lbs	8 lbs	9 lbs
Grapefruit or orange sections	Hot	Pints or Quarts	8 minutes	6 lbs	7 lbs	8 lbs	9 lbs
	Raw	Pints	8 minutes	6 lbs	7 lbs	8 lbs	9 lbs
	Raw	Quarts	10 minutes	6 lbs	7 lbs	8 lbs	9 lbs
Peaches, apricots, or nectarines	Hot or Raw	Pints or Quarts	10 minutes	6 lbs	7 lbs	8 lbs	9 lbs
Pears	Hot	Pints or Quarts	10 minutes	6 lbs	7 lbs	8 lbs	9 lbs
Plums	Hot or Raw	Pints or Quarts	10 minutes	6 lbs	7 lbs	8 lbs	9 lbs
Rhubarb	Hot	Pints or Quarts	8 minutes	6 lbs	7 lbs	8 lbs	9 lbs

*After the process is complete, turn off the heat and remove the canner lid. Wait five to 10 minutes before removing jars.

Process Times for Fruits and Fruit Products in a Weighted-Gauge Pressure Canner*

Type of Fruit	Style of Pack	Jar Size	Process Time	Canner Pressure (PSI) at Altitudes of:	
				0–1,000 ft	Above 1,000 ft
Applesauce	Hot	Pints	8 minutes	5 lbs	10 lbs
	Hot	Quarts	10 minutes	5 lbs	10 lbs
Apples, sliced	Hot	Pints or Quarts	8 minutes	5 lbs	10 lbs

Type of Fruit	Style of Pack	Jar Size	Process Time	Canner Pressure (PSI) at Altitudes of:	
				0–1,000 ft	Above 1,000 ft
Berries, whole	Hot	Pints or Quarts	8 minutes	5 lbs	10 lbs
	Raw	Pints	8 minutes	5 lbs	10 lbs
	Raw	Quarts	10 minutes	5 lbs	10 lbs
Cherries, sour or sweet	Hot	Pints	8 minutes	5 lbs	10 lbs
	Hot	Quarts	10 minutes	5 lbs	10 lbs
	Raw	Pints or Quarts	10 minutes	5 lbs	10 lbs
Fruit purées	Hot	Pints or Quarts	8 minutes	5 lbs	10 lbs
Grapefruit or orange sections	Hot	Pints or Quarts	8 minutes	5 lbs	10 lbs
	Raw	Pints	8 minutes	5 lbs	10 lbs
	Raw	Quarts	10 minutes	5 lbs	10 lbs
Peaches, apricots, or nectarines	Hot or Raw	Pints or Quarts	10 minutes	5 lbs	10 lbs
Pears	Hot	Pints or Quarts	10 minutes	5 lbs	10 lbs
Plums	Hot or Raw	Pints or Quarts	10 minutes	5 lbs	10 lbs
Rhubarb	Hot	Pints or Quarts	8 minutes	5 lbs	10 lbs

*After the process is complete, turn off the heat and remove the canner lid. Wait five to 10 minutes before removing jars.

Syrups

Adding syrup to canned fruit helps to retain its flavor, color, and shape, although jars still need to be processed to prevent spoilage. Follow the chart below for syrups of varying sweetness. Light corn syrups or mild-flavored honey may be used to replace up to half the table sugar called for in syrups.

Directions

1. Bring water and sugar to a boil in a medium saucepan.
2. Pour over raw fruits in jars.

> For hot packs, bring water and sugar to boil, add fruit, reheat to boil, and fill into jars immediately.

Sugar and Water in Syrup

Syrup Type	Approx. % Sugar	Measures of Water and Sugar				Fruits Commonly Packed in Syrup
		For 9-Pt Load*		For 7-Qt Load		
		Cups Water	Cups Sugar	Cups Water	Cups Sugar	
Very Light	10	6 ½	¾	10 ½	1 ¼	Approximates natural sugar levels in most fruits and adds the fewest calories.
Light	20	5 ¾	1 ½	9	2 ¼	Very sweet fruit. Try a small amount the first time to see if your family likes it.
Medium	30	5 ¼	2 ¼	8 ¼	3 ¾	Sweet apples, sweet cherries, berries, grapes.
Heavy	40	5	3 ¼	7 ¾	5 ¼	Tart apples, apricots, sour cherries, gooseberries, nectarines, peaches, pears, plums.
Very Heavy	50	4 ¼	4 ¼	6 ½	6 ¾	Very sour fruit. Try a small amount the first time to see if your family likes it.

*This amount is also adequate for a four-quart load.

Canning Without Sugar

In canning regular fruits without sugar, it is very important to select fully ripe but firm fruits of the best quality. It is generally best to can fruit in its own juice, but blends of unsweetened apple, pineapple, and white grape juice are also good for pouring over solid fruit pieces. Adjust headspaces and lids and use the processing recommendations for regular fruits. Add sugar substitutes, if desired, when serving.

Apple Juice

The best apple juice is made from a blend of varieties. If you don't have your own apple press, try to buy fresh juice from a local cider maker within 24 hours after it has been pressed.

Directions

1. Refrigerate juice for 24 to 48 hours.
2. Without mixing, carefully pour off clear liquid and discard sediment. Strain the clear liquid through a paper coffee filter or double layers of damp cheesecloth.
3. Heat quickly in a saucepan, stirring occasionally, until juice begins to boil.
4. Fill immediately into sterile pint or quart jars or into clean half-gallon jars, leaving ¼-inch headspace.

5. Adjust lids and process. See below for recommended times for a boiling-water canner.

Process Times for Apple Juice in a Boiling-Water Canner*

Style of Pack	Jar Size	Process Time at Altitudes of:		
		0–1,000 ft	1,001– 6,000 ft	Above 6,000 ft
Hot	Pints or Quarts	5 min	10	15
	Half-Gallons	10	15	20

*After the process is complete, turn off the heat and remove the canner lid. Wait five minutes before removing jars.

Apple Butter

The best apple varieties to use for apple butter include Jonathan, Winesap, Stayman, Golden Delicious, and Macintosh apples, but any of your favorite varieties will work. Don't bother to peel the apples, as you will strain the fruit before cooking it anyway. This recipe will yield eight to nine pints.

Applesauce

Besides being delicious on its own or paired with dishes like pork chops or latkes, applesauce can be used as a butter substitute in many baked goods. Select apples that are sweet, juicy, and crisp. For a tart flavor, add one to two pounds of tart apples to each three pounds of sweeter fruit.

Quantity

1. An average of 21 pounds of apples is needed per canner load of seven quarts.
2. An average of 13 ½ pounds of apples is needed per canner load of nine pints.
3. A bushel weighs 48 pounds and yields 14 to 19 quarts of sauce—an average of three pounds per quart.

Directions

1. Wash, peel, and core apples. Slice apples into water containing a little lemon juice to prevent browning.
2. Place drained slices in an 8- to 10-quart pot. Add ½ cup water. Stirring occasionally to prevent burning, heat quickly until tender (5 to 20 minutes, depending on maturity and variety).
3. Press through a sieve or food mill, or skip the pressing step if you prefer chunky-style sauce. Sauce may be packed without sugar, but if desired, sweeten to taste (start with ⅛ cup sugar per quart of sauce).
4. Reheat sauce to boiling. Fill jars with hot sauce, leaving ½-inch headspace. Adjust lids and process.

Ingredients

8 lbs apples
2 cups cider
2 cups vinegar
2 ¼ cups white sugar
2 ¼ cups packed brown sugar
2 tbsp ground cinnamon
1 tbsp ground cloves

Directions

1. Wash, stem, quarter, and core apples.
2. Cook slowly in cider and vinegar until soft. Press fruit through a colander, food mill, or strainer.
3. Cook fruit pulp with sugar and spices, stirring frequently. To test for doneness, remove a spoonful and hold it away from steam for 2 minutes. If the butter remains mounded on the spoon, it is done. If you're still not sure, spoon a small quantity onto a plate. When a rim of liquid does not separate around the edge of the butter, it is ready for canning.
4. Fill hot into sterile half-pint or pint jars, leaving ¼-inch headspace. Quart jars need not be pre-sterilized.

Process Times for Apple Butter in a Boiling-Water Canner*

		Process Time at Altitudes of:		
Style of Pack	**Jar Size**	0–1,000 ft	1,001–6,000 ft	Above 6,000 ft
Hot	Half-pints or Pints	5 minutes	10 minutes	15 minutes
	Quarts	10 minutes	15 minutes	20 minutes

*After the process is complete, turn off the heat and remove the canner lid. Wait five minutes before removing jars.

Process Times for Applesauce in a Boiling-Water Canner*

Style of Pack	Jar Size	Process Time at Altitudes of:			
		0–1,000 ft	1,001–3,000 ft	3,001–6,000 ft	Above 6,000 ft
Hot	Pints	15 minutes	20 minutes	20 minutes	25 minutes
	Quarts	20 minutes	25 minutes	30 minutes	35 minutes

*After the process is complete, turn off the heat and remove the canner lid. Wait five minutes before removing jars.

Process Times for Applesauce in a Dial-Gauge Pressure Canner*

Style of Pack	Jar Size	Process Time	Canner Pressure (PSI) at Altitudes of:			
			0–2,000 ft	2,001–4,000 ft	4,001–6,000 ft	6,001–8,000 ft
Hot	Pints	8 minutes	6 lb	7 lb	8 lb	9 lb
	Quarts	10 minutes	6 lb	7 lb	8 lb	9 lb

*After the canner is completely depressurized, remove the weight from the vent port or open the petcock. Wait 10 minutes; then unfasten the lid and remove it carefully. Lift the lid with the underside away from you so that the steam coming out of the canner does not burn your face.

Process Times for Applesauce in a Weighted-Gauge Pressure Canner*

Style of Pack	Jar Size	Process Time	Canner Pressure (PSI) at Altitudes of:	
			0–1,000 ft	Above 1,000 ft
Hot	Pints	8 minutes	5 lb	10 lb
	Quarts	10 minutes	5 lb	10 lb

*After the canner is completely depressurized, remove the weight from the vent port or open the petcock. Wait 10 minutes; then unfasten the lid and remove it carefully. Lift the lid with the underside away from you so that the steam coming out of the canner does not burn your face.

Spiced Apple Rings

12 lbs firm tart apples (maximum diameter 2-½ inches)
12 cups sugar
6 cups water
1-¼ cups white vinegar (5%)
3 tbsp whole cloves
¾ cup red hot cinnamon candies or 8 cinnamon sticks
1 tsp red food coloring (optional)
Yield: About 8 to 9 pints

Directions

1. Wash apples. To prevent discoloration, peel and slice one apple at a time. Immediately cut crosswise into ½-inch slices, remove core area with a melon baller and immerse in ascorbic acid solution.

2. To make flavored syrup, combine sugar water, vinegar, cloves, cinnamon candies, or cinnamon sticks and food coloring in a 6-qt saucepan. Stir, heat to boil, and simmer 3 minutes.

3. Drain apples, add to hot syrup, and cook 5 minutes. Fill jars (preferably wide-mouth) with apple rings and hot flavored syrup, leaving ½-inch headspace. Adjust lids and process according to the chart below.

Process time for spiced apple rings in a boiling-water canner.

Table 1. Recommended process time for Spiced Apple Rings in a boiling-water canner.				
		Process Time at Altitudes of		
Style of Pack	**Jar Size**	**0–1,000 ft**	**1,001– 6,000 ft**	**Above 6,000 ft**
Hot	Half-pints or Pints	**10 min**	15	20

Apricots, Halved or Sliced

Apricots are excellent in baked goods, stuffing, chutney, or on their own. Choose firm, well-colored mature fruit for best results.

Quantity

- An average of 16 pounds is needed per canner load of seven quarts.
- An average of 10 pounds is needed per canner load of nine pints.
- A bushel weighs 50 pounds and yields 20 to 25 quarts—an average of 2 ¼ pounds per quart.

Directions

1. Dip fruit in boiling water for 30 to 60 seconds until skins loosen. Dip quickly in cold water and slip off skins.
2. Cut in half, remove pits, and slice if desired. To prevent darkening, keep peeled fruit in water with a little lemon juice.
3. Prepare and boil a very light, light, or medium syrup (see page 148) or pack apricots in water, apple juice, or white grape juice.

Process Times for Halved or Sliced Apricots in a Dial-Gauge Pressure Canner*

			Canner Pressure (PSI) at Altitudes of:			
Style of Pack	**Jar Size**	**Process Time**	0–2,000 ft	2,001–4,000 ft	4,001–6,000 ft	6,001–8,000 ft
Hot or Raw	Pints or Quarts	10 minutes	6 lbs	7 lbs	8 lbs	9 lbs

*After the process is complete, turn off the heat and remove the canner lid. Wait five minutes before removing jars.

Process Times for Halved or Sliced Apricots in a Weighted-Gauge Pressure Canner*

			Canner Pressure (PSI) at Altitudes of:	
Style of Pack	**Jar Size**	**Process Time**	0–1,000 ft	Above 1,000 ft
Hot or Raw	Pints or Quarts	10 minutes	5 lbs	10 lbs

*After the process is complete, turn off the heat and remove the canner lid. Wait five minutes before removing jars.

Berries, Whole

Preserved berries are perfect for use in pies, muffins, pancakes, or in poultry or pork dressings. Nearly every berry preserves well, including blackberries, blueberries, currants, dewberries, elderberries, gooseberries, huckleberries, loganberries, mulberries, and raspberries. Choose ripe, sweet berries with uniform color.

Quantity

- An average of 12 pounds is needed per canner load of seven quarts.
- An average of 8 pounds is needed per canner load of nine pints.
- A 24-quart crate weighs 36 pounds and yields 18 to 24 quarts—an average of 1 ¾ pounds per quart.

Directions

1. Wash 1 or 2 quarts of berries at a time. Drain, cap, and stem if necessary. For gooseberries, snip off heads and tails with scissors.

2. Prepare and boil preferred syrup, if desired (see page 148). Add ½ cup syrup, juice, or water to each clean jar.

Hot pack—(Best for blueberries, currants, elderberries, gooseberries, and huckleberries) Heat berries in boiling water for 30 seconds and drain. Fill jars and cover with hot juice, leaving ½-inch headspace.

Raw pack—Fill jars with any of the raw berries, shaking down gently while filling. Cover with hot syrup, juice, or water, leaving ½-inch headspace.

Recommended Process Times for Whole Berries in a Boiling-Water Canner*

		Process Time at Altitudes of:			
Style of Pack	Jar Size	0–1,000 ft	1,001–3,000 ft	3,001–6,000 ft	Above 6,000 ft
Hot	Pints or Quarts	15 minutes	20 minutes	20 minutes	25 minutes
Raw	Pints	15 minutes	20 minutes	20 minutes	25 minutes
	Quarts	20 minutes	25 minutes	30 minutes	35 minutes

*After the process is complete, turn off the heat and remove the canner lid. Wait five minutes before removing jars.

Process Times for Whole Berries in a Dial-Gauge Pressure Canner*

			Canner Pressure (PSI) at Altitudes of:			
Style of Pack	Jar Size	Process Time	0–2,000 ft	2,001–4,000 ft	4,001–6,000 ft	6,001–8,000 ft
Hot	Pints or Quarts	8 minutes	6 lbs	7 lbs	8 lbs	9 lbs
Raw	Pints	8 minutes	6 lbs	7 lbs	8 lbs	9 lbs
Raw	Quarts	10 minutes	6 lbs	7 lbs	8 lbs	9 lbs

*After the process is complete, turn off the heat and remove the canner lid. Wait five minutes before removing jars.

Process Times for Whole Berries in a Weighted-Gauge Pressure Canner*

Style of Pack	Jar Size	Process Time	Canner Pressure (PSI) at Altitudes of:	
			0–1,000 ft	Above 1,000 ft
Hot	Pints or Quarts	8 minutes	5 lbs	10 lbs
Raw	Pints	8 minutes	5 lbs	10 lbs
Raw	Quarts	10 minutes	5 lbs	10 lbs

*After the process is complete, turn off the heat and remove the canner lid. Wait five minutes before removing jars.

Berry Syrup

Juices from fresh or frozen blueberries, cherries, grapes, raspberries (black or red), and strawberries are easily made into toppings for use on ice cream and pastries. For an elegant finish to cheesecakes or pound cakes, drizzle a thin stream in a zigzag across the top just before serving. Berry syrups are also great additions to smoothies or milkshakes. This recipe makes about nine half-pints.

Directions

1. Select 6 ½ cups of fresh or frozen berries of your choice. Wash, cap, and stem berries and crush in a saucepan.

2. Heat to boiling and simmer until soft (5 to 10 minutes). Strain hot through a colander placed in a large pan and drain until cool enough to handle.

3. Strain the collected juice through a double layer of cheesecloth or jelly bag. Discard the dry pulp. The yield of the pressed juice should be about 4 ½ to 5 cups.

4. Combine the juice with 6 ¾ cups of sugar in a large saucepan, bring to a boil, and simmer 1 minute.

5. Fill into clean half-pint or pint jars, leaving ½-inch headspace. Adjust lids and process.

To make syrup with whole berries, rather than crushed, save 1 or 2 cups of the fresh or frozen fruit, combine these with the sugar, and simmer until soft. Remove from heat, skim off foam, and fill into clean jars, following processing directions for regular berry syrup.

Process Times for Berry Syrup in a Boiling-Water Canner*

Style of Pack	Jar Size	Process Time at Altitudes of:		
		0–1,000 ft	1,001–6,000 ft	Above 6,000 ft
Hot	Half-pints or Pints	10 minutes	15 minutes	20 minutes

*After the process is complete, turn off the heat and remove the canner lid. Wait five minutes before removing jars.

Fruit Purées

Almost any fruit can be puréed for use as baby food, in sauces, or just as a nutritious snack. Puréed prunes and apples can be used as a butter replacement in many baked goods. Use this recipe for any fruit except figs and tomatoes.

Directions

1. Stem, wash, drain, peel, and remove pits if necessary. Measure fruit into large saucepan, crushing slightly if desired.

2. Add 1 cup hot water for each quart of fruit. Cook slowly until fruit is soft, stirring frequently. Press through sieve or food mill. If desired, add sugar to taste.

3. Reheat pulp to boil, or until sugar dissolves (if added). Fill hot into clean jars, leaving ¼-inch headspace. Adjust lids and process.

Process Times for Fruit Purées in a Boiling-Water Canner*

Style of Pack	Jar Size	Process Time at Altitudes of:		
		0–1,000 ft	1,001–6,000 ft	Above 6,000 ft
Hot	Pints or Quarts	15 minutes	20 minutes	25 minutes

*After the process is complete, turn off the heat and remove the canner lid. Wait five minutes before removing jars.

Process Times for Fruit Purées in a Dial-Gauge Pressure Canner*

Style of Pack	Jar Size	Process Time	Canner Pressure (PSI) at Altitudes of:			
			0–2,000 ft	2,001–4,000 ft	4,001–6,000 ft	6,001–8,000 ft
Hot	Pints or Quarts	8 minutes	6 lbs	7 lbs	8 lbs	9 lbs

*After the canner is completely depressurized, remove the weight from the vent port or open the petcock. Wait 10 minutes; then unfasten the lid and remove it carefully. Lift the lid with the underside away from you so that the steam coming out of the canner does not burn your face.

Process Times for Fruit Purées in a Weighted-Gauge Pressure Canner*

Style of Pack	Jar Size	Process Time (Min)	Canner Pressure (PSI) at Altitudes of:	
			0–1,000 ft	Above 1,000 ft
Hot	Pints or Quarts	8 minutes	5 lbs	10 lbs

*After the canner is completely depressurized, remove the weight from the vent port or open the petcock. Wait 10 minutes; then unfasten the lid and remove it carefully. Lift the lid with the underside away from you so that the steam coming out of the canner does not burn your face.

Grape Juice

Purple grapes are full of antioxidants and help to reduce the risk of heart disease, cancer, and Alzheimer's disease. For juice, select sweet, well-colored, firm, mature fruit.

Quantity
- An average of 24 ½ pounds is needed per canner load of seven quarts.
- An average of 16 pounds per canner load of nine pints.
- A lug weighs 26 pounds and yields seven to nine quarts of juice—an average of 3 ½ pounds per quart.

Directions

1. Wash and stem grapes. Place grapes in a saucepan and add boiling water to cover. Heat and simmer slowly until skin is soft.
2. Strain through a damp jelly bag or double layers of cheesecloth, and discard solids. Refrigerate juice for 24 to 48 hours.
3. Without mixing, carefully pour off clear liquid and save; discard sediment. If desired, strain through a paper coffee filter for a clearer juice.
4. Add juice to a saucepan and sweeten to taste. Heat and stir until sugar is dissolved. Continue heating with occasional stirring until juice begins to boil. Fill into jars immediately, leaving ¼-inch headspace. Adjust lids and process.

Process Times for Grape Juice in a Boiling-Water Canner*

Style of Pack	Jar Size	Process Time at Altitudes of:		
		0–1,000 ft	1,001–6,000 ft	Above 6,000 ft
Hot	Pints or Quarts	5 minutes	10 minutes	15 minutes
	Half-gallons	10 minutes	15 minutes	20 minutes

*After the process is complete, turn off the heat and remove the canner lid. Wait five minutes before removing jars.

Peaches, Halved or Sliced

Peaches are delicious in cobblers, crisps, and muffins, or grilled for a unique cake topping. Choose ripe, mature fruit with minimal bruising.

Quantity

- An average of 17 ½ pounds is needed per canner load of seven quarts.
- An average of 11 pounds is needed per canner load of nine pints.
- A bushel weighs 48 pounds and yields 16 to 24 quarts—an average of 2 ½ pounds per quart.

Directions

1. Dip fruit in boiling water for 30 to 60 seconds until skins loosen. Dip quickly in cold water and slip off skins. Cut in half, remove pits, and slice if desired. To prevent darkening, keep peeled fruit in ascorbic acid solution.

2. Prepare and boil a very light, light, or medium syrup or pack peaches in water, apple juice, or white grape juice. Raw packs make poor quality peaches.

   Hot pack—In a large saucepan, place drained fruit in syrup, water, or juice and bring to boil. Fill jars with hot fruit and cooking liquid, leaving ½-inch headspace. Place halves in layers, cut side down.

   Raw pack—Fill jars with raw fruit, cut side down, and add hot water, juice, or syrup, leaving ½-inch headspace.

3. Adjust lids and process.

Process Times for Halved or Sliced Peaches in a Boiling-Water Canner*

Style of Pack	Jar Size	Process Time at Altitudes of:			
		0–1,000 ft	1,001–3,000 ft	3,001–6,000 ft	Above 6,000 ft
Hot	Pints	20 minutes	25 minutes	30 minutes	35 minutes
	Quarts	25 minutes	30 minutes	35 minutes	40 minutes
Raw	Pints	25 minutes	30 minutes	35 minutes	40 minutes
	Quarts	30 minutes	35 minutes	40 minutes	45 minutes

*After the process is complete, turn off the heat and remove the canner lid. Wait five minutes before removing jars.

Process Times for Halved or Sliced Peaches in a Dial-Gauge Pressure Canner*

Style of Pack	Jar Size	Process Time	Canner Pressure (PSI) at Altitudes of:			
			0–2,000 ft	2,001–4,000 ft	4,001–6,000 ft	6,001–8,000 ft
Hot or Raw	Pints or Quarts	10 minutes	6 lbs	7 lbs	8 lbs	9 lbs

*After the canner is completely depressurized, remove the weight from the vent port or open the petcock. Wait 10 minutes; then unfasten the lid and remove it carefully. Lift the lid with the underside away from you so that the steam coming out of the canner does not burn your face.

Process Times for Halved or Sliced Peaches in a Weighted-Gauge Pressure Canner*

Style of Pack	Jar Size	Process Time	Canner Pressure (PSI) at Altitudes of:	
			0–1,000 ft	Above 1,000 ft
Hot or Raw	Pints or Quarts	10 minutes	5 lbs	10 lbs

*After the canner is completely depressurized, remove the weight from the vent port or open the petcock. Wait 10 minutes; then unfasten the lid and remove it carefully. Lift the lid with the underside away from you so that the steam coming out of the canner does not burn your face.

Pears, Halved

Choose ripe, mature fruit for best results. For a special treat, filled halved pears with a mixture of chopped dried apricots, pecans, brown sugar, and butter; bake or microwave until warm and serve with vanilla ice cream.

Quantity

- An average of 17 ½ pounds is needed per canner load of seven quarts.
- An average of 11 pounds is needed per canner load of nine pints.
- A bushel weighs 50 pounds and yields 16 to 25 quarts—an average of 2 ½ pounds per quart.

Directions

1. Wash and peel pears. Cut lengthwise in halves and remove core. A melon baller or metal measuring spoon works well for coring pears. To prevent discoloration, keep pears in water with a little lemon juice.

2. Prepare a very light, light, or medium syrup (see page 148) or use apple juice, white grape juice, or water. Raw packs make poor quality pears. Boil drained pears 5 minutes in syrup, juice, or water. Fill jars with hot fruit and cooking liquid, leaving ½-inch headspace. Adjust lids and process.

Process Times for Halved Pears in a Boiling-Water Canner*

Style of Pack	Jar Size	Process Time at Altitudes of:			
		0–1,000 ft	1,001–3,000 ft	3,001–6,000 ft	Above 6,000 ft
Hot	Pints	20 minutes	25 minutes	30 minutes	35 minutes
	Quarts	25 minutes	30 minutes	35 minutes	40 minutes

*After the process is complete, turn off the heat and remove the canner lid. Wait five minutes before removing jars.

Process Times for Halved Pears in a Dial-Gauge Pressure Canner*

Style of Pack	Jar Size	Process Time	Canner Pressure (PSI) at Altitudes of:			
			0–2,000 ft	2,001–4,000 ft	4,001–6,000 ft	6,001–8,000 ft
Hot	Pints or Quarts	10 minutes	6 lbs	7 lbs	8 lbs	9 lbs

*After the canner is completely depressurized, remove the weight from the vent port or open the petcock. Wait 10 minutes; then unfasten the lid and remove it carefully. Lift the lid with the underside away from you so that the steam coming out of the canner does not burn your face.

Process Times for Halved Pears in a Weighted-Gauge Pressure Canner*

Style of Pack	Jar Size	Process Time	Canner Pressure (PSI) at Altitudes of:	
			0–1,000 ft	Above 1,000 ft
Hot	Pints or Quarts	10 minutes	5 lbs	10 lbs

*After the canner is completely depressurized, remove the weight from the vent port or open the petcock. Wait 10 minutes; then unfasten the lid and remove it carefully. Lift the lid with the underside away from you so that the steam coming out of the canner does not burn your face.

Rhubarb, Stewed

Rhubarb in the garden is a sure sign that spring has sprung and summer is well on its way. But why not enjoy rhubarb all year round? The brilliant red stalks make it as appropriate for a holiday table as for an early summer feast. Rhubarb is also delicious in crisps, cobblers, or served hot over ice cream. Select young, tender, well-colored stalks from the spring or, if available, late fall crop.

Quantity

- An average of 10 ½ pounds is needed per canner load of seven quarts.
- An average of 7 pounds is needed per canner load of nine pints.
- A lug weighs 28 pounds and yields 14 to 28 quarts—an average of 1 ½ pounds per quart.

Directions

1. Trim off leaves. Wash stalks and cut into ½-inch to 1-inch pieces.
2. Place rhubarb in a large saucepan, and add ½ cup sugar for each quart of fruit. Let stand until juice appears. Heat gently to boiling. Fill jars without delay, leaving ½-inch headspace. Adjust lids and process.

Process Times for Stewed Rhubarb in a Boiling-Water Canner*

Style of Pack	Jar Size	Process Time at Altitudes of:		
		0–1,000 ft	1,001–6,000 ft	Above 6,000 ft
Hot	Pints or Quarts	15 minutes	20 minutes	25 minutes

*After the process is complete, turn off the heat and remove the canner lid. Wait five minutes before removing jars.

Process Times for Stewed Rhubarb in a Dial-Gauge Pressure Canner*

Style of Pack	Jar Size	Process Time	Canner Pressure (PSI) at Altitudes of			
			0–2,000 ft	2,001–4,000 ft	4,001–6,000 ft	6,001–8,000 ft
Hot	Pints or Quarts	8 minutes	6 lbs	7 lbs	8 lbs	9 lbs

*After the canner is completely depressurized, remove the weight from the vent port or open the petcock. Wait 10 minutes; then unfasten the lid and remove it carefully. Lift the lid with the underside away from you so that the steam coming out of the canner does not burn your face.

Process Times for Stewed Rhubarb in a Weighted-Gauge Pressure Canner*

Style of Pack	Jar Size	Process Time	Canner Pressure (PSI) at Altitudes of:	
			0–1,000 ft	Above 1,000 ft
Hot	Pints or Quarts	8 minutes	5 lbs	10 lbs

*After the canner is completely depressurized, remove the weight from the vent port or open the petcock. Wait 10 minutes; then unfasten the lid and remove it carefully. Lift the lid with the underside away from you so that the steam coming out of the canner does not burn your face.

Canned Pie Fillings

Using a pre-made pie filling will cut your pie preparation time by more than half, but most commercially produced fillings are oozing with high fructose corn syrup and all manner of artificial coloring and flavoring. (Food coloring is not at all necessary, but if you're really concerned about how the inside of your pie will look, appropriate amounts are added to each recipe as an optional ingredient.) Making and preserving your own pie fillings means that you can use your own fresh ingredients and adjust the sweetness to your taste. Because some folks like their pies rich and sweet and others prefer a natural tart flavor, you might want to first make a single quart, make a pie with it, and see how you like it. Then you can adjust the sugar and spices in the recipe to suit your personal preferences before making a large batch. Experiment with combining fruits or adding different spices, but the amount of lemon juice should not be altered, as it aids in controlling the safety and storage stability of the fillings.

These recipes use Clear Jel® (sometimes sold as Clear Jel A®), a chemically modified cornstarch that produces excellent sauce consistency even after fillings are canned and baked. By using Clear Jel® you can lower the sugar content of your fillings without sacrificing safety, flavor, or texture. (Note: Instant Clear Jel® is not meant to be cooked and should not be used for these recipes. Sure-Gel® is a natural fruit pectin and is not a suitable substitute for Clear Jel®. Cornstarch, tapioca starch, or arrowroot starch can be used in place of Clear Jel®, but the finished product is likely to be runny.) One pound of Clear Jel® costs less than five dollars and is enough to make fillings for about 14 pies. It will keep for at least a year if stored in a cool, dry place. Clear Jel® is increasingly available among canning and freezing supplies in some stores. Alternately, you can order it by the pound at any of the following online stores:

- www.barryfarm.com
- www.kitchenkrafts.com
- www.theingredientstore.com

Apple Pie Filling

Use firm, crisp apples, such as Stayman, Golden Delicious, or Rome varieties for the best results. If apples lack tartness, use an additional ¼ cup of lemon juice for each six quarts of slices. Ingredients are included for a one-quart (enough for one 8-inch pie) or a seven-quart recipe.

Ingredients

	1 Quart	7 Quarts
Blanched, sliced fresh apples	3 ½ cups	6 quarts
Granulated sugar	¾ cup + 2 tbsp	5 ½ cups
Clear Jel®	¼ cup	1 ½ cup
Cinnamon	½ tsp	1 tbsp
Cold water	½ cup	2 ½ cups
Apple juice	¾ cup	5 cups
Bottled lemon juice	2 tbsp	¾ cup
Nutmeg (optional)	⅛ tsp	1 tsp

Directions

1. Wash, peel, and core apples. Prepare slices ½ inch wide and place in water containing a little lemon juice to prevent browning.

2. For fresh fruit, place 6 cups at a time in 1 gallon of boiling water. Boil each batch 1 minute after the water returns to a boil. Drain, but keep heated fruit in a covered bowl or pot.

3. Combine sugar, Clear Jel®, and cinnamon in a large kettle with water and apple juice. Add nutmeg, if desired. Stir and cook on medium-high heat until mixture thickens and begins to bubble.

4. Add lemon juice and boil 1 minute, stirring constantly. Fold in drained apple slices immediately and fill jars with mixture without delay, leaving 1-inch headspace. Adjust lids and process immediately.

> When using frozen cherries and blueberries, select unsweetened fruit. If sugar has been added, rinse it off while fruit is frozen. Thaw fruit, then collect, measure, and use juice from fruit to partially replace the water specified in the recipe.

Process Times for Apple Pie Filling in a Boiling-Water Canner*

Style of Pack	Jar Size	Process Time at Altitudes of:			
		0–1,000 ft	1,001–3,000 ft	3,001–6,000 ft	Above 6,000 ft
Hot	Pints or Quarts	25 minutes	30 minutes	35 minutes	40 minutes

*After the process is complete, turn off the heat and remove the canner lid. Wait five minutes before removing jars.

Blueberry Pie Filling

Select fresh, ripe, and firm blueberries. Unsweetened frozen blueberries may be used. If sugar has been added, rinse it off while fruit is still frozen. Thaw fruit, then collect, measure, and use juice from fruit to partially replace the water specified in the recipe. Ingredients are included for a one-quart (enough for one 8-inch pie) or seven-quart recipe.

Ingredients

	1 Quart	7 Quarts
Fresh or thawed blueberries	3 ½ cups	6 quarts
Granulated sugar	¾ cup + 2 tbsp	6 cups
Clear Jel®	¼ cup + 1 tbsp	2 ¼ cup
Cold water	1 cup	7 cups
Bottled lemon juice	3 ½ cups	½ cup
Blue food coloring (optional)	3 drops	20 drops
Red food coloring (optional)	1 drop	7 drops

Directions

1. Wash and drain blueberries. Place 6 cups at a time in 1 gallon boiling water. Allow water to return to a boil and cook each batch for 1 minute. Drain but keep heated fruit in a covered bowl or pot.
2. Combine sugar and Clear Jel® in a large kettle. Stir. Add water and food coloring if desired. Cook on medium-high heat until mixture thickens and begins to bubble.
3. Add lemon juice and boil 1 minute, stirring constantly. Fold in drained berries immediately and fill jars with mixture without delay, leaving 1-inch headspace. Adjust lids and process immediately.

Process Times for Blueberry Pie Filling in a Boiling-Water Canner*

Style of Pack	Jar Size	Process Time at Altitudes of:			
		0–1,000 ft	1,001–3,000 ft	3,001–6,000 ft	Above 6,000 ft
Hot	Pints or Quarts	30 minutes	35 minutes	40 minutes	45 minutes

*After the process is complete, turn off the heat and remove the canner lid. Wait five minutes before removing jars.

Cherry Pie Filling

Select fresh, very ripe, and firm cherries. Unsweetened frozen cherries may be used. If sugar has been added, rinse it off while the fruit is still frozen. Thaw fruit, then collect, measure, and use juice from fruit to partially replace the water specified in the recipe. Ingredients are included for a one-quart (enough for one 8-inch pie) or seven-quart recipe.

Ingredients

	1 Quart	7 Quarts
Fresh or thawed sour cherries	3 ⅓ cups	6 quarts
Granulated sugar	1 cup	7 cups
Clear Jel®	¼ cup + 1 tbsp	1-¾ cups
Cold water	1 ⅓ cups	9 ⅓ cups
Bottled lemon juice	1 tbsp + 1 tsp	½ cup
Cinnamon (optional)	⅛ tsp	1 tsp
Almond extract (optional)	¼ tsp	2 tsp
Red food coloring (optional)	6 drops	¼ tsp

Directions

1. Rinse and pit fresh cherries, and hold in cold water. To prevent stem end browning, use water with a little lemon juice. Place 6 cups at a time in 1 gallon boiling water. Boil each batch 1 minute after the water returns to a boil. Drain but keep heated fruit in a covered bowl or pot.

2. Combine sugar and Clear Jel® in a large saucepan and add water. If desired, add cinnamon, almond extract, and food coloring. Stir mixture and cook over medium-high heat until mixture thickens and begins to bubble.

3. Add lemon juice and boil 1 minute, stirring constantly. Fold in drained cherries immediately and fill jars with mixture without delay, leaving 1-inch headspace. Adjust lids and process immediately.

Process Times for Cherry Pie Filling in a Boiling-Water Canner*

Style of Pack	Jar Size	Process Time at Altitudes of:			
		0–1,000 ft	1,001–3,000 ft	3,001–6,000 ft	Above 6,000 ft
Hot	Pints or Quarts	30 minutes	35 minutes	40 minutes	45 minutes

*After the process is complete, turn off the heat and remove the canner lid. Wait five minutes before removing jars.

Festive Mincemeat Pie Filling

Mincemeat pie originated as "Christmas Pie" in the eleventh century, when the English crusaders returned from the Holy Land bearing oriental spices. They added three of these spices—cinnamon, cloves, and nutmeg—to their meat pies to represent the three gifts that the magi brought to the Christ child. Mincemeat pies are traditionally small and are perfect paired with a mug of hot buttered rum. Walnuts or pecans can be used in place of meat if preferred. This recipe yields about seven quarts.

Ingredients

2 cups finely chopped suet

4 lbs ground beef or 4 lbs ground venison and 1 lb sausage

5 qts chopped apples

2 lbs dark seedless raisins

1 lb white raisins

2 qts apple cider

2 tbsp ground cinnamon

2 tsp ground nutmeg

½ tsp cloves

5 cups sugar

2 tbsp salt

Directions

1. Cook suet and meat in water to avoid browning. Peel, core, and quarter apples. Put suet, meat, and apples through food grinder using a medium blade.

2. Combine all ingredients in a large saucepan, and simmer 1 hour or until slightly thickened. Stir often.

3. Fill jars with mixture without delay, leaving 1-inch headspace. Adjust lids and process.

Process Times for Festive Mincemeat Pie Filling in a Dial-Gauge Pressure Canner*

Style of Pack	Jar Size	Process Time	Canner Pressure (PSI) at Altitudes of:			
			0–2,000 ft	2,001–4,000 ft	4,001–6,000 ft	6,000–8,000 ft
Hot	Quarts	90 minutes	11 lb	12 lb	13 lb	14 lb

*After the canner is completely depressurized, remove the weight from the vent port or open the petcock. Wait 10 minutes; then unfasten the lid and remove it carefully. Lift the lid with the underside away from you so that the steam coming out of the canner does not burn your face.

Process Times for Festive Mincemeat Pie Filling in a Weighted-Gauge Pressure Canner*

Style of Pack	Jar Size	Process Time	Canner Pressure (PSI) at Altitudes of:	
			0–1,000 ft	Above 1,000 ft
Hot	Quarts	90 minutes	10 lb	15 lb

*After the canner is completely depressurized, remove the weight from the vent port or open the petcock. Wait 10 minutes; then unfasten the lid and remove it carefully. Lift the lid with the underside away from you so that the steam coming out of the canner does not burn your face.

Jams, Jellies, and Other Fruit Spreads

Homemade jams and jellies have lots more flavor than store-bought, over-processed varieties. The combinations of fruits and spices are limitless, so have fun experimenting with these recipes. If you can bear to part with your creations when you're all done, they make wonderful gifts for any occasion.

Pectin is what makes jams and jellies thicken and gel. Many fruits, such as crab apples, citrus fruits, sour plums, currants, quinces, green apples, or Concord grapes, have plenty of their own natural pectin, so there's no need to add more pectin to your recipes. You can use less sugar when you don't add pectin, but you will have to boil the fruit for longer. Still, the process is relatively simple and you don't have to worry about having store-bought pectin on hand.

To use fresh fruits with a low pectin content or canned or frozen fruit juice, powdered or liquid pectin must be added for your jams and jellies to thicken and set properly.

Jelly or jam made with added pectin requires less cooking and generally gives a larger yield. These products have more natural fruit flavors, too. In addition, using added pectin eliminates the need to test hot jellies and jams for proper gelling.

Beginning this section are descriptions of the differences between methods and tips for success with whichever you use.

Making Jams and Jellies without Added Pectin

> **TIP**
>
> If you are not sure if a fruit has enough of its own pectin, combine 1 tablespoon of rubbing alcohol with 1 tablespoon of extracted fruit juice in a small glass. Let stand 2 minutes. If the mixture forms into one solid mass, there's plenty of pectin. If you see several weak blobs, you need to add pectin or combine with another high-pectin fruit.

Jelly without Added Pectin

Making jelly without added pectin is not an exact science. You can add a little more or less sugar according to your taste, substitute honey for up to ½ of the sugar, or experiment with combining small amounts of low-pectin fruits with other high-pectin fruits. The Ingredients table below shows you the basics for common high-pectin fruits. Use it as a guideline as you experiment with other fruits.

As fruit ripens, its pectin content decreases, so use fruit that has recently been picked, and mix ¾ ripe fruit with ¼

under-ripe. Cooking cores and peels along with the fruit will also increase the pectin level. Avoid using canned or frozen fruit as they contain very little pectin. Be sure to wash all fruit thoroughly before cooking. One pound of fruit should yield at least 1 cup of clear juice.

> **TIP**
>
> Commercially frozen and canned juices may be low in natural pectins and make soft textured spreads.

Ingredients

Fruit	Water to be Added per Pound of Fruit	Minutes to Simmer Fruit before Extracting Juice	Ingredients Added to Each Cup of Strained Juice		Yield from 4 Cups of Juice (Half-pints)
			Sugar (Cups)	Lemon Juice (Tsp)	
Apples	1 cup	20 to 25	¾	1 ½ (opt)	4 to 5
Blackberries	None or ¼ cup	5 to 10	¾ to 1	None	7 to 8
Crab apples	1 cup	20 to 25	1	None	4 to 5
Grapes	None or ¼ cup	5 to 10	¾ to 1	None	8 to 9
Plums	½ cup	15 to 20	¾	None	8 to 9

Temperature test—Use a jelly or candy thermometer and boil until mixture reaches the following temperatures:

Sea Level	1,000 ft	2,000 ft	3,000 ft	4,000 ft	5,000 ft	6,000 ft	7,000 ft	8,000 ft
220°F	218°F	216°F	214°F	212°F	211°F	209°F	207°F	205°F

Sheet or spoon test—Dip a cool metal spoon into the boiling jelly mixture. Raise the spoon about 12 inches above the pan (out of steam). Turn the spoon so the liquid runs off the side. The jelly is done when the syrup forms two drops that flow together and sheet or hang off the edge of the spoon.

Process Times for Jelly without Added Pectin in a Boiling Water Canner*

		Process Time at Altitudes of:		
Style of Pack	Jar Size	0–1,000 ft	1,001–6,000 ft	Above 6,000 ft
Hot	Half-pints or pints	5 minutes	10 minutes	15 minutes

*After the process is complete, turn off the heat and remove the canner lid. Wait five minutes before removing jars.

Directions

1. Crush soft fruits or berries; cut firmer fruits into small pieces (there is no need to peel or core the fruits, as cooking all the parts adds pectin).

2. Add water to fruits that require it, as listed in the Ingredients table above. Put fruit and water in large saucepan and bring to a boil. Then simmer according to the times below until fruit is soft, while stirring to prevent scorching.

3. When fruit is tender, strain through a colander, then strain through a double layer of cheesecloth or a jelly bag. Allow juice to drip through, using a stand or colander to hold the bag. Avoid pressing or squeezing the bag or cloth as it will cause cloudy jelly.

4. Using no more than 6 to 8 cups of extracted fruit juice at a time, measure fruit juice, sugar, and lemon juice according to the Ingredients table, and heat to boiling.

5. Stir until the sugar is dissolved. Boil over high heat to the jellying point. To test jelly for doneness, use one of the following methods:

6. Remove from heat and quickly skim off foam. Fill sterile jars with jelly. Use a measuring cup or ladle the jelly through a wide-mouthed funnel, leaving ¼-inch headspace. Adjust lids and process.

Preventing spoilage

Even though sugar helps preserve jellies and jams, molds can grow on the surface of these products. Research now indicates that the mold which people usually scrape off the surface of jellies may not be as harmless as it seems. Mycotoxins have been found in some jars of jelly having surface mold growth. Mycotoxins are known to cause cancer in animals; their effects on humans are still being researched. Because of possible mold contamination, paraffin or wax seals are no longer recommended for any sweet spread, including jellies. To prevent growth of molds and loss of good flavor or color, fill products hot into sterile Mason jars, leaving ¼-inch headspace, seal with self-sealing lids, and process 5 minutes in a boiling-water canner. Correct process time at higher elevations by adding 1 additional minute per 1,000 ft above sea level. If unsterile jars are used, the filled jars should be processed 10 minutes. Use of sterile jars is preferred, especially when fruits are low in pectin, since the added 5-minute process time may cause weak gels.

Lemon Curd

Lemon curd is a rich, creamy spread that can be used on (or in) a variety of teatime treats—crumpets, scones, cake fillings, tartlets, or meringues are all enhanced by its tangy-sweet flavor. Follow the recipe carefully, as variances in ingredients, order, and temperatures may lead to a poor texture or flavor. For Lime Curd, use the same recipe but substitute 1 cup bottled lime juice and ¼ cup fresh lime zest for the lemon juice and zest. This recipe yields about three to four half-pints.

Ingredients

2 ½ cups superfine sugar*

½ cup lemon zest (freshly zested), optional

1 cup bottled lemon juice** ***

¾ cup unsalted butter, chilled, cut into approximately ¾-inch pieces

7 large egg yolks

4 large whole eggs

Directions

1. Wash 4 half-pint canning jars with warm, soapy water. Rinse well; keep hot until ready to fill. Prepare canning lids according to manufacturer's directions.

2. Fill boiling water canner with enough water to cover the filled jars by 1 to 2 inches. Use a thermometer to preheat the water to 180°F by the time filled jars are ready to be added. **Caution:** Do not heat the water in the canner to more than 180°F before jars are added. If the water in the canner is too hot when jars are added, the process time will not be long enough. The time it takes for the canner to reach boiling after the jars are added is ex-

pected to be 25 to 30 minutes for this product. Process time starts after the water in the canner comes to a full boil over the tops of the jars.

3. Combine the sugar and lemon zest in a small bowl, stir to mix, and set aside about 30 minutes. Pre-measure the lemon juice and prepare the chilled butter pieces.

4. Heat water in the bottom pan of a double boiler until it boils gently. The water should not boil vigorously or touch the bottom of the top double boiler pan or bowl in which the curd is to be cooked. Steam produced will be sufficient for the cooking process to occur.

5. In the top of the double boiler, on the counter top or table, whisk the egg yolks and whole eggs together until thoroughly mixed. Slowly whisk in the sugar and zest, blending until well mixed and smooth. Blend in the lemon juice and then add the butter pieces to the mixture.

6. Place the top of the double boiler over boiling water in the bottom pan. Stir gently but continuously with a silicone spatula or cooking spoon, to prevent the mixture from sticking to the bottom of the pan. Continue cooking until the mixture reaches a temperature of 170°F. Use a food thermometer to monitor the temperature.

7. Remove the double boiler pan from the stove and place on a protected surface, such as a dishcloth or towel on the counter top. Continue to stir gently until the curd thickens (about 5 minutes). Strain curd through a mesh strainer into a glass or stainless steel bowl; discard collected zest.

8. Fill hot strained curd into the clean, hot half-pint jars, leaving ½-inch headspace. Remove air bubbles and adjust headspace if needed. Wipe rims of jars with a dampened, clean paper towel; apply two-piece metal canning lids. Process. Let cool, undisturbed, for 12 to 24 hours and check for seals.

* If superfine sugar is not available, run granulated sugar through a grinder or food processor for 1 minute, let settle, and use in place of superfine sugar. Do not use powdered sugar.

** Bottled lemon juice is used to standardize acidity. Fresh lemon juice can vary in acidity and is not recommended.

*** If a double boiler is not available, a substitute can be made with a large bowl or saucepan that can fit partway down into a saucepan of a smaller diameter. If the bottom pan has a larger diameter, the top bowl or pan should have a handle or handles that can rest on the rim of the lower pan.

Process Times for Lemon Curd in a Boiling-Water Canner*

Style of Pack	Jar Size	Process Time at Altitudes of:		
		0–1,000 ft	1,001–6,000 ft	Above 6,000 ft
Hot	Half-pints	15 minutes	20 minutes	25 minutes

*After the process is complete, turn off the heat and remove the canner lid. Wait five minutes before removing jars.

Jam without Added Pectin

Making jam is even easier than making jelly, as you don't have to strain the fruit. However, you'll want to be sure to remove all stems, skins, and pits. Be sure to wash and rinse all fruits thoroughly before cooking, but don't let them soak. For best flavor, use fully ripe fruit. Use the Ingredients table below as a guideline as you experiment with less common fruits.

Ingredients

Fruit	Quantity (Crushed)	Sugar	Lemon Juice	Yield (Half-pints)
Apricots	4 to 4 ½ cups	4 cups	2 tbsp	5 to 6
Berries*	4 cups	4 cups	None	3 to 4
Peaches	5 ½ to 6 cups	4 to 5 cups	2 tbsp	6 to 7

* Includes blackberries, boysenberries, dewberries, gooseberries, loganberries, raspberries, and strawberries.

1. Remove stems, skins, seeds, and pits; cut into pieces and crush. For berries, remove stems and blossoms and crush. Seedy berries may be put through a sieve or food mill. Measure crushed fruit into large saucepan using the ingredient quantities specified above.

2. Add sugar and bring to a boil while stirring rapidly and constantly. Continue to boil until mixture thickens. Use one of the following tests to determine when jams and jellies are ready to fill. Remember that the jam will thicken as it cools.

3. Remove from heat and skim off foam quickly. Fill sterile jars with jam. Use a measuring cup or ladle the jam through a wide-mouthed funnel, leaving ¼-inch headspace. Adjust lids and process.

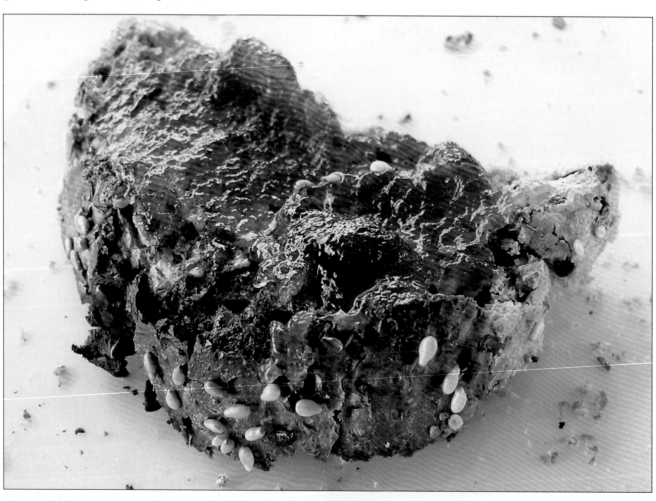

Temperature test—Use a jelly or candy thermometer and boil until mixture reaches the temperature for your altitude.

Sea Level	1,000 ft	2,000 ft	3,000 ft	4,000 ft	5,000 ft	6,000 ft	7,000 ft	8,000 ft
220°F	218°F	216°F	214°F	212°F	211°F	209°F	207°F	205°F

Refrigerator test—Remove the jam mixture from the heat. Pour a small amount of boiling jam on a cold plate and put it in the freezer compartment of a refrigerator for a few minutes. If the mixture gels, it is ready to fill.

Process Times for Jams without Added Pectin in a Boiling-Water Canner*

		Process Time at Altitudes of:		
Style of Pack	Jar Size	0–1,000 ft	1,001–6,000 ft	Above 6,000 ft
Hot	Half-pints	5 minutes	10 minutes	15 minutes

*After the process is complete, turn off the heat and remove the canner lid. Wait five minutes before removing jars.

Jams and Jellies with Added Pectin

To use fresh fruits with a low pectin content or canned or frozen fruit juice, powdered or liquid pectin must be added for your jams and jellies to thicken and set properly. Jelly or jam made with added pectin requires less cooking and generally gives a larger yield. These products have more natural fruit flavors, too. In addition, using added pectin eliminates the need to test hot jellies and jams for proper gelling.

Commercially produced pectin is a natural ingredient, usually made from apples and available at most grocery stores. There are several types of pectin now commonly available; liquid, powder, low-sugar, and no-sugar pectins each have their own advantages and downsides. Pomona's Universal Pectin® is a citrus pectin that allows you to make jams and jellies with little or no sugar. Because the order of combining ingredients depends on the type of pectin used, it is best to follow the common jam and jelly recipes that are included right on most pectin packages. How ever, if you want to try something a little different, follow one of the following recipes for mixed fruit and spiced fruit jams and jellies.

TIPS

- Adding ½ teaspoon of butter or margarine with the juice and pectin will reduce foaming. How- ever, these may cause off-flavor in a long-term storage of jellies and jams.
- Purchase fresh fruit pectin each year. Old pectin may result in poor gels.
- Be sure to use mason canning jars, self-sealing two-piece lids, and a five-minute process (cor- rected for altitude, as necessary) in boiling water.

Process Times for Jams and Jellies with Added Pectin in a Boiling-Water Canner*

		Process Time at Altitudes of:		
Style of Pack	Jar Size	0–1,000 ft	1,001–6,000 ft	Above 6,000 ft
Hot	Half-pints	5 minutes	10 minutes	15 minutes

*After the process is complete, turn off the heat and remove the canner lid. Wait five minutes before removing jars.

Pear-Apple Jam

This is a delicious jam perfect for making at the end of autumn, just before the frost gets the last apples. For a warming, spicy twist add a teaspoon of fresh grated ginger along with the cinnamon. This recipe yields seven to eight half-pints.

Ingredients

2 cups peeled, cored, and finely chopped pears (about 2 lbs)

1 cup peeled, cored, and finely chopped apples

¼ tsp ground cinnamon

6 ½ cups sugar

⅓ cup bottled lemon juice

6 oz liquid pectin

Directions

1. Peel, core, and slice apples and pears into a large saucepan and stir in cinnamon. Thoroughly mix sugar and lemon juice with fruits and bring to a boil over high heat, stirring constantly and crushing fruit with a potato masher as it softens.

2. Once boiling, immediately stir in pectin. Bring to a full rolling boil and boil hard 1 minute, stirring constantly.

3. Remove from heat, quickly skim off foam, and fill sterile jars, leaving ¼ inch headspace. Adjust lids and process.

Process Times for Pear-Apple Jam in a Boiling Water Canner*

Style of Pack	Jar Size	Process Time at Altitudes of:		
		0–1,000 ft	1,001–6,000 ft	Above 6,000 ft
Hot	Half-pints	5 minutes	10 minutes	15 minutes

*After the process is complete, turn off the heat and remove the canner lid. Wait five minutes before removing jars.

Strawberry-Rhubarb Jelly

Strawberry-rhubarb jelly will turn any ordinary piece of bread into a delightful treat. You can also spread it on short-cake or pound cake for a simple and unique dessert. This recipe yields about seven half-pints.

Ingredients

1 ½ lbs red stalks of rhubarb
1 ½ qts ripe strawberries
½ tsp butter or margarine to reduce foaming (optional)
6 cups sugar
6 oz liquid pectin

Directions

1. Wash and cut rhubarb into 1-inch pieces and blend or grind. Wash, stem, and crush strawberries, one layer at a time, in a saucepan. Place both fruits in a jelly bag or double layer of cheesecloth and gently squeeze juice into a large measuring cup or bowl.

2. Measure 3 ½ cups of juice into a large saucepan. Add butter and sugar, thoroughly mixing into juice. Bring to a boil over high heat, stirring constantly.

3. As soon as mixture begins to boil, stir in pectin. Bring to a full rolling boil and boil hard 1 minute, stirring constantly. Remove from heat, quickly skim off foam, and fill sterile jars, leaving ¼-inch headspace. Adjust lids and process.

Process Times for Strawberry-Rhubarb Jelly in a Boiling-Water Canner*

Style of Pack	Jar Size	Process Time at Altitudes of:		
		0–1,000 ft	1,001–6,000 ft	Above 6,000 ft
Hot	Half-pints or pints	5 minutes	10 minutes	15 minutes

*After the process is complete, turn off the heat and remove the canner lid. Wait five minutes before removing jars.

Blueberry-Spice Jam

This is a summery treat that is delicious spread over waffles with a little butter. Using wild blueberries results in a stronger flavor, but cultivated blueberries also work well. This recipe yields about five half-pints.

Ingredients

2 ½ pints ripe blueberries
1 tbsp lemon juice
½ tsp ground nutmeg or cinnamon
¾ cup water
5 ½ cups sugar
1 box (1 ¾ oz) powdered pectin

Directions

1. Wash and thoroughly crush blueberries, adding one layer at a time, in a saucepan. Add lemon juice, spice, and water. Stir pectin and bring to a full, rolling boil over high heat, stirring frequently.

2. Add the sugar and return to a full rolling boil. Boil hard for 1 minute, stirring constantly. Remove from heat, quickly skim off foam, and fill sterile jars, leaving ¼-inch headspace. Adjust lids and process.

Process Times for Blueberry-Spice Jam in a Boiling-Water Canner*

Style of Pack	Jar Size	Process Time at Altitudes of:		
		0–1,000 ft	1,001–6,000 ft	Above 6,000 ft
Hot	Half-pints or pints	5 minutes	10 minutes	15 minutes

*After the process is complete, turn off the heat and remove the canner lid. Wait five minutes before removing jars.

Grape-Plum Jelly

If you think peanut butter and jelly sandwiches are only for kids, try grape-plum jelly spread with a natural nut butter over a thick slice of whole wheat bread. You'll change your mind. This recipe yields about 10 half-pints.

Ingredients

3 ½ lbs ripe plums

3 lbs ripe Concord grapes

8 ½ cups sugar

1 cup water

½ tsp butter or margarine to reduce foaming (optional)

1 box (1 ¾ oz) powdered pectin

Directions

1. Wash and pit plums; do not peel. Thoroughly crush the plums and grapes, adding one layer at a time, in a saucepan with water. Bring to a boil, cover, and simmer 10 minutes.

2. Strain juice through a jelly bag or double layer of cheesecloth. Measure sugar and set aside. Combine 6 ½ cups of juice with butter and pectin in large saucepan. Bring to a hard boil over high heat, stirring constantly.

3. Add the sugar and return to a full rolling boil. Boil hard for 1 minute, stirring constantly. Remove from heat, quickly skim off foam, and fill sterile jars, leaving ¼-inch headspace. Adjust lids and process.

Process Times for Grape-Plum Jelly in a Boiling-Water Canner*

Style of Pack	Jar Size	Process Time at Altitudes of:		
		0–1,000 ft	1,001–6,000 ft	Above 6,000 ft
Hot	Half-pints or pints	5 minutes	10 minutes	15 minutes

*After the process is complete, turn off the heat and remove the canner lid. Wait five minutes before removing jars.

Making Reduced-Sugar Fruit Spreads

A variety of fruit spreads may be made that are tasteful, yet lower in sugars and calories than regular jams and jellies. The most straightforward method is probably to buy low-sugar pectin and follow the directions on the package, but the recipes below show alternate methods of using gelatin or fruit pulp as thickening agents. Gelatin recipes should not be processed and should be refrigerated and used within four weeks.

Peach-Pineapple Spread

This recipe may be made with any combination of peaches, nectarines, apricots, and plums. You can use no sugar, up to two cups of sugar, or a combination of sugar and another sweetener (such as honey, Splenda®, or agave nectar). Note that if you use aspartame, the spread may lose its sweetness within three to four weeks. Add cinnamon or star anise if desired. This recipe yields five to six half-pints.

Ingredients

4 cups drained peach pulp (follow directions below)

2 cups drained unsweetened crushed pineapple

¼ cup bottled lemon juice

2 cups sugar (optional)

Directions

1. Thoroughly wash 4 to 6 pounds of firm, ripe peaches. Drain well. Peel and remove pits. Grind fruit flesh with a medium or coarse blade, or crush with a fork (do not use a blender).

2. Place ground or crushed peach pulp in a 2-quart saucepan. Heat slowly to release juice, stirring constantly, until fruit is tender. Place cooked fruit in a jelly bag or strainer lined with four layers of cheesecloth. Allow juice to drip about 15 minutes. Save the juice for jelly or other uses.

3. Measure 4 cups of drained peach pulp for making spread. Combine the 4 cups of pulp, pineapple, and lemon juice in a 4-quart saucepan. Add up to 2 cups of sugar or other sweetener, if desired, and mix well.

4. Heat and boil gently for 10 to 15 minutes, stirring enough to prevent sticking. Fill jars quickly, leaving ¼-inch headspace. Adjust lids and process.

Process Times for Peach-Pineapple Spread in a Boiling-Water Canner*

Style of Pack	Jar Size	Process Time at Altitudes of:			
		0–1,000 ft	1,001–3,000 ft	3,001–6,000 ft	Above 6,000 ft
Hot	Half-pints	15 minutes	20 minutes	20 minutes	25 minutes
	Pints	20 minutes	25 minutes	30 minutes	35 minutes

*After the process is complete, turn off the heat and remove the canner lid. Wait five minutes before removing jars.

Refrigerated Apple Spread

This recipe uses gelatin as a thickener, so it does not require processing but it should be refrigerated and used within four weeks. For spiced apple jelly, add two sticks of cinnamon and four whole cloves to mixture before boiling. Remove both spices before adding the sweetener and food coloring (if desired). This recipe yields four half-pints.

Ingredients

2 tbsp unflavored gelatin powder

1 qt bottle unsweetened apple juice

2 tbsp bottled lemon juice

2 tbsp liquid low-calorie sweetener (e.g., sucralose, honey, or 1–2 tsp liquid stevia)

Directions

1. In a saucepan, soften the gelatin in the apple and lemon juices. To dissolve gelatin, bring to a full rolling boil and boil 2 minutes. Remove from heat.

2. Stir in sweetener and food coloring (if desired). Fill jars, leaving ¼-inch headspace. Adjust lids. Refrigerate (do not process or freeze).

Refrigerated Grape Spread

This is a simple, tasty recipe that doesn't require processing. Be sure to refrigerate and use within four weeks. This recipe makes three half-pints.

Ingredients

2 tbsp unflavored gelatin powder

1 bottle (24 oz) unsweetened grape juice

2 tbsp bottled lemon juice

2 tbsp liquid low-calorie sweetener (e.g., sucralose, honey, or 1–2 tsp liquid stevia)

Directions

1. In a saucepan, heat the gelatin in the grape and lemon juices until mixture is soft. Bring to a full rolling boil to dissolve gelatin. Boil 1 minute and remove from heat. Stir in sweetener.

2. Fill jars quickly, leaving ¼-inch headspace. Adjust lids. Refrigerate (do not process or freeze).

Remaking Soft Jellies

Sometimes jelly just doesn't turn out right the first time. Jelly that is too soft can be used as a sweet sauce to drizzle over ice cream, cheesecake, or angel food cake, but it can also be re-cooked into the proper consistency.

To Remake with Powdered Pectin

1. Measure jelly to be re-cooked. Work with no more than 4 to 6 cups at a time. For each quart (4 cups) of jelly, mix ¼ cup sugar, ½ cup water, 2 tablespoons bottled lemon juice, and 4 teaspoons powdered pectin. Bring to a boil while stirring.

2. Add jelly and bring to a rolling boil over high heat, stirring constantly. Boil hard ½ minute. Remove from heat, quickly skim foam off jelly, and fill sterile jars, leaving ¼-inch headspace. Adjust new lids and process as recommended (see page 171).

To Remake with Liquid Pectin

1. Measure jelly to be re-cooked. Work with no more than 4 to 6 cups at a time. For each quart (4 cups) of jelly, measure into a bowl ¾ cup sugar, 2 tablespoons bottled lemon juice, and 2 tablespoons liquid pectin.

2. Bring jelly only to boil over high heat, while stirring. Remove from heat and quickly add the sugar, lemon

juice, and pectin. Bring to a full rolling boil, stirring constantly. Boil hard for 1 minute. Quickly skim off foam and fill sterile jars, leaving ¼-inch headspace. Adjust new lids and process as recommended (see page 171)

To Remake without Added Pectin

1. For each quart of jelly, add 2 tablespoons bottled lemon juice. Heat to boiling and continue to boil for 3 to 4 minutes.

2. To test jelly for doneness, use one of the following methods:

3. Remove from heat, quickly skim off foam, and fill sterile jars, leaving ¼-inch headspace. Adjust new lids and process.

Temperature test—Use a jelly or candy thermometer and boil until mixture reaches the following temperatures at the altitudes below:

Sea Level	1,000 ft	2,000 ft	3,000 ft	4,000 ft	5,000 ft	6,000 ft	7,000 ft	8,000 ft
220°F	218°F	216°F	214°F	212°F	211°F	209°F	207°F	205°F

Sheet or spoon test—Dip a cool metal spoon into the boiling jelly mixture. Raise the spoon about 12 inches above the pan (out of steam). Turn the spoon so the liquid runs off the side. The jelly is done when the syrup forms two drops that flow together and sheet or hang off the edge of the spoon.

Process Times for Remade Soft Jellies in a Boiling-Water Canner

Style of Pack	Jar Size	Process Time at Altitudes of:		
		0–1,000 ft	1,001–6,000 ft	Above 6,000 ft
Hot	Half-pints or pints	5 minutes	10 minutes	15 minutes

*After the process is complete, turn off the heat and remove the canner lid. Wait five minutes before removing jars.

Vegetables, Pickles, and Tomatoes

Beans or Peas, Shelled or Dried (All Varieties)

Shelled or dried beans and peas are inexpensive and easy to buy or store in bulk, but they are not very convenient when it comes to preparing them to eat. Hydrating and canning beans or peas enable you to simply open a can and use them rather than waiting for them to soak. Sort and discard discolored seeds before rehydrating.

Directions

1. Place dried beans or peas in a large pot and cover with water. Soak 12 to 18 hours in a cool place. Drain water. To quickly hydrate beans, you may cover sorted and washed beans with boiling water in a saucepan. Boil 2 minutes, remove from heat, soak 1 hour, and drain.

2. Cover beans soaked by either method with fresh water and boil 30 minutes. Add ½ teaspoon of salt per pint or 1 teaspoon per quart to each jar, if desired. Fill jars with beans or peas and cooking water, leaving 1-inch headspace. Adjust lids and process.

Quantity

- An average of five pounds is needed per canner load of seven quarts.
- An average of 3 ¼ pounds is needed per canner load of nine pints—an average of ¾ pounds per quart.

Process Times for Beans or Peas in a Dial-Gauge Pressure Canner*

Style of Pack	Jar Size	Process Time	Canner Pressure (PSI) at Altitudes of:			
			0–2,000 ft	2,001–4,000 ft	4,001–6,000 ft	6,001–8,000 ft
Hot	Pints	75 minutes	11 lbs	12 lbs	13 lbs	14 lbs
	Quarts	90 minutes	11 lbs	12 lbs	13 lbs	14 lbs

*After the canner is completely depressurized, remove the weight from the vent port or open the petcock. Wait 10 minutes; then unfasten the lid and remove it carefully. Lift the lid with the underside away from you so that the steam coming out of the canner does not burn your face.

Process Times for Beans or Peas in a Weighted-Gauge Pressure Canner*

Style of pack	Jar Size	Process Time	Canner Pressure (PSI) at Altitudes of:	
			0–1,000 ft	Above 1,000 ft
Hot	Pints	75 minutes	10 lbs	15 lbs
	Quarts	90 minutes	10 lbs	15 lbs

*After the canner is completely depressurized, remove the weight from the vent port or open the petcock. Wait 10 minutes; then unfasten the lid and remove it carefully. Lift the lid with the underside away from you so that the steam coming out of the canner does not burn your face.

Baked Beans

Baked beans are an old New England favorite, but every cook has his or her favorite variation. Two recipes are included here, but feel free to alter them to your own taste.

Quantity

- An average of five pounds of beans is needed per canner load of seven quarts.
- An average of 3 ¼ pounds is needed per canner load of nine pints—an average of ¾ pounds per quart.

Directions

1. Sort and wash dry beans. Add 3 cups of water for each cup of dried beans. Boil 2 minutes, remove from heat, soak 1 hour, and drain.

2. Heat to boiling in fresh water, and save liquid for making sauce. Make your choice of the following sauces:

Tomato Sauce—Mix 1 quart tomato juice, 3 tablespoons sugar, 2 teaspoons salt, 1 tablespoon chopped onion, and ¼ teaspoon each of ground cloves, allspice, mace, and cayenne pepper. Heat to boiling. Add 3 quarts cooking liquid from beans and bring back to boiling.

Molasses Sauce—Mix 4 cups water or cooking liquid from beans, 3 tablespoons dark molasses, 1 tablespoon vinegar, 2 teaspoons salt, and ¾ teaspoon powdered dry mustard. Heat to boiling.

3. Place seven ¾-inch pieces of pork, ham, or bacon in an earthenware crock, a large casserole, or a pan. Add beans and enough molasses sauce to cover beans.

4. Cover and bake 4 to 5 hours at 350ºF. Add water as needed—about every hour. Fill jars, leaving 1-inch headspace. Adjust lids and process.

Process Times for Baked Beans in a Dial-Gauge Pressure Canner*

Style of Pack	Jar Size	Process Time	Canner Pressure (PSI) at Altitudes of:			
			0–2,000 ft	2,001–4,000 ft	4,001–6,000 ft	6,001–8,000 ft
Hot	Pints	65 minutes	11 lbs	12 lbs	13 lbs	14 lbs
	Quarts	75 minutes	11 lbs	12 lbs	13 lbs	14 lbs

*After the canner is completely depressurized, remove the weight from the vent port or open the petcock. Wait 10 minutes; then unfasten the lid and remove it carefully. Lift the lid with the underside away from you so that the steam coming out of the canner does not burn your face.

Process Times for Baked Beans in a Weighted-Gauge Pressure Canner*

Style of pack	Jar Size	Process Time	Canner Pressure (PSI) at Altitudes of:	
			0–1,000 ft	Above 1,000 ft
Hot	Pints	65 minutes	10 lbs	15 lbs
	Quarts	75 minutes	10 lbs	15 lbs

*After the canner is completely depressurized, remove the weight from the vent port or open the petcock. Wait 10 minutes; then unfasten the lid and remove it carefully. Lift the lid with the underside away from you so that the steam coming out of the canner does not burn your face.

Green Beans

This process will work equally well for snap, Italian, or wax beans. Select filled but tender, crisp pods, removing any diseased or rusty pods.

Quantity

- An average of 14 pounds is needed per canner load of seven quarts.
- An average of nine pounds is needed per canner load of nine pints.
- A bushel weighs 30 pounds and yields 12 to 20 quarts—an average of 2 pounds per quart.

Directions

1. Wash beans and trim ends. Leave whole, or cut or break into 1-inch pieces.

   Hot pack—Cover with boiling water; boil 5 minutes. Fill jars loosely, leaving 1-inch headspace.
   Raw pack—Fill jars tightly with raw beans, leaving 1-inch headspace. Add 1 teaspoon of salt per quart to each jar, if desired. Add boiling water, leaving 1-inch headspace.

2. Adjust lids and process.

Process Times for Green Beans in a Dial-Gauge Pressure Canner*

Style of Pack	Jar Size	Process Time	Canner Pressure (PSI) at Altitudes of:			
			0–2,000 ft	2,001–4,000 ft	4,001–6,000 ft	6,001–8,000 ft
Hot or Raw	Pints	20 minutes	11 lb	12 lb	13 lb	14 lb
	Quarts	25 minutes	11 lb	12 lb	13 lb	14 lb

*After the canner is completely depressurized, remove the weight from the vent port or open the petcock. Wait 10 minutes; then unfasten the lid and remove it carefully. Lift the lid with the underside away from you so that the steam coming out of the canner does not burn your face.

Process Times for Green Beans in a Weighted-Gauge Pressure Canner*

Style of Pack	Jar Size	Process Time	Canner Pressure (PSI) at Altitudes of:	
			0–1,000 ft	Above 1,000 ft
Hot or Raw	Pints	20 minutes	10 lbs	15 lbs
	Quarts	25 minutes	10 lbs	15 lbs

*After the canner is completely depressurized, remove the weight from the vent port or open the petcock. Wait 10 minutes; then unfasten the lid and remove it carefully. Lift the lid with the underside away from you so that the steam coming out of the canner does not burn your face.

Beets

You can preserve beets whole, cubed, or sliced, according to your preference. Beets that are 1 to 2 inches in diameter are the best, as larger ones tend to be too fibrous.

Quantity

- An average of 21 pounds (without tops) is needed per canner load of seven quarts.
- An average of 13 ½ pounds is needed per canner load of nine pints.
- A bushel (without tops) weighs 52 pounds and yields 15 to 20 quarts—an average of three pounds per quart.

Directions

1. Trim off beet tops, leaving an inch of stem and roots to reduce bleeding of color. Scrub well. Cover with boiling water. Boil until skins slip off easily, about 15 to 25 minutes depending on size.

2. Cool, remove skins, and trim off stems and roots. Leave baby beets whole. Cut medium or large beets into ½-inch cubes or slices. Halve or quarter very large slices. Add 1 teaspoon of salt per quart to each jar, if desired.

3. Fill jars with hot beets and fresh hot water, leaving 1-inch headspace. Adjust lids and process.

Process Times for Beets in a Dial-Gauge Pressure Canner*

Style of Pack	Jar Size	Process Time	Canner Pressure (PSI) at Altitudes of:			
			0–2,000 ft	2,001–4,000 ft	4,001–6,000 ft	6,001–8,000 ft
Hot	Pints	30 minutes	11 lbs	12 lbs	13 lbs	14 lbs
	Quarts	35 minutes	11 lbs	12 lbs	13 lbs	14 lbs

*After the canner is completely depressurized, remove the weight from the vent port or open the petcock. Wait 10 minutes; then unfasten the lid and remove it carefully. Lift the lid with the underside away from you so that the steam coming out of the canner does not burn your face.

Process Times for Beets in a Weighted-Gauge Pressure Canner*

Style of Pack	Jar Size	Process Time	Canner Pressure (PSI) at Altitudes of:	
			0–1,000 ft	Above 1,000 ft
Hot or Raw	Pints	30 minutes	10 lbs	15 lbs
	Quarts	35 minutes	10 lbs	15 lbs

*After the canner is completely depressurized, remove the weight from the vent port or open the petcock. Wait 10 minutes; then unfasten the lid and remove it carefully. Lift the lid with the underside away from you so that the steam coming out of the canner does not burn your face.

Carrots

Carrots can be preserved sliced or diced according to your preference. Choose small carrots, preferably 1 to 1 ¼ inches in diameter, as larger ones are often too fibrous.

Quantity

- An average of 17 ½ pounds (without tops) is needed per canner load of seven quarts.
- An average of 11 pounds is needed per canner load of nine pints.
- A bushel (without tops) weighs 50 pounds and yields 17 to 25 quarts—an average of 2 ½ pounds per quart.

Directions

1. Wash, peel, and rewash carrots. Slice or dice.

   Hot pack—Cover with boiling water; bring to boil and simmer for 5 minutes. Fill jars with carrots, leaving 1-inch headspace.

   Raw pack—Fill jars tightly with raw carrots, leaving 1-inch headspace.

2. Add 1 teaspoon of salt per quart to the jar, if desired. Add hot cooking liquid or water, leaving 1-inch headspace. Adjust lids and process.

Process Times for Carrots in a Dial-Gauge Pressure Canner*

Style of Pack	Jar Size	Process Time	Canner Pressure (PSI) at Altitudes of:			
			0–2,000 ft	2,001–4,000 ft	4,001–6,000 ft	6,001–8,000 ft
Hot or Raw	Pints	25 minutes	11 lb	12 lb	13 lb	14 lb
	Quarts	30 minutes	11 lb	12 lb	13 lb	14 lb

*After the canner is completely depressurized, remove the weight from the vent port or open the petcock. Wait 10 minutes; then unfasten the lid and remove it carefully. Lift the lid with the underside away from you so that the steam coming out of the canner does not burn your face.

Process Times for Carrots in a Weighted-Gauge Pressure Canner*

Style of Pack	Jar Size	Process Time	Canner Pressure (PSI) at Altitudes of:	
			0–1,000 ft	Above 1,000 ft
Hot or Raw	Pints	25 minutes	10 lb	15 lb
	Quarts	30 minutes	10 lb	15 lb

*After the canner is completely depressurized, remove the weight from the vent port or open the petcock. Wait 10 minutes; then unfasten the lid and remove it carefully. Lift the lid with the underside away from you so that the steam coming out of the canner does not burn your face.

Corn, Cream Style

The creamy texture comes from scraping the corncobs thoroughly and including the juices and corn pieces with the kernels. If you want to add milk or cream, butter, or other ingredients, do so just before serving (do not add dairy products before canning). Select ears containing slightly immature kernels for this recipe.

Quantity

- An average of 20 pounds (in husks) of sweet corn is needed per canner load of nine pints.
- A bushel weighs 35 pounds and yields 12 to 20 pints—an average of 2 ¼ pounds per pint.

Directions

1. Husk corn, remove silk, and wash ears. Cut corn from cob at about the center of kernel. Scrape remaining corn from cobs with a table knife.

   Hot pack—To each quart of corn and scrapings in a saucepan, add 2 cups of boiling water. Heat to boiling. Add ½ teaspoon salt to each jar, if desired. Fill pint jars with hot corn mixture, leaving 1-inch headspace.

   Raw pack—Fill pint jars with raw corn, leaving 1-inch headspace. Do not shake or press down. Add ½ teaspoon salt to each jar, if desired. Add fresh boiling water, leaving 1-inch headspace.

2. Adjust lids and process.

Process Times for Cream Style Corn in a Dial-Gauge Pressure Canner

Style of pack	Jar Size	Process Time	Canner Pressure (PSI) at Altitudes of:			
			0–2,000 ft	2,001–4,000 ft	4,001–6,000 ft	6,001–8,000 ft
Hot	Pints	85 minutes	11 lbs	12 lbs	13 lbs	14 lbs
Raw	Pints	95 minutes	11 lbs	12 lbs	13 lbs	14 lbs

*After the canner is completely depressurized, remove the weight from the vent port or open the petcock. Wait 10 minutes; then unfasten the lid and remove it carefully. Lift the lid with the underside away from you so that the steam coming out of the canner does not burn your face.

Process Times for Cream Style Corn in a Weighted-Gauge Pressure Canner*

Style of Pack	Jar Size	Process Time	Canner Pressure (PSI) at Altitudes of:	
			0–1,000 ft	Above 1,000 ft
Hot	Pints	85 minutes	10 lb	15 lb
Raw	Pints	95 minutes	10 lb	15 lb

*After the canner is completely depressurized, remove the weight from the vent port or open the petcock. Wait 10 minutes; then unfasten the lid and remove it carefully. Lift the lid with the underside away from you so that the steam coming out of the canner does not burn your face.

Corn, Whole Kernel

Select ears containing slightly immature kernels. Canning of some sweeter varieties or kernels that are too immature may cause browning. Try canning a small amount to test color and flavor before canning large quantities.

Quantity

- An average of 31 ½ pounds (in husks) of sweet corn is needed per canner load of seven quarts.
- An average of 20 pounds is needed per canner load of nine pints.
- A bushel weighs 35 pounds and yields 6 to 11 quarts—an average of 4 ½ pounds per quart.

Directions

1. Husk corn, remove silk, and wash. Blanch 3 minutes in boiling water. Cut corn from cob at about three-fourths the depth of kernel. Do not scrape cob, as it will create a creamy texture.

   Hot pack —To each quart of kernels in a saucepan, add 1 cup of hot water, heat to boiling, and simmer 5 minutes. Add 1 teaspoon of salt per quart to each jar, if desired. Fill jars with corn and cooking liquid, leaving 1-inch headspace.

   Raw pack—Fill jars with raw kernels, leaving 1-inch headspace. Do not shake or press down. Add 1 teaspoon of salt per quart to the jar, if desired.

2. Add fresh boiling water, leaving 1-inch headspace. Adjust lids and process.

Process Times for Whole Kernel Corn in a Dial-Gauge Pressure Canner*

Style of Pack	Jar Size	Process Time	Canner Pressure (PSI) at Altitudes of:			
			0–2,000 ft	2,001–4,000 ft	4,001–6,000 ft	6,001–8,000 ft
Hot or Raw	Pints	55 minutes	11 lbs	12 lbs	13 lbs	14 lbs
	Quarts	85 minutes	11 lbs	12 lbs	13 lbs	14 lbs

*After the canner is completely depressurized, remove the weight from the vent port or open the petcock. Wait 10 minutes; then unfasten the lid and remove it carefully. Lift the lid with the underside away from you so that the steam coming out of the canner does not burn your face.

Process Times for Whole Kernel Corn in a Weighted-Gauge Pressure Canner*

Style of Pack	Jar Size	Process Time	Canner Pressure (PSI) at Altitudes of:	
			0–1,000 ft	Above 1,000 ft
Hot or Raw	Pints	55 minutes	10 lbs	15 lbs
	Quarts	85 minutes	10 lbs	15 lbs

*After the canner is completely depressurized, remove the weight from the vent port or open the petcock. Wait 10 minutes; then unfasten the lid and remove it carefully. Lift the lid with the underside away from you so that the steam coming out of the canner does not burn your face.

Mixed Vegetables

Use mixed vegetables in soups, casseroles, pot pies, or as a quick side dish. You can change the suggested proportions or substitute other favorite vegetables, but avoid leafy greens, dried beans, cream-style corn, winter squash, and sweet potatoes, as they will ruin the consistency of the other vegetables. This recipe yields about seven quarts.

Ingredients

6 cups sliced carrots

6 cups cut, whole-kernel sweet corn

6 cups cut green beans

6 cups shelled lima beans

4 cups diced or crushed tomatoes

4 cups diced zucchini

Directions

1. Carefully wash, peel, de-shell, and cut vegetables as necessary. Combine all vegetables in a large pot or kettle, and add enough water to cover pieces.

2. Add 1 teaspoon salt per quart to each jar, if desired. Boil 5 minutes and fill jars with hot pieces and liquid, leaving 1-inch headspace. Adjust lids and process.

Process Times for Mixed Vegetables in a Dial-Gauge Pressure Canner*

Style of Pack	Jar Size	Process Time	Canner Pressure (PSI) at Altitudes of:			
			0–2,000 ft	2,001–4,000 ft	4,001–6,000 ft	6,001–8,000 ft
Hot	Pints	75 minutes	11 lbs	12 lbs	13 lbs	14 lbs
	Quarts	90 minutes	11 lbs	12 lbs	13 lbs	14 lbs

*After the canner is completely depressurized, remove the weight from the vent port or open the petcock. Wait 10 minutes; then unfasten the lid and remove it carefully. Lift the lid with the underside away from you so that the steam coming out of the canner does not burn your face.

Process Times for Mixed Vegetables in a Weighted-Gauge Pressure Canner*

Style of Pack	Jar Size	Process Time	Canner Pressure (PSI) at Altitudes of:	
			0–1,000 ft	Above 1,000 ft
Hot	Pints	75 minutes	10 lbs	15 lbs
	Quarts	90 minutes	10 lbs	15 lbs

*After the canner is completely depressurized, remove the weight from the vent port or open the petcock. Wait 10 minutes; then unfasten the lid and remove it carefully. Lift the lid with the underside away from you so that the steam coming out of the canner does not burn your face.

Peas, Green or English, Shelled

Green and English peas preserve well when canned, but sugar snap and Chinese edible pods are better frozen. Select filled pods containing young, tender, sweet seeds, and discard any diseased pods.

Quantity
- An average of 31 ½ pounds (in pods) is needed per canner load of seven quarts.
- An average of 20 pounds is needed per canner load of nine pints.
- A bushel weighs 30 pounds and yields 5 to 10 quarts—an average of 4 ½ pounds per quart.

Directions

1. Shell and wash peas. Add 1 teaspoon of salt per quart to each jar, if desired.

Hot pack—Cover with boiling water. Bring to a boil in a saucepan, and boil 2 minutes. Fill jars loosely with hot peas, and add cooking liquid, leaving 1-inch headspace.

Raw pack—Fill jars with raw peas, and add boiling water, leaving 1-inch headspace. Do not shake or press down peas.

2. Adjust lids and process.

Process Times for Peas in a Dial-Gauge Pressure Canner*

			Canner Pressure (PSI) at Altitudes of:			
Style of Pack	Jar Size	Process Time	0–2,000 ft	2,001–4,000 ft	4,001–6,000 ft	6,001–8,000 ft
Hot or Raw	Pints or Quarts	40 minutes	11 lbs	12 lbs	13 lbs	14 lbs

*After the canner is completely depressurized, remove the weight from the vent port or open the petcock. Wait 10 minutes; then unfasten the lid and remove it carefully. Lift the lid with the underside away from you so that the steam coming out of the canner does not burn your face.

Process Times for Peas in a Weighted-Gauge Pressure Canner*

			Canner Pressure (PSI) at Altitudes of:	
Style of Pack	Jar Size	Process Time	0–1,000 ft	Above 1,000 ft
Hot or Raw	Pints or Quarts	40 minutes	10 lbs	15 lbs

*After the canner is completely depressurized, remove the weight from the vent port or open the petcock. Wait 10 minutes; then unfasten the lid and remove it carefully. Lift the lid with the underside away from you so that the steam coming out of the canner does not burn your face.

Potatoes, Sweet

Sweet potatoes can be preserved whole, in chunks, or in slices, according to your preference. Choose small to medium-sized potatoes that are mature and not too fibrous. Can within one to two months after harvest.

Quantity
- An average of 17 ½ pounds is needed per canner load of seven quarts.
- An average of 11 pounds is needed per canner load of nine pints.
- A bushel weighs 50 pounds and yields 17 to 25 quarts—an average of 2 ½ pounds per quart.

Directions

1. Wash potatoes and boil or steam until partially soft (15 to 20 minutes). Remove skins. Cut medium potatoes, if needed, so that pieces are uniform in size. Do not mash or purée pieces.
2. Fill jars, leaving 1-inch headspace. Add 1 teaspoon salt per quart to each jar, if desired. Cover with your choice of fresh boiling water or syrup, leaving 1-inch headspace. Adjust lids and process.

Process Times for Sweet Potatoes in a Dial-Gauge Pressure Canner*

			Canner Pressure (PSI) at Altitudes of:			
Style of Pack	Jar Size	Process Time	0–2,000 ft	2,001–4,000 ft	4,001–6,000 ft	6,001–8,000 ft
Hot	Pints	65 minutes	11 lbs	12 lbs	13 lbs	14 lbs
	Quarts	90 minutes	11 lbs	12 lbs	13 lbs	14 lbs

*After the canner is completely depressurized, remove the weight from the vent port or open the petcock. Wait 10 minutes; then unfasten the lid and remove it carefully. Lift the lid with the underside away from you so that the steam coming out of the canner does not burn your face.

Process Times for Sweet Potatoes in a Weighted-Gauge Pressure Canner*

Style of Pack	Jar Size	Process Time	Canner Pressure (PSI) at Altitudes of:	
			0–1,000 ft	Above 1,000 ft
Hot	Pints	65 minutes	10 lbs	15 lbs
	Quarts	90 minutes	10 lbs	15 lbs

*After the canner is completely depressurized, remove the weight from the vent port or open the petcock. Wait 10 minutes; then unfasten the lid and remove it carefully. Lift the lid with the underside away from you so that the steam coming out of the canner does not burn your face.

Pumpkin and Winter Squash

Pumpkin and squash are great to have on hand for use in pies, soups, quick breads, or as side dishes. They should have a hard rind and stringless, mature pulp. Small pumpkins (sugar or pie varieties) are best. Before using for pies, drain jars and strain or sieve pumpkin or squash cubes.

Quantity

- An average of 16 pounds is needed per canner load of seven quarts.
- An average of 10 pounds is needed per canner load of nine pints—an average of 2 ¼ pounds per quart.

Directions

1. Wash, remove seeds, cut into 1-inch-wide slices, and peel. Cut flesh into 1-inch cubes. Boil 2 minutes in water. Do not mash or purée.
2. Fill jars with cubes and cooking liquid, leaving 1-inch headspace. Adjust lids and process.

Process Times for Pumpkin and Winter Squash in a Dial-Gauge Pressure Canner*

Style of Pack	Jar Size	Process Time	Canner Pressure (PSI) at Altitudes of:			
			0–2,000 ft	2,001–4,000 ft	4,001–6,000 ft	6,001–8,000 ft
Hot	Pints	55 minutes	11 lbs	12 lbs	13 lbs	14 lbs
	Quarts	90 minutes	11 lbs	12 lbs	13 lbs	14 lbs

*After the canner is completely depressurized, remove the weight from the vent port or open the petcock. Wait 10 minutes; then unfasten the lid and remove it carefully. Lift the lid with the underside away from you so that the steam coming out of the canner does not burn your face.

Process Times for Pumpkin and Winter Squash in a Weighted-Gauge Pressure Canner*

Style of Pack	Jar Size	Process Time	Canner Pressure (PSI) at Altitudes of:	
			0–1,000 ft	Above 1,000 ft
Hot	Pints	55 minutes	10 lbs	15 lbs
	Quarts	90 minutes	10 lbs	15 lbs

*After the canner is completely depressurized, remove the weight from the vent port or open the petcock. Wait 10 minutes; then unfasten the lid and remove it carefully. Lift the lid with the underside away from you so that the steam coming out of the canner does not burn your face.

Succotash

To spice up this simple, satisfying dish, add a little paprika and celery salt before serving. It is also delicious made into a pot pie, with or without added chicken, turkey, or beef. This recipe yields seven quarts.

Ingredients

1 lb unhusked sweet corn or 3 qts cut whole kernels

14 lbs mature green podded lima beans or 4 qts shelled lima beans

2 qts crushed or whole tomatoes (optional)

Directions

1. Husk corn, remove silk, and wash. Blanch 3 minutes in boiling water. Cut corn from cob at about three-fourths the depth of kernel. Do not scrape cob, as it will create a creamy texture. Shell lima beans and wash thoroughly.

    Hot pack—Combine all prepared vegetables in a large kettle with enough water to cover the pieces. Add 1 teaspoon salt to each quart jar, if desired. Boil gently 5 minutes and fill jars with pieces and cooking liquid, leaving 1-inch headspace.

    Raw pack—Fill jars with equal parts of all prepared vegetables, leaving 1-inch headspace. Do not shake or press down pieces. Add 1 teaspoon salt to each quart jar, if desired. Add fresh boiling water, leaving 1-inch headspace.

2. Adjust lids and process.

Process Times for Succotash in a Dial-Gauge Pressure Canner*

Style of Pack	Jar Size	Process Time	Canner Pressure (PSI) at Altitudes of:			
			0–2,000 ft	2,001–4,000 ft	4,001–6,000 ft	6,001–8,000 ft
Hot or Raw	Pints	60 minutes	11 lbs	12 lbs	13 lbs	14 lbs
	Quarts	85 minutes	11 lbs	12 lbs	13 lbs	14 lbs

*After the canner is completely depressurized, remove the weight from the vent port or open the petcock. Wait 10 minutes; then unfasten the lid and remove it carefully. Lift the lid with the underside away from you so that the steam coming out of the canner does not burn your face.

Process Times for Succotash in a Weighted-Gauge Pressure Canner*

Style of Pack	Jar Size	Process Time	Canner Pressure (PSI) at Altitudes of:	
			0–1,000 ft	Above 1,000 ft
Hot or Raw	Pints	60 minutes	10 lbs	15 lbs
	Quarts	85 minutes	10 lbs	15 lbs

*After the canner is completely depressurized, remove the weight from the vent port or open the petcock. Wait 10 minutes; then unfasten the lid and remove it carefully. Lift the lid with the underside away from you so that the steam coming out of the canner does not burn your face.

Soups

Vegetable, dried bean or pea, meat, poultry, or seafood soups can all be canned. Add pasta, rice, or other grains to soup just prior to serving, as grains tend to get soggy when canned. If dried beans or peas are used, they *must* be fully rehydrated first. Dairy products should also be avoided in the canning process.

Directions

1. Select, wash, and prepare vegetables.
2. Cook vegetables. For each cup of dried beans or peas, add 3 cups of water, boil 2 minutes, remove from heat, soak 1 hour, and heat to boil. Drain and combine with meat broth, tomatoes, or water to cover. Boil 5 minutes.
3. Salt to taste, if desired. Fill jars halfway with solid mixture. Add remaining liquid, leaving 1-inch headspace. Adjust lids and process.

Process Times for Soups in a Dial-Gauge Pressure Canner*

Style of Pack	Jar Size	Process Time	Canner Pressure (PSI) at Altitudes of:			
			0–2,000 ft	2,001–4,000 ft	4,001–6,000 ft	6,001–8,000 ft
Hot	Pints	60* minutes	11 lbs	12 lbs	13 lbs	14 lbs
	Quarts	75* minutes	11 lbs	12 lbs	13 lbs	14 lbs

***Caution: Process 100 minutes if soup contains seafood.**

*After the canner is completely depressurized, remove the weight from the vent port or open the petcock. Wait 10 minutes; then unfasten the lid and remove it carefully. Lift the lid with the underside away from you so that the steam coming out of the canner does not burn your face.

Process Times for Soups in a Weighted-Gauge Pressure Canner*

Style of Pack	Jar Size	Process Time	Canner Pressure (PSI) at Altitudes of:	
			0–1,000 ft	Above 1,000 ft
Hot	Pints	60* minutes	10 lbs	15 lbs
	Quarts	75* minutes	10 lbs	15 lbs

***Caution: Process 100 minutes if soup contains seafood.**

*After the canner is completely depressurized, remove the weight from the vent port or open the petcock. Wait 10 minutes; then unfasten the lid and remove it carefully. Lift the lid with the underside away from you so that the steam coming out of the canner does not burn your face.

Meat Stock (Broth)

"Good broth will resurrect the dead," says a South American proverb. Bones contain calcium, magnesium, phosphorus, and other trace minerals, while cartilage and tendons hold glucosamine, which is important for joints and muscle health. When simmered for extended periods, these nutrients are released into the water and broken down into a form that our bodies can absorb. Not to mention that good broth is the secret to delicious risotto, reduction sauces, gravies, and dozens of other gourmet dishes.

Beef

1. Saw or crack fresh trimmed beef bones to enhance extraction of flavor. Rinse bones and place in a large stockpot or kettle, cover bones with water, add pot cover, and simmer 3 to 4 hours.

2. Remove bones, cool broth, and pick off meat. Skim off fat, add meat removed from bones to broth, and reheat to boiling. Fill jars, leaving 1-inch headspace. Adjust lids and process.

Chicken or Turkey

1. Place large carcass bones in a large stockpot, add enough water to cover bones, cover pot, and simmer 30 to 45 minutes or until meat can be easily stripped from bones.

2. Remove bones and pieces, cool broth, strip meat, discard excess fat, and return meat to broth. Reheat to boiling and fill jars, leaving 1-inch headspace. Adjust lids and process.

Process Times for Meat Stock in a Dial-Gauge Pressure Canner*

			Canner Pressure (PSI) at Altitudes of:			
Style of Pack	Jar Size	Process Time	0–2,000 ft	2,001–4,000 ft	4,001–6,000 ft	6,001–8,000 ft
Hot	Pints	20 minutes	11 lbs	12 lbs	13 lbs	14 lbs
	Quarts	25 minutes	11 lbs	12 lbs	13 lbs	14 lbs

*After the canner is completely depressurized, remove the weight from the vent port or open the petcock. Wait 10 minutes; then unfasten the lid and remove it carefully. Lift the lid with the underside away from you so that the steam coming out of the canner does not burn your face.

Process Times for Meat Stock in a Weighted-Gauge Pressure Canner*

			Canner Pressure (PSI) at Altitudes of:	
Style of Pack	Jar Size	Process Time	0–1,000 ft	Above 1,000 ft
Hot	Pints	20 minutes	10 lbs	15 lbs
	Quarts	25 minutes	10 lbs	15 lbs

*After the canner is completely depressurized, remove the weight from the vent port or open the petcock. Wait 10 minutes; then unfasten the lid and remove it carefully. Lift the lid with the underside away from you so that the steam coming out of the canner does not burn your face.

Fermented Foods and Pickled Vegetables

Pickled vegetables play a vital role in Italian antipasto dishes, Chinese stir-fries, British piccalilli, and much of Russian and Finnish cuisine. And, of course, the Germans love their sauerkraut, kimchee is found on nearly every Korean dinner table, and many an American won't eat a sandwich without a good strong dill pickle on the side.

Fermenting vegetables is not complicated, but you'll want to have the proper containers, covers, and weights ready before you begin. For containers, keep the following in mind:

- A one-gallon container is needed for each 5 pounds of fresh vegetables. Therefore, a five-gallon stone crock is of ideal size for fermenting about 25 pounds of fresh cabbage or cucumbers.
- Food-grade plastic and glass containers are excellent substitutes for stone crocks. Other one- to three-gallon non-food-grade plastic containers may be used if lined inside with a clean food-grade plastic bag. **Caution: Be certain that foods contact only food-grade plastics. Do not use garbage bags or trash liners.**
- Fermenting sauerkraut in quart and half-gallon mason jars is an acceptable practice, but may result in more spoilage losses.

Some vegetables, like cabbage and cucumbers, need to be kept 1 to 2 inches under brine while fermenting. If you find them floating to top of the container, here are some suggestions:

- After adding prepared vegetables and brine, insert a suitably sized dinner plate or glass pie plate inside the fermentation container. The plate must be slightly smaller than the container opening, yet large enough to cover most of the shredded cabbage or cucumbers.
- To keep the plate under the brine, weight it down with two to three sealed quart jars filled with water. Covering the container opening with a clean, heavy bath towel helps to prevent contamination from insects and molds while the vegetables are fermenting.

- Fine quality fermented vegetables are also obtained when the plate is weighted down with a very large, clean, plastic bag filled with three quarts of water containing 4 ½ tablespoons of salt. Be sure to seal the plastic bag. Freezer bags sold for packaging turkeys are suitable for use with five-gallon containers.

Be sure to wash the fermentation container, plate, and jars in hot sudsy water, and rinse well with very hot water before use.

Regular dill pickles and sauerkraut are fermented and cured for about 3 weeks. Refrigerator dills are fermented for about 1 week. During curing, colors and flavors change and acidity increases. Fresh-pack or quick-process pickles are not fermented; some are brined several hours or overnight, then drained and covered with vinegar and seasonings. Fruit pickles usually are prepared by heating fruit in a seasoned syrup acidified with either lemon juice or vinegar. Relishes are made from chopped fruits and vegetables that are cooked with seasonings and vinegar.

Be sure to remove and discard a ¹⁄₁₆-inch slice from the blossom end of fresh cucumbers. Blossoms may contain an enzyme which causes excessive softening of pickles.

Caution: The level of acidity in a pickled product is as important to its safety as it is to taste and texture.

- **Do not alter vinegar, food, or water proportions in a recipe or use a vinegar with unknown acidity.**
- **Use only recipes with tested proportions of ingredients.**
- **There must be a minimum, uniform level of acid throughout the mixed product to prevent the growth of botulinum bacteria.**

Ingredients

Select fresh, firm fruits or vegetables free of spoilage. Measure or weigh amounts carefully, because the proportion of fresh food to other ingredients will affect flavor and, in many instances, safety.

Use canning or pickling salt. Noncaking material added to other salts may make the brine cloudy. Since flake salt varies in density, it is not recommended for making pickled and fermented foods. White granulated and brown sugars

are most often used. Corn syrup and honey, unless called for in reliable recipes, may produce undesirable flavors. White distilled and cider vinegars of 5 percent acidity (50 grain) are recommended. White vinegar is usually preferred when light color is desirable, as is the case with fruits and cauliflower.

Pickles with reduced salt content

In the making of fresh-pack pickles, cucumbers are acidified quickly with vinegar. Use only tested recipes formulated to produce the proper acidity. While these pickles may be prepared safely with reduced or no salt, their quality may be noticeably lower. Both texture and flavor may be slightly, but noticeably, different than expected. You may wish to make small quantities first to determine if you like them.

However, the salt used in making fermented sauerkraut and brined pickles not only provides characteristic flavor but also is vital to safety and texture. In fermented foods, salt favors the growth of desirable bacteria while inhibiting the growth of others. **Caution: Do not attempt to make sauerkraut or fermented pickles by cutting back on the salt required.**

Preventing spoilage

Pickle products are subject to spoilage from microorganisms, particularly yeasts and molds, as well as enzymes that may affect flavor, color, and texture. Processing the pickles in a boiling-water canner will prevent both of these problems. Standard canning jars and self-sealing lids are recommended. Processing times and procedures will vary according to food acidity and the size of food pieces.

Dill Pickles

Feel free to alter the spices in this recipe, but stick to the same proportion of cucumbers, vinegar, and water. Check the label of your vinegar to be sure it contains 5 percent acetic acid. Fully fermented pickles may be stored in the original container for about four to six months, provided they are refrigerated and surface scum and molds are removed regularly, but canning is a better way to store fully fermented pickles.

Ingredients

Use the following quantities for each gallon capacity of your container:

4 lbs of 4-inch pickling cucumbers
2 tbsp dill seed or 4 to 5 heads fresh or dry dill weed
½ cup salt
¼ cup vinegar (5 percent acetic acid)
8 cups water and one or more of the following ingredients:
2 cloves garlic (optional)
2 dried red peppers (optional)
2 tsp whole mixed pickling spices (optional)

Directions

1. Wash cucumbers. Cut 1/16-inch slice off blossom end and discard. Leave ¼ inch of stem attached. Place half of dill and spices on bottom of a clean, suitable container (see page 198 for suggestions on containers, lids, and weights).

2. Add cucumbers, remaining dill, and spices. Dissolve salt in vinegar and water and pour over cucumbers. Add suitable cover and weight. Store where temperature is between 70 and 75°F for about 3 to 4 weeks while fermenting. Temperatures of 55 to 65°F are acceptable, but the fermentation will take 5 to 6 weeks. Avoid temperatures above 80°F, or pickles will become too soft during fermentation. Fermenting pickles cure slowly. Check the container several times a week and promptly remove surface scum or mold. **Caution: If the pickles become soft, slimy, or develop a disagreeable odor, discard them.**

3. Once fully fermented, pour the brine into a pan, heat slowly to a boil, and simmer 5 minutes. Filter brine through paper coffee filters to reduce cloudiness, if desired. Fill jars with pickles and hot brine, leaving ½-inch headspace. Adjust lids and process in a boiling water canner, or use the low-temperature pasteurization treatment described below.

Low-Temperature Pasteurization Treatment

The following treatment results in a better product texture but must be carefully managed to avoid possible spoilage.

1. Place jars in a canner filled halfway with warm (120 to 140°F) water. Then, add hot water to a level 1 inch above jars.

2. Heat the water enough to maintain 180 to 185°F water temperature for 30 minutes. Check with a candy or jelly thermometer to be certain that the water temperature is at least 180°F during the entire 30 minutes. Temperatures higher than 185°F may cause unnecessary softening of pickles.

Process Times for Dill Pickles in a Boiling-Water Canner*

Style of Pack	Jar Size	Process Time at Altitudes of:		
		0–1,000 ft	1,001–6,000 ft	Above 6,000 ft
Raw	Pints	10 minutes	15 minutes	20 minutes
	Quarts	15 minutes	20 minutes	25 minutes

*After the process is complete, turn off the heat and remove the canner lid. Wait five minutes before removing jars.

Sauerkraut

For the best sauerkraut, use firm heads of fresh cabbage. Shred cabbage and start kraut between 24 and 48 hours after harvest. This recipe yields about nine quarts.

Ingredients

25 lbs cabbage
¾ cup canning or pickling salt

Directions

1. Work with about 5 pounds of cabbage at a time. Discard outer leaves. Rinse heads under cold running water and drain. Cut heads in quarters and remove cores. Shred or slice to the thickness of a quarter.

2. Put cabbage in a suitable fermentation container (see page 198 for suggestions on containers, lids, and weights), and add 3 tablespoons of salt. Mix thoroughly, using clean hands. Pack firmly until salt draws juices from cabbage.

3. Repeat shredding, salting, and packing until all cabbage is in the container. Be sure it is deep enough so that its rim is at least 4 or 5 inches above the cabbage. If juice does not cover cabbage, add boiled and cooled brine (1 ½ tablespoons of salt per quart of water).

4. Add plate and weights; cover container with a clean bath towel. Store at 70 to 75°F while fermenting. At temperatures between 70 and 75°F, kraut will be fully fermented in about 3 to 4 weeks; at 60° to 65°F, fermentation may take 5 to 6 weeks. At temperatures lower than 60°F, kraut may not ferment. Above 75°F, kraut may become soft.

Note: If you weigh the cabbage down with a brine-filled bag, do not disturb the crock until normal fermentation is completed (when bubbling ceases). If you use jars as weight, you will have to check the kraut 2 to 3 times each week and remove scum if it forms. Fully fermented kraut may be kept tightly covered in the refrigerator for several months or it may be canned as follows:

Hot pack—Bring kraut and liquid slowly to a boil in a large kettle, stirring frequently. Remove from heat and fill jars rather firmly with kraut and juices, leaving ½-inch headspace.

Raw pack—Fill jars firmly with kraut and cover with juices, leaving ½-inch headspace.

5. Adjust lids and process.

Process Times for Sauerkraut in a Boiling-Water Canner*

Style of Pack	Jar Size	Process Time at Altitudes of:			
		0–1,000 ft	1,001-3,000 ft	3,001-6,000 ft	Above 6,000 ft
Hot	Pints	10 minutes	15 minutes	15 minutes	20 minutes
	Quarts	15 minutes	20 minutes	20 minutes	25 minutes
Raw	Pints	20 minutes	25 minutes	30 minutes	35 minutes
	Quarts	25 minutes	30 minutes	35 minutes	40 minutes

*After the process is complete, turn off the heat and remove the canner lid. Wait five minutes before removing jars.

Pickled Three-Bean Salad

This is a great side dish to bring to a summer picnic or potluck. Feel free to add or adjust spices to your taste. This recipe yields about five to six half-pints.

Ingredients

1 ½ cups cut and blanched green or yellow beans (prepared as below)

1 ½ cups canned, drained red kidney beans

1 cup canned, drained garbanzo beans

½ cup peeled and thinly sliced onion (about 1 medium onion)

½ cup trimmed and thinly sliced celery (1 ½ medium stalks)

½ cup sliced green peppers (½ medium pepper)

½ cup white vinegar (5 percent acetic acid)

¼ cup bottled lemon juice

¾ cup sugar

1 ¼ cups water

¼ cup oil

½ tsp canning or pickling salt

Directions

1. Wash and snap off ends of fresh beans. Cut or snap into 1- to 2-inch pieces. Blanch 3 minutes and cool immediately. Rinse kidney beans with tap water and drain again. Prepare and measure all other vegetables.

2. Combine vinegar, lemon juice, sugar, and water and bring to a boil. Remove from heat. Add oil and salt and mix well. Add beans, onions, celery, and green pepper to solution and bring to a simmer.

3. Marinate 12 to 14 hours in refrigerator, then heat entire mixture to a boil. Fill clean jars with solids. Add hot liquid, leaving ½-inch headspace. Adjust lids and process.

Process Times for Pickled Three-Bean Salad in a Boiling Water Canner*

		Process Time at Altitudes of:		
Style of Pack	Jar Size	0–1,000 ft	1,001–6,000 ft	Above 6,000 ft
Hot	Half-pints or Pints	15 minutes	20 minutes	25 minutes

Pickled Horseradish Sauce

Select horseradish roots that are firm and have no mold, soft spots, or green spots. Avoid roots that have begun to sprout. The pungency of fresh horseradish fades within one to two months, even when refrigerated, so make only small quantities at a time. This recipe yields about two half-pints.

Ingredients

2 cups (¾ lb) freshly grated horseradish

1 cup white vinegar (5 percent acetic acid)

½ tsp canning or pickling salt

¼ tsp powdered ascorbic acid

Directions

1. Wash horseradish roots thoroughly and peel off brown outer skin. Grate the peeled roots in a food processor or cut them into small cubes and put through a food grinder.

2. Combine ingredients and fill into sterile jars, leaving ¼-inch headspace. Seal jars tightly and store in a refrigerator.

Marinated Peppers

Any combination of bell, Hungarian, banana, or jalapeño peppers can be used in this recipe. Use more jalapeño peppers if you want your mix to be hot, but remember to wear rubber or plastic gloves while handling them or wash hands thoroughly with soap and water before touching your face. This recipe yields about nine half-pints.

Ingredients

4 lbs firm peppers

1 cup bottled lemon juice

2 cups white vinegar (5 percent acetic acid)

1 tbsp oregano leaves

1 cup olive or salad oil

½ cup chopped onions

2 tbsp prepared horseradish (optional)

2 cloves garlic, quartered (optional)

2 ¼ tsp salt (optional)

Directions

1. Select your favorite pepper. Peppers may be left whole or quartered. Wash, slash two to four slits in each pepper, and blanch in boiling water or blister in order to peel tough-skinned hot peppers. Blister peppers using one of the following methods:

Oven or broiler method—Place peppers in a hot oven (400°F) or broiler for 6 to 8 minutes or until skins blister.

Range-top method—Cover hot burner, either gas or electric, with heavy wire mesh. Place peppers on burner for several minutes until skins blister.

2. Allow peppers to cool. Place in pan and cover with a damp cloth. This will make peeling the peppers easier. After several minutes of cooling, peel each pepper. Flatten whole peppers.

3. Mix all remaining ingredients except garlic and salt in a saucepan and heat to boiling. Place ¼ garlic clove (optional) and ¼ teaspoon salt in each half-pint or ½ teaspoon per pint. Fill jars with peppers, and add hot, well-mixed oil/pickling solution over peppers, leaving ½-inch headspace. Adjust lids and process.

Process Times for Marinated Peppers in a Boiling-Water Canner*

Style of Pack	Jar Size	Process Time at Altitudes of:			
		0–1,000 ft	1,001–3,000 ft	3,001–6,000 ft	Above 6,000 ft
Raw	Half-pints and pints	15 minutes	20 minutes	20 minutes	25 minutes

*After the process is complete, turn off the heat and remove the canner lid. Wait five minutes before removing jars.

Piccalilli

Piccalilli is a nice accompaniment to roasted or braised meats and is common in British and Indian meals. It can also be mixed with mayonnaise or crème fraîche as the basis of a French remoulade. This recipe yields nine half-pints.

Ingredients

6 cups chopped green tomatoes

1 ½ cups chopped sweet red peppers

1 ½ cups chopped green peppers

2 ¼ cups chopped onions

7 ½ cups chopped cabbage

½ cup canning or pickling salt

3 tbsp whole mixed pickling spice

4 ½ cups vinegar (5 percent acetic acid)

3 cups brown sugar

Directions

1. Wash, chop, and combine vegetables with salt. Cover with hot water and let stand 12 hours. Drain and press in a clean white cloth to remove all possible liquid.

2. Tie spices loosely in a spice bag and add to combined vinegar and brown sugar and heat to a boil in a saucepan. Add vegetables and boil gently 30 minutes or until the volume of the mixture is reduced by one-half. Remove spice bag.

3. Fill hot sterile jars with hot mixture, leaving ½-inch headspace. Adjust lids and process.

Process Times for Piccalilli in a Boiling-Water Canner

		Process Time at Altitudes of:		
Style of Pack	Jar Size	0–1,000 ft	1,001–6,000 ft	Above 6,000 ft
Hot	Half-pints or Pints	5 minutes	10 minutes	15 minutes

*After the process is complete, turn off the heat and remove the canner lid. Wait five minutes before removing jars.

Bread-and-Butter Pickles

These slightly sweet, spiced pickles will add flavor and crunch to any sandwich. If desired, slender (1 to 1 ½ inches in diameter) zucchini or yellow summer squash can be substituted for cucumbers. After processing and cooling, jars should be stored four to five weeks to develop ideal flavor. This recipe yields about eight pints.

Ingredients

6 lbs of 4- to 5-inch pickling cucumbers
8 cups thinly sliced onions (about 3 pounds)
½ cup canning or pickling salt
4 cups vinegar (5 percent acetic acid)
4 ½ cups sugar
2 tbsp mustard seed
1 ½ tbsp celery seed
1 tbsp ground turmeric
1 cup pickling lime (optional—for use in variation below for making firmer pickles)

Directions

1. Wash cucumbers. Cut ¹⁄₁₆ inch off blossom end and discard. Cut into ³⁄₁₆-inch slices. Combine cucumbers and onions in a large bowl. Add salt. Cover with 2 inches crushed or cubed ice. Refrigerate 3 to 4 hours, adding more ice as needed.
2. Combine remaining ingredients in a large pot. Boil 10 minutes. Drain cucumbers and onions, add to pot, and slowly reheat to boiling. Fill jars with slices and cooking syrup, leaving ½-inch headspace.
3. Adjust lids and process in boiling-water canner, or use the low-temperature pasteurization treatment described below.

Low-Temperature Pasteurization Treatment

The following treatment results in a better product texture but must be carefully managed to avoid possible spoilage.

1. Place jars in a canner filled halfway with warm (120 to 140°F) water. Then, add hot water to a level 1 inch above jars.
2. Heat the water enough to maintain 180 to 185°F water temperature for 30 minutes. Check with a candy or jelly thermometer to be certain that the water temperature is at least 180°F during the entire 30 minutes. Temperatures higher than 185°F may cause unnecessary softening of pickles.

Variation for firmer pickles: Wash cucumbers. Cut ¹⁄₁₆ inch off blossom end and discard. Cut into ³⁄₁₆-inch slices. Mix 1 cup pickling lime and ½ cup salt to 1 gallon water in a 2- to 3-gallon crock or enamelware container. Avoid inhaling lime dust while mixing the lime-water solution. Soak cucumber slices in lime water for 12 to 24 hours, stirring occasionally. Remove from lime solution, rinse, and resoak 1 hour in fresh cold water. Repeat the rinsing and soaking steps two more times. Handle carefully, as slices will be brittle. Drain well.

Process Times for Bread-and-Butter Pickles in a Boiling-Water Canner*

		Process Time at Altitudes of:		
Style of Pack	Jar Size	0–1,000 ft	1,001–6,000 ft	Above 6,000 ft
Hot	Pints or Quarts	10 minutes	15 minutes	20 minutes

*After the process is complete, turn off the heat and remove the canner lid. Wait five minutes before removing jars.

Quick Fresh-Pack Dill Pickles

For best results, pickle cucumbers within twenty-four hours of harvesting, or immediately after purchasing. This recipe yields seven to nine pints.

Ingredients

8 lbs of 3- to 5-inch pickling cucumbers
2 gallons water
1 ¼ to 1 ½ cups canning or pickling salt
1 ½ qts vinegar (5 percent acetic acid)
¼ cup sugar
2 to 2 ¼ quarts water
2 tbsp whole mixed pickling spice
3 to 5 tbsp whole mustard seed (2 tsp to 1 tsp per pint jar)

14 to 21 heads of fresh dill (1 ½ to 3 heads per pint jar) *or*
4 ½ to 7 tbsp dill seed (1-½ tsp to 1 tbsp per pint jar)

Directions

1. Wash cucumbers. Cut ¹⁄₁₆-inch slice off blossom end and discard, but leave ¼-inch of stem attached. Dissolve ¾ cup salt in 2 gallons water. Pour over cucumbers and let stand 12 hours. Drain.
2. Combine vinegar, ½ cup salt, sugar and 2 quarts water. Add mixed pickling spices tied in a clean white cloth. Heat to boiling. Fill jars with cucumbers. Add 1 tsp mustard seed and 1 ½ heads fresh dill per pint.
3. Cover with boiling pickling solution, leaving ½-inch headspace. Adjust lids and process.

Process Times for Quick Fresh-Pack Dill Pickles in a Boiling-Water Canner*

		Process Time at Altitudes of:		
Style of Pack	Jar Size	0–1,000 ft	1,001–6,000 ft	Above 6,000 ft
Raw	Pints	10 minutes	15 minutes	20 minutes
	Quarts	15 minutes	20 minutes	25 minutes

*After the process is complete, turn off the heat and remove the canner lid. Wait five minutes before removing jars.

Pickle Relish

A food processor will make quick work of chopping the vegetables in this recipe. Yields about nine pints.

Ingredients

3 qts chopped cucumbers
3 cups each of chopped sweet green and red peppers
1 cup chopped onions
¾ cup canning or pickling salt
4 cups ice
8 cups water
4 tsp each of mustard seed, turmeric, whole allspice, and whole cloves
2 cups sugar
6 cups white vinegar (5 percent acetic acid)

Directions

1. Add cucumbers, peppers, onions, salt, and ice to water and let stand 4 hours. Drain and re-cover vegetables with fresh ice water for another hour. Drain again.
2. Combine spices in a spice or cheesecloth bag. Add spices to sugar and vinegar. Heat to boiling and pour mixture over vegetables. Cover and refrigerate 24 hours.
3. Heat mixture to boiling and fill hot into clean jars, leaving ½-inch headspace. Adjust lids and process.

Process Times for Pickle Relish in a Boiling-Water Canner*

Style of Pack	Jar Size	Process Time at Altitudes of:		
		0–1,000 ft	1,001–6,000 ft	Above 6,000 ft
Hot	Half-pints or Pints	10 minutes	15 minutes	20 minutes

*After the process is complete, turn off the heat and remove the canner lid. Wait five minutes before removing jars.

Quick Sweet Pickles

Quick and simple to prepare, these are the sweet pickles to make when you're short on time. After processing and cooling, jars should be stored four to five weeks to develop ideal flavor. If desired, add two slices of raw whole onion to each jar before filling with cucumbers. This recipe yields about seven to nine pints.

Ingredients

8 lbs of 3- to 4-inch pickling cucumbers
⅓ cup canning or pickling salt
4 ½ cups sugar
3 ½ cups vinegar (5 percent acetic acid)
2 tsp celery seed
1 tbsp whole allspice
2 tbsp mustard seed
1 cup pickling lime (optional)

Directions

1. Wash cucumbers. Cut ¹⁄₁₆ inch off blossom end and discard, but leave ¼ inch of stem attached. Slice or cut in strips, if desired.
2. Place in bowl and sprinkle with salt. Cover with 2 inches of crushed or cubed ice. Refrigerate 3 to 4 hours. Add more ice as needed. Drain well.
3. Combine sugar, vinegar, celery seed, allspice, and mustard seed in 6-quart kettle. Heat to boiling.

   Hot pack—Add cucumbers and heat slowly until vinegar solution returns to boil. Stir occasionally to make sure mixture heats evenly. Fill sterile jars, leaving ½-inch headspace.

   Raw pack—Fill jars, leaving ½-inch headspace.
4. Add hot pickling syrup, leaving ½-inch headspace. Adjust lids and process.

> **Variation for firmer pickles:** Wash cucumbers. Cut ¹⁄₁₆ inch off blossom end and discard, but leave ¼ inch of stem attached. Slice or strip cucumbers. Mix 1 cup pickling lime and ⅓ cup salt with 1 gallon water in a 2- to 3-gallon crock or enamelware container. **Caution: Avoid inhaling lime dust while mixing the lime-water solution.** Soak cucumber slices or strips in lime-water solution for 12 to 24 hours, stirring occasionally. Remove from lime solution, rinse, and soak 1 hour in fresh cold water. Repeat the rinsing and soaking two more times. Handle carefully, because slices or strips will be brittle. Drain well.

Process Times for Quick Sweet Pickles in a Boiling-Water Canner*

Style of Pack	Jar Size	Process Time at Altitudes of:		
		0–1,000 ft	1,001-6,000 ft	Above 6,000 ft
Hot	Pints or Quarts	5 minutes	10 minutes	15 minutes
Raw	Pints	10 minutes	15 minutes	20 minutes
	Quarts	15 minutes	20 minutes	25 minutes

*After the process is complete, turn off the heat and remove the canner lid. Wait five minutes before removing jars.

Reduced-Sodium Sliced Sweet Pickles

Whole allspice can be tricky to find. If it's not available at your local grocery store, it can be ordered at www.spice-barn.com or at www.gourmetsleuth.com. This recipe yields about four to five pints.

Ingredients

4 lbs (3- to 4-inch) pickling cucumbers
Canning syrup:
1 ⅔ cups distilled white vinegar (5 percent acetic acid)
3 cups sugar
1 tbsp whole allspice
2 ¼ tsp celery seed
Brining solution:
1 qt distilled white vinegar (5 percent acetic acid)

1 tbsp canning or pickling salt
1 tbsp mustard seed
½ cup sugar

Directions

1. Wash cucumbers and cut ¹⁄₁₆ inch off blossom end, and discard. Cut cucumbers into ¼-inch slices. Combine all ingredients for canning syrup in a saucepan and bring to boiling. Keep syrup hot until used.
2. In a large kettle, mix the ingredients for the brining solution. Add the cut cucumbers, cover, and simmer until the cucumbers change color from bright to dull green (about 5 to 7 minutes). Drain the cucumber slices.
3. Fill jars, and cover with hot canning syrup leaving ½-inch headspace. Adjust lids and process.

Process Times for Reduced-Sodium Sliced Sweet Pickles in a Boiling-Water Canner*

Style of Pack	Jar Size	Process Time at Altitudes of:		
		0–1,000 ft	1,001–6,000 ft	Above 6,000 ft
Hot	Pints	10 minutes	15 minutes	20 minutes

*After the process is complete, turn off the heat and remove the canner lid. Wait five minutes before removing jars.

Tomatoes

Canned tomatoes should be a staple in every cook's pantry. They are easy to prepare and, when made with garden-fresh produce, make ordinary soups, pizza, or pastas into five-star meals. Be sure to select only disease-free, preferably vine-ripened, firm fruit. Do not can tomatoes from dead or frost-killed vines.

Green tomatoes are more acidic than ripened fruit and can be canned safely with the following recommendations.

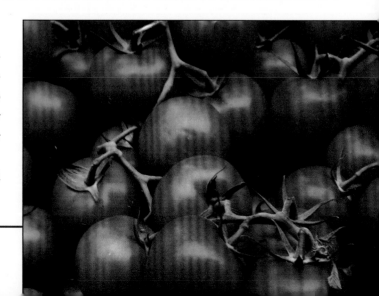

- To ensure safe acidity in whole, crushed, or juiced tomatoes, add two tablespoons of bottled lemon juice or ½ teaspoon of citric acid per quart of tomatoes. For pints, use one tablespoon bottled lemon juice or ¼ teaspoon citric acid.

- Acid can be added directly to the jars before filling with product. Add sugar to offset acid taste, if desired. Four tablespoons of 5 percent acidity vinegar per quart may be used instead of lemon juice or citric acid. However, vinegar may cause undesirable flavor changes.

- Using a pressure canner will result in higher quality and more nutritious canned tomato products. If your pressure canner cannot be operated above 15 PSI, select a process time at a lower pressure.

Tomato Juice

Tomato juice is a good source of vitamin A and C and is tasty on its own or in a cocktail. It's also the secret ingredient in some very delicious cakes. If desired, add carrots, celery, and onions, or toss in a few jalapeños for a little kick.

Directions

1. Wash tomatoes, remove stems, and trim off bruised or discolored portions. To prevent juice from separating,

---

Quantity

- An average of 23 pounds is needed per canner load of seven quarts, or an average of 14 pounds per canner load of nine pints.
- A bushel weighs 53 pounds and yields 15 to 18 quarts of juice—an average of 3 ¼ pounds per quart.

---

quickly cut about 1 pound of fruit into quarters and put directly into saucepan. Heat immediately to boiling while crushing.

2. Continue to slowly add and crush freshly cut tomato quarters to the boiling mixture. Make sure the mixture boils constantly and vigorously while you add the remaining tomatoes. Simmer 5 minutes after you add all pieces.

3. Press heated juice through a sieve or food mill to remove skins and seeds. Add bottled lemon juice or citric acid to jars (see page 135). Heat juice again to boiling.

4. Add 1 teaspoon of salt per quart to the jars, if desired. Fill jars with hot tomato juice, leaving ½-inch headspace. Adjust lids and process.

Process Times for Tomato Juice in a Boiling-Water Canner*

Style of Pack	Jar Size	Process Time at Altitudes of:			
		0–1,000 ft	1,001–3,000 ft	3,001–6,000 ft	Above 6,000 ft
Hot	Pints	35 minutes	40 minutes	45 minutes	50 minutes
	Quarts	40 minutes	45 minutes	50 minutes	55 minutes

*After the process is complete, turn off the heat and remove the canner lid. Wait five minutes before removing jars.

Process Times for Tomato Juice in a Dial-Gauge Pressure Canner*

Style of Pack	Jar Size	Process Time	Canner Gauge Pressure (PSI) at Altitudes of:			
			0–2,000 ft	2,001–4,000 ft	4,001–6,000 ft	6,001–8,000 ft
Hot	Pints or Quarts	20 minutes	6 lbs	7 lbs	8 lbs	9 lbs
		15 minutes	11 lbs	12 lbs	13 lbs	14 lbs

*After the canner is completely depressurized, remove the weight from the vent port or open the petcock. Wait 10 minutes; then unfasten the lid and remove it carefully. Lift the lid with the underside away from you so that the steam coming out of the canner does not burn your face.

Process Times for Tomato Juice in a Weighted-Gauge Pressure Canner*

Style of Pack	Jar Size	Process Time	Canner Gauge Pressure (PSI) at Altitudes of:	
			0–1,000 ft	Above 1,000 ft
Hot	Pints or Quarts	20 minutes	5 lbs	10 lbs
		15 minutes	10 lbs	15 lbs

Crushed Tomatoes with No Added Liquid

Crushed tomatoes are great for use in soups, stews, thick sauces, and casseroles. Simmer crushed tomatoes with kidney beans, chili powder, sautéed onions, and garlic to make an easy pot of chili.

Directions

1. Wash tomatoes and dip in boiling water for 30 to 60 seconds or until skins split. Then dip in cold water, slip off skins, and remove cores. Trim off any bruised or discolored portions and quarter.
2. Heat ⅙ of the quarters quickly in a large pot, crushing them with a wooden mallet or spoon as they are added to the pot. This will exude juice. Continue heating the tomatoes, stirring to prevent burning.
3. Once the tomatoes are boiling, gradually add remaining quartered tomatoes, stirring constantly. These remaining tomatoes do not need to be crushed; they will soften with heating and stirring. Continue until all tomatoes are added. Then boil gently 5 minutes.
4. Add bottled lemon juice or citric acid to jars (see page 135). Add 1 teaspoon of salt per quart to the jars, if desired. Fill jars immediately with hot tomatoes, leaving ½-inch headspace. Adjust lids and process.

> **Quantity**
> - An average of 22 pounds is needed per canner load of seven quarts.
> - An average of 14 fresh pounds is needed per canner load of nine pints.
> - A bushel weighs 53 pounds and yields 17 to 20 quarts of crushed tomatoes—an average of 2 ¾ pounds per quart.

Process Times for Crushed Tomatoes in a Dial-Gauge Pressure Canner*

Style of Pack	Jar Size	Process Time	Canner Gauge Pressure (PSI) at Altitudes of:			
			0–2,000 ft	2,001–4,000 ft	4,001–6,000 ft	6,001–8,000 ft
Hot	Pints or Quarts	20 minutes	6 lbs	7 lbs	8 lbs	9 lbs
		15 minutes	11 lbs	12 lbs	13 lbs	14 lbs

*After the canner is completely depressurized, remove the weight from the vent port or open the petcock. Wait 10 minutes; then unfasten the lid and remove it carefully. Lift the lid with the underside away from you so that the steam coming out of the canner does not burn your face.

Process Times for Crushed Tomatoes in a Weighted-Gauge Pressure Canner*

Style of Pack	Jar Size	Process Time	Canner Gauge Pressure (PSI) at Altitudes of:	
			0–1,000 ft	Above 1,000 ft
Hot	Pints or Quarts	20 minutes	5 lbs	10 lbs
		15 minutes	10 lbs	15 lbs

*After the canner is completely depressurized, remove the weight from the vent port or open the petcock. Wait 10 minutes; then unfasten the lid and remove it carefully. Lift the lid with the underside away from you so that the steam coming out of the canner does not burn your face.

Process Times for Crushed Tomatoes in a Boiling-Water Canner*

Style of Pack	Jar Size	Process Time at Altitudes of:			
		0–1,000 ft	1,001–3,000 ft	3,001–6,000 ft	Above 6,000 ft
Hot	Pints	35 minutes	40 minutes	45 minutes	50 minutes
	Quarts	45 minutes	50 minutes	55 minutes	60 minutes

*After the process is complete, turn off the heat and remove the canner lid. Wait five minutes before removing jars.

Tomato Sauce

This plain tomato sauce can be spiced up before using in soups or in pink or red sauces. The thicker you want your sauce, the more tomatoes you'll need.

Directions

1. Prepare and press as for making tomato juice (see page 211). Simmer in a large saucepan until sauce reaches desired consistency. Boil until volume is reduced by about one-third for thin sauce, or by one-half for thick sauce.

2. Add bottled lemon juice or citric acid to jars (see page 135). Add 1 teaspoon of salt per quart to the jars, if desired. Fill jars, leaving ¼-inch headspace. Adjust lids and process.

Quantity

For thin sauce:

- An average of 35 pounds is needed per canner load of seven quarts.
- An average of 21 pounds is needed per canner load of nine pints.
- A bushel weighs 53 pounds and yields 10 to 12 quarts of sauce—an average of five pounds per quart.

For thick sauce:

- An average of 46 pounds is needed per canner load of seven quarts.
- An average of 28 pounds is needed per canner load of nine pints.
- A bushel weighs 53 pounds and yields seven to nine quarts of sauce—an average of 6 ½ pounds per quart.

Process Times for Tomato Sauce in a Boiling-Water Canner*

Style of Pack	Jar Size	Process Time at Altitudes of:			
		0–1,000 ft	1,001–3,000 ft	3,001–6,000 ft	Above 6,000 ft
Hot	Pints	35 minutes	40 minutes	45 minutes	50 minutes
	Quarts	40 minutes	45 minutes	50 minutes	55 minutes

*After the process is complete, turn off the heat and remove the canner lid. Wait five minutes before removing jars.

Process Times for Tomato Sauce in a Dial-Gauge Pressure Canner*

Style of Pack	Jar Size	Process Time	Canner Gauge Pressure (PSI) at Altitudes of:			
			0–2,000 ft	2,001–4,000 ft	4,001–6,000 ft	6,001–8,000 ft
Hot	Pints or Quarts	20 minutes	6 lbs	7 lbs	8 lbs	9 lbs
		15 minutes	11 lbs	12 lbs	13 lbs	14 lbs

*After the canner is completely depressurized, remove the weight from the vent port or open the petcock. Wait 10 minutes; then unfasten the lid and remove it carefully. Lift the lid with the underside away from you so that the steam coming out of the canner does not burn your face.

Process Times for Tomato Sauce in a Weighted-Gauge Pressure Canner*

| Style of Pack | Jar Size | Process Time | Canner Gauge Pressure (PSI) at Altitudes of: | |
			0–1,000 ft	Above 1,000 ft
Hot	Pints or Quarts	20 minutes	5 lbs	10 lbs
		15 minutes	10 lbs	15 lbs

*After the canner is completely depressurized, remove the weight from the vent port or open the petcock. Wait 10 minutes; then unfasten the lid and remove it carefully. Lift the lid with the underside away from you so that the steam coming out of the canner does not burn your face.

Tomatoes, Whole or Halved, Packed in Water

Whole or halved tomatoes are used for scalloped tomatoes, savory pies (baked in a pastry crust with parmesan cheese, mayonnaise, and seasonings), or stewed tomatoes.

Directions

1. Wash tomatoes. Dip in boiling water for 30 to 60 seconds or until skins split; then dip in cold water. Slip off skins and remove cores. Leave whole or halve.

2. Add bottled lemon juice or citric acid to jars (see page 135). Add 1 teaspoon of salt per quart to the jars, if desired. For hot pack products, add enough water to cover the tomatoes and boil them gently for 5 minutes.

Quantity
- An average of 21 pounds is needed per canner load of seven quarts.
- An average of 13 pounds is needed per canner load of nine pints.
- A bushel weighs 53 pounds and yields 15 to 21 quarts—an average of three pounds per quart.

3. Fill jars with hot tomatoes or with raw peeled tomatoes. Add the hot cooking liquid to the hot pack, or hot water for raw pack to cover, leaving ½-inch headspace. Adjust lids and process.

Process Times for Water-Packed Whole Tomatoes in a Boiling-Water Canner*

| Style of Pack | Jar Size | Process Time at Altitudes of: | | | |
		0–1,000 ft	1,001–3,000 ft	3,001–6,000 ft	Above 6,000 ft
Hot or Raw	Pints	40 minutes	45 minutes	50 minutes	55 minutes
	Quarts	45 minutes	50 minutes	55 minutes	60 minutes

*After the process is complete, turn off the heat and remove the canner lid. Wait five minutes before removing jars.

Process Times for Water-Packed Whole Tomatoes in a Dial-Gauge Pressure Canner*

| Style of Pack | Jar Size | Process Time | Canner Gauge Pressure (PSI) at Altitudes of: | | | |
			0–2,000 ft	2,001–4,000 ft	4,001–6,000 ft	6,001–8,000 ft
Hot or Raw	Pints or Quarts	15 minutes	6 lbs	7 lbs	8 lbs	9 lbs
		10 minutes	11 lbs	12 lbs	13 lbs	14 lbs

*After the canner is completely depressurized, remove the weight from the vent port or open the petcock. Wait 10 minutes; then unfasten the lid and remove it carefully. Lift the lid with the underside away from you so that the steam coming out of the canner does not burn your face.

Process Times for Water-Packed Whole Tomatoes in a Weighted-Gauge Pressure Canner*

Style of Pack	Jar Size	Process Time	Canner Gauge Pressure (PSI) at Altitudes of:	
			0–1,000 ft	Above 1,000 ft
Hot or Raw	Pints or Quarts	15 minutes	5 lbs	10 lbs
		10 minutes	10 lbs	15 lbs

*After the canner is completely depressurized, remove the weight from the vent port or open the petcock. Wait 10 minutes; then unfasten the lid and remove it carefully. Lift the lid with the underside away from you so that the steam coming out of the canner does not burn your face.

Spaghetti Sauce Without Meat

Homemade spaghetti sauce is like a completely different food than store-bought varieties—it tastes fresher, is more flavorful, and is far more nutritious. Adjust spices to taste, but do not increase proportions of onions, peppers, or mushrooms. This recipe yields about nine pints.

Ingredients

30 lbs tomatoes

1 cup chopped onions

5 cloves garlic, minced

1 cup chopped celery or green pepper

1 lb fresh mushrooms, sliced (optional)

4 ½ tsp salt

2 tbsp oregano

4 tbsp minced parsley

2 tsp black pepper

¼ cup brown sugar

¼ cup vegetable oil

Directions

1. Wash tomatoes and dip in boiling water for 30 to 60 seconds or until skins split. Dip in cold water and slip off skins. Remove cores and quarter tomatoes. Boil 20 minutes, uncovered, in large saucepan. Put through food mill or sieve.

2. Sauté onions, garlic, celery or peppers, and mushrooms (if desired) in vegetable oil until tender. Combine sautéed vegetables and tomatoes and add spices, salt, and sugar. Bring to a boil.

3. Simmer uncovered, until thick enough for serving. Stir frequently to avoid burning. Fill jars, leaving 1-inch headspace. Adjust lids and process.

Process Times for Spaghetti Sauce Without Meat in a Dial-Gauge Pressure Canner*

Style of Pack	Jar Size	Process Time	Canner Gauge Pressure (PSI) at Altitudes of:			
			0–2,000 ft	2,001–4,000 ft	4,001–6,000 ft	6,001–8,000 ft
Hot	Pints	20 minutes	11 lbs	12 lbs	13 lbs	14 lbs
	Quarts	25 minutes	11 lbs	12 lbs	13 lbs	14 lbs

*After the canner is completely depressurized, remove the weight from the vent port or open the petcock. Wait 10 minutes; then unfasten the lid and remove it carefully. Lift the lid with the underside away from you so that the steam coming out of the canner does not burn your face.

Process Times for Spaghetti Sauce Without Meat in a Weighted-Gauge Pressure Canner*

Style of Pack	Jar Size	Process Time	Canner Gauge Pressure (PSI) at Altitudes of:	
			0–1,000 ft	Above 1,000 ft
Hot	Pints	20 minutes	10 lbs	15 lbs
	Quarts	25 minutes	10 lbs	15 lbs

*After the canner is completely depressurized, remove the weight from the vent port or open the petcock. Wait 10 minutes; then unfasten the lid and remove it carefully. Lift the lid with the underside away from you so that the steam coming out of the canner does not burn your face.

Tomato Ketchup

Ketchup forms the base of several condiments, including Thousand Island dressing, fry sauce, and barbecue sauce. And, of course, it's an American favorite in its own right. This recipe yields six to seven pints.

Ingredients

24 lbs ripe tomatoes
3 cups chopped onions
¾ tsp ground red pepper (cayenne)
4 tsp whole cloves
3 sticks cinnamon, crushed
1-½ tsp whole allspice
3 tbsp celery seeds
3 cups cider vinegar (5 percent acetic acid)
1-½ cups sugar
¼ cup salt

Directions

1. Wash tomatoes. Dip in boiling water for 30 to 60 seconds or until skins split. Dip in cold water. Slip off skins and remove cores. Quarter tomatoes into 4-gallon stockpot or a large kettle. Add onions and red pepper. Bring to boil and simmer 20 minutes, uncovered.

2. Combine remaining spices in a spice bag and add to vinegar in a 2-quart saucepan. Bring to boil. Turn off heat and let stand until tomato mixture has been cooked 20 minutes. Then, remove spice bag and combine vinegar and tomato mixture. Boil about 30 minutes.

3. Put boiled mixture through a food mill or sieve. Return to pot. Add sugar and salt, boil gently, and stir frequently until volume is reduced by one-half or until mixture rounds up on spoon without separation. Fill pint jars, leaving ⅛-inch headspace. Adjust lids and process.

Process Times for Tomato Ketchup in a Boiling-Water Canner*

		Process Time at Altitudes of:		
Style of Pack	Jar Size	0–1,000 ft	1,001-6,000 ft	Above 6,000 ft
Hot	Pints	15 minutes	20 minutes	25 minutes

*After the process is complete, turn off the heat and remove the canner lid. Wait five minutes before removing jars.

Chile Salsa (Hot Tomato-Pepper Sauce)

For fantastic nachos, cover corn chips with chile salsa, add shredded Monterey jack or cheddar cheese, bake under broiler for about five minutes, and serve with guacamole and sour cream. Be sure to wear rubber gloves while handling chiles or wash hands thoroughly with soap and water before touching your face. This recipe yields six to eight pints.

Ingredients

5 lbs tomatoes
2 lbs chile peppers
1 lb onions
1 cup vinegar (5 percent)
3 tsp salt
½ tsp pepper

Directions

1. Wash and dry chiles. Slit each pepper on its side to allow steam to escape. Peel peppers using one of the following methods:

Oven or broiler method: Place chiles in oven (400°F) or broiler for 6 to 8 minutes until skins blister. Cool and slip off skins.

Range-top method: Cover hot burner, either gas or electric, with heavy wire mesh. Place chiles on burner for several minutes until skins blister. Allow peppers to cool. Place in a pan and cover with a damp cloth. This will make peeling the peppers easier. After several minutes, peel each pepper.

2. Discard seeds and chop peppers. Wash tomatoes and dip in boiling water for 30 to 60 seconds or until skins split. Dip in cold water, slip off skins, and remove cores.

3. Coarsely chop tomatoes and combine chopped peppers, onions, and remaining ingredients in a large saucepan. Heat to boil, and simmer 10 minutes. Fill jars, leaving ½-inch headspace. Adjust lids and process.

Process Times for Chile Salsa in a Boiling-Water Canner*

		Process Time at Altitudes of:		
Style of Pack	Jar Size	0–1,000 ft	1,001–6,000 ft	Above 6,000 ft
Hot	Pints	15 minutes	20 minutes	25 minutes

*After the process is complete, turn off the heat and remove the canner lid. Wait five minutes before removing jars.

CHEESE

There are endless varieties of cheese you can make, but they all fall into two main categories: soft and hard. Soft cheeses (like cream cheese) are easier to make because they don't require a cheese press. The curds in hard cheeses (like cheddar) are pressed together to form a solid block or wheel, which requires more time and effort, but hard cheeses will keep longer than soft cheeses, and generally have a much stronger flavor.

Cheese is basically curdled milk and is made by adding an enzyme (typically rennet) to milk, allowing curds to form, heating the mixture, straining out the whey, and finally pressing the curds together. Cheeses such as *queso fresco* or *queso blanco* (traditionally eaten in Latin American countries) and *paneer* (traditionally eaten in India), are made with an acid such as vinegar or lemon juice instead of bacterial cultures or rennet.

You can use any kind of milk to make cheese, including cow's milk, goat's milk, sheep's milk, and even buffalo's milk (used for traditional mozzarella). For the richest flavor, try to get raw milk from a local farmer. If you don't know of one near you, visit www.realmilk.com/where.html for a listing of raw milk suppliers in your state. You can use homogenized milk, but it will produce weaker curds and a milder flavor. If your milk is pasteurized, you'll need to "ripen" it by heating it in a double boiler until it reaches 86°F and then adding 1 cup of unpasteurized, preservative-free cultured buttermilk per gallon of milk and letting it stand 30 minutes to three hours (the longer you leave it, the sharper the flavor will be). If you cannot find unpasteurized buttermilk, diluting ⅛ teaspoon calcium chloride (available from online cheesemaker suppliers) in ¼ cup of water and adding it to your milk will create a similar effect.

Rennet (also called rennin or chymosin) is sold online at cheesemaking sites in tablet or liquid form. You may also be able to find Junket rennet tablets near the pudding and gelatin in your grocery store. One teaspoon of liquid rennet is the equivalent of one rennet tablet, which is enough to turn 5 gallons of milk into cheese (estimate four drops of liquid rennet per gallon of milk). Microbial rennet is a vegetarian alternative that is available for purchase online.

Preparation

It's important to keep your hands clean and all equipment sterile when making cheese.

1. Wash hands and all equipment with soapy detergent before and after use.
2. Rinse all equipment with clean water, removing all soapy residue.
3. Boil all cheesemaking equipment between uses.
4. For best-quality cheese, use new cheesecloth each time you make cheese. (Sterilize cheesecloth by first washing, then boiling.)
5. Squeaky clean is clean. If you can feel a residue on the equipment, it is not clean.

Yogurt Cheese

This soft cheese has a flavor similar to sour cream and a texture like cream cheese. A pint of yogurt will yield approximately ¼ pound of cheese. The yogurt cheese has a shelf life of approximately 7 to 14 days when wrapped and placed in the refrigerator and kept at less than 40°F. Add a little salt and pepper and chopped fresh herbs for variety.

Plain, whole-milk yogurt

1. Line a large strainer or colander with cheesecloth.
2. Place the lined strainer over a bowl and pour in the yogurt. Do not use yogurt made with the addition of gelatin, as gelatin will inhibit whey separation.
3. Let yogurt drain overnight, covered with plastic wrap. Empty the whey from the bowl.
4. Fill a strong plastic storage bag with some water, seal, and place over the cheese to weigh it down. Let the cheese stand another 8 hours and then enjoy!

Queso Blanco

Queso blanco is a white, semi-hard cheese made without culture or rennet. It is eaten fresh and may be flavored with peppers, herbs, and spices. It is considered a "frying cheese," meaning it does not melt and may be deep-fried or grilled. *Queso blanco* is best eaten fresh, so try this small recipe the first time you make it. If it disappears quickly, next time double or triple the recipe. This recipe will yield about ½ cup of cheese.

2 cups milk
4 teaspoons white vinegar
Salt
Minced jalapeño, black pepper, chives, or other herbs to taste

1. Heat milk to 176°F for 20 minutes.
2. Add vinegar slowly to the hot milk until the whey is semi-clear and the curd particles begin to form stretchy clumps. Stir for 5 to 10 minutes. When it's ready you should be able to stretch a piece of curd about ⅓ inch before it breaks.
3. Allow to cool, and strain off the whey by filtering through a cheesecloth-lined colander or a cloth bag.
4. Work in salt and spices to taste.
5. Press the curd in a mold or simply leave in a ball.
6. *Queso blanco* may keep for several weeks if stored in a refrigerator, but is best eaten fresh.

Ricotta Cheese

Making ricotta is very similar to making *queso blanco*, though it takes a bit longer. Start the cheese in the morning for use at dinner, or make a day ahead. Use it in lasagna, in desserts, or all on its own.

1 gallon milk
¼ teaspoon salt
⅓ cup plus 1 teaspoon white vinegar

1. Pour milk into a large pot, add salt, and heat slowly while stirring until the milk reaches 180°F.
2. Remove from heat and add vinegar. Stir for 1 minute as curds begin to form.
3. Cover and allow to sit undisturbed for 2 hours.
4. Pour mixture into a colander lined with cheesecloth, and allow to drain for two or more hours.
5. Store in a sealed container for up to a week.

Mozzarella

This mild cheese will make your homemade pizza especially delicious. Or slice it and eat with fresh tomatoes and basil from the garden. Fresh cheese can be stored in salt water but must be eaten within two days.

1 gallon 2 percent milk
¼ cup fresh, plain yogurt (see recipe on page 294)
One tablet rennet or 1 teaspoon liquid rennet dissolved in ½ cup tap water
Brine: use 2 pounds of salt per gallon of water

1. Heat milk to 90°F and add yogurt. Stir slowly for 15 minutes while keeping the temperature constant.
2. Add rennet mixture and stir for 3 to 5 minutes.
3. Cover, remove from heat, and allow to stand until coagulated, about 30 minutes.
4. Cut curd into ½-inch cubes. Allow to stand for 15 minutes with occasional stirring.
5. Return to heat and slowly increase temperature to 118°F over a period of 45 minutes. Hold this temperature for an additional 15 minutes.
6. Drain off the whey by transferring the mixture to a cheesecloth-lined colander. Use a spoon to press the liquid out of the curds. Transfer the mat of curd to a flat pan that can be kept warm in a low oven. Do not cut mat, but turn it over every 15 minutes for a 2-hour period. Mat should be tight when finished.
7. Cut the mat into long strips 1 to 2 inches wide and place in hot water (180°F). Using wooden spoons, tumble and stretch it under water until it becomes elastic, about 15 minutes.
8. Remove curd from hot water and shape it by hand into a ball or a loaf, kneading in the salt. Place cheese in cold water (40°F) for approximately 1 hour.
9. Store in a solution of 2 teaspoons salt to 1 cup water.

Cheddar Cheese

Cheddar is a New England and Wisconsin favorite. The longer you age it, the sharper the flavor will be. Try a slice with a wedge of homemade apple pie.

Ingredients

1 gallon milk
¼ cup buttermilk
1 tablet rennet, or 1 teaspoon liquid rennet
1 ½ teaspoon salt

Directions

1. Combine milk and buttermilk and allow the mixture to ripen overnight.
2. The next day, heat milk to 90ºF in a double boiler and add rennet.
3. After about 45 minutes, cut curds into small cubes and let sit 15 minutes.
4. Heat very slowly to 100ºF and cook for about an hour or until a cooled piece of curd will keep its shape when squeezed.
5. Drain curds and rinse out the double boiler.
6. Place a rack lined with cheesecloth inside the double boiler and spread the curds on the cloth. Cover and reheat at about 98° 30 to 40 minutes. The curds will become one solid mass.
7. Remove the curds, cut them into 1-inch wide strips, and return them to the pan. Turn the strips every 15 to 20 minutes for one hour.
8. Cut the strips into cubes and mix in salt.

9. Let the curds stand for 10 minutes, place them in cheesecloth, and press in a cheese press with 15 pounds for 10 minutes, then with 30 pounds for an hour.

10. Remove the cheese from the press, unwrap it, dip in warm water, and fill in any cracks.

11. Wrap again in cheesecloth and press with 40 pounds for 24 hours.

12. Remove from the press and let the cheese dry about five days in a cool, well-ventilated area, turning the cheese twice a day and wiping it with a clean cloth. When a hard skin has formed, rub with oil or seal with wax. You can eat the cheese after six weeks, but for the strongest flavor, allow cheese to age for six months or more.

Make Your Own Simple Cheese Press

1. Remove both ends of a large coffee can or thoroughly cleaned paint can, saving one end. Use an awl or a hammer and long nail to pierce the sides in several places, piercing from the inside out.

2. Place the can on a cooling rack inside a larger basin. Leave the bottom of the can in place.

3. Use a saw to cut a ¾-inch-thick circle of wood to create a "cheese follower." It should be small enough in diameter to fit easily in the can.

4. Place cheese curds in the can, and top with the cheese follower. Place several bricks wrapped in cloth or foil on top of the cheese follower to weigh down curds.

5. Once the cheese is fully pressed, remove the bricks and bottom of the can. Use the cheese follower to push the cheese out of the can.

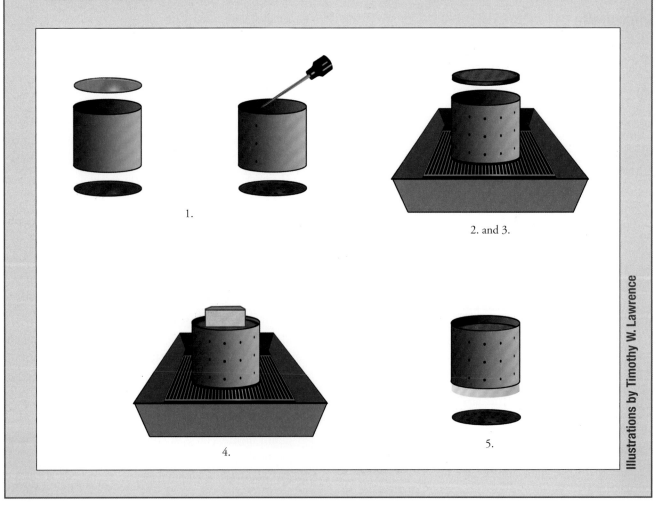

1.

2. and 3.

4.

5.

Illustrations by Timothy W. Lawrence

Homemade Crackers

Crackers are very easy to make and can be varied end-lessly by adding seasonings of your choice. Try sprinkling a coarse sea salt and dried orega-no or cinnamon and sugar over the crackers just before baking. Serve with homemade cheese.

1 ½ cups all-purpose flour
1 ½ cups whole wheat flour
1 teaspoon salt
1 cup warm water
⅓ cup olive oil
Herbs, seeds, spices, or coarse sea salt as desired

1. Stir together the dry ingre-dients in a mixing bowl. Add the water and olive oil and knead until dough is elastic and not too sticky (about 5 minutes in an electric mixer with a dough attachment or 10 minutes by hand).
2. Allow dough to rest at room temperature for about half an hour. Preheat oven to 450°F.
3. Flour a clean, dry surface and roll dough to about ⅛ inch thick. Cut into squares or shapes (using a cookie cutter) and place on a cookie sheet. Sprin-kle with desired topping. Bake for about 5 minutes or until crackers are golden brown.

CIDER

There's little more refreshing than a glass of cold, crisp hard cider. And when you make your own, you have the ability to control the flavor and strength of your cider to create a brew that's uniquely yours.

Apples and Fermentables

Not every apple is good for cider, and very rarely will you receive a blend with all the characteristics you desire from using one variety of apple. Cider makers generally use a mix of apples, choosing bland ones as the base since they generously provide plenty of juice and fermentables. Tart apples are often added to balance the sweetness of other sugars involved in the process of brewing, while sweet apples compensate for the acidity of the base apple.

There are four basic types of apples used to make ciders:

Sweet: Baldwin, Cortland, Delicious, Rome Beauty

Acidic: (medium to high acidity) Northern Spy, Winesap, Greening, and Pippin

Aromatic: MacIntosh and Russet

Astringent: wild and crabapples

If you've got all of these types at your disposal, make your first brew a blend of 50 percent sweet, 35 percent acidic, 10 percent aromatic, and 5 percent astringent. After you've made and tested one batch, you can vary the flavor to your liking by experimenting with different ratios.

A lot depends on how hard you want your cider. If your goal is a sweet, light cider, use only apples. If you're in the mood for a stronger one, you'll need to add additional fermentables, like cane and corn sugar, which are cheap, or honey and maple sugar, which offer more flavor and body to the cider (If you end up adding the latter two, it's helpful to add some sort of yeast nutrient that will improve the efficiency of fermentation.). Adding adjunct sugars will change the amount of fermentables, and they must be accounted for when you're scheduling the fermentation process. How many you add depends on the goal taste: for a sweet and more full-bodied taste, use honey and don't let it ferment all the way out; for a dryer taste, use cane or corn sugar; and you can achieve a "dark" cider by adding brown sugar or molasses.

You can add sulfite to the "must" in order to reduce unwanted bacteria and wild yeast. One or two Campden tablets per gallon are sufficient. Afterward, make sure to aerate

the "must" for at least 24 hours before pitching the yeast. Champagne yeast is the best option, especially if your goal is a sparkling cider. It's also a good idea to add a bit of acid to increase the tartness; do so by throwing in a quarter to a half teaspoon of citric acid or winemaker's acid.

How To Brew Hard Cider

You can buy a grinder (starting at around $450) or simply wash, cut, and grind your apples with a food processor and/ or juicer. To do the latter, wash your apples outside with a big bucket and a garden hose. Make sure to remove any leftover leaves or branches sticking to the apple and to discard apples that are brown or beginning to rot. The simple rule of, "if you wouldn't eat it, you shouldn't drink it" sums it up nicely.

Chop large apples into small pieces, removing any bad sections as you go. Once chopped, pour the bits into the bowl of your processor, screw the lid on tightly, and turn the power on. Do this in smaller sections so as to not overload the processor and don't worry that the apples are turning brown due to the quick oxidization.

Bring the pulp to the press (often a basket screw press— but even if that isn't the case, there will be a cheesecloth or other straining material that will line the press, retaining the solid and allowing the liquid to run through) and put a small amount of pulp in, a few inches deep at most. While this may seem excessively careful, it will reward you with more juice; with a larger amount of pulp, the juice in the center cannot easily run out through the compressed pulp surrounding it.

Use an empty container to collect the juice. Tighten the screw and wait in between each tightening to maximize the amount of juice collected. Add sugars and other fermentables, if desired, and the "must" is ready for fermentation.

Closed fermentation is recommended, and the equipment you'll need is the same as for brewing beer: plastic buckets, glass carboys, and airlocks. After putting it in the container and locking it, the cider can work for 3-4 months. Cider improves with aging, but it's more difficult to achieve sparkling cider if the yeast has mostly dropped out. The 4-month-old cider is then primed with corn sugar (⅞ of a cup for 5 gallons) and bottled in champagne bottles (or otherwise), usually corked rather than capped. If you opt for beer bottles, choose the ones that have caps that do not screw off—the process of fermenting can blow those off. The bottled cider is stowed away to age and condition for at least six months at room temperature.

COOKING

Cooking is a very personal and creative act. Every cook has their own ideas of how to peel a potato, the best way to chop herbs, how long a steak should cook, when pasta is "done," what goes into tomato sauce . . . Chances are, if you've ever cooked with another person, you're well aware of the wide differences between one cook's methods and another's. Thus, rather than offering you much in the way of techniques or instructions, I'm simply offering a variety of classic recipes for your inspiration. Modify, modernize, or do with them whatever you will—cook's have been preparing many of these dishes for their families for centuries, and you'd be hard-pressed to find two who do any one dish exactly alike.

Breakfast

Eggs

POACHED EGGS

Use a skillet, or muffin-rings placed in a pan of water, not too deep. The water should barely cover the eggs. Bring the water to the boiling point, drop in the eggs carefully, one at a time, and remove from the fire immediately. Cover the pan and let stand until cooked. A teaspoon of lemon-juice or vinegar in the water will keep the whites firm and preserve the shape of the eggs. Poached eggs are usually served on thin slices of buttered toast. Take up with a skimmer and let drain thoroughly before placing on the toast. Sprinkle with salt and pepper. As every other writer who has given directions for poaching eggs has said that "the beauty of a poached egg is for the yolk to be seen blushing through the veiled white," the author of this book will make no allusion to it.

SCRAMBLED EGGS

> 2 Tbs Butter
> 6 eggs
> Salt and pepper, to taste

Put the butter into a frying pan. When it sizzles, break into it with the eggs and mix thoroughly with a silver spoon for 2 minutes without stopping. Season with salt and pepper and a slight grating of nutmeg, if desired. Scrambled eggs should be thick and creamy. Alternatively, beat the eggs thoroughly before putting into the pan, adding one teaspoon of water or milk for each egg you use. The cooking process is the same.

CHEESE SCRAMBLE

½ cup grated cheese

6 eggs, well-beaten

1 tsp water or milk for each egg

Cook as you would for scrambled eggs, but mix the cheese with the eggs before cooking. Add fresh chopped herbs, if desired.

EGGS A LA PAYSANNE

½ cup cream

6 eggs

Salt and pepper, to taste

Dash of sweet green pepper

Dash of parsley, minced

Put the cream into a baking dish, and break the eggs into it. Place the dish in the oven until the eggs are set. Sprinkle with the seasonings.

SPANISH EGGS

1 cup tomato, stewed and strained

1 garlic clove, finely minced

1 onion, chopped

2 sweet green peppers, seeded and chopped

1 egg

Slices of toast

Cook the tomatoes, garlic, onion, and peppers gently, until reduced one half. Spread on thin slices of toast and lay a fried or poached egg on each slice.

DEVILED EGGS

10 hard-boiled eggs

3 Tbs vinegar

½ tsp salt

½ tsp pepper

Pinch sugar (if desired)

Take the hard-boiled eggs and cut each in half, removing the yolks from the whites. Place all yolks in a mixing bowl and mash them. Add the vinegar, salt, pepper, and sugar (if desired). Put the mixture in the whites.

BREAKFAST OMELET

1 cup milk

1 cup bread crumbs

6 eggs

Pinch salt

Pinch pepper

Bring the milk to a boil. Once it is boiling, pour it into a deep dish, stirring in the breadcrumbs as it cooks. Add eggs and stir them in, but do not beat them. Add the salt and pepper, and fry in a hot, well-buttered frying pan over medium-high heat.

SPANISH OMELET

1 slice ham, diced

1 chili pepper, diced

1 small onion, diced

6 eggs

Dice the ham, pepper, and onion and sauté in a frying pan. Cook until the onions are brown, stirring constantly. Add the eggs and let the mixture set. Fold up as you would for a plain omelet.

OYSTER OMELET

8 oysters, chopped fine

6 eggs

1 cup flour

¼ cup milk

Salt and pepper, to taste

Beat the eggs lightly and add the oysters and the flour, which must be mixed to a paste with the milk. Season with the salt and pepper and fry it in hot butter, without turning it. As soon as it is done, slip it on a dish and serve it hot.

EGGS AU BECHAMEL

1 Tbs butter

½ cup thin cream

3 hardboiled eggs

1 Tbs flour

½ cup chicken broth

Salt and pepper, to taste

Mix butter and flour together until smooth. Add the cream, chicken broth, and seasoning, cook until thick, and pour the sauce over brown toast and then the hard boiled eggs, which have been put through a colander, and then the minced bacon on top. Serve hot.

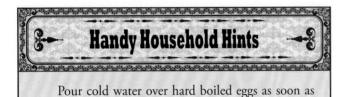

Handy Household Hints

Pour cold water over hard boiled eggs as soon as taken from the kettle, and they will not be discolored.

HAM OMELET

1½ Tbs corn starch
1 Tbs butter
½ cup minced ham
1 cup milk
3 eggs

Stir corn starch in the milk. Add the beaten eggs and ham. Beat all together. Melt butter in an omelet pan and cook. Serve hot.

Breakfast Favorites

FRENCH TOAST

Bread slices
2 eggs
1 cup milk
1 Tbs melted butter
Grated lemon or orange peel

Make the batter, beating the eggs well. Dip the trimmed slices of bread in the batter, and fry in butter until brown.

MRS. HARDING'S WAFFLES

2 eggs
2 Tbs sugar
2 Tbs butter
1 tsp salt

1 pt. milk
Flour to make thin batter
2 Tbs baking powder

Beat the yolks of the eggs. Melt the butter. Add sugar, salt, and the melted butter, then the milk and the flour. Just before ready to bake, add the beaten egg whites and baking powder. Bake on a hot waffle iron.

RICE WAFFLES

1¾ cup flour
⅔ cup cold, cooked rice broken up with a fork
1½ cup warm milk
1 egg
2 Tbs sugar
4 tsp Royal Baking powder
1 Tbs melted butter

Mix and sift dry ingredients; add rice, milk, well-beaten egg yolk, butter, and egg white beaten stiff.

OATMEAL

1 cup oats or oatmeal
½ tsp salt
3 cup boiling water

Stir oats and salt gradually in boiling water. If cooked over direct heat, cook slowly, (stirring occasionally) about 40 minutes, or cook in a double boiler for 1-2 hours. It may be placed in a fireless cook overnight.

CREAM OF WHEAT

½ cup cream of wheat
½ tsp salt
2½ cup water

Add salt to boiling water, stir in cream of wheat, and continue stirring for a few moments to prevent lumping. Cook slowly over direct heat for 20 minutes or more or in a double boiler for 1 hour or more. It can be placed in a fireless cooker for 3 or 4 hours. Serve with cream and sugar.

Salads

Dressings

MAYONNAISE DRESSING

2 egg yolks

1 tsp mustard

1 tsp powdered sugar

1 tsp salt

¼ tsp cayenne pepper

1 pt. olive oil

¼ cup vinegar

¼ cup lemon juice

Add the dry ingredients to the yolks of the eggs, and beat well before adding the oil. Add the oil slowly while beating vigorously. Once the dressing is thick, thin it with vinegar, then alternate adding oil and vinegar, and lastly, add the lemon juice. Less oil can be used if less dressing is required. The eggs can make a foundation for almost any quantity of mayonnaise.

FRENCH DRESSING

1 Tbs vinegar

3 Tbs olive oil

½ tsp salt

1 saltspoon pepper

Put the salt and pepper in a bowl and gradually add the oil, mixing until the salt is thoroughly dissolved. Then, slowly add the vinegar, and stir for one minute. Keep the mixture cold.

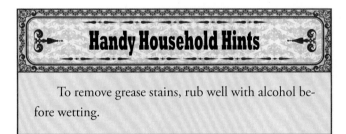

Handy Household Hints

To remove grease stains, rub well with alcohol before wetting.

THOUSAND ISLAND DRESSING

1 bottle small pickled onions

1 box pimentos

8 hard-boiled eggs

Paprika

Salt

Cayenne pepper

Grind eggs in meat chopper, slice pimentos, and finely cut up the onions. Mix them all together, adding salt, paprika, and cayenne to taste.

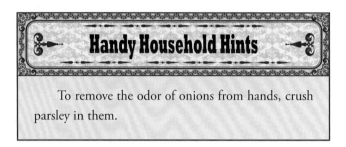

Handy Household Hints

To remove the odor of onions from hands, crush parsley in them.

Leafy Salads

WALDORF SALAD

1 head of celery, sliced

3 apples, sliced

½ cup chopped walnuts

Lettuce

Slice the celery and the apples and mix with the walnuts. Arrange over lettuce leaves and pour any dressing you desire over it.

Other Salads

CHICKEN SALAD

Chicken, cooked and chopped (white meat, preferably)

Oil and vinegar

Salt and pepper, to taste

Celery, diced

Mayonnaise dressing (above)

Cut the cooked chicken into small pieces and let marinate in a bowl with the oil and vinegar: 3 parts vinegar to 1 part oil seasoned with salt and pepper. Let it marinate for two hours, and once done, pour off anything that has not been absorbed. Mix the chicken with equal parts of diced celery and then stir it together with the mayonnaise dressing. Garnish with a sliced pickle, olives, capers, anything.

Waldorf Salad

FRENCH POTATO SALAD

Cold potatoes, sliced thin
1 small onion, chopped
Salt and pepper, to taste
1 cup sour cream
¼ cup vinegar

Prepare the potatoes and onion and mix well together and season. Combine the sour cream and vinegar, stirring well, and pour over the potatoes. Don't allow the potatoes to get too moist.

POTATO SALAD

12 large white potatoes
2 large white onions, chopped
4 boiled eggs, chopped
2 Tbs sugar
1 Tbs salt
2 Tbs Coleman's ground mustard, English imported
1 Tbs celery seed
4 Tbs imported olive oil
2 cup cider vinegar

Boil the potatoes, then take out of the pot and let cool. Once cool, chop them up and add them to the chopped onions and boiled eggs. Add the rest of the ingredients and stir well.

LOBSTER SALAD

1 lg. lobster
3 Tbs French mustard, OR
2 tsp common mixed mustard
½ cup vinegar
½ cup oil
5 hard-boiled eggs
Salt, to taste
1 tsp cayenne pepper
The inside leaves of 2 heads of cabbage lettuce

Cut the lobster and the lettuce into small pieces. Boil the eggs, cut them open, and remove the yolks with a spoon. Mash them with a spoon and add enough oil to them to create a smooth paste, and then add the vinegar, mustard,

pepper, and salt. Mix that thoroughly with the lobster and lettuce, and serve before the salad wilts.

Handy Household Hints

The gaily-colored nasturtium offers delights to the palate. It may be used as a filling for sandwiches or mixed judiciously with other materials in salad. Its delicious pungency appeals to the epicure, while physicians say it aids digestion.

CLAM SALAD

Small Little Neck clams
Head of lettuce
Lemon juice
White or red pepper
Mayonnaise or Tabasco sauce

The very small neck clams may be served in a salad. Put them on crisp lettuce. Season with a little lemon juice and white or red pepper. Mayonnaise or Tabasco sauce can be used.

JELLIED APPLE SALAD

1 cup apples, diced
¼ cup celery, diced
¼ cup figs, chopped
2 Tbs gelatin
¼ cup cold water
1½ cup boiling water
2 Tbs sugar
⅓ cup lemon juice
¼ tsp salt
2 red pimentos, cut in strips

Mix ingredients with mayonnaise. Soften gelatin with cold water. Dissolve with boiling water. Add the ingredients to the gelatin mixture. Pour into mold. When congealed, serve with Mayonnaise.

FRUIT SALAD

- 1 can pineapple
- 1 grapefruit
- 1 can white cherries
- ½ box gelatin
- 1 stalk celery
- 1 cup pecan meats
- 2 lemons (juice)

Cut celery, pineapple, and grapefruit in pieces. Add nuts chopped fine, cherries stoned, and the juice of the lemons. Use juice of pineapple and cherries. Put gelatin ½ cup cold water, dissolve by setting cup in hot water. Stir into fruit and congeal. Serve with mayonnaise.

PIMENTO MOUSSE SALAD

- 1 pt. cream
- 2 pimentos
- 1 cup cold water
- 1 package gelatin
- Red pepper
- Salt
- Grated cheese

Add gelatin to a cup of cold water and let stand for half an hour. Dissolve by setting in a pan of boiling water. Whip the cream stiff, add the pimentos, finely mashed, and add salt and red pepper to taste. Stir dissolved gelatin in mixture and set on ice until congealed. Serve with grated cheese and mayonnaise.

CREAM DRESSING

- ¼ Tbs salt
- ½ Tbs mustard
- ¾ Tbs sugar
- 1 egg, slightly beaten
- 2½ Tbs melted butter
- ¾ cup cream
- ¼ cup vinegar

Mix ingredients in the order listed, adding very slowly. Cook over boiling water, stirring constantly until the mixture thickens. Strain and cool.

MARSHMALLOW SALAD

- ½ lb. marshmallows
- 1 lb. almonds
- 1 lb. can pineapple
- 3 egg yolks
- ½ pt. cream
- ½ tsp salt
- 2½ Tbs vinegar
- ½ Tbs butter
- 1 Tbs sugar

Blanch almonds; cut in small pieces. Cut marshmallows in quarters. Cut pineapple in small pieces. Mix together in bowl and let stand. Beat yolks of eggs lightly and add salt, vinegar, butter, and sugar; boil till thick, stirring the entire time. Remove from heat, and when almost cold, stir in the bowl of prepared marshmallows, almonds, and pineapple. Put in cold place, and just before serving, fold cream (previously whipped and on ice) in and serve on crisp lettuce leaves.

Soups

Light Soups

CREAM OF CHICKEN SOUP

- 1 Tbs butter
- 1 Tbs flour
- 2 qt. chicken stock
- 1 egg, beaten
- 1 pt. cream

Brown the butter and flour in a pot. Add the chicken stock, and boil. Beat the egg, and then add the cream to it. Gradually add the mixture to the stock. Don't boil after adding the egg and cream.

CORN SOUP

- 1 qt. corn, fresh or canned (if from cob, scrape off any sweet bits)
- 1 qt. hot water
- 2 Tbs butter
- 1 Tbs flour
- 1 pt. milk, boiled

1 cup cream

Cayenne pepper and salt, to taste

To the corn, add the hot water. Boil for an hour or longer. Once finished, strain it with a colander. Put butter into a saucepan, and once it has melted, add the flour. Cook for a moment, stirring constantly, and then gradually add the leftover corn pulp. Season the mixture, and then add the milk and cream.

PEA SOUP

1 can of peas

1 qt. chicken stock

1 cup cream or milk

2 Tbs butter

2 Tbs flour

1 onion, chopped

Salt and pepper, to taste

Cook the onion, peas, and stock together for 20 minutes. Remove the onion and put the peas and stock through a sieve. Return it to the stew pan and let it simmer for 10 more minutes. Mix the butter and flour to a cream, and gradually add a ½ C of the soup to the cream. Pour the mixture into the stew pan. Add the salt and pepper and the cream. Boil for 3 minutes, and then serve.

CREAMY TOMATO SOUP

½ can tomatoes

¼ small onion, chopped

1 bay leaf

1 cardamom seed

⅛ tsp cayenne

1 tsp salt

1 Tbs cornstarch thickened in 1 pt. hot milk

Combine all the ingredients and let it stand for half an hour. Then, boil for 10 minutes, strain, and add the cornstarch mixture.

FRENCH SOUP

1 qt. tomatoes

1 onion, sliced

2 potatoes, sliced

1 Tbs butter

Salt and pepper, to taste

2 cup hot water

1 tsp Wyandotte soda

1 cup milk, boiled

1 tsp flour

1 cup canned corn

Slice the onions and potatoes and place them and the rest of the ingredients in a pot. Let simmer for a few hours, covered. Boil the milk and add the flour and canned corn. After the few hours of simmering, add the milk mixture and the Wyandotte soda to the soup. If, at any point, the soup boils away, add hot water.

Heavy Soups and Stews

SAVORY STEW

1½ lb. beef or mutton

1 qt. cold water

1 small head of celery

1 carrot

1 small onion

¼ cup of pearled barley

1 turnip

2 bay leaves

3-4 potatoes, chopped

Cut the meat into bite-size portions, and put it in a double-boiler with the cold water. Put it over low heat and cook for eight hours without allowing it to boil. After six hours, add the vegetables and seasoning, and the resulting stew will be a very nutritious dinner.

Chicken Soup

GUMBO

¼ lb sliced ham, diced

1 clove of garlic, chopped

1 small onion, chopped

Fresh red pepper or cayenne pepper, to taste

3 cup tomatoes

1 pt. oyster (shrimp or crabs can be substituted)

Season with fresh savory herbs in summer or dried herbs in winter

Sauté the ham, garlic, onion, and pepper in butter or oil. Once the onions have browned and the ham is cooked, add the tomatoes and let it simmer. Strain out the liquid, and just before serving, add the oysters to the mixture. Season with the herbs and serve.

CHICKEN SOUP

A large chicken

Sweet herbs, chopped

2 onions, chopped

2 stalks celery, chopped

2 carrots, chopped

2 potatoes, chopped

1 bell pepper (any color), chopped

Parsley, finely minced

Salt and pepper, to taste

Wash the chicken, put it in a pot, and cover it with water and a little salt. Bring it to a boil and add the herbs and vegetables. When the chicken is tender, add a small bunch of finely minced parsley. Let it boil a few minutes, and then serve. Season with salt and pepper as needed.

*Noodles, dumplings, or rice may be susbtituted for potatoes, or used in addition.

FRENCH STEW

2 lbs. beef

6 onions, chopped

2 cup water

2 tomatoes, chopped

Salt and pepper, to taste

1 oz. butter, possibly rolled in flour

Flour, to thicken

Cut the beef into 1-inch squares, and put the meat in a stew-pan with the onions and tomatoes. Season it with salt and pepper and add a little flour. Add the ingredients in layers, alternating between the meat and vegetable combination and the seasonings. Once all the ingredients are in, add the water and let it simmer until the beef is tender. Halfway through, if the gravy seems thin, add the butter rolled in flour, but if the stew seems thick enough, add the butter without the flour. If the gravy isn't thick enough at the end, add flour mixed with cold water to thicken it.

Handy Household Hints

To Prevent Lumps in Gravy: Gravy, soups and thickenings of any sort will not be lumpy if the salt is mixed with the flour before wetting. Stir with a fork instead of a spoon. Better yet, use a small egg beater and the sauce will be perfectly smooth within two minutes.

LAMB STEW WITH ONIONS

Onions, peeled and sliced

Lamb chops

Salt and pepper, to taste

Butter rolled in flour

Peel and slice the onions, and place them in your stew-pan. Cut off the ends of the chops, pound them, and lay them in the pan with the onions and some salt and pepper. Pour in as much water as will cook them. Let them stew slowly until they are tender, then add the butter rolled in flour to thicken the gravy.

Setting Table and Serving Meals

Lay the tablecloth with the crease exactly down the middle of the table. Arrange plates, right side up, at equal distances around the table, one inch from edge of table. Place knives at right of plates, sharp edge toward plates, with handle ends one inch from edge of table. Put forks at left of plates, tines up, one inch from table edge. Place spoons at right of knives. Place napkin, neatly folded, at left of forks. Place tumblers at tip of knives, butter plates at tip of forks.

Arrange neatly inside of the place settings the steadies (the usual dinner accompaniments), the salt and pepper, vinegar and any oils needed, sugar and cream, milk and water.

Place coffee pot and teapot at right of hostess with cups and saucers before her.

Be sure that everything is on the table before beginning to serve meals.

If possible have some little decoration for the center of the table, either a bunch of flowers in season or a little green plant.

All hot food should be served in hot dishes. All dishes should be offered at the left of the guest, if the guest is to help themselves. Dishes placed for the guest must be placed from the right side.

Remove dishes from the right of the guest. Never reach across the guest to place or remove any dish.

If a dessert is served, remove everything from the previous courses before serving the dessert.

After the meal have a time for rest and pleasant intercourse. Cleanliness, good taste, well-cooked food, and pleasant manners will greatly aid digestion.

Sides

Vegetables

BOILED POTATOES

10 potatoes
1 oz. butter
¼ cup or cream

Wash the potatoes, and boil them in water with a little salt. Once they are soft, peel them and serve. Alternatively, once they're peeled, mash them, adding butter, milk or cream, and more salt, if necessary.

FRIED SWEET POTATOES

Boil some sweet potatoes until they are soft enough to pass the prong of a fork through them. Peel them, and when they get cold, slice them. Season with salt and pepper, dredge flour over, and fry them in hot oil or butter until light brown.

ROASTED POTATOES

Wash them, and put them in a pan in a moderate oven. When they can be easily pierced by a fork, they are cooked all the way through. Serve them with the skins on.

POTATO CAKES

6 potatoes
3 hard-boiled eggs, finely chopped
Salt and pepper, to taste
1 Tbs milk or cream

Boil the potatoes, and then mash them. Add to it the three hard-boiled eggs, salt and pepper, and the milk or cream. Form them into small cakes, flour them on both sides, and fry them in a skillet until they are a delicate brown.

Handy Household Hints

To Prevent Rust in Deep Pans of Any Sort: Put a few drops of olive or cottonseed oil in them after they have been used and brush this along the seams with a small brush, so that every part of the surface is washed with the oil.

POTATO KALE

6 potatoes
Half a head of cabbage
2 oz. butter
½ cup cream
Salt and pepper, to taste

Boil the cabbage with a little salt in the water. When it is nearly done, pare the potatoes and drop them in with the cabbage. Once the potatoes are soft, take them out—drain the cabbage—wipe the pot they were in, and put the potatoes and cabbage back in. Mash them both very fine, adding the butter, cream, and salt and pepper. Set the pot over fire and stir it until the potatoes are hot.

STEWED TOMATOES

If they are not very ripe, pour boiling water over them and let them stand a few minutes; the skin will peel off very easily. Then, cut them up and put them in a pan without any water. Cook them until they are soft. If they prove too juicy, dip some of the water out and mash them fine. Season with butter, cayenne pepper, and salt. If you prefer them thicker, use bread crumbs or crushed crackers.

FRIED EGGPLANT

Slice the eggplant in slices ⅓ of an inch thick, pare, put into a deep dish, and cover with well-salted cold water. Soak for one hour. Drain, wipe, dip in egg and crumbs, and fry brown.

BROILED MUSHROOMS

Choose large, firm mushrooms. Remove the stems, peel, wash, and wipe dry. Rub with melted butter and broil. Serve with a sauce made of melted butter, lemon juice, and minced parsley.

FRIED MUSHROOMS

Prepare as above, dip in egg and crumbs, and fry in hot oil or fat. Or, alternatively, sauté in butter in a frying pan. Breaded mushrooms may be broiled if dipped in melted butter or oil before broiling.

BAKED MUSHROOMS

Prepare as above. Place in a shallow baking dish, hollow side up, sprinkle with salt and pepper, and place a small piece of butter on each. Baste with melted butter and a few drops of lemon juice. Serve very hot on buttered toast.

GRILLED MUSHROOMS

Cut off the stalks, peel, and score lightly on the underside of large, firm, fresh mushrooms. Sprinkle with pepper and salt and soak a few moments in oil. Drain and broil. Serve with lemon quarters and garnish with parsley.

CORN FRITTERS

1 cup milk
3 eggs
1 can corn
A pinch of salt
As much flour as will form a batter

Beat the eggs, the yolks, and whites separate. To the yolks of the eggs add the corn, salt, milk, and flour enough to form a batter, beat the whole very hard, then stir in the whites, and drop the batter, a spoonful at a time, into hot lard, and fry them on both sides until it's a light brown color.

CORN OYSTERS

2 cup green corn, grated
½ cup milk
2 eggs
1 tsp salt
1 Tbs butter
1 Tbs lard

Beat the yolks of the eggs, add the milk and then the flour and salt. Beat to a smooth batter, add the corn, then beat again, adding the well-beaten whites of the eggs last. Put the lard and butter into a frying pan, and when very hot, put in the batter by small spoonfuls. Brown on one side, then turn. If the batter is too thick, add a little more milk. The thinner the batter, the more delicate and tender the oysters will be. Canned corn may be used, if it is chopped very fine, but it is not as good. By scoring deeply with a sharp knife each row of kernels on an ear of corn, the pulp may be pressed out with a knife. The corn may be cut from the cob and chopped, but the better way is to press out the pulp.

SWEET POTATO SOUFFLE

2 cup sweet potatoes

1 cup hot milk

2 Tbs sugar

½ tsp salt

2 Tbs butter

2 eggs

1 tsp nutmeg

½ cup seeded raisins

½ cup walnut meats

½ lb. marshmallows

Mash potatoes well. Heat milk and dissolve sugar and salt in it. Add butter, stirring until melted. Add this mixture to potatoes and beat until light and fluffy. Add the nutmeg, raisings, and nuts. Fold in well-beaten egg whites. Pour the mixture into a buttered baking dish; arrange marshmallows one-half inch apart on top. Bake in a moderate oven until the soufflé is set and the marshmallows a delicate brown. Serve immediately.

SWEET POTATO DELIGHT

2 medium-sized potatoes

1 cup sugar

1 cup nuts

1 box seedless raisins

1 box shredded coconut

20 marshmallows

Add potatoes, peeled and sliced thin, to a sauce pan. Cover with sugar and water; cook until tender, placing alternately layers of potatoes, coconut, and raisins sprinkled with nuts. Add syrup in which potatoes were candied and cover with marshmallows. Cook in a moderate oven until a golden brown. Serve in the dish in which it is cooked.

Noodles and Such

TO MAKE NOODLES

1 egg

Flour

Beat an egg and add as much flour as will make a very stiff dough. Roll it out in a thin sheet, flour it, and roll it up closely, as you would do a sheet of paper. Then, with a sharp knife, cut it in shavings like cabbage for slaw. Flour these cuttings to prevent them from adhering to each other. Add them to a pot of boiling water and boil for ten minutes.

DUMPLINGS

2 cup flour

2 tsp baking powder

½ tsp salt

½ cup milk

Water or chicken or beef broth

Sift the flour, baking powder, and salt together and stir in the milk to make dough. Drop the dough from a spoon into boiling water or broth (enough so that you will not need to replenish it while the dumplings are cooking). Boil for about 15 minutes without a lid, and then for 5 minutes with one.

MACARONI AND CHEESE

1 package macaroni (or ziti)

2 cup grated cheese

3 Tbs butter

Salt and pepper, to taste

2 cup crackers, crushed

2 cup milk

Cook macaroni as directed, until tender, and drain. Add the butter, cheese, milk, salt, and pepper, and stir well. Pour into a baking dish and sprinkle the crackers on top. Bake until the surface is a nice golden brown.

To Keep the Lid on a Boiling Pot: If you will drop a teaspoonful of butter into the water in which you are boiling dry beans or other starchy vegetables, you will not be annoyed by having the lid of the pot jump off, as it will otherwise do. The butter acts as oil on the troubled water, and keeps it calm and manageable.

RISOTTO

1 package rice, cooked
1 medium onion, chopped
2 cup pure white bouillon
Powdered saffron or curry, to taste
Parmesan cheese

Cook the rice as directed. Brown the onion in a frying pan with butter, and then add the rice. Stir carefully with a wooden spoon, gradually adding the bouillon. Let it simmer for a few moments. Add the seasoning, sprinkle with cheese, and serve as hot as possible.

Meats

Pork

BAKED HAM

Ham (any size)
1 cup vinegar
1 cup hot water
1 tsp English mustard
Brown sugar

Cover the ham in cold water and simmer gently, just long enough to loosen the skin so that it can be easily pulled off, between two and three hours. Dissolve the mustard in the hot water. Once the ham has been skinned, put it in a pan and pour the vinegar and mustard-water solution. Bake this for two hours, basting occasionally with the liquid. Take the ham out of the oven and cover it with brown sugar, about an inch thick. Press the sugar down firmly and do not baste again until the sugar has formed a thick crust. Bake for an hour after covering the ham with sugar, or until it becomes a rich, golden-brown. When done, put the ham on a dish to cool.

MACARONI AND HAM

1 cup macaroni
1 cup ham, cooked and chopped
1 egg
½ cup sweet milk
Cheese
Pepper
Butter

Cook macaroni as directed. Put alternating layers of ham and macaroni in a baking dish. Sprinkle each layer with pepper and bits of butter. When the baking dish is full, pour the milk over the top and let stand for a few hours. Then, grate as much cheese as you desire, sprinkle it over the top, and bake.

PORK STEAKS

Pork steaks, sliced thin
Cayenne pepper, salt, and rubbed sage, to taste

Cut the steaks in thin slices. Season them with cayenne pepper, salt, and rubbed sage. They may be broiled and buttered or fried in hot lard, with a gravy thickened with a little flour and poured over them.

SAUSAGE MEAT

66 lbs. lean pork
33 lbs. fat pork
$\frac{5}{12}$ cup black pepper
$\frac{1}{3}$ cup red pepper
½ cup salt
3 ⅓ cup sage

Cut meat in oblong strips, 1-2 inches thick, 3-5 inches long. In grinding, alternate fat and lean meat in order to secure thorough mixing. Work the seasoning well through the meat, or same can be sprinkled over meat before grinding. Casings of unbleached domestic should be made 3-5 inches wide and 10-15 inches long, sewed up all but at one end. Fill with sausage and tie with a strong cord. One pound of casings stuffs 50 lbs. of sausage.

Beef and Veal

HAMBURGER STEAK

1 lb ground steak
1 egg, beaten
4 salted crackers, crushed
1 onion, diced
1 tomato, cooked

Add all the ingredients, except for the crackers, to the meat and mix in. Season to taste and form into patties. Roll the patties in cracker crumbs and then fry slowly in hot butter.

PLAIN FRIED VEAL

Veal, sliced thin
Salt and pepper, to taste
Flour

Cut the meat into thin slices, then pound and wash them. Season them with salt and pepper and fry them in a pan with hot oil or butter until they are brown on both sides. When the meat is done, stir flour into the fat and pour in some water. Let it come to a boil once, then pour it over the veal, and serve.

FRIED VEAL WITH TOMATOES

Veal, sliced thin
Salt and pepper, to taste
Stewed tomatoes

Cut the veal in thin slices, season it with salt and pepper, and fry it in a pan with butter or oil until it turns a nice brown color. Have ready tomatoes which have been stewed very dry. Pass them through a sieve to take out the seeds, and then put them into the pan in which the meat was fried. Add enough butter to make a rich gravy. Pour them over the veal and serve it.

BAKED VEAL CUTLETS

Cutlets, cut ¾ of an inch thick
1 egg, beaten
Bread or cracker crumbs
Salt and pepper, to taste
Bacon, as needed

Cut the cutlets, flatten them, and then brush them over with the beaten egg. Sprinkle with bread or cracker crumbs. Next, season with the salt and pepper, and then place on a greased baking pan in between layers of bacon. Bake for about an hour.

MEAT BALLS

1 lb. hamburger meat
1 tsp salt
1 tsp pepper
1 egg
1 cup milk
Butter or oil, for frying

Loosen the meat with your hands or a meat mallet for a few minutes. Season with the salt and pepper, and then add the egg and milk. Mix together using your hands or a mixer. When all the ingredients are thoroughly incorporated into the meat, grease a frying pan, and drop the meat by spoonfuls into the pan. Brown the meat and serve hot.

ROAST BEEF

2 ribs of fine beef
Salt, to taste
Flour

Season the beef with salt, and place it in a roaster before a clear bright fire. Don't set it too close at first. The cook has to judge the amount of time the beef will roast depending on the taste of those eating. Rare beef will take an hour and a half, and two hours will cook two large ribs sufficiently, but if you desire well-done meat, it must be done longer. While the beef is roasting, baste it frequently with its own gravy. When nearly finished, dredge flour lightly over it so as to brown it. When the meat is taken out, skim the fat off the top of the gravy, pour the remainder in a pan, add a little flour, salt, and water, and boil it. Serve the gravy with the roast beef.

SMOTHERED STEAK

12 lg. onions
1 beef steak
Salt and pepper, to taste
Hot beef dripping

Take the onions and boil them in very little water until they are tender. Meanwhile, pound and wash a beef steak, seasoning it with salt and pepper, and place it in a pan with some hot beef dripping. Fry until it is done, then take it out and put it on a dish where it will stay hot. Once the onions are soft, drain and mash them in the pan with the steak gravy. Add salt and pepper, put it on the stove, and as soon as it is hot, pour it over the steak and serve.

VEAL CROQUETTES

2 cup cooked veal
1 egg
1 cup white sauce
Season with onion, salt, and pepper, to taste.

Mix all together, forming into balls. Dip in beaten eggs, and then in cracker crumbs. Fry in hot lard.

Lamb and Mutton

ROAST LEG OF LAMB

1 leg of lamb
Salt and pepper, to taste

Dressing

1 cup bread crumbs
1 tsp salt
1 tsp pepper
1 tsp sweet marjoram or summer savory
Butter, as much as the crumbs need to stick together

Put together the dressing. Cut deep incisions around the bone and in the flesh, and fill them with the dressing. Season the meat with salt and pepper. Put it on the spit and roast it before a clear fire. When nearly finished, dredge flour over and baste it with gravy. Skim the fat off the gravy, add a little flour and water, bring it to a boil once, and serve alongside in a gravy boat.

MUTTON CHOPS

Mutton chops
Butter
Salt and pepper, to taste
Slices of lemon, if desired

Trim the mutton chops, taking off the loose fat. Heat the grill, grease the grates, and put the chops on top over clear coals. Turn them frequently, and when done, put them in a dish, butter them well, and then season with salt and pepper. They may be served with slices of lemon.

Poultry

MRS. HARDING'S CHICKEN PIE

1 good-sized chicken
1 qt. flour
Lard size of egg
6 small potatoes
1 onion
5 tsp baking powder
1 tsp salt
Milk

Boil chicken gently until it falls from bones; cut in small pieces. Cook potatoes and onion in chicken broth. Make a pastry of flour, lard, salt, and baking powder—add enough milk to make a soft dough. Line the baking dish with the pasty and bake in a hot oven. Then fill this with the chicken, potatoes, and a small amount of broth. Cover the pastry and brown in the oven. Thicken the remaining broth and serve over the pie.

CREAMED CHICKEN

½ cup cold chicken, diced
1 cup rich white sauce
½ cup rich chicken stock
2 hard boiled eggs, diced
½ of a canned pimento, cut in strips
1 cup celery, diced

For seasoning, use salt, pepper, (white), and paprika. Heat the ingredients in the white sauce.

CHICKEN CROQUETTES A LA PIEDMONT

1 lb. ground, cooked chicken

1 cup milk

1 Tbs flour

2 Tbs butter

1 egg yolk

Cream the butter and the flour together, add the yolk and then the milk. Cook all until thick. Mix with the chicken—season with salt and pepper, to taste. Shape into balls—roll in flour—next in beaten egg and then in cracker crumbs. Fry in deep fat.

Fish

Boiled

SPICED SHAD

1 lg. shad

2 Tbs salt

3 tsp cayenne pepper

2 Tbs whole allspice

Vinegar

Split the shad open, rub over it the salt, and let it stand for a several hours. Have a pot of boiling water ready with enough water to cover the shad, allowing a teaspoon of salt to every quart of water. Boil it for 20 minutes, drain it, and bruise the allspice just so as to crack the grains. Sprinkle the allspice and pepper, and then cover it with cold vinegar.

STEWED CLAMS

1 pt. clams

2 pt. water

1 oz. butter rolled in flour

Cayenne pepper and salt, to taste

½ cup cream

Wash the clams, put them in a pot with a lid, and set them near heat. As soon as they begin to open, take them out of their shell. Drain them, and add the water, butter, and cayenne pepper. Let them stew for ten minutes, and just before they are to be served, add the cream.

CREAMED CODFISH

2 cup shredded codfish

3 cup milk

1 egg yolk

1 Tbs butter

2 Tbs flour

2 qt. water

Salt and pepper, to taste

Cover the fish with water in a pot and set it over a low fire. When it boils, drain it, and cover with the milk. Bring it to a boil again. Have the butter and flour rubbed smooth with a little cold milk and add to the boiling milk. Stir steadily till it thickens, then add the beaten egg yolk and cook five minutes longer. Season with salt, pepper, and minced parsley, if desired.

Fried

FRIED SHAD

1 large shad

Cayenne pepper and salt, to taste

Flour

Hot lard

Cut the shad in half, wash it, and wipe it dry. Score it, and season with the cayenne pepper and salt. Dredge some flour over it, and then fry it in hot lard. When nicely browned, put the two halves together, so it looks like a whole fish, and serve.

HALIBUT

1 large halibut

Cayenne pepper and salt, to taste

1 egg

Breadcrumbs

Cut the halibut into thin slices, less than a quarter of an inch thick. Wash and dry them, and then season with cayenne pepper and salt. Beat an egg and dip the cutlets into the egg and then cover them in breadcrumbs. Use a frying pan and butter and fry until golden brown.

CATFISH

1 large catfish
Cayenne pepper and salt, to taste
1 egg

Cut the fish into two parts, down the back and stomach, and take out the upper part of the back bone near the head. Wash and wipe the pieces dry. Season them with cayenne pepper and salt. Dip into a beaten egg and fry in a pan.

FRIED CLAMS

1 pt. clams
Cayenne pepper and salt, to taste

Wash the clams, place them in a vessel without any water, and watch them closely. As soon as they open, take them out of the shell. Dry them in a napkin, season them with the cayenne pepper and salt, and fry them in butter. Alternatively, they can be dipped in egg and bread crumbs and fried.

Baked

BAKED SMOKED HADDOCK

Put the haddock into a baking pan, cover with boiling water, drain, dot with butter, sprinkle with black pepper, and bake in a hot oven for ten minutes. Serve very hot.

BAKED OYSTERS

3 pt. oysters
1 pt. tomato catsup
1 cup cream
Butter size of an egg
1 tsp Worcestershire Sauce
Salt and pepper, to taste

Open oyster shells and boil oysters slightly in their own liquid; then combine this with sauce made of the remaining ingredients, place in oyster shells, and bake in oven, adding sauce from time to time, until all possible is used. Care must be taken not to have the oven too hot, or juice will burn on edges. Fancy shells may be used or ramekins.

DRYING

Drying fruits, vegetables, herbs, and even meat is a great way to preserve foods for longer-term storage, especially if your pantry or freezer space is limited. Dried foods take up much less space than their fresh, frozen, or canned counterparts. Drying requires relatively little preparation time, and is simple enough that kids will enjoy helping. Drying with a food dehydrator will ensure the fastest, safest, and best quality results. However, you can also dry produce in the sunshine, in your oven, or strung up over a woodstove.

For more information on food drying, check out *So Easy to Preserve, 5th ed.* from the Cooperative Extension Service, the University of Georgia. Much of the information that follows is adapted from this excellent source.

Drying with a Food Dehydrator

Food dehydrators use electricity to produce heat and have a fan and vents for air circulation. Dehydrators are efficiently designed to dry foods fast at around 140ºF. Look for food dehydrators in discount department stores, mail-order catalogs, the small appliance section of a department store, natural food stores, and seed or garden supply catalogs. Costs vary depending on features. Some models are expandable and additional trays can be purchased later. Twelve square feet of drying space dries about a half-bushel of produce.

Dehydrator Features to Look For

- Double-wall construction of metal or high-grade plastic. Wood is not recommended, because it is a fire hazard and is difficult to clean.
- Enclosed heating elements
- Countertop design
- An enclosed thermostat from 85 to 160ºF
- Fan or blower
- Four to ten open mesh trays made of sturdy, lightweight plastic for easy washing
- Underwriters Laboratory (UL) seal of approval

- A one-year guarantee
- Convenient service
- A dial for regulating temperature
- A timer. Often the completed drying time may occur during the night, and a timer turns the dehydrator off to prevent scorching.

Types of Dehydrators

There are two basic designs for dehydrators. One has horizontal air flow and the other has vertical air flow. In units with horizontal flow, the heating element and fan are located on the side of the unit. The major advantages of horizontal flow are: it reduces flavor mixture so several different foods can be dried at one time; all trays receive equal heat penetration; and juices or liquids do not drip down into the heating element. Vertical air flow dehydrators have the heating element and fan located at the base. If different foods are dried, flavors can mix and liquids can drip into the heating element.

Fruit Drying Procedures

Apples—Select mature, firm apples. Wash well. Pare, if desired, and core. Cut in rings or slices ⅛ to ¼ inch thick or cut in quarters or eighths. Soak in ascorbic acid, vinegar, or lemon juice for 10 minutes. Remove from solution and drain well. Arrange in single layer on trays, pit side up. Dry until soft, pliable, and leathery; there should be no moist area in center when cut.

Apricots—Select firm, fully ripe fruit. Wash well. Cut in half and remove pit. Do not peel. Soak in ascorbic acid, vinegar, or lemon juice for 10 minutes. Remove from solution and drain well. Arrange in single layer on trays, pit side up with cavity popped up to expose more flesh to the air. Dry until soft, pliable, and leathery; there should be no moist area in center when cut.

Bananas—Select firm, ripe fruit. Peel. Cut in ⅛-inch slices. Soak in ascorbic acid, vinegar, or lemon juice for 10 minutes. Remove and drain well. Arrange in single layer on trays. Dry until tough and leathery.

Berries—Select firm, ripe fruit. Wash well. Leave whole or cut in half. Dip in boiling water 30 seconds to crack skins. Arrange on drying trays not more than two berries deep. Dry until hard and berries rattle when shaken on trays.

Cherries—Select fully ripe fruit. Wash well. Remove stems and pits. Dip whole cherries in boiling water 30 seconds to crack skins. Arrange in single layer on trays. Dry until tough, leathery, and slightly sticky.

Citrus peel—Select thick-skinned oranges with no signs of mold or decay and no color added to skin. Scrub oranges well with brush under cool running water. Thinly peel outer ¹⁄₁₆ to ⅛ inch of the peel; avoid white bitter part. Soak in ascorbic acid, vinegar, or lemon juice for 10 minutes. Remove from solution and drain well. Arrange in single layers on trays. Dry at 130°F for 1 to 2 hours; then at 120°F until crisp.

Figs—Select fully ripe fruit. Wash or clean well with damp towel. Peel dark-skinned varieties if desired. Leave whole if small or partly dried on tree; cut large figs in halves or slices. If drying whole figs, crack skins by dipping in boiling water for 30 seconds. For cut figs, soak in ascorbic acid, vinegar, or lemon juice for 10 minutes. Remove and drain well. Arrange in single layers on trays. Dry until leathery and pliable.

Grapes and black currants—Select seedless varieties. Wash, sort, and remove stems. Cut in half or leave whole. If drying whole, crack skins by dipping in boiling water for 30 seconds. If halved, dip in ascorbic acid or other antimicrobial solution for 10 minutes. Remove and drain well. Dry until pliable and leathery with no moist center.

Melons—Select mature, firm fruits that are heavy for their size; cantaloupe dries better than watermelon. Scrub outer surface well with brush under cool running water. Remove outer skin, any fibrous tissue, and seeds. Cut into ¼- to ½-inch-thick slices. Soak in ascorbic acid, vinegar, or lemon juice for 10 minutes. Remove and drain well. Arrange in single layer on trays. Dry until leathery and pliable with no pockets of moisture.

Nectarines and peaches—Select ripe, firm fruit. Wash and peel. Cut in half and remove pit. Cut in quarters or slices if desired. Soak in ascorbic acid, vinegar, or lemon juice for 10 minutes. Remove and drain well. Arrange in single layer on trays, pit side up. Turn halves over when visible juice disappears. Dry until leathery and somewhat pliable.

Pears—Select ripe, firm fruit. Bartlett variety is recommended. Wash fruit well. Pare, if desired. Cut in half lengthwise and core. Cut in quarters, eighths, or slices ⅛ to ¼ inch thick. Soak in ascorbic acid, vinegar, or lemon juice for 10 minutes. Remove and drain. Arrange in single layer on trays, pit side up. Dry until springy and suede-like with no pockets of moisture.

Plums and prunes—Wash well. Leave whole if small; cut large fruit into halves (pit removed) or slices. If left whole, crack skins in boiling water 1 to 2 minutes. If cut in half, dip in ascorbic acid or other antimicrobial solution for 10 minutes. Remove and drain. Arrange in single layer on trays, pit side up, cavity popped out. Dry until pliable and leathery; in whole prunes, pit should not slip when squeezed.

Fruit Leathers

Fruit leathers are a tasty and nutritious alternative to store-bought candies that are full of artificial sweeteners and preservatives. Blend the leftover fruit pulp from making jelly or use fresh, frozen, or drained canned fruit. Ripe or slightly overripe fruit works best.

Chances are the fruit leather will get eaten before it makes it into the cupboard, but it can keep up to one month at room temperature. For storage up to one year, place tightly wrapped rolls in the freezer.

Ingredients

2 cups fruit
2 tsp lemon juice or ⅛ tsp ascorbic acid (optional)
¼ to ½ cup sugar, corn syrup, or honey (optional)

Directions

1. Wash fresh fruit or berries in cool water. Remove peel, seeds, and stem.
2. Cut fruit into chunks. Use 2 cups of fruit for each 13 x 15-inch inch fruit leather. Purée fruit until smooth.
3. Add 2 teaspoons of lemon juice or ⅛ teaspoon ascorbic acid (375 mg) for each 2 cups light-colored fruit to prevent darkening.
4. Optional: To sweeten, add corn syrup, honey, or sugar. Corn syrup or honey is best for longer storage because

these sweeteners prevent crystals. Sugar is fine for immediate use or short storage. Use ¼ to ½ cup sugar, corn syrup, or honey for each 2 cups of fruit. Avoid aspartame sweeteners as they may lose sweetness during drying.

> - Applesauce can be dried alone or added to any fresh fruit purée as an extender. It decreases tartness and makes the leather smoother and more pliable.
> - To dry fruit in the oven, a 13 X 15-inch cookie pan with edges works well. Line pan with plastic wrap, being careful to smooth out wrinkles. Do not use waxed paper or aluminum foil.

5. Pour the leather. Fruit leathers can be poured into a single large sheet (13 X 15 inches) or into several smaller

Spices, Flavors, and Garnishes

To add interest to your fruit leathers, include spices, flavorings, or garnishes.

- **Spices to try**—Allspice, cinnamon, cloves, coriander, ginger, mace, mint, nutmeg, or pumpkin pie spice. Use sparingly; start with ⅛ teaspoon for each 2 cups of purée.
- **Flavorings to try**—Almond extract, lemon juice, lemon peel, lime juice, lime peel, orange extract, orange juice, orange peel, or vanilla extract. Use sparingly; try ⅛ to ¼ teaspoon for each 2 cups of purée.
- **Delicious additions to try**—Shredded coconut, chopped dates, other dried chopped fruits, granola, miniature marshmallows, chopped nuts, chopped raisins, poppy seeds, sesame seeds, or sunflower seeds.
- **Fillings to try**—Melted chocolate, softened cream cheese, cheese spreads, jam, preserves, marmalade, marshmallow cream, or peanut butter. Spread one or more of these on the leather after it is dried and then roll. Store in refrigerator.

sizes. Spread purée evenly, about ⅛ inch thick, onto drying tray. Avoid pouring purée too close to the edge of the cookie sheet.

6. Dry the leather. Dry fruit leathers at 140ºF. Leather dries from the outside edge toward the center. Larger fruit leathers take longer to dry. Approximate drying times are 6 to 8 hours in a dehydrator, up to 18 hours in an oven, and 1 to 2 days in the sun. Test for dryness by touching center of leather; no indentation should be evident. While warm, peel from plastic and roll, allow to cool, and rewrap the roll in plastic. Cookie cutters can be used to cut out shapes that children will enjoy. Roll, and wrap in plastic.

Vegetable Leathers

Pumpkin, mixed vegetables, and tomatoes make great leathers. Just purée cooked vegetables, strain, spread on a tray lined with plastic wrap, and dry. Spices can be added for flavoring.

Mixed Vegetable Leather

2 cups cored, cut-up tomatoes
1 small onion, chopped
¼ cup chopped celery
Salt to taste

Combine all ingredients in a covered saucepan and cook over low heat 15 to 20 minutes. Purée or force through a sieve or colander. Return to saucepan and cook until thickened. Spread on a cookie sheet or tray lined with plastic wrap. Dry at 140ºF.

Pumpkin Leather

2 cups canned pumpkin or 2 cups fresh pumpkin, cooked and puréed
½ cup honey
¼ tsp cinnamon
⅛ tsp nutmeg
⅛ tsp powdered cloves
Blend ingredients well. Spread on tray or cookie sheet lined with plastic wrap. Dry at 140ºF.

Tomato Leather

Core ripe tomatoes and cut into quarters. Cook over low heat in a covered saucepan, 15 to 20 minutes. Purée or force through a sieve or colander and pour into electric fry pan or shallow pan. Add salt to taste and cook over low heat until thickened. Spread on a cookie sheet or tray lined with plastic wrap. Dry at 140ºF.

Vine Drying

One method of drying outdoors is vine drying. To dry beans (navy, kidney, butter, great northern, lima, lentils, and soybeans) leave bean pods on the vine in the garden until the beans inside rattle. When the vines and pods are dry and shriveled, pick the beans and shell them. No pretreatment is necessary. If beans are still moist, the drying process is not complete and the beans will mold if not more thoroughly dried. If needed, drying can be completed in the sun, an oven, or a dehydrator.

How to Make a Woodstove Food Dehydrator

1. Collect pliable wire mesh or screens (available at hardware stores) and use wire cutters to trim to squares 12 to 16 inches on each side. The trays should be of the same size and shape. Bend up the edges of each square to create a half-inch lip (see illustrations on page 254).

2. Attach one S hook from the hardware store or a large paperclip to each side of each square (four clips per tray) to attach the trays together.

3. Cut four equal lengths of chain or twine that will reach from the ceiling to the level of the top tray. Use a wire or metal loop to attach the four pieces together at the top and secure to a hook in the ceiling above the woodstove. Attach the chain or twine to the hooks on the top tray.

4. To use, fill trays with food to dry, starting with the top tray. Link trays together using the S hooks or strong paperclips. When the foods are dried, remove the entire stack and disassemble. Remove the dried food and store.

Herbs

Drying is the easiest method of preserving herbs. Simply expose the leaves, flowers, or seeds to warm, dry air. Leave the herbs in a well-ventilated area until the moisture evaporates. Sun drying is not recommended because the herbs can lose flavor and color.

The best time to harvest most herbs for drying is just before the flowers first open when they are in the bursting bud stage. Gather the herbs in the early morning after the dew has evaporated to minimize wilting. Avoid bruising the leaves. They should not lie in the sun or unattended after harvesting. Rinse herbs in cool water and gently shake to remove excess moisture. Discard all bruised, soiled, or imperfect leaves and stems.

Dehydrator drying is another fast and easy way to dry high-quality herbs because temperature and air circulation can be controlled. Preheat dehydrator with the thermostat set to 95°F to 115°F. In areas with higher humidity, temperatures as high as 125°F may be needed. After rinsing under cool, running water and shaking to remove excess moisture, place the herbs in a single layer on dehydrator trays. Drying times may vary from one to four hours. Check periodically. Herbs are dry when they crumble, and stems break when bent. Check your dehydrator instruction booklet for specific details.

Less tender herbs—The more sturdy herbs, such as rosemary, sage, thyme, summer savory, and parsley, are the easiest to dry without a dehydrator. Tie them into small bundles and hang them to air dry. Air drying outdoors is often possible; however, better color and flavor retention usually results from drying indoors.

Tender-leaf herbs—Basil, oregano, tarragon, lemon balm, and the mints have a high moisture content and will mold if not dried quickly. Try hanging the tender-leaf herbs or those with seeds inside paper bags to dry. Tear or punch holes in the sides of the bag. Suspend a small bunch (large amounts will mold) of herbs in a bag and close the top with a rubber band. Place where air currents will circulate through the bag. Any leaves and seeds that fall off will be caught in the bottom of the bag.

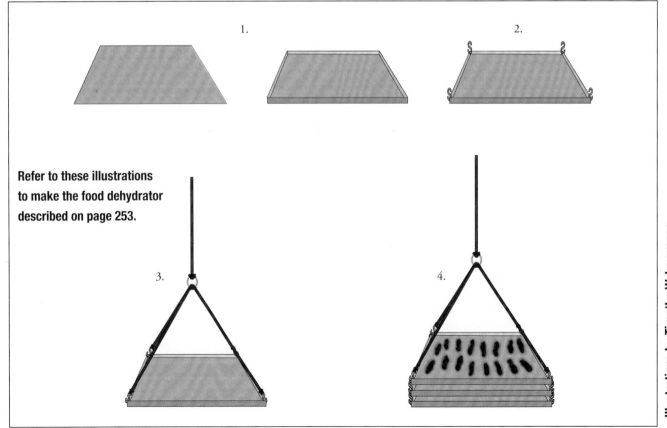

Refer to these illustrations to make the food dehydrator described on page 253.

1.

2.

3.

4.

Illustrations by Timothy W. Lawrence

Another method, especially nice for mint, sage, or bay leaf, is to dry the leaves separately. In areas of high humidity, it will work better than air drying whole stems. Remove the best leaves from the stems. Lay the leaves on a paper towel, without allowing leaves to touch. Cover with another towel and layer of leaves. Five layers may be dried at one time using this method. Dry in a very cool oven. The oven light of an electric range or the pilot light of a gas range furnishes enough heat for overnight drying. Leaves dry flat and retain a good color.

Microwave ovens are a fast way to dry herbs when only small quantities are to be prepared. Follow the directions that come with your microwave oven.

When the leaves are crispy, dry, and crumble easily between the fingers, they are ready to be packaged and stored. Dried leaves may be left whole and crumbled as used, or coarsely crumbled before storage. Husks can be removed from seeds by rubbing the seeds between the hands and blowing away the chaff. Place herbs in airtight containers and store in a cool, dry, dark area to protect color and fragrance.

Dried herbs are usually three to four times stronger than the fresh herbs. To substitute dried herbs in a recipe that calls for fresh herbs, use ¼ to ⅓ of the amount listed in the recipe.

Jerky

Jerky is great for hiking or camping because it supplies protein in a very lightweight form—plus it can be very tasty. A pound of meat or poultry weighs about four ounces after being made into jerky. In addition, because most of the moisture is removed, it can be stored for one to two months without refrigeration.

Jerky has been around since the ancient Egyptians began drying animal meat that was too big to eat all at once. Native Americans mixed ground dried meat with dried fruit or suet to make pemmican. *Biltong* is dried meat or game used in many African countries. The English word *jerky* came from the Spanish word *charque*, which means, "dried salted meat."

Drying is the world's oldest and most common method of food preservation. Enzymes require moisture in order to react with food. By removing the moisture, you prevent this biological action.

Jerky can be made from ground meat, which is often less expensive than strips of meat and allows you to combine different kinds of meat if desired. You can also make it into any shape you want! As with strips of meat, an internal temperature of 160ºF is necessary to eliminate disease-causing bacteria such as *E. coli*, if present.

Food Safety

The USDA Meat and Poultry Hotline's current recommendation for making jerky safely is to heat meat to 160°F and poultry to 165°F before the dehydrating process. This ensures that any bacteria present are destroyed by heat. If your food dehydrator doesn't heat up to 160°F, it's important to cook meat slightly in the oven or by steaming before drying. After heating, maintain a constant dehydrator temperature of 130 to 140°F during the drying process.

According to the USDA, you should always:

- Wash hands thoroughly with soap and water before and after working with meat products.
- Use clean equipment and utensils.
- Keep meat and poultry refrigerated at 40°F or slightly below; use or freeze ground beef and poultry within two days, and whole red meats within three to five days.
- Defrost frozen meat in the refrigerator, not on the kitchen counter.
- Marinate meat in the refrigerator. Don't save marinade to re-use. Marinades are used to tenderize and flavor the jerky before dehydrating it.
- If your food dehydrator doesn't heat up to 160°F (or 165°F for poultry), steam or roast meat before dehydrating it.
- Dry meats in a food dehydrator that has an adjustable temperature dial and will maintain a temperature of at least 130 to 140°F throughout the drying process.

Preparing the Meat

1. Partially freeze meat to make slicing easier. Slice meat across the grain ⅛ to ¼ inch thick. Trim and discard all fat, gristle, and membranes or connective tissue.
2. Marinate the meat in a combination of oil, salt, spices, vinegar, lemon juice, teriyaki, soy sauce, beer, or wine.

Marinated Jerky

¼ cup soy sauce

1 tbsp Worcestershire sauce

1 tsp brown sugar

¼ tsp black pepper

½ tsp fresh ginger, finely grated

1 tsp salt

1 ½ to 2 lbs of lean meat strips

(beef, pork, or venison)

1. Combine all ingredients except the strips, and blend. Add meat, stir, cover, and refrigerate at least one hour.

2. If your food dehydrator doesn't heat up to 160°F, bring strips and marinade to a boil and cook for 5 minutes.

3. Drain meat in a colander and absorb extra moisture with clean, absorbent paper towels. Arrange strips in a single layer on dehydrator trays, or on cake racks placed on baking sheets for oven drying.

4. Place the racks in a dehydrator or oven preheated to 140°F, or 160°F if the meat wasn't precooked. Dry until a test piece cracks but does not break when it is bent (10 to 24 hours for samples not heated in marinade, 3 to 6 hours for preheated meat). Use paper towel to pat off any excess oil from strips, and pack in sealed jars, plastic bags, or plastic containers.

EDIBLE WILD PLANTS

Wild Vegetables, Fruits, and Nuts

Agave

Description: Agave plants have large clusters of thick leaves that grow around one stalk. They grow close to the ground and only flower once before dying.

Location: Agave like dry, open areas and are found in the deserts of the American west.

Edible Parts and Preparing: Only agave flowers and buds are edible. Boil these before consuming. The juice can be collected from the flower stalk for drinking.

Other Uses: Most agave plants have thick needles on the tips of their leaves that can be used for sewing.

Asparagus

Description: When first growing, asparagus looks like a collection of green fingers. Once mature, the plant has fern-like foliage and red berries (which are toxic if eaten). The

flowers are small and green and several species have sharp, thornlike projections.

Location: It can be found growing wild in fields and along fences. Asparagus is found in temperate areas in the United States.

Edible Parts and Preparing: It is best to eat the young stems, before any leaves grow. Steam or boil them for 10 to 15 minutes before consuming. The roots are a good source of starch, but don't eat any part of the plant raw, as it could cause nausea or diarrhea.

Beech

Description: Beech trees are large forest trees. They have smooth, light gray bark, very dark leaves, and clusters of prickly seedpods.

Location: Beech trees prefer to grow in moist, forested areas. These trees are found in the Temperate Zone in the eastern United States.

Edible Parts and Preparing: Eat mature beechnuts by breaking the thin shells with your fingers and removing the sweet, white kernel found inside. These nuts can also be used as a substitute for coffee by roasting them until the kernel turns hard and golden brown. Mash up the kernel and boil or steep in hot water.

Blackberry and Raspberry

Description: These plants have prickly stems that grow upright and then arch back toward the ground. They have alternating leaves and grow red or black fruit.

Location: Blackberry and raspberry plants prefer to grow in wide, sunny areas near woods, lakes, and roads. They grow in temperate areas.

Edible Parts and Preparing: Both the fruits and peeled young shoots can be eaten. The leaves can be used to make tea.

Burdock

Description: Burdock has wavy-edged, arrow-shaped leaves. Its flowers grow in burrlike clusters and are purple or pink. The roots are large and fleshy.

Location: This plant prefers to grow in open waste areas during the spring and summer. It can be found in the Temperate Zone in the north.

Edible Parts and Preparing: The tender leaves growing on the stalks can be eaten raw or cooked. The roots can be boiled or baked.

Cattail

Description: These plants are grasslike and have leaves shaped like straps. The male flowers grow above the female flowers, have abundant, bright yellow pollen, and die off quickly. The female flowers become the brown cattails.

Location: Cattails like to grow in full-sun areas near lakes, streams, rivers, and brackish water. They can be found all over the country.

Edible Parts and Preparing: The tender, young shoots can be eaten either raw or cooked. The rhizome (rootstalk) can be pounded and made into flour. When the cattail is immature, the female flower can be harvested, boiled, and eaten like corn on the cob.

Other Uses: The cottony seeds of the cattail plant are great for stuffing pillows. Burning dried cattails helps repel insects.

Chicory

Description: This is quite a tall plant, with clusters of leaves at the base of the stem and very few leaves on the stem itself. The flowers are sky blue in color and open only on sunny days. It produces a milky juice.

Location: Chicory grows in fields, waste areas, and alongside roads. It grows primarily as a weed all throughout the country.

Edible Parts and Preparing: The entire plant is edible. The young leaves can be eaten in a salad. The leaves and roots may also be boiled as you would regular vegetables. Roast

the roots until they are dark brown, mash them up, and use them as a substitute for coffee.

Cranberry

Description: The cranberry plant has tiny, alternating leaves. Its stems crawl along the ground and it produces red berry fruits.

Location: Cranberries only grow in open, sunny, wet areas. They thrive in the colder areas in the northern states.

Edible Parts and Preparing: The berries can be eaten raw, though they are best when cooked in a small amount of water, adding a little bit of sugar if desired.

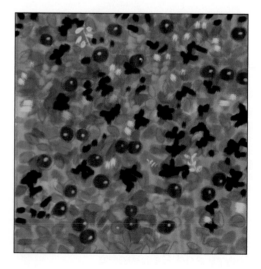

Dandelion

Description: These plants have jagged leaves and grow close to the ground. They have bright yellow flowers.

Location: Dandelions grow in almost any open, sunny space in the United States.

Edible Parts and Preparing: All parts of this plant are edible. The leaves can be eaten raw or cooked and the roots boiled. Roasted and ground roots can make a good substitute for coffee.

Other Uses: The white juice in the flower stem can be used as glue.

Elderberry

Description: This shrub has many stems containing opposite, compound leaves. Its flower is white, fragrant, and grows in large clusters. Its fruits are berry-shaped and are typically dark blue or black.

Location: Found in open, wet areas near rivers, ditches, and lakes, the elderberry grows mainly in the eastern states.

Edible Parts and Preparing: The flowers can be soaked in water for eight hours and then the liquid can be drunk. The fruit is also edible but don't eat any other parts of the plant—they are poisonous.

Hazelnut

Description: The nuts grow on bushes in very bristly husks.

Location: Hazelnut grows in dense thickets near streambeds and in open areas and can be found all over the United States.

Edible Parts and Preparing: In the autumn, the hazelnut ripens and can be cracked open and the kernel eaten. Eating dried nuts is also tasty.

Juniper

Description: Also known as cedar, this shrub has very small, scaly leaves that are densely crowded on the branches.

Berrylike cones on the plant are usually blue and are covered with a whitish wax.

Location: They grow in open, dry, sunny places throughout the country.

Edible Parts and Preparing: Both berries and twigs are edible. The berries can be consumed raw or the seeds may be roasted to make a substitute for coffee. Dried and crushed berries are good to season meat. Twigs can be made into tea.

Lotus

Description: This plant has large, yellow flowers and leaves that float on or above the surface of the water. The lotus fruit has a distinct, flattened shape and possesses around 20 hard seeds.

Location: Found on fresh water in quiet areas, the lotus plant is native to North America.

Edible Parts and Preparing: All parts of the lotus plant are edible, raw or cooked. Bake or boil the fleshy parts that grow underwater and boil young leaves. The seeds are quite nutritious and can be eaten raw or they can be ground into flour.

Marsh Marigold

Description: Marsh marigold has round, dark green leaves and a short stem. It also has bright yellow flowers.

Location: The plant can be found in bogs and lakes in the northeastern states.

Edible Parts and Preparing: All parts can be boiled and eaten. Do not consume any portion raw.

Mulberry

Description: The mulberry tree has alternate, lobed leaves with rough surfaces and blue or black seeded fruits.

Location: These trees are found in forested areas and near roadsides in temperate and tropical regions of the United States.

Edible Parts and Preparing: The fruit can be consumed either raw or cooked and it can also be dried. Make sure the fruit is ripe or it can cause hallucinations and extreme nausea.

Nettle

Description: Nettle plants grow several feet high and have small flowers. The stems, leafstalks, and undersides of the leaves all contain fine, hairlike bristles that cause a stinging sensation on the skin.

Location: This plant grows in moist areas near streams or on the edges of forests. It can be found throughout the United States.

Edible Parts and Preparing: The young shoots and leaves are edible. To eat, boil the plant for 10 to 15 minutes.

Oak

Description: These trees have alternating leaves and acorns. Red oaks have bristly leaves and smooth bark on the upper part of the tree and their acorns need two years to reach maturity. White oaks have leaves with no bristles and rough bark on the upper part of the tree. Their acorns only take one year to mature.

Location: Found in various locations and habitats throughout the country.

Edible Parts and Preparing: All parts of the tree are edible, but most are very bitter. Shell the acorns and soak them in water for one or two days to remove their tannic acid. Boil the acorns to eat or grind them into flour for baking.

Palmetto Palm

Description: This is a tall tree with no branches and has a continual leaf base on the trunk. The leaves are large, simple, and lobed and it has dark blue or black fruits that contain a hard seed.

Location: This tree is found throughout the southeastern coast.

Edible Parts and Preparing: The palmetto palm fruit can be eaten raw. The seeds can also be ground into flour, and the heart of the palm is a nutritious source of food, but the top of the tree must be cut down in order to reach it.

Pine

Description: Pine trees have needlelike leaves that are grouped into bundles of one to five needles. They have a very pungent, distinguishing odor.

Location: Pines grow best in sunny, open areas and are found all over the United States.

Edible Parts and Preparing: The seeds are completely edible and can be consumed either raw or cooked. Also, the young male cones can be boiled or baked and eaten. Peel the bark off of thin twigs and chew the juicy inner bark. The needles can be dried and brewed to make tea that's high in vitamin C.

Other Uses: Pine tree resin can be used to waterproof items. Collect the resin from the tree, put it in a container, heat it, and use it as glue or, when cool, rub it on items to waterproof them.

Plantain

Description: The broad-leafed plantain grows close to the ground and the flowers are situated on a spike that rises from the middle of the leaf cluster. The narrow-leaf species has leaves covered with hairs that form a rosette. The flowers are very small.

Location: Plantains grow in lawns and along the side of the road in the northern Temperate Zone.

Edible Parts and Preparing: Young, tender leaves can be eaten raw and older leaves should be cooked before consumption. The seeds may also be eaten either raw or roasted. Tea can also be made by boiling 1 ounce of the plant leaves in a few cups of water.

Pokeweed

Description: A rather tall plant, pokeweed has elliptical leaves and produces many large clusters of purple fruits in the late spring.

Location: Pokeweed grows in open and sunny areas in fields and along roadsides in the eastern United States.

Edible Parts and Preparing: If cooked, the young leaves and stems are edible. Be sure to boil them twice and discard

the water from the first boiling. The fruit is also edible if cooked. Never eat any part of this plant raw, as it is poisonous.

Prickly Pear Cactus

Description: This plant has flat, pad-like green stems and round, furry dots that contain sharp-pointed hairs.

Location: Found in arid regions and in dry, sandy areas in wetter regions, it can be found throughout the United States.

Edible Parts and Preparing: All parts of this plant are edible. To eat the fruit, peel it or crush it to make a juice. The seeds can be roasted and ground into flour.

Reindeer Moss

Description: This is a low plant that does not flower. However, it does produce bright red structures used for reproduction.

Location: It grows in dry, open areas in much of the country.

Edible Parts and Preparing: While having a crunchy, brittle texture, the whole plant can be eaten. To remove some of the bitterness, soak it in water and then dry and crush it, adding it to milk or other foods.

Sassafras

Description: This shrub has different leaves—some have one lobe, others two lobes, and others have none at all. The flowers are small and yellow and appear in the early spring. The plant has dark blue fruit.

Location: Sassafras grows near roads and forests in sunny, open areas. It is common throughout the eastern states.

Edible Parts and Preparing: The young twigs and leaves can be eaten either fresh or dried—add them to soups. Dig out the underground portion of the shrub, peel off the bark, and dry it. Boil it in water to make tea.

Other Uses: Shredding the tender twigs will make a handy toothbrush.

Spatterdock

Description: The leaves of this plant are quite long and have a triangular notch at the base. Spatterdock has yellow flowers that become bottle-shaped fruits, which are green when ripe.

Location: Spatterdock is found in fresh, shallow water throughout the country.

Edible Parts and Preparing: All parts of the plant are edible and the fruits have brown seeds that can be roasted and ground into flour. The rootstock can be dug out of the mud, peeled, and boiled.

Strawberry

Description: This is a small plant with a three-leaved pattern. Small white flowers appear in the springtime and the fruit is red and very fleshy.

Location: These plants prefer sunny, open spaces, are commonly planted, and appear in the northern Temperate Zone.

Edible Parts and Preparing: The fruit can be eaten raw, cooked, or dried. The plant leaves may also be eaten or dried to make tea.

Thistle

Description: This plant may grow very high and has long-pointed, prickly leaves.

Location: Thistle grows in woods and fields all over the country.

Edible Parts and Preparing: Peel the stalks, cut them into smaller sections, and boil them to consume. The root may be eaten raw or cooked.

Walnut

Description: Walnuts grow on large trees and have divided leaves. The walnut has a thick outer husk that needs to be removed before getting to the hard, inner shell.

Location: The black walnut tree is common in the eastern states.

Edible Parts and Preparing: Nut kernels become ripe in the fall and the meat can be obtained by cracking the shell.

Water Lily

Description: With large, triangular leaves that float on water, these plants have fragrant flowers that are white or red. They also have thick rhizomes that grow in the mud.

Location: Water lilies are found in many temperate areas.

Edible Parts and Preparing: The flowers, seeds, and rhizomes can be eaten either raw or cooked. Peel the corky rind off of the rhizome and eat it raw or slice it thinly, dry it, and grind into flour. The seeds can also be made into flour after drying, parching, and grinding.

Wild Grapevine

Description: This vine will climb on tendrils, and most of these plants produce deeply lobed leaves. The grapes grow in pyramidal bunches and are black-blue, amber, or white when ripe.

Location: Climbing over other vegetation on the edges of forested areas, they can be found in the eastern and southwestern parts of the United States.

Edible Parts and Preparing: Only the ripe grape and the leaves can be eaten.

Wild Onion and Garlic

Description: These are recognized by their distinctive odors.

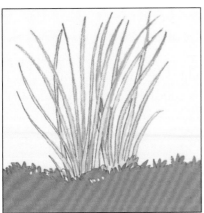

Location: They are found in open areas that get lots of sun throughout temperate areas.

Edible Parts and Preparing: The bulbs and young leaves are edible and can be consumed either raw or cooked.

Wild Rose

Description: This shrub has alternating leaves and sharp prickles. It has red, pink, or yellow flowers and fruit (rose hip) that remains on the shrub all year.

Location: These shrubs occur in dry fields throughout the country.

Edible Parts and Preparing: The flowers and buds are edible raw or boiled. Boil fresh, young leaves to make tea. The rose hips can be eaten once the flowers fall and they can be crushed once dried to make flour.

Violets

Violets can be candied and used to decorate cakes, cookies, or pastries. Pick the flowers with a tiny bit of stem, wash, and allow to dry thoroughly on a paper towel or a rack. Heat ½ cup water, 1 cup sugar, and ¼ teaspoon almond extract in a saucepan. Use tweezers to carefully dip each flower in the hot liquid. Set on wax paper and dust with sugar until every flower is thoroughly coated. If desired, snip off remaining stems with small scissors. Allow flowers to dry for a few hours in a warm, dry place.

FERMENTING

Fermenting is the process of preserving certain food products using either bacteria or yeast and keeping exposure to air at a minimum. Fermentation is used to help preserve certain foods and to create new flavors. This process is used to create products such as kefir, sauerkraut, and kombucha.

Kefir

Kefir is slightly fermented milk. It is similar to yogurt, except it is milder in flavor and of a thinner consistency.

To make kefir you will need to obtain kefir "grains"— these are made up of yeast and bacteria. These can be purchased online, or at select health-food stores (you may have to look hard, as they are not common). The grains will vary in size, anywhere from rice to walnut size. They resemble pieces of cauliflower.

Rinse the grains. Add about ½ cup grains to one quart of milk. The milk should be cold, preferably directly from the fridge. The milk can be store-bought, and skim or low-fat milk works best. Cover the milk, but do not seal it tightly, and leave it to sit at room temperature for 24-48 hours. Stir the milk once a day. Test the milk occasionally after the 24 hour period- finished kefir should have a mildly acidic taste and be slightly carbonated. Strain the grains from the kefir and refrigerate it. Rinse the grains.

The grains will continue to grow and multiply as they are used. Occasionally some grains will have to be removed from the milk mixture to ensure it does not become too thick. Remove the extra grains, wash them in cold water and allow them to dry completely between two pieces of cheesecloth.

To revive either dried or new grains place them in one cup of milk for 24 hours. Drain out the grains, rinse them and add them back to the milk, with another cup of fresh milk added. After two days you can add enough milk to make one quart and create kefir using the normal process.

Sauerkraut

Sauerkraut is the simple process of fermenting cabbage. The only things necessary are salt and a non-metal crock to put the cabbage in.

To make sauerkraut—Mix shredded cabbage and salt in a large pot and let it sit for 15 minutes. The usual ratio is 5 lbs. young cabbage to 3 tablespoons of salt. Then, pack the mixture into a clean crock, or other non-metal container. Use a wooden spoon to press the cabbage tightly into the crock. Make sure that the juices from the cabbage and salt mixture cover the cabbage, if they do not some water may be added. Keep about six inches of the crock above the pressed cabbage clear. Cover the cabbage with cheesecloth, tucking it down the edges of the crock. Use a heavy, tight-fitting lid to weigh down the cabbage. The seal should be airtight and the cabbage should not be exposed to the air. Ferment the cabbage for five to six weeks at room temperature. Check the cabbage periodically, skimming off any scum that develops and changing the cheesecloth and lid if they develop scum. The fermentation process is complete when bubbles no longer rise to the surface.

To process the fermented sauerkraut, place the crock in a hot water bath and bring the sauerkraut to a simmer, making sure not to completely boil. Pack the sauerkraut in clean, hot jars, being sure to leave ½ inch at the top. Place the jars in a boiling water bath; 15, minutes for pint sized jars and 20 minutes for quart sized ones.

handled, separated like two pancakes, and moved to another container. The yeast in the tapped liquid will continue to survive.

It is important to check your mushroom for any signs of mold. Any green spots mean that the mushroom has been compromised—it is best at this point to discard both the kombucha and the mushroom and begin again.

Kombucha

Kombucha is a medicinal tea, reputed to have healing benefits. It can be brewed by fermenting tea using a visible, solid mass of yeast and bacteria which forms the kombucha culture , called the "mushroom".

To brew kombucha: Start with any tea of choice. Black tea is typically used, but green tea is fine, too. Steep 5 tea bags (or 2 to 3 teaspoons loose leaf tea) in 3 quarts of purified water. Allow to steep for at least 15 minutes, remove tea bags or leaves, add 1 cup of sugar (necessary for the bacteria to survive), and allow to cool. The tea being sweetened is important, as the bacteria will use the sugar for food. Add the culture (purchased online or secured from a fellow kombucha-maker) and 1 ½ to 2 cups of previously-brewed kombucha tea (your starter) to the tea and cover it with a cloth; the cloth keeps out dust while still allowing for air flow, so be sure the cloth is porous and not too thick. Let the kombucha sit for one to two weeks at room temperature. During this time the mushroom will form at the top of the liquid. The longer the kombucha sits the more acidic it will become.

When the kombucha has become the taste you wish it to be the liquid can be "tapped." Always reserve some of the liquid from the tapped jar; it needs to stay with its mushroom to maintain a balance within the jar when new tea is added. In each batch, the mushroom culture will produce a "daughter," which can be directly

Chilled kefir soup makes a refreshing summer lunch or light dinner.

FREEZING

Many foods preserve well in the freezer and can make preparing meals easy when you are short on time. If you make a big pot of soup, serve it for dinner, put a small container in the refrigerator for lunch the next day, and then stick the rest in the freezer. A few weeks later you'll be ready to eat it again and it will only take a few minutes to thaw out and serve. Many fruits also freeze well and are perfect for use in smoothies and desserts, or served with yogurt for breakfast or dessert. Vegetables frozen shortly after harvesting keep many of the nutrients found in fresh vegetables and will taste delicious when cooked.

they make the frozen contents much easier to get out. They are also generally reusable and make it easier to stack foods in the refrigerator. When using rigid containers, be sure to leave headspace so that the container won't explode when the contents expand with freezing. Covers for rigid containers should fit tightly. If they do not, reinforce the seal with freezer tape. Freezer tape is specially designed to stick at freezing temperatures. Freezer bags or aluminum foil are good for meats, breads and baked goods, or fruits and vegetables that don't contain much liquid. Be sure to remove as much air as possible from bags before closing.

Containers for Freezing

The best packaging materials for freezing include rigid containers such as jars, bottles, or Tupperware, and freezer bags or aluminum foil. Sturdy containers with rigid sides are especially good for liquids such as soup or juice because

Headspace to Allow Between Packed Food and Closure

Headspace is the amount of empty air left between the food and the lid. Headspace is necessary because foods expand when frozen.

Type of Pack	Container with Wide Opening		Container with Narrow Opening	
	Pint	Quart	Pint	Quart
Liquid pack*	½ inch	1 inch	¾ inch	1 ½ inch
Dry pack**	½ inch	½ inch	½ inch	½ inch
Juices	½ inch	1 inch	1 ½ inch	1 ½ inch

*Fruit packed in juice, sugar syrup, or water; crushed or puréed fruit.
**Fruit or vegetable packed without added sugar or liquid.

Foods That Do Not Freeze Well

Food	Usual Use	Condition After Thawing
Cabbage*, celery, cress, cucumbers*, endive, lettuce, parsley, radishes	As raw salad	Limp, waterlogged; quickly develops oxidized color, aroma, and flavor
Irish potatoes, baked or boiled	In soups, salads, sauces or with butter	Soft, crumbly, waterlogged, mealy
Cooked macaroni, spaghetti, or rice	When frozen alone for later use	Mushy, tastes warmed over
Egg whites, cooked	In salads, creamed foods, sandwiches, sauces, gravy or desserts	Soft, tough, rubbery, spongy
Meringue	In desserts	Soft, tough, rubbery, spongy
Icings made from egg whites	Cakes, cookies	Frothy, weeps
Cream or custard fillings	Pies, baked goods	Separates, watery, lumpy
Milk sauces	For casseroles or gravies	May curdle or separate
Sour cream	As topping, in salads	Separates, watery
Cheese or crumb toppings	On casseroles	Soggy
Mayonnaise or salad dressing	On sandwiches (not in salads)	Separates
Gelatin	In salads or desserts	Weeps
Fruit jelly	Sandwiches	May soak bread
Fried foods	All except French fried potatoes and onion rings	Lose crispness, become soggy

* Cucumbers and cabbage can be frozen as marinated products such as "freezer slaw" or "freezer pickles." These do not have the same texture as regular slaw or pickles.

Effect of Freezing on Spices and Seasonings

- Pepper, cloves, garlic, green pepper, imitation vanilla and some herbs tend to get strong and bitter.
- Onion and paprika change flavor during freezing.
- Celery seasonings become stronger.
- Curry develops a musty off-flavor.
- Salt loses flavor and has the tendency to increase rancidity of any item containing fat.
- When using seasonings and spices, season lightly before freezing, and add additional seasonings when reheating or serving.

Frozen berries are perfect for smoothies and milkshakes.

How to Freeze Vegetables

Because many vegetables contain enzymes that will cause them to lose color when frozen, you may want to blanche your vegetables before putting them in the freezer. To do this, first wash the vegetables thoroughly, peel if desired, and chop them into bite-size pieces. Then pour them into boiling water for a couple of minutes (or cook longer for very dense vegetables, such as beets), drain, and immediately dunk the vegetables in ice water to stop them from cooking further. Use a paper towel or cloth to absorb excess water from the vegetables, and then pack in resealable airtight bags or plastic containers.

Blanching Times for Vegetables

Artichokes	3-6 minutes
Asparagus	2-3 minutes
Beans	2-3 minutes
Beets	30-40 minutes
Broccoli	3 minutes
Brussels sprouts	4-5 minutes
Cabbage	3-4 minutes
Carrots	2-5 minutes
Cauliflower	6 minutes
Celery	3 minutes
Corn (off the cob)	2-3 minutes
Eggplant	4 minutes
Okra	3-4 minutes
Peas	1-2 minutes
Peppers	2-3 minutes
Squash	2-3 minutes
Turnips or Parsnips	2 minutes

How to Freeze Fruits

Many fruits freeze easily and are perfect for use in baking, smoothies, or sauces. Wash, peel, and core fruit before freezing. To easily peel peaches, nectarines, or apricots, dip them in boiling water for 15 to 20 seconds to loosen the skins. Then chill and remove the skins and stones.

Berries should be frozen immediately after harvesting and can be frozen in a single layer on a paper towel-lined tray or cookie sheet to keep them from clumping together. Allow them to freeze until hard (about 3 hours) and then pour them into a resealable plastic bag for long-term storage.

Some fruits have a tendency to turn brown when frozen. To prevent this, you can add ascorbic acid (crush a vitamin C in a little water), citrus juice, plain sugar, or a sweet syrup (1 part sugar and 2 parts water) to the fruit before freezing. Apples, pears, and bananas are best frozen with ascorbic acid or citrus juice, while berries, peaches, nectarines, apricots, pineapple, melons, and berries are better frozen with a sugary syrup.

How to Freeze Meat

Be sure your meat is fresh before freezing. Trim off excess fats and remove bones, if desired. Separate the meat into portions that will be easy to use when preparing meals and wrap in foil or place in resealable plastic bags or plastic containers. Refer to the chart to determine how long your meat will last at best quality in your freezer.

Meat	Months
Bacon and sausage	1 to 2
Ham, hotdogs, and lunchmeats	1 to 2
Meat, uncooked roasts	4 to 12
Meat, uncooked steaks or chops	4 to 12
Meat, uncooked ground	3 to 4
Meat, cooked	2 to 3
Poultry, uncooked whole	12
Poultry, uncooked parts	9
Poultry, uncooked giblets	3 to 4
Poultry, cooked	4
Wild game, uncooked	8 to 12

GINGERBREAD HOUSES

Making gingerbread houses is a fun and creative way to spend a winter afternoon. You can keep things simple or get very elaborate, gather a group of friends or dive into the project single-handedly. Here are some directions to get you started from gingerbread masters Danette and Katie Morris.

Gingerbread House Recipe

- 2 ½ C packed dark brown sugar
- 1 ½ C heavy or whipping cream
- 1 ¼ C molasses
- 9 ½ C flour
- 2 T baking soda
- 1 T ground ginger

1. In a very large bowl, with wire whisk beat brown sugar, cream and molasses until sugar lumps dissolve and mixture is smooth. In medium bowl, combined flour, soda and ginger. With spoon, stir flour mixture into cream mixture in 3 additions until dough is smooth.

2. Divide dough into 4 equal portions; flatten each into a rectangle to speed chilling. Wrap each piece well with plastic wrap and refrigerate at least 4 hours or overnight, until dough is firm enough to roll.

Roll dough

1. Grease and flour large cookie sheets. Roll out dough, 1 rectangle at a time, on each cookie sheet to about ⅛ inch thickness (put damp paper towels under cookie sheets to keep them from shifting while you roll). If dough is too crumbly, add water a tablespoon at a time until it is elastic.

2. Trim excess dough from sheet; wrap and reserve in refrigerator. Chill rolled dough on cookie sheet for 10 minutes.

3. Place floured cardboard patterns on dough and cut pieces making sure to leave 1 ¼ inches between pieces for expansion. Combine and reroll as necessary.

Bake dough

1. Preheat oven to 350°F.
2. Brush pieces lightly with water before baking. Bake 25 to 30 minutes or until pieces are firm to the touch—don't overbake.
3. While gingerbread is warm, place pattern pieces on top and re-trim if necessary.

Momma's note: I basically ignore the "Roll dough" section and just roll it out on a counter (cold soapstone helps here), cut the pieces, transfer and reshape on the sheet and bake. I think the key is to let them harden for a day before constructing.

Royal Icing

> 1 lb. confectioner's sugar
> ½ tsp cream of tartar
> 3 egg whites

In bowl of electric mixer mix all ingredients until combined. Set speed on high and beat at least 7 minutes. Store in airtight container.

Construct the House

While the tree house seen above was constructed with the help of two civil engineering degrees, a simple gingerbread house is easy to build and a lot of fun to make, especially with extra hands. If you do happen to attempt the architectural feat alone, it's a good idea to grab some canned goods (or anything, really) to prop up the pieces while the icing dries.

1. Choose some sort of solid base for the house: a cookie sheet, a piece of cardboard, anything that is flat and sturdy. Some people decide to line their base with aluminum foil or wax paper, but that is entirely up to you.
2. As you build, make sure to prop your pieces for a sufficient amount of time so that the icing can dry. If the different sides of the house are weak, they're more likely to fall over, obviously. The more times you have to fix it, the more likely it is that you'll break one of the pieces. If one of the pieces does happen to break, you can always mortar it back together with the royal icing. Make sure to wait at least an hour for the walls of the house to dry before attempting to place the roof.

3. Once the royal icing has dried, leaving the base structure sturdy, you can build add the roof. Once you've done this, it's a good idea to let the house stand for at least another hour before you add the decorations.

4. The decorations on a house depend solely on what you want the house to look like and which candies and snacks you would like eating the most while decorating. Be creative when you decorate the house and the yard around it. Think about: roof tiles, pathways, walls, windowsills, shrubs and other plants. Perhaps you even want to attempt a rope swing. The possibilities are endless.

ICE CREAM

Is there anything more summery and delicious than homemade ice cream? There are many electric ice cream makers on the market that make the process incredibly quick and simple. But if you want to make your own simple ice cream maker as well as the ice cream itself, give this a try.

Supplies

1-pound coffee can
3-pound coffee can
Duct tape
Ice
1 cup salt

Ingredients

2 cups half and half
½ cup sugar
1 teaspoon vanilla

Directions

1. Mix all the ingredients in the 1-pound coffee can. Cover the lid with duct tape to ensure it is tightly sealed.
2. Place the smaller can inside the larger can and fill the space between the two with ice and salt.
3. Cover the large can and seal with duct tape. Roll the can back and forth for 15 minutes. To reduce noise, place a towel on your working surface, or work on a rug.
4. Dump out ice and water. Stir contents of small can. Store ice cream in a glass or plastic container (if you leave it in the can it may take on a metallic flavor).

If desired, add cocoa powder, coffee granules, crushed peppermint sticks or other candy, or fruit.

MAPLE SYRUP

Making maple syrup is a time-consuming process, but if you've ever tasted the sweet amber liquid straight from the boiling pan, you'll understand why North Americans have been doing it since the Native Americans first figured out how to draw the sap from the trees. The equipment for a small-scale sugaring operation is minimal; in fact, you probably already own what you need. The sap begins running when the nights are below freezing and the days are above freezing, which generally happens from mid-February through early April. Scope out your land for maple trees and get your supplies ready ahead of time so you'll be ready to go as soon as the snow begins to thaw. You can expect each tap to yield an average of 10 gallons of sap per season, which will give you about a quart of syrup.

Terms You Should Know

Tap: To drill a hole in the tree—without causing any permanent damage—and to insert a spile into the hole to capture the sap that collects there. A bucket is connected to the spile, its purpose being to amass the dripping sap so it can be evaporated.

Spile: A spout.

Evaporate: The process of boiling the sap to make maple syrup. When the sap is collected, it's comprised of water mostly (98%, in fact), when finished maple syrup should be 33% water and 67% sugar. It takes about forty gallons of sap to make one gallon of syrup.

Sugar sand: Minerals and nutrients that condense as the excess water boils away. The sugar sap must be filtered out, otherwise the syrup will look cloudy.

What You'll Need

1. Sugar Maple trees (also known as Hard Maples or Rock Maples) are the best to tap because they have the most sugar in their sap, but Red Maples, Silver Maples, and Ash-leafed Maples can also be tapped. Their sap is not as sweet, however, and the sap doesn't flow as long. Trees should be at least 10 inches in diameter and 4 ½ feet

above the ground. For every additional 8 inches, another tap can be added. If your tree has a lot of branches, it's likely that more sap will be produced. Tapping a tree when it's too young or adding too many taps to any tree can permanently damage it.

2. A drill with a $\frac{7}{16}$ inch diameter drill bit.

3. A spile (normally metal or plastic) for each taphole. You can buy these at feed stores or farm supply stores. Making one of your own is an option, as well. If you can acquire an elderberry stem a little larger than the hole you'll be making in the tree, then you can cut it until it is 4 or 5 inches long. Sharpen the end that will go in the tap hole, and then use a skinny rod to push the heart wood out of the center.

4. A rust-proof bucket or container with a cover. You need a cover so rainwater doesn't dilute the sap and to prevent bark, insects, and other uninvited guests from falling into the sap.

5. Storage or collection containers for the sap before it gets processed. You could use clean, galvanized or plastic trash cans, large pails, or something similar.

6. A pan with high sides for boiling sap—stainless steel works best.

7. A thermometer calibrated to at least 230 degrees Fahrenheit.

8. A stove of sorts, preferably outdoors so the steam doesn't damage anything in your home (warning: the steam will cause your wallpaper to peel off the walls if you decide to boil inside). You could build an outdoor fire pit, or use a camp stove or even an outdoor grill.

9. A filter or cheesecloth for filtering hot, finished syrup.

10. Containers for storing the finished maple syrup; you can use metal containers or glass jars that will seal. Canning jars are perfect for this.

How to Collect Sap

1. Find your tree and locate a good spot to tap. Look for unblemished bark, and if there were taps there before, do not put yours directly above or under the old holes. Keep your hole at least 4 inches from the side of the former taphole. Once you've found the spot, drill, with your $\frac{7}{16}$ inch in diameter drill bit, a hole in the trunk that is 2-3 inches deep.

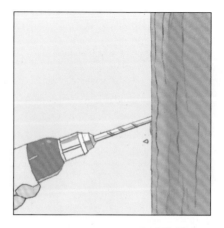

2. Clean any loose wood out of the hole and lightly tap a spile into it with a hammer. You want to make sure the spile is tight enough that it cannot be pulled out by hand, but you do not want to hit it too hard so that the tree splits.

3. Hang your container from the spile so it can catch the sap and place the lid on top.

4. When the containers fill, the sap should be collected and refrigerated until you are ready to boil it. It should be processed as soon as possible.

Once it's collected, it's time to process the sap.

Boiling

1. Depending on what kind of stove you decide to employ, you should use dry, fast-burning wood to provide the heat that is necessary for boiling.
2. Get your sap from the refrigerator and pour it into your pan. You should fill the pan, but not all the way to the top, because you don't want it to boil over. Rubbing butter or margarine on the rim of the pan also helps prevent this.
3. Take the temperature of the sap when it first comes to a boil. This will help you know when it is finished.
4. As the water boils away, add more sap to the pan. Never have less than an inch of sap in the pan.
5. You can transport the sap indoors to finish it off, if desired. When you've run out of sap to add to the boiling pan (or you're ready to quit for the day), allow the sap to thicken until it's more like syrup than water and carefully carry the pan to your kitchen stove.

6. The syrup will have the right concentration of sugar when it is 7 degrees Fahrenheit hotter than when it was first boiled. At or near sea level, the temperature will be about 219 degrees for finished syrup. Make sure to watch the syrup keenly at this point to prevent scorching or boiling over.
7. Once finished, strain the syrup through the cheesecloth or filter to remove any bits of sugar sand that are left over.
8. After you've filtered, the syrup should be reheated to at least 180 degrees or almost to boiling before pouring it into containers for storage.
9. Fill the containers so that there is little air left in them and screw on the covers tightly while the bottles are still warm. Since the sugar content is so high, your syrup can be stored at room temperature without spoiling and in the refrigerator or freezer without freezing.

You could sell your maple syrup, keep it for yourself, or, by boiling the sap a little longer, make maple cream, maple sugar, or maple candy.

A sugar shack in the fall.

MUSHROOMS

A walk through the woods will likely reveal several varieties of mushrooms, and chances are that some are the types that are edible. However, because some mushrooms are very poisonous, it is important never to try a mushroom of which you are unsure. Never eat a mushroom with gills, or, for that matter, any mushroom that you cannot positively identify as edible. Also, never eat mushrooms that appear wilted, damaged, or rotten.

Here are some common edible mushroom that you can easily identify and enjoy.

Chanterelles

These trumpet-shaped mushrooms have wavy edges and interconnected blunt-ridged gills under the caps. They are varied shades of yellow and have a fruity fragrance. They grow in summer and fall on the ground of hardwood forests. Because chanterelles tend to be tough, they are best when slowly sautéed or added to stews or soups.

Notes: Beware of Jack O'Lantern mushrooms, which look and smell similarly to chanterelles. Jack O'Lanterns have sharp knifelike gills instead of the blunt gills of chanterelles, and generally grow in large clusters at the base of trees or on decaying wood.

Chanterelles

Morel mushroom

Coral Fungi

These fungi are aptly named for their bunches of up-ward-facing branching stems, which look strikingly like coral. They are whitish, tan, yellowish, or sometimes pinkish or purple. They may reach 8 inches in height. They grow in the summer and fall in shady, wooded areas.

Notes: Avoid coral fungi that are bitter, have soft, gelatinous bases, or turn brown when you poke or squeeze them. These may have a laxative effect, though are not life-threatening.

Coral fungi

Morels

Morels are sometimes called sponge, pine cone, or honeycomb mushrooms because of the pattern of pits and ridges that appears on the caps. They can be anywhere from 2 to 12 inches tall. They may be yellow, brown, or black and grow in spring and early summer in wooded areas and on river bottoms. To cook, cut in half to check for insects, wash, and sauté, bake, or stew.

Notes: False morels can be poisonous and appear similar to morels because of their brainlike irregularly shaped caps. However, they can be distinguished from true morels because false morel caps bulge inward instead of outward. The caps have lobes, folds, flaps, or wrinkles, but not pits and ridges like a true morel.

Morel

Puffballs

These round or pear-shaped mushrooms are often mistaken for golf balls or eggs. They are always whitish, tan, or gray and sometimes have a thick stem. Young puffballs tend to be white and older ones yellow or brown. Fully matured puffballs have dark spores scattered over the caps. Puffballs are generally found in late summer and fall on lawns, in the woods, or on old tree stumps. To eat, peel off the outer skin and eat raw or batter-fried.

Notes: Slice each puffball open before eating to be sure it is completely white inside. If there is any yellow, brown, or

Puffball

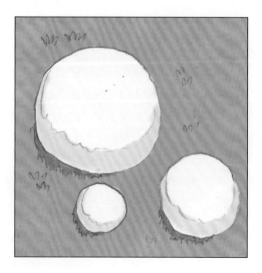

black, or **if there is a developing mushroom inside with a stalk, gills, and cap, do not eat!** Amanitas, which are very poisonous, can appear similar to puffballs when they are young. Do not eat if the mushroom gives off an unpleasant odor.

Shaggy Mane Mushrooms

This mushroom got its name from its cap, which is a white cylinder with shaggy, upturned, brownish scales. As the mushroom matures, the bottom outside circumference of the cap becomes black. Shaggy manes are generally 4 to 6 inches tall and grow in all the warm seasons in fields and on lawns.

Shaggy manes are tastiest eaten when young, but they're easiest to identify once the bottoms of the caps begin to turn black. They are delicious sautéed in butter or olive oil and lightly seasoned with salt, garlic, or nutmeg.

Shaggy Mane Mushroom

Amanita mushrooms are very poisonous. Do not eat any mushroom that resembles an amanita.

289

ROOT BEER

The primary ingredients that give root beer its distinctive flavor are sassafrass, vanilla, and wintergreen. Other optional ingredients include ginger, licorice, and sarsaparilla. Sassafras contains the chemical known as safrole, which has fairly recently been shown to be a carcinogen in laboratory animals and has been banned by the USDA. Though the amount of safrole you would consume in home brewed root beer is probably not enough to cause worry, it can't hurt to use sassafrass extract that has had the safrole removed (available online and from some health food stores).

If you're gathering your roots, be sure to rinse them thoroughly and cut thicker ones in half. When using leaves, be sure to use only the leaves and not the stems, roots, or flowers. Leaves should be rinsed thoroughly and then dried.

Root Beer

¼ ounce wintergreen leaves

¼ ounce sarsaparilla root

½ ounce sassafras root bark or 8 to 10 drops sassafras extract

1 1-inch piece ginger root, unpeeled and thinly sliced

½ ounce burdock root

¼ ounce licorice root

⅛ teaspoon active dry yeast

1 cup molasses

1 cup sugar

1 teaspoon vanilla extract

1 gallon water

1. Place roots and leaves in 2 quarts water in a large pot and bring to a boil. Remove from heat and allow to steep for 2 hours.

2. Strain through a cheesecloth-lined strainer into a clean pot. Discard roots and leaves. Add remaining 2 quarts of water to infused liquid.

3. Add yeast, molasses, vanilla, and sugar to liquid, cover, and set aside to ferment for about 20 minutes.

4. Use a funnel to fill 4 1-liter plastic soda bottles. Leave at least 2 inches of space at the top of each bottle. Screw on lids and store at room temperature for 12 hours.

5. Chill root beer in the refrigerator for at least 2 days, or until it reaches the desired fizziness. After 4 or 5 days the root beer will become slightly alcoholic. Remove caps slowly to allow gas to escape gradually.

Sassafras leaves

291

Plant a Row for the Hungry (PAR) Program

One in ten households in the United States experiences hunger or the risk of hunger every year. This is an astounding number of adults and children who are not receiving the proper foods, nutrients, and sustenance to maintain a healthy lifestyle. What can you do as a gardener to help provide fresh produce to the hungry in your community? Churches, schools, and other local organizations try to provide food for those who are hungry by giving to local food banks and shelters and by establishing soup kitchens and other programs. Often these food drives can only accept nonperishable items, but, recognizing the need for fresh fruits and vegetables in a healthy diet, many organizations are beginning to take produce on a conditional basis. If they can use the produce in their prepared meals or distribute them to needy families before the food begins to perish, they will accept it.

Started in 1995 in Anchorage, Alaska, by Jeff Lowenfels as a public service program of the Garden Writers Association (GWA) and Foundation, PAR is a way for those who garden and grow their own vegetables and fruit to help combat hunger and poverty in their local communities. PAR encourages local gardeners to plant one extra row of produce in their gardens and then to donate that harvest to neighbors or others in the community who struggle to feed themselves and their families. Donating excess vegetables and fruit to your local food bank, soup kitchen, shelter, or other food agency to help feed the hungry in your local area is a wonderful way to share your love of gardening and to be an active and important member of your community.

Anyone in the U.S. or Canada can participate in the PAR program, and members grow and donate over one million tons of food a year to help fight hunger in their local communities. Some people establish a community garden that solely grows food to give to local pantries and soup kitchens. Others simply grow additional crops and give individually to those in need.

If you want to become involved in the PAR program, it is best to call or e-mail the GWA in order to receive more information about PAR programs in your area. It is also beneficial to call or visit local food agencies and see what crops they need most, or what types of fruits and vegetables they would like to provide for their customers. Most agencies seek out food that can be shelved for a few days if necessary and also foods that are high in nutritional value. They also accept fresh herbs that they can then use when making foods, such as soups. Flowers are also acceptable donations.

When thinking about growing excess vegetables and fruit in your garden for donation, it is good to select easily grown crops that will help to maximize your harvest and will encourage you to be enthusiastic about growing and tending the plants. Some easily grown and highly desirable plants are: beans, cucumbers, peas, radishes, summer squash, and tomatoes. Then, once your produce is ready for harvesting, gather up the excess and take it to a selected PAR drop-off site or, if acceptable, directly to the food agency.

There are some guidelines for growing fruits and vegetables that are acceptable for donation. Always be sure to contact your local food banks and soup kitchens for information on what they need in terms of fresh produce. Be sure to space your plantings apart so that your growing season and harvesting season are extended over a longer period of time. That way, you'll be able to donate longer. Choose to grow produce that lasts well and stays fresher longer. Pick your ripe produce promptly and be sure to clean it of any dirt (but don't wash it). And, of course, don't ever give away overripe or spoiled produce.

How to Start Your Own PAR Program

What if your local community does not yet have a PAR program? It is easy to establish your own program by contacting PAR (either by e-mail, by visiting the GWA website, or by calling toll free; see contact information below). This way, you will be able to gain access to information on creating a successful PAR program. Next, you should try to recruit volunteers (neighbors, community gardeners, garden clubs, garden centers, and nurseries) to be a part of the program. It is important to establish a local coordinator who can be a "go-to" person for any questions volunteers may have about the program, growing tips, and drop-off sites.

In addition to gathering volunteers, it is essential to find a food distribution agency partner who wants to collect the donated produce. This may be a food bank, food pantry, soup kitchen, or local shelter. It is also wise to find someone to market and publicize your local program. Getting the word out about the program will help gather support and more gardeners who want to help. There should also be someone appointed to help coordinate and oversee the drop-off sites and to take food to the designated area if need be.

For a PAR program to be a success, it is a good idea to reach out to your local extension services; community, church, and school gardens; businesses; and local food agencies to see if a specific PAR garden can be established. It is also wise to ask farmers to donate their unsold produce from farmers' markets and trucks in the area. All of these various elements will help you to create a successful fresh food donation program in your area and will ensure that those who are hungry in your community will not be lacking in fresh produce in their diets.

For more information on Plant a Row for the Hungry, please call, e-mail, or visit the Garden Writers Association at:

1-877-GWAA-PAR

par@gwaa.org

www.gwaa.org

To find a local PAR contact in your area, please call the Foodchain at: 1-800-845-3008.

YOGURT

Yogurt is basically fermented milk. You can make it by adding the active cultures *Streptococcus thermophilus* and *Lactobacillus bulgaricus* to heated milk, which will produce lactic acid, creating yogurt's tart flavor and thick consistency. Yogurt is simple to make and is delicious on its own, as a dessert, in baked goods, or in place of sour cream.

Yogurt is thought to have originated many centuries ago among the nomadic tribes of Eastern Europe and Western Asia. Milk stored in animal skins would acidify and coagulate. The acid helped preserve the milk from further spoilage and from the growth of pathogens (disease-causing microorganisms).

Ingredients

Makes 4 to 5 cups of yogurt

- **1 quart milk** (cream, whole, low-fat, or skim)—In general the higher the milk fat level in the yogurt, the creamier and smoother it will taste. **Note:** If you use home-produced milk it *must* be pasteurized before preparing yogurt. See page 295 for tips on pasteurizing milk.

- **Nonfat dry milk powder**—Use ⅓ cup powder when using whole or low-fat milk, or use ⅔ cup powder when using skim milk. The higher the milk solids, the firmer the yogurt will be. For even more firmness add gelatin (directions below).

- **Commercial, unflavored, cultured yogurt**—Use ¼ cup. Be sure the product label indicates that it contains a live culture. Also note the content of the culture. *L. bulgaricus* and *S. thermophilus* are required in yogurt, but some manufacturers may in addition add *L. acidophilus* or *B. bifidum*. The latter two are used for slight variations in flavor, but more commonly for health reasons attributed to these organisms. All culture variations will make a successful yogurt.

- **2 to 4 tablespoons sugar or honey (optional)**

- **1 teaspoon unflavored gelatin (optional)**—For a thick, firm yogurt, swell 1 teaspoon gelatin in a little milk for 5 minutes. Add this to the milk and nonfat dry milk mixture before cooking.

Supplies

- **Double boiler or regular saucepan**—1 to 2 quarts in capacity larger than the volume of yogurt you wish to make.
- **Cooking or jelly thermometer**—A thermometer that can clip to the side of the saucepan and remain in the milk works best. Accurate temperatures are critical for successful processing.
- **Mixing spoon**
- **Yogurt containers**—cups with lids or canning jars with lids.
- **Incubator**—a yogurt-maker, oven, heating pad, or warm spot in your kitchen. To use your oven, place yogurt containers into deep pans of 110°F water. Water should come at least halfway up the containers. Set oven temperature at lowest point to maintain water temperature at 110°F. Monitor temperature throughout incubation, making adjustments as necessary.

Processing

> ### How to Pasteurize Raw Milk
>
> If you are using fresh milk that hasn't been processed, you can pasteurize it yourself. Heat water in the bottom section of a double boiler and pour milk into the top section. Cover the milk and heat to 165°F while stirring constantly for uniform heating. Cool immediately by setting the top section of the double boiler in ice water or cold running water. Store milk in the refrigerator in clean containers until ready for making yogurt.

1. **Combine ingredients and heat.** Heating the milk is necessary in order to change the milk proteins so that they set together rather than form curds and whey. Do not substitute this heating step for pasteurization. Place cold, pasteurized milk in top of a double boiler and stir in nonfat dry milk powder. Adding nonfat dry milk to heated milk will cause some milk proteins to coagulate and form strings. Add sugar or honey if a sweeter, less tart yogurt is desired. Heat milk to 200°F, stirring gently

and hold for 10 minutes for thinner yogurt, or hold 20 minutes for thicker yogurt. Do not boil. Be careful and stir constantly to avoid scorching if not using a double boiler.

2. **Cool and inoculate.** Place the top of the double boiler in cold water to cool milk rapidly to 112 to 115°F. Remove 1 cup of the warm milk and blend it with the yogurt starter culture. Add this to the rest of the warm milk. The temperature of the mixture should now be 110 to 112°F.

3. **Incubate.** Pour immediately into clean, warm containers; cover and place in prepared incubator. Close the incubator and incubate about 4 to 7 hours at 110°F, ± 5°F. Yogurt should set firm when the proper acid level is achieved (pH 4.6). Incubating yogurt for several hours past the time after the yogurt has set will produce more acidity. This will result in a more tart or acidic flavor and eventually cause the whey to separate.

4. **Refrigerate.** Rapid cooling stops the development of acid. Yogurt will keep for about 10 to 21 days if held in the refrigerator at 40°F or lower.

Yogurt Types

Set yogurt: A solid set where the yogurt firms in a container and is not disturbed.

Stirred yogurt: Yogurt made in a large container then spooned or otherwise dispensed into secondary serving containers. The consistency of the "set" is broken and the texture is less firm than set yogurt. This is the most popular form of commercial yogurt.

Drinking yogurt: Stirred yogurt into which additional milk and flavors are mixed. Add fruit or fruit syrups to taste. Mix in milk to achieve the desired thickness. The shelf life of this product is 4 to 10 days, since the pH is raised by the addition of fresh milk. Some whey separation will occur and is natural. Commercial products recommend a thorough shaking before consumption.

Fruit yogurt: Fruit, fruit syrups, or pie filling can be added to the yogurt. Place them on top, on bottom, or stir them into the yogurt.

Troubleshooting

- If milk forms some clumps or strings during the heating step, some milk proteins may have jelled. Take the solids out with a slotted spoon or, in difficult cases, after cooking pour the milk mixture through a clean colander or cheesecloth before inoculation.
- When yogurt fails to coagulate (set) properly, it's because the pH is not low enough. Milk proteins will coagulate when the pH has dropped to 4.6. This is done by the culture growing and producing acids. Adding culture to very hot milk (+115°F) can kill bacteria. Use a thermometer to carefully control temperature.
- If yogurt takes too long to make, it may be because the temperature is off. Too hot or too cold of an incubation temperature can slow down culture growth. Use a thermometer to carefully control temperature.
- If yogurt just isn't working, it may be because the starter culture was of poor quality. Use a fresh, recently purchased culture from the grocery store each time you make yogurt.
- If yogurt tastes or smells bad, it's likely because the starter culture is contaminated. Obtain new culture for the next batch.
- Yogurt has over-set or incubated too long. Refrigerate yogurt immediately after a firm coagulum has formed.
- If yogurt tastes a little odd, it could be due to overheating or boiling of the milk. Use a thermometer to carefully control temperature.
- When whey collects on the surface of the yogurt, it's called syneresis. Some syneresis is natural. Excessive separation of whey, however, can be caused by incubating yogurt too long or by agitating the yogurt while it is setting.

Storing Your Yogurt

- Always pasteurize milk or use commercially pasteurized milk to make yogurt.
- Discard batches that fail to set properly, especially those due to culture errors.
- Yogurt generally has a 10- to 21-day shelf life when made and stored properly in the refrigerator below 40°F.
- Always use clean and sanitized equipment and containers to ensure a long shelf life for your yogurt. Clean equipment and containers in hot water with detergent, then rinse well. Allow to air dry.

PART THREE

CARPENTRY AND WOODWORKING

BASIC TOOLS

Woodworking is one of the most basic and most useful skills a person dedicated to self-sufficiency can master. Once you understand the basics of what tools to use and how to use them, there's almost no limit to what you can construct. Here you will find a wide selection of the hand tools, materials, and processes used in woodworking, as well as diagrams and instructions for dozens of specific projects. Much of this section is adapted from the writings and artwork of Paul N. Hasluck, whose wealth of knowledge on the subject has served as the foundation for generations of woodworkers.

Classifying Tools

Tools can be categorized by their functions, as follows:

1. Geometrical tools for laying off and testing work: they include rules, straight-edges, gauges, etc.

2. Tools for holding and supporting work: such tools are benches, vices, stools, etc.

3. Paring or shaving tools, like chisels, spokeshaves, planes, etc.

4. Saws.

5. Percussion or impelling tools, such as hammers, mallets, screwdrivers, and (combined with cutting) hatchets and axes.

6. Boring tools, including gimlets and brace-bites, etc.

7. Abrading and scraping tools, such as rasps, scrapers, glasspaper, and implements such as whetstones, for sharpening edged tools.

The most useful tools in these categories are discussed in the following pages.

GEOMETRICAL TOOLS

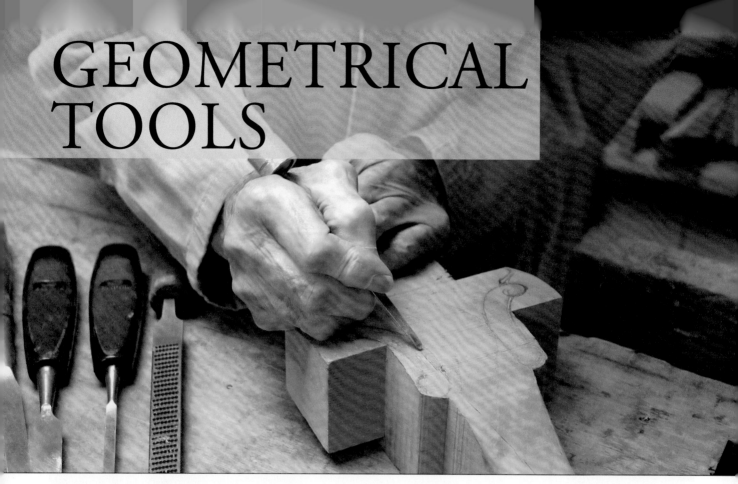

Tools for Marking and Scribing

The simplest of these is the pencil. Sharpen a flat oval section to a chisel edge; if sharpened to a point, the pencil wears away quickly and will only mark a fine, solid line for a few minutes. The greater surface area of lead in the chisel edge makes it last for longer before it requires re-sharpening. Steel scribing and marking tools are illustrated by Figures 1 to 3. The chisel end marking awl (Figure 1) and the striking knife (Figure 2) are used for all purposes of scribing and marking smooth work, where an indented line works better than a black line, the scratch providing a good starting point for edge tools. A striking knife (Figure 3) can be made by grinding down an old table knife.

Figure 1—Chisel-end Marking Awl

Figure 2—Striking Knife and Marking Awl

Straight-edge

A straight-edge 15 feet long, 6 inches wide, and 1¼ inches thick is large enough for all practical purposes of the joiner, mason, bricklayer, engineer, etc. If you're making your own, the best material is pine, as it is the least affected (permanently) by change of temperature or weather. The pine board must be cut from a straight-grown tree, since a board from a crooked trunk will not keep parallel and straight for any length of time, owing to the grain crossing and recrossing its (thickness) edge. Straight-edges are made from all parts of boards cut from whole logs, but you cannot assume they will keep perfectly straight and true for any length of time.

To test the truth of a straight-edge this size, get a clean board 1 foot longer and about 7 inches or 8 inches wide. Lay

Figure 3—Homemade Striking Knife

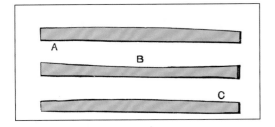

Figure 4—Whitworth Method of Testing Straight-edges

the straight strip at about the center of the board, and with a sharp pencil draw a line on the board along the trued edge of the strip, keeping the side close to the board, and making the line as fine as possible. Now turn the strip of the line, and if the trued edge is perfectly straight the line also will appear so. If the line is wavy, the edge must be planed until only one line is made when marked and tested from each side; mark a fresh line for each test, otherwise you can become confused and inaccurate. With one edge now perfectly true, you can proceed with the other edge. Set a sharp gauge to the required width, and mark the second edge lightly on each side of the rule, working the gauge from the true edge; then the wood is planed off to the gauge marks, and the second edge tested as to its being true with the first one, using the pencil line, as before. An even more precise test than the gauge line for parallelism can be done by using a pair of calipers. The points of the calipers are drawn along the edges, and if they are perfectly parallel, there will be no easy or hard places, the presence of which might possibly not be detected by the pencil line. If the edges will stand both these tests, the strip is perfectly straight and parallel.

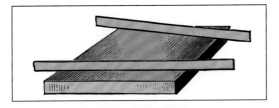

Figure 5—Testing Surface with Straight-edges

Whitworth Method of Testing Straight-edges

Sir J. Whitworth's famous method of trueing engineers' straight-edges should interest any woodworker. Three straight-edges are prepared individually, and each is brought to a moderate state of accuracy; two of them, A and B (Figure 4), are compared with each other by placing them edge to edge, and any irregularities found are removed. The process is then repeated until A and B fit each other perfectly. The third straight-edge, C, now is compared with both A and B, and when it fits both perfectly, then there is no doubt whatsoever that the three are straight. Why this is the case becomes obvious when you remember that though A may be rounded instead of being straight, and B may be hollow sufficiently to make them fit each other perfectly, it is impossible for C to fit both the rounded and the hollow straight-edge.

Testing Surfaces with Straight-edges

How surfaces are tested for winding with straight-edges is shown by Figure 5, from which it is obvious that if the work has warped ever so slightly a true straight-edge must disclose the fact, as it could not then lie flat on its edge across the work. If the board is in winding, each straight-edge will magnify the error. If the winding is wavy, the edges will touch at certain points, and in other places light will be seen between them and the work. Taking a sight from one straight-edge to the other is another test you can perform.

Rules

A 2-feet four-fold boxwood rule (Figure 6) is the best for the all-round purposes of the joiner; and for those who can use the slide rule, the tool shown in Figure 7 would be handy. A simple 2-feet two-fold rule (Figure 8) is useful, but the rule with double arch joints shown by Figure 9 will be your best bet. The average worker will find a simple rule preferable to an elaborate one. Figure 10 shows a combined rule and spirit level, the rule joint also being set out to serve as a protractor.

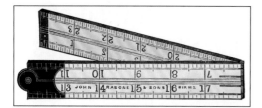

Figure 6—Two-feet Four-fold Rule

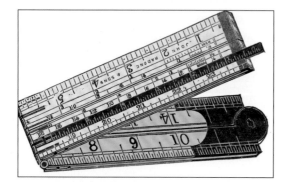

Figure 7—Rule with Brass Slide

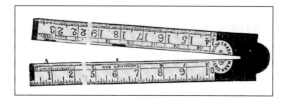

Figure 8—Two-feet Two-fold Rule

This tool may prove useful in special circumstances, but its use as a spirit level is not recommended. It is preferable to have rule and level as two distinct tools.

Dividing a Board with a Rule

A simple way to divide a board of any width into any number of parts is illustrated by Figure 11. Suppose a board of 9 inches is to be cut into six equal parts; place the 1-foot rule so that its ends touch the opposite edges of board, as shown in Figure 11; draw a line right across, and upon this line mark off from the rule every 2 inches, as 2, 4, 6, etc. Remove the rule, and draw lines parallel with the edge of the board, intersecting with the marks upon the oblique line, thus obtaining six parts, each really 1½ inches wide.

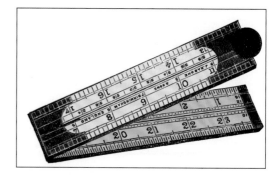

Figure 9—Rule with Double Arch Joints

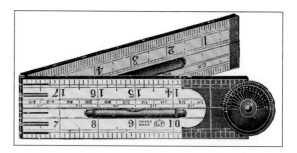

Figure 10—Rule with Spirit Level

The principle of this is simple: 2 inches is the one-sixth part of 1 foot, and whatever be the slant of the rule across the board (and the narrower the board, the greater the slant) each 2-inch mark must denote a one-sixth less than 24 inches length or width is to be divided into eight parts; then as 3 inches is one-eighth of 2 feet, use a 2-foot rule in the same way as before, and mark off at every 3 inches.

Squares and Bevels

Woodworkers constantly use squares for setting out and testing work. The simplest is the try square. A combination try and miter square is shown by Figure 12. This has an iron stock hollowed out to lower its weight to that of a wooden one. This is a useful and cheap tool, very unlikely to get out of truth. A patent adjustable try square is illustrated by Figure 13. The set screw clamps the blade in the stock just where it may be most convenient for awkward work such as puttingbutts, locks, and other fittings on doors and windows. The graduated blade is very useful. The sliding bevel is handy for setting off angles in duplicate, since by using the set screw, the blade can be made to assume any angle

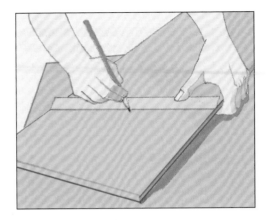

Figure 11—Dividing with Rule

with the stock. Figure 14 shows a bevel with a simple ebony stock, and Figure 15 shows one with an ebony stock framed in brass, this protection keeping the edges true for an almost unlimited period. The joiner's steel square is a mere right angle of steel, sometimes nickel plated, graduated in inches, ¼ inch, and ¹⁄₁₆ inch.

Testing and Correcting Try square

A carpenter's try square that is thought to be untrue may be tested in the following way. Get a piece of board with an edge that has been proven to be straight, apply the square as shown at A (Figure 16), and draw a line; then turn the square as at B, and if it is true the blade should fit the line; if it is less than a right angle it will be as seen in C and D (Figure 16), and if more than a right angle the defect will be as indicated at E and F (Figure 16). If the blade has moved or has been knocked out of truth through a fall, it should

be knocked back into its proper position, and, when true, the rivets should be tightened by careful hammering. If the blade is too fast in the stock to be knocked back, it should be filed true.

Crenellated Squares

A crenellated square has a tongue in which there is a series of crenellations or notches at the graduations. It is especially useful in marking off mortises, though it is available for all other ordinary applications. Three sides of a piece of timber can be set out without moving the work. To use this square, say in marking out a mortise or tenon, take it in the left hand and lay its tongue upon the surface of the work, as in Figure 17. The lower end of the main arm is lowered for

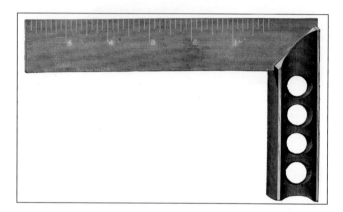

Figure 12—Iron Frame Try Square

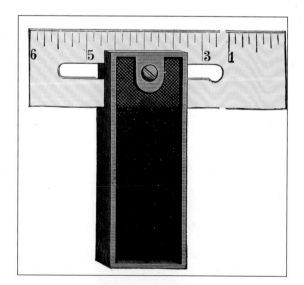

Figure 13—Adjustable Try Square

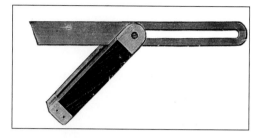

Figure 14—Ordinary Sliding Bevel

2 inches or so from the surface to get a better purchase, and then an awl, held in the right hand, is placed in a notch at the correct distance from the edge to mark the left-hand edge of the mortise or left hand face of the tenon as the case may be. Then push the square forward, pressing it down gently on the work to make one mark. Next, replace the square and place the awl in another notch at the thickness of the tenon or width of the mortise to make a second mark. Horizontals are drawn by means of the smooth edge of the tongue, as shown in Figure 17.

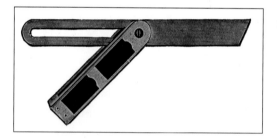

Figure 15—Brass Frame Sliding Bevel

Marking Work for Sawing

The chalk line, pencil and rule, and scribe are all used for marking lines to guide the saw. The chalk line is used for long pieces of timber; pencil and rule for ordinary and roughly approximate work, and lastly, the scribe is used for the most accurate sawing. Lining off a plank or board for ripping, when rough on the edges, is commonly done with a straightedge or chalk line. If it is square-edged, it can be done by the rule and pencil, which is explained in Figure 18. Hold the rule in your left hand, measuring off on the board the breadth to be ripped, and place your forefinger against the edge to act as a fence. The pencil is held in your right hand and placed at the end of the rule on the board. Then move

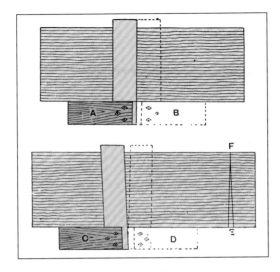

Figure 16—Testing Try Squares in Truth

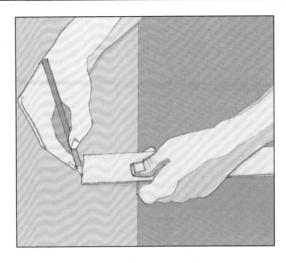

Figure 18—Lining Board with Rule and Pencil

both hands simultaneously, and you can trace the required line backward or forward, as may be the case. Lines for cross-cutting, when square across or at right angles to the edge, are easily obtained by the square, as long as you keep its blade flat on the board or plank and the stock hard to the edge (see Figure 19). Use the miter square for lines at an angle of 45 degrees to the edge (Figure 20), and for other angles, set and apply the bevel-stock in the same way (see Figure 21). The bevel-stock differs from the square only in that it has a moveable blade and is capable of being adjusted at any desired angle with the stock by using a screw. To use the chalk line method of marking (Figure 22), whiten a piece of fine cord with chalk and pull it taut between two points whose positions are marked to correspond with the end of the cutting

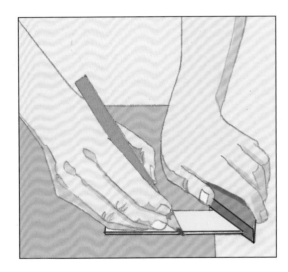

Figure 19—Squaring Line on Board

Figure 17—Marking Mortise with Crenellated Square

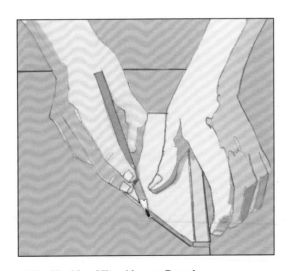

Figure 20—Marking Miter Line on Board

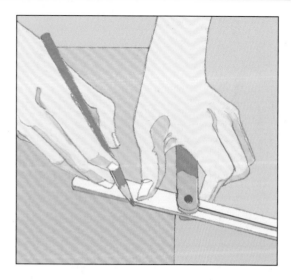

Figure 21—Using Sliding Bevel

Figure 22—Using Chalk Line

line. Then lift the chalk line vertically at or near the center, and as it's suddenly released, it chalks a perfectly straight and fine line on the timber, and furnishes a correct guide for you to saw. Lines are marked with the timber scribe in such cases as squaring the ends of planed stuff or in marking dovetails and tenons. The saw can then be made to cut close outside the scribed line, allowing for sufficient margin of material to be removed with the plane; or the saw can pass along the scribed line, as in cutting dovetails and tenons, with no after-finish required. In either case, use the scribed line over the pencil-marked one, because the cutting can be done much more accurately in the first case than the latter. Also, when the end of a piece of timber has to be squared with the plane,

there is, besides having greater accuracy, much less risk of spalting or breaking out of the grain occurring with scribed lines than with pencil-marked lines.

Marking and Cutting Gauges

The carpenter draws a line at a short distance from, and parallel to, the edge of a board by using a rule and pencil, the method previously described and seen in Figure 18. The use of the pencil or marking gauge is an advantage over this method. Figures 23 and 24 show that there are two ways to make the pencil gauge. It can be made of any hard wood, such as beech. The stem can be round (Figure 23) or square (Figure 24), and the head may be round or octagonal. Make sure the head can slide up and down the stem easily, but without sideplay. The gauge may be made to use up odd pieces of lead pencil, and these should be sharpened (with a chisel)

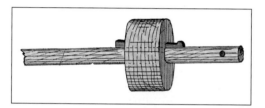

Figure 23—Pencil Gauge with Round Stem

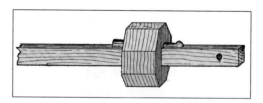

Figure 24—Pencil Gauge with Square Stem

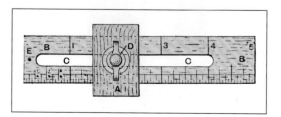

Figure 25—Rule Pencil and Cutting Gauge

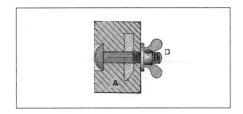

Figure 26—Section through Rule Gauge

Figure 27—Improved Pencil Gauge

to a wedge-shaped point. Figures 25 and 26 show a pencil gauge made from a broken rule fitted into a block so as to run easily, and secured at any distance (as indicated by the rule's edge) using a thumbscrew. A represents a block of birch wood, 1½ inches by 1 inch by 1 inch, mortised so it receives the rule. B is a 5-inch length of an ordinary rule, with a slot C just large enough to fit screw D, which is fixed to block A.

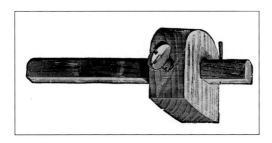

Figure 28—Ordinary Marking Gauge

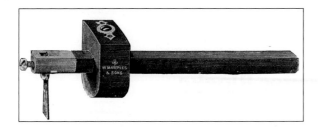

Figure 29—Cutting Gauge

The thickness of the wood between the washer and the rule should be only 1⅛ inches, to allow a little pliability. You can make a cutting or scratch gauge similarly by inserting a pin at E, exactly over the first 1/16 inch, as that distance is always allowed for. Store-bought marking and cutting gauges are illustrated by Figures 27 to 32. A beechwood pencil gauge is shown by Figure 27, a marking gauge with a steel point by Figure 28, an improved cutting gauge for scribing deep lines in Figure 29, and mortise gauges for scribing mortise holes and tenons by Figures 30 to 32. The mortise gauges are of ebony and brass, the one illustrated by Figure 32 having a stem of brass. The ordinary marking gauge is shown in Figure 33.

Panel Gauges

A panel gauge (Figure 34) is used to mark a line parallel to the true edge of a panel, or any piece of wood that is too wide for the ordinary gauge to take inches. The stock is of maple, beech, or similar wood. It is 1 inch thick, and has a ⅜-inch by ⅜-inch rebate at the bottom. A mortise is made for the stem to pass through, and another is made at the side for the wedge. The wedge should be made of box-wood or ebony if possible, and is a bare ¼ inches thick. The taper of the mortise in the stock must be made to correspond with it. The stem should be about 2 feet 6 inches long, and may be made of a piece of straight-grained mahogany. It should not

Figure 30—Square Mortise Gauge

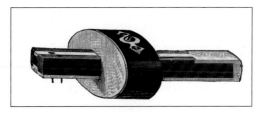

Figure 31—Oval Mortise Gauge

Figure 32—Brass Stem Mortise Gauge

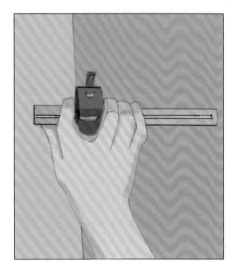

Figure 33—Using Marking Gauge

Figure 34—Panel Gauge

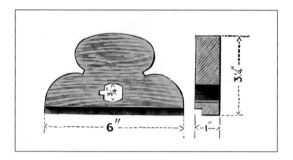

Figures 35 and 36—Elevation and Section of Panel Gauge Stock

fit the mortise too tightly, just so that it can be moved with the hands without tapping, and is held in position by the wedge when set. A piece is dovetailed in the end, as shown, to bring the marking point level with the bottom of the rebate. The stem may be made square if possible, or if the rounded mortise presents a difficulty. Make sure the stock is well finished and nicely polished.

Compasses, Dividers, and Calipers

Joiners and cabinet workers have a multitude of uses for the above tools, which are of the simplest construction. The ordinary form of wing compasses is such that the wing (the curved side projection) forms one with the left leg, while the right leg has a slot that lets it slide up and down the wing, and the set screw is tightened when the legs need to be fixed at a certain distance apart. For very accurate work, compasses with the sensitive adjustment at side, can be very

useful. They are used for stepping off a number of equal distances, for transferring measurements, and for scribing. Calipers are used for measuring diameters of round pins, circular recesses, etc.; for the former purpose use outside calipers and inside calipers for the latter purpose.

Appliances for Mitering

The technical term *miter* is applied usually to the angle between any two pieces of wood or molding where they join or intersect, for example the angles in a picture frame. In this instance, the joint would be a true miter—that is to say, it would be 45° or half the right angle (90°) formed by the two inner edges of the frame. Although the term miter is generally understood to apply to a right angle, any angle, acute or obtuse, can also be called a miter.

Miter Blocks

There are various appliances employed in cutting miters, but the simplest is the miter block. Lay the project on

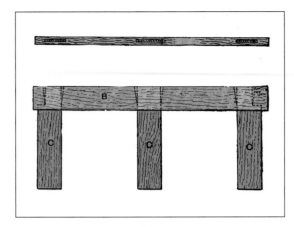

Figures 37 and 38—Frame of Improved Shooting Board

the rebate, shown by c (Figure 41), and use the saw kerfs A and B as a guide for the tenon saw. The best form of miter block is made from a piece of dry beech, about 16 inches long, 6 inches wide, and 3 inches thick. Then cut a rebate c to about the size shown in Figure 41. Make sure that the angle is perfectly true. Lines A and B are set out to an angle of 45°, and then square down the rebate and back of the block. Cut the lines down with a tenon saw, and remember that the accuracy of the saw depends on the value of the finished miters. Figure 42 is a section of the miter block commonly used by joiners. This is simply two pieces of wood planed up true and screwed or nailed together. This plan answers very well, as when it becomes worn and out of truth another can be made inexpensively. The block shown by Figure 43 has a ledge on the bottom as shown; owning to the inward slant the project is more easily held.

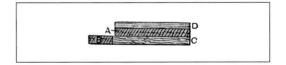

Figure 39—End Elevation of Improved Shooting Board

Miter Box

Figure 44 shows a miter box which serves the same purpose as the block. This is made with three pieces of deal about 1 inch thick, nailed together at the bottom as shown. Miter boxes for heavy work require a strengthening piece on top

to hold the sides together (see Figure 45). Sometimes three pieces may be necessary (see Figure 46). Both these illustrations show pieces of molding in position for miter-cutting.

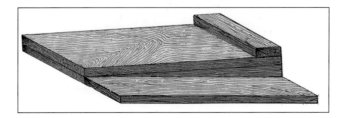

Figure 40—Shooting Board Giving Oblique Planing

Miter Shooting Block

Figures 47 and 48 show a miter shooting block for shooting or planing the edges of objects sawn in the miter block or box. In Figure 47 the bottom piece is of dry red deal, 2 feet 6 inches long, and rebated. For the top piece, select a hard material, such as mahogany or beech, or whichever you prefer. Make sure it is planed perfectly true, and cut at the ends to a "true miter" (45°); Then firmly screw it to the bottom piece. It is better to fix the ledger pieces across the bottom, to keep the board from warping. Figure 62 shows that the bottom piece is made of two separate boards.

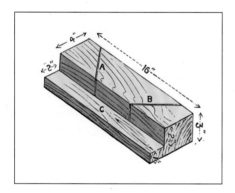

Figure 41—Miter Sawing Block

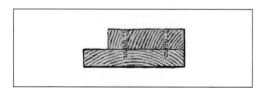

Figure 42—Section of Miter Block

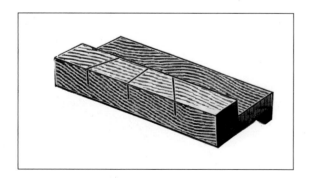

Figure 43—Inclined Miter Sawing Block

Combination Shooting Board

Notwithstanding the large number of patented mitering machines in the market, skilled jointers, when any particularly good piece of work is in hand, still prefer to use the ordinary homemade wooden shoot. The machines, while new and in good condition, are without a doubt faster, but if they are carelessly used, they are apt to get out of order, and then their work is far from satisfactory; while the wood shoot will stand a deal of rough usage, and is also easily repaired, Figure 49 illustrates several improvements on the old form of shoot. A miter-shoot, square-shoot, and joint-shoot are combined in one board, which will prove very handy when you only want to use these appliances occasionally. The shoot consists of a top board of seasoned yellow deal 3 feet by 9 inches by 1½ inches, slot-screwed to an under-frame of teak, made up of the plane bed B, 3 feet by 2¾ inches by 1½ inches, into which are framed three cross rails 2¼ inches by 1 inch

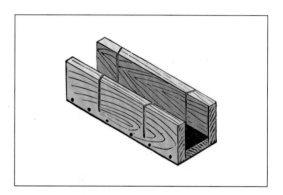

Figure 44—Miter Box

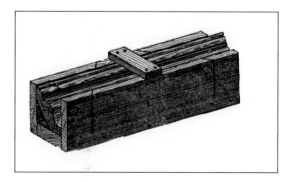

Figure 45—Miter Box with Strengthening Piece

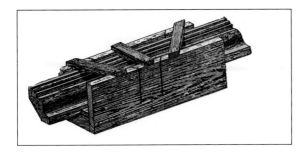

Figure 46—Miter Box with Strengthening Pieces

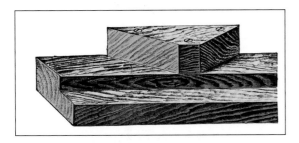

Figure 47—Miter Shooting Block with Solid Base

flush on the under-side. The center of the top board is the miter-block, a piece of dry oak 2 inches thick cut with two of its sides exactly square with each other, and at an angle of 45° with the third. This block, instead of being fixed in the general way, should be mounted on a pivot in its center, as

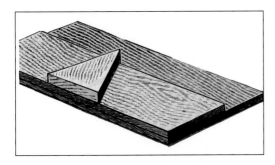

Figure 48—Miter Shooting Block with made-up Base

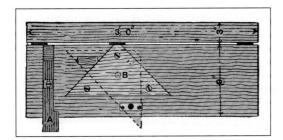

Figure 49—Combination Shooting Board

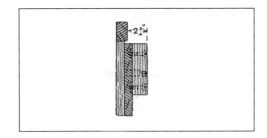

Figure 50—End Elevation of Combination Board

shown at B (Figures 49 and 50), and is capable of adjustment either as a miter-shoot, as shown in the full lines, or as a square-shoot, as indicated by the dotted lines in Figure 49; it is firmly secured in either position by means of three ½-inch by 3-inch square-head screw bolts. Be sure that the grain of the block runs parallel with the plane bed so shrinkage will not alter its shape. The board should be arranged for jointing by removing the miter-block and working the boards against the adjustable stop A. This stop works in an undercut groove, and is secured in any required position by the screw bolt; the projection at the end prevents the boards from slipping while being planed.

Miter Templates

You will use miter templates constantly as an aid in cutting miters. They are made from a piece of hard wood, in the form shown by Figure 54; it is usually about 4 inches long, 3 inches wide either way, and ½ inch thick. You can make a true miter by planing up true a square block of hard wood, cutting out a rebate B, and on each end make a "true miter" (45°) as shown. If an ordinary cupboard framing is examined at the junction of the rail with the jambs, it will be seen that each of the molded edges has been mitered as shown in

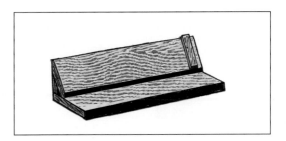

Figure 51—Donkey's ear Shooting Block

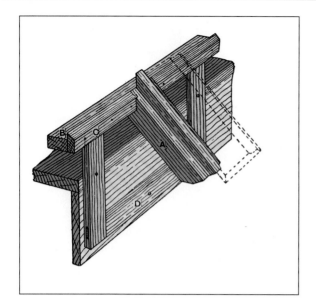

Figure 52—Donkey's-ear Block for Shooting Wide Surfaces

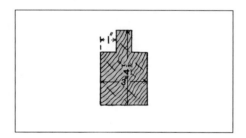

Figure 53—Rest of Donkey's-ear Block

Figure 54—Miter Template

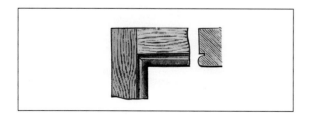

Figure 55—Molding with Mitered Joint

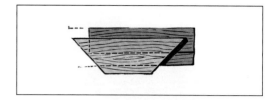

Figure 56—Application of Miter Template

Figure 57—Using Chisel with Miter Template

Figure 55. To obtain this miter, you can use the template. Figure 56 and Figure 57 show the template applied to the edge being held by the left hand, while the right guides the chisel A.

Spirit Level

The spirit level is used to determine the plane of the horizon—that is the plane forming a right angle to the vertical plane. A frame holds a closed glass tube nearly filled with anhydrous ether or with a mixture of ether and alcohol. Make sure your spirit level comes with a graduated scale engraved on the glass tube or on a brass or steel rule attached to the frame beside it, so as to mark the position of the bubble, the tube being so shaped that when the level is lying on a flat and horizontal surface the bubble occupies the center of the tube. Many levels have provision for altering the length of the bubble. Figure 58 is a view of an ordinary spirit level, and its construction is made quite clear by the sectional view, Figure 59. To use the spirit level, apply it to the work twice, reversing it at the second application, and find the mean of the two indications when you are finished. Spirit levels are made in many shapes and sizes, but the method of construction is always the same. A serviceable tool has a narrow shape, about

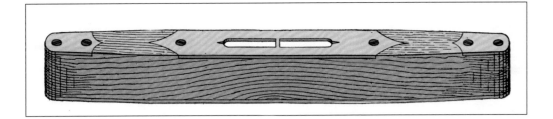

Figure 58—Spirit Level

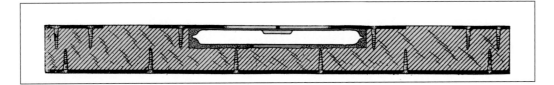

Figure 59—Section of Spirit Level

10 inches long and its greatest breadth being 1 ¹⁄₁₆ inches, and diminishing to ½ inch at the ends.

The frame is of any hard, tough wood, like box, ebony, lignum-vitae, birch, beech, walnut, or oak. At the back of the tube should be silvering, which shows up the bubble and enables side lights to be dispensed with. The tube is set in plaster-of-paris, and has a brass cover. Store-bought spirit levels are constructed generally in ebony or rosewood, with better qualities having a metal protection for the edges and faces. This will preserve the truth of the instruments for a longer time. A serviceable American level has a mount entirely of steel, which is hexagonal in section, and has rounded ends. Another handy form is the one mounted wholly in brass; this has a revolving protector over the bulb opening, and there is provision for adjustment should the level after a time wear out of truth. A very convenient form of level is the one with a graduated screw slide that allows the fall per foot to be shown at a glance.

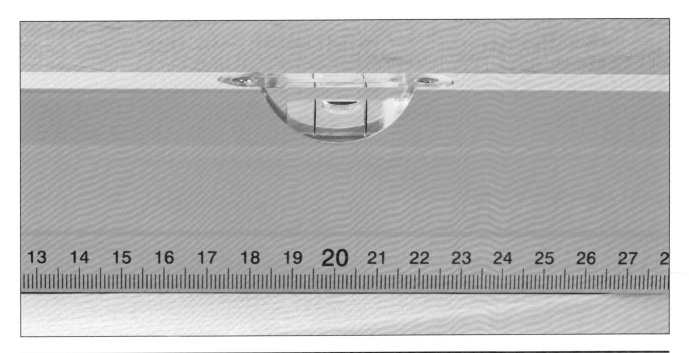

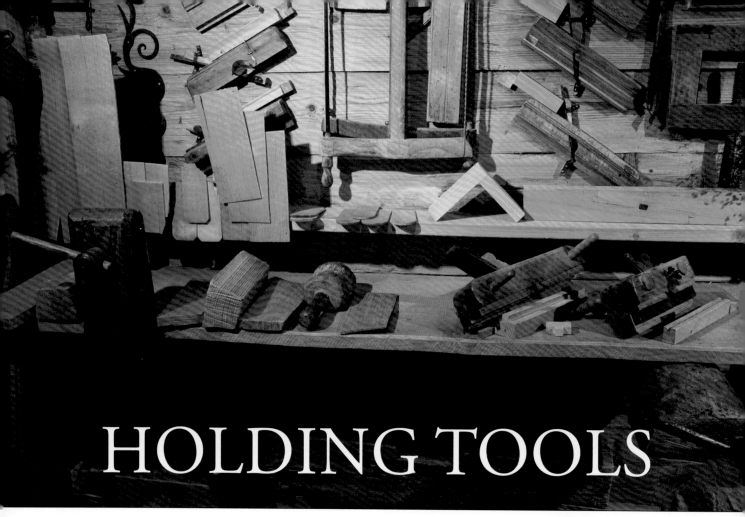

HOLDING TOOLS

Benches

Y ou'll need a work bench before you begin almost any
sizable project. For general manual work, the ordinary bench is used by joiners as it is the most serviceable; it
should be fitted with two wood bench screws and wood vice
cheeks, one at each left-hand corner of the bench, to accommodate two workers. If possible, place the bench so that light
falls directly on both ends—in other words, face the window
while you work.

There are many good benches on the market, but do not
get one that is too low, and make sure the height is right for
the kind of work you intend to perform on it. The smaller
benches sold at the tool-shops are not high enough for an
adult—from 33 to 34 inches is great for an adult, while 26 to
30 inches will suit younger workers. Choose a bench that is
high enough to give you the most command over your tools.
You should be able to work conveniently without having to
stoop; but the height of the bench should not prevent you

Figure 60—Workman and Bench

from standing well over your work (see Figure 60). A height of 2 feet 6 inches might be just right for occasional use, but too low to work at for any length of time. A simple method of raising it slightly from the floor is to put a piece of quarter-ing under each pair of legs. For heavy work the bench may have to be fixed to the quartering, and the quartering to the floor, for which purpose stout screws or screw bolts will do. Figure 60 shows the relative heights of worker and bench.

Various Kinds of Benches

Figure 61 shows a general view of a simple bench with side and tail vices. This form is extremely useful for cabinet making and similar work, where it is helpful to hold pieces of material that may have to be planed, molded, chamfered,

Figure 62—Double Bench with Vice at each end

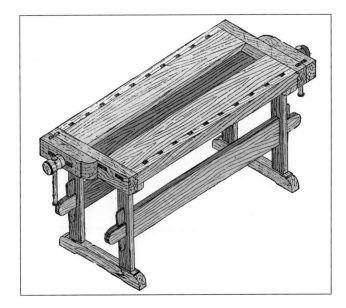

Figure 61—Bench with Side and Tail Vices

Figure 63—Folding Bench in use

mortised, grooved, etc., without using a bench knife or similar method of fixing. The material could be held between stops, one being inserted in one of the holes in the top of the bench and another in the hole made in the cheek of the tail vice. The following dimensions are only suggestions and the bench may be made longer or shorter, narrower or wider, to meet your needs: Top, 5 feet by 1 foot, 9 inches, and 2 inches thick. Height, 2 feet, 7 inches. Distance between legs, 3 feet, 2 inches lengthwise, and 1 foot, 3 inches sideways. Legs, 3 inches by 3 inches. Your bench should be constructed of hard wood, such as beech or birch, and in any case it will be best to have hard wood for all the parts forming the top, side cheeks, and cheeks of vices, as these are the main parts of the bench; consider choosing red deal for the framing of

Figure 64—Folding Bench not in use

the legs, rails, etc. A simple bench is illustrated by Figure 62; this is a suitable model for general carpentry and joinery. The framework is made of a thoroughly seasoned dry spruce fir or red pine, and the top of birch or yellow pine.

The folding bench illustrated in Figure 63 will be found useful in cases where a portable bench is required for occasional use only. When not using the bench, the screw, screw cheek, and runner can be taken out, the legs folded on to the wall, and the top and side folded down as seen in Figure 64. You will find a more elaborate bench for cabinet work in Figure 65; it consists of two principle parts, the underneath framework and support, and the top. The former has two standards joined by two bars. On the feet of the standards rests a board which serves to hold heavy tools and other articles. There is a rack for small tools and underneath this a band, tacked at short intervals, for other tools. The front rail has holes on its top face 1 inches by ¾ inches for holding bench stops, while in the front face of the rail are round holes for holding other pins, illustrated by т, 1½ inches square at one end, but made round at the other end to fit tightly into the holes. The pin т and the block v (Figure 65), screwed on the end of the moveable jaw of the vice, serve to hold wood during the process of edge planing. Holes in the back rail receive pins w which are convenient for cramping up joints. A kitchen table bench is shown in Figure 66. The end of the table employed is not the one containing the usual drawer. Two blocks of wood A B, 3 inches square, are attached to the table top by two cramps embedded in the ends of one of the pieces. Through mortise holes c c are inserted slats glued and wedged to block A, but running loosely in holes in B. s is a screw, and the two parts of the bench serve the purpose also of vice cheeks; though if desired the two blocks can be screwed together solid.

Bench Stops

Generally, store-bought benches are provided with holes for the reception of stops that allow for work to be held for planing, etc. These stops are made of iron and shaped as in Figure 67, and have springs at their sides by means of which they are held tightly and at any required height in the bench holes. You will find an adjustable stop for screwing to the bench in Figure 68. For a temporary stop some workers drive

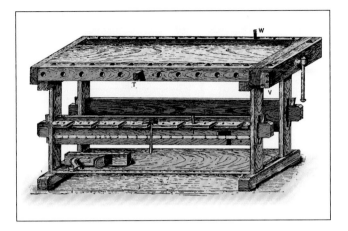

Figure 65—Cabinet-worker's Bench

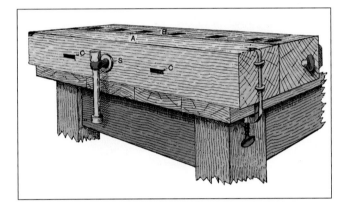

Figure 66—Kitchen Table Bench

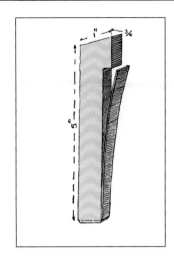

Figure 67—Iron Bench Vice

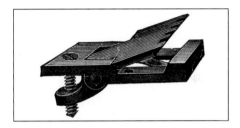

Figure 68—Adjustable Bench Stop

Figure 69—Hinge used as Bench Stop

Figure 70—Bench Screw Vice

a few nails into the bench end, leaving the heads projecting enough to hold the wood. A better substitute can be made out of an ordinary butt hinge, one end of which should be filed into teeth so as to hold the wood better. Be sure to leave this end loose, and screw the other end down tightly to the bench end as shown by Figure 69. A long, light screw through the middle hole in the loose side will afford sufficient adjustment for thin or thick work. When you are finished, the hinge can be taken up and put away.

Common Bench Screw Vice

A common form of joiner's bench screw is shown in general view by Figure 70 and Figure 71 is a sectional view. D is the side or cheek of the bench to which the wooden nut A is screwed. The box B, which accurately fits the runner

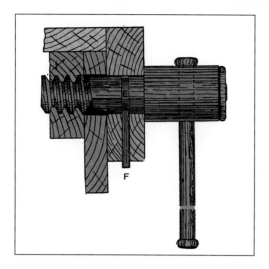

Figure 71—Section through Screw Vice

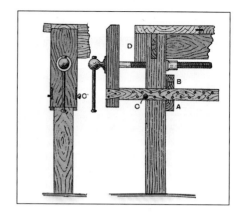

Figures 72 and 73—Kitchen Table Screw Vice

shown inside it, is fixed to the top rail connecting the legs, and to the top and side of the bench. Be careful to keep the runner at right angles to the vice cheeks. To fasten the vice outer cheek and screw together so that upon turning the cheek, the screw will follow, cut a groove shown by Figure 71. Then from the under edge of the cheek, make a mortise and drive in a hardwood key, F, so that it fits fairly tightly into the mortise and its end enters E. The screw cheek is usually about 1 foot, 9 inches long, 9 inches wide, and 2 inches to 3 inches thick. The runner is about 3 inches by 3 inches and 2 feet long. You can also buy the wooden screws and nuts ready made.

Screw Vice for Kitchen Table

Figures 72 to 74 show a simple device for fixing a screw vice to a kitchen table, the vice being detachable for removal as required. The device illustrated does not cause the least degree of damage to the table. A hole is made in the table leg for the screw to pass through, the nut or box of which is fixed to the back of the leg as shown. Two hardwood runners, 2 inches by ¾ inch by 1 foot, 8 inches, should be made and dovetailed into the screw cheek, which is 2¼ inches thick, 1 foot, 3 inches long, and has its breadth regulated by the size of the leg. The distance between runners should be the same as the thickness of the leg. The runners are kept in position

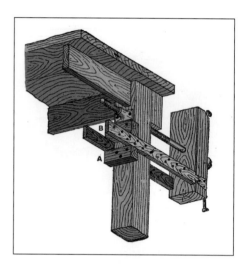

Figure 74—Side General View of Kitchen Table Vice

by two blocks A and B, which are screwed to the back of the leg. An adjustable pin C made from a piece of ½-inch round iron, will be required, and must be sufficiently long to pass through both runners. Screw a block, D (Figure 73), to the leg, the face of the block being flush with the front edge of the top.

Sawing Stools or Trestles

Figure 75 shows a standard sawing stool, Figure 76 is a side elevation, and Figure 77 an end elevation. Suggestive sizes are figured on the drawings. The thickness of the material can, of course, be increased or decreased according to requirements. The simplest sawing stool, but the least reliable, is the one with three legs shown by Figure 78, but this is of

Figure 75—Common Sawing Stool

Cramps

Cramps are used to hold work on the bench, to hold together work in course construction, to facilitate the making of articles in which tight and accurate joints are

little service and almost useless. Better and more usual forms are shown by Figures 79 and 80, these being about 20 inches high, firmly and stiffly made. In Figure 79 all the parts are mortised and tenoned together, and strutted to give strength, but in Figure 80 the legs are simply shouldered and bolted into the sides of the top. The cross stretchers are slightly shouldered back and screwed or bolted to the legs.

Figure 78—Three-legged Sawing Stool

Figure 79—Braced Sawing Stool

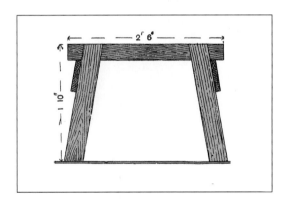

Figure 76—Front Elevation of Sawing Stool

Figure 80—Bolted Sawing Stool

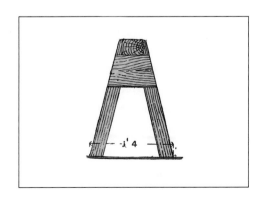

Figure 77—End Elevation of Sawing Stool

Figure 81—Sawing Horse

Figure 82—Bench Holdfast

Figure 84—G-cramp

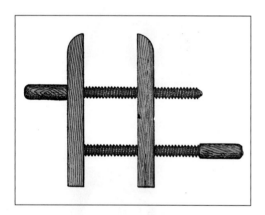

Figure 83—Wooden Holdfast Cramp

Figure 85—Hammer's G-cramp

essential, and to hold together glued joints until the glue is dry and hard. A holdfast for temporarily securing work to the bench is shown in Figure 82. This ranges in length from 12 inches to 16 inches. The old-fashioned holdfast cramp is illustrated by Figure 83; this is made entirely of wood, and the cheeks of the cramp range in length from 6 inches to 16 inches. Iron cramps are shown by Figures 84 and 85, where Figure 84 is the ordinary G-cramp. Figure 85 shows one of Hammer's G-cramps with instantaneous adjustment, this being an improved appliance of some merit. The screw is merely pushed until it is tight on a work held in the cramp,

and a slight turn of the winged head then tightens up the screw sufficiently. The sliding pattern G-cramp is illustrated by Figure 86, this possessing an advantage similar to, but not as great as, that of Hammer's cramp. Sash cramps and jointers' cramps (non-patent) resemble Figure 87. There are several makes and many differences in detail, but Figure 87 illustrates the type. There are a number of patent cramps for sashes and general joinery, but Crampton's appliance (Figure 88) is typically sufficient. You can set the right-hand jaw at any position on the rack. After you insert the work, push the right-hand saw against it tightly and the

Figure 86—Sliding G-cramp

lever handle adjusts instantly. When you joint up a thin project with an ordinary cramp, there is a great risk of the material buckling up and the joint being broken. You can easily avoid this risk by using a cramp which is sufficiently explanatory when it is said that the cross pieces slide upon the side pieces, one sliding bar being made immovable by iron pins placed in holes in the side pieces. In cramping very thin projects, place a weight upon it before finally tightening the hand screw.

Rope and Block Cramp

To cramp up boards with ropes and blocks, place the wood blocks A about 4 inches long, and 1½ inches square, on the edges of the boards B, and place a rope around them twice and make a knot. Then place a small piece of wood between the two strands of rope and twist it around. This twisting draws the rope tighter on the blocks, thereby cramping the boards together. Three of these sets would be sufficient to cramp a number of long boards.

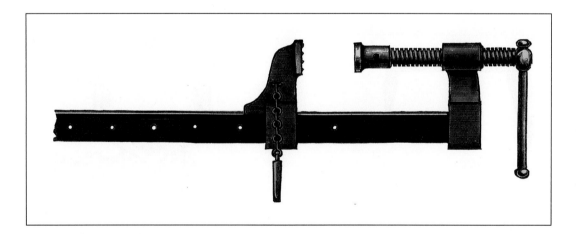

Figure 87—Sash Cramp

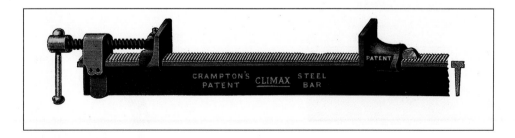

Figure 88—Crampton's Patent Cramp

Cramping Floor Boards

Floor boards are commonly cramped or tightened up using "dogs," of which two forms are shown respectively by Figures 89 and 90. The boards being already close together, the dog is inserted across at a right angle to the line of joint, one point being in one board and one in the other. The further you hammer in the dog, the closer the boards are cramped together. Floor boards can be tightened up without using a floor dog with the method illustrated in Figure 91.

The board next to the wall should be well secured to the joists, and then three or four boards can be laid down and tightened up by means of wedges, as shown. The following is the method of procedure: Place a piece of quartering about 2 inches by 3 inches next to the floor board, as at C. Cut a wedge, and place it at B, then nail down a piece of batten to the joists, as at A (both this and the wedge can be cut out of odd pieces of floor board). The wedge B should be driven with a large hammer or axe until the joints of the board are quite close.

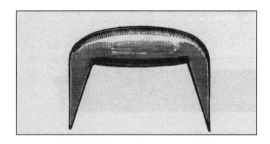

Figure 89—Dog, Round Section

Figure 90—Dog, Square Section

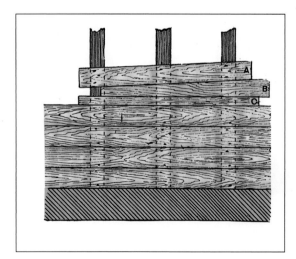

Figure 91—Wedge Cramp for Floor Boards

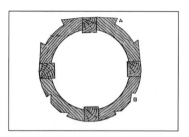

Figure 92—Circular Seat with Cut Cramping Pieces

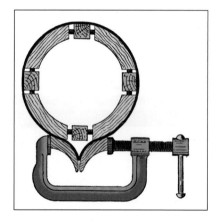

Figure 93—Circular Seat with Flexible Cramp

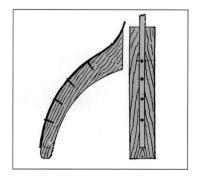

Figures 94 and 95—Wood Horn of Flexible Cramp

Pincers

Pincers are used for extracting and beheading nails, and for other purposes where a form of hand vice is wanted for momentary use. There is little variety in their shape, and they range in size from 5 inches to 9 inches Usually one handle ends in a small cone (see Figure 96) or ball (See Figure 97), and the other in a claw for levering out nails, etc. Figure 96 shows Lancashire pincers, and Figure 97 Tower Pincers.

Figure 96—Lancashire Pincers

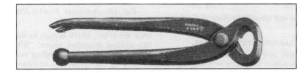

Figure 97—Tower Pincers

BEDROOM FURNITURE

Shaving Cabinet

The shaving cabinet illustrated in Figure 1 is arranged so that when you open the doors the light is reflected on each side of the face. You should begin by making the back of the cabinet. It is of ½-inch material, 2 feet, 7 inches by 2 feet, 5 inches, and is either jointed up, as shown in section by Figure 2, or framed, a ¼-inch board forming its center. Cut it out with a bow-saw; the ornamental panels at the top have the outlines cut in about ⅛ inch, the centers being beveled down and punched with the point of a wire nail. The top and bottom boards at the back are ½ inch thick and 2 feet, 6½ inches by 7½ inches The ends form an angle of 45° with the back. The front edge has a length of 1 foot, 3½ inches These boards are 1 foot apart, and are fixed with screws through the back as at a (Figure 3). Frame up the front 1 foot, 3 inches by 1 foot, the rails and stiles being 1¼ inches by ¾ inch, rebated for the glass, and the back edge of the stiles being beveled off at an angle of 67.5°. Cut and fit the brackets as shown in Figure 2. They are glued and screwed through the bottom board and back. Get out the ornamental rails at the top and bottom, miter

them, and glue and brad them in place, ½ inch within each edge, and they will then cover up the screws used to secure the front, as at c (Figure 3). Fasten a slip cut to the section of B in each angle at the back; with these slips the doors will come flush. Frame the doors 1 foot by 9 ⅛ inches of similar section to the front, with the hinge stile beveled to 67.5°, and hand the door in place. This can be fitted either with locks or with hanging handles and catches. Stain and polish, or otherwise finish to your taste, and, finally, insert the mirrors. These should be securely wedged like at D (Figure 3), and covered at the back with thin wood to protect them from scratches.

Boot and Shoe Rack

Figures 4 and 5 show a rack for boots, shoes, and slippers, which you will find useful in almost any part of your house. The rack can be made of ¾-inch yellow pine or sound red deal, and requires two ends, 2 feet, 9 inches by 11 inches by ¾ inch; one top, 4 feet by 11¾ inches by ¾ inch two braces, 3 feet, 6 inches by 3 inches by ½ inch; six rails for

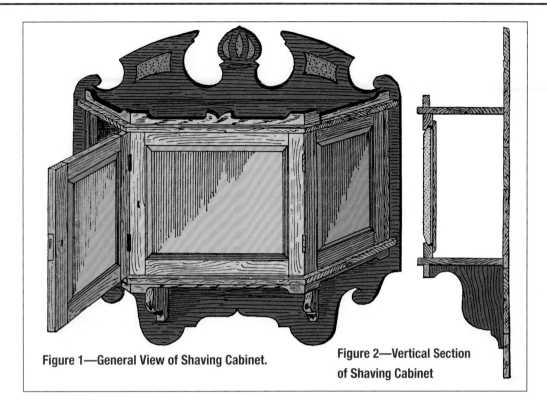

Figure 1—General View of Shaving Cabinet.

Figure 2—Vertical Section of Shaving Cabinet

shelves, 3 feet, 7 inches by 2 inches by 1 inch; and one molding, 6 feet by 1½ inches by ⅞ inch. First prepare the rails, then plane them square, and take off the sharp corner to form a slight chamfer. Shoulder and tenon each end ready for the ends, which can then be prepared with the mortises. Space them out as shown in Figure 6, the top two rails for children's boots and slippers. After you cut the mortises, gently drive on the ends and wedge the rail tenons. Across the back, screw two braces b (Figure 6), and then screw on the top from the under-side. Make sure to gouge the ends and braces for the screws. At the front, immediately below the top, place a ⁵⁄₁₆-inch diameter iron rod I R (Figure 6) suspended by brass hooks, on which to fix a curtain c (Figures 4 and 6) to cover the boots. The finishing molding along the top can be fixed next, and the rack is then often ready for painting, or staining and varnishing. Rails are preferable to solid shelves, as they admit of a free current of air to dry the soles of the boots.

Figure 3—Joints of Shaving Cabinet

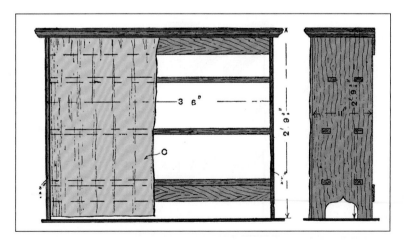

Figures 4 and 5—Elevations of Boot and Shoe Rack

Bed Rest

For the bed rest illustrated by Figure 7 the following materials will be required: For the frame a (Figure 7), two pieces 2 feet, 1 inch by 3 inches by 1 inch, planed to 2¾ inches by ⅞ inch. For the frame b, two pieces 1 foot, 4½ inches, planed to 1¾ inches by 1⅛ inches. For the frame c, two pieces 10½ inches by 1¾ inches by 1¼ inches, planed to 1½ inches by 1 inch; and one piece 11½ inches by 1¾ inches by 1¼ inches, planed to 1½ inches by 1 inch. There will also be required between 5 yd. and 6 yd. of webbing, one pair of 1¼-inches back flaps, some ¾-inch screws, and two coach-screws, 3 inches by ¼ inch. The wood must first be planed to true. Begin by planing true the face side and face edge of each piece of wood, afterwards gauging to the several widths and thickness, taking care that in each measurement the gauge is set quite 1/32 inch full to allow for smoothing off when glued up. When this is true, lay it on the bench, face up, and as the other pieces are planed lay them on the prepared piece, and if each piece does not lie perfectly flat, plane off the part where it touches, that being the highest part. The face edge of each piece can be done in a similar manner. To test with the eye alone, hold up a piece of wood to your line of sight, then slowly turn up the further edge into your line

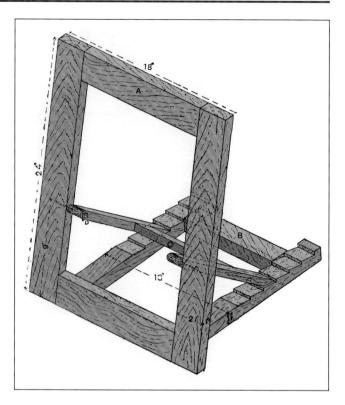

Figure 7—Bed Rest

of sight. Notice which corner comes into view first—that will be the highest part. Plane this side down, and the opposite corner of the diagonal. This is because if one corner curls up ⅛ inch, taking 1/16 inch off the opposite corner will make it equally true and the material will hold thicker that if ⅛ inch was planed off one corner. Having planed all the material on the face sides and edges, set the gauge for the width. As each piece is gauged, plane down to the line, taking care that the edge is exactly square before the line is reached; if two gauges are to hand, then the other gauge can be set to the thickness, and the planing of each piece; plan down to the gauge line on each edge, testing for level by placing the back edge of the try-square across the grain.

Setting out Bed Rest

Now proceed to set out, remembering that in setting out work the face sides and face edges must always pair. For example, take the frame marked a (Figure 7). Take one of the sides, 2 feet, 1 inch long, 2¾ inches by ⅞ inch, and, keeping the face side to the worker and the face edge uppermost, find

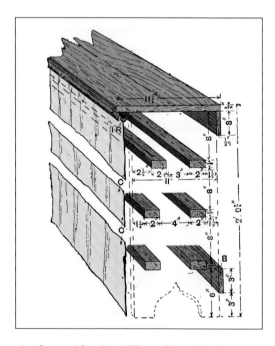

Figure 6—General Sectional View of Boot Rack

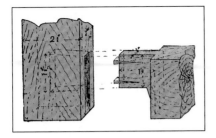

Figure 8—Joint at Corner of Bed Rest Frame

the center of length; mark off 1 foot (half the height) on each side of this, and square lines across the edge. Then from these lines mark 2¾ inches inwards, being the width of rail at top and bottom of frame, and set out 1⅝ inches width of tenon. Note that the tenon does not go through, only entering 1⅝ inches, which is quite sufficient, the joint being fastened by fox-wedging, which is here shown by Figure 8, but which has already been described. Thus, setting-out lines will only be needed on the face edge. Having set out one side of frame a, take the other side and place it beside the one set out, so that the two sides pair—that is, both face sides are outside and both face edges uppermost. Having squared lines across, set the mortise gauge to the chisel that is nearest one-third the thickness of material, adjusting the one teeth of the mortise gauge so that the chisel just rests comfortably between the points of the teeth, then set the gauge head so that the teeth are exactly in the center, trying first one side and then the other by pricking in the points of the teeth until the points coincide; then tighten up, and holding the stock of the gauge close to the face side, mark the mortises and continue gauge lines for the haunching, running the lines on to the end. Next proceed to set out top and bottom rail of frame a (Figure 7). Take one rail, find the center of its length, and set off

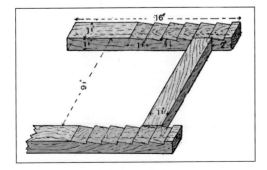

Figure 9—Horizontal Frame of Bed Rest

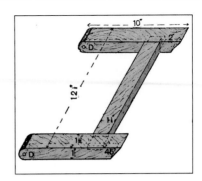

Figure 10—Bed Rest Strut

half-width of frame less 2¾ inches (width of stile) on each side of center. Square all these lines, taking great care, for it is very important that these, being shoulder lines, should be marked true. Note that the try-square should always be applied to either a face side or face edge. The lines must next be squared to the other rail, it having been first noted in placing the rails together that they pair. Then with the mortise gauge mark the tenons all round the end from shoulder to shoulder. Now set out frame b (Figure 7). The material should be ½ inch longer than the finished lengths. Set out one side, starting ¼ inch from one end, and marking off 1 foot, 4 inches, thus leaving a little waste to be sawn off, and enabling a square end to be cut, which can also be shot just before the gluing together. From the first line set off 2 inches (Figure 9). On the face side of one of the 1-foot, 6-inches pieces, commencing ¾ inch from the rail end, set out successively six spaces of 1½ inches each; place the other side so that they pair. Square lines across the two pieces by means of the try-square; you should also square down each edge a little way, and with a marking gauge set to ¼ inch just make a mark on the line; mark the slope with a striking knife, using the edge of the try-square as a straight-edge; then set out the rails 9 ½ inches between shoulders, the tenons 1 inch longer, prepared for fox-wedging, as in frame A. Frame c (Figure 7)

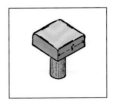

Figure 11—Head of Coach-screw

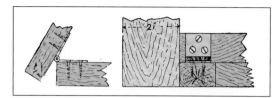

Figures 12 and 13—Side and Back Views of Corner of Bed Rest Frame

comes next, taking one side and proceeding to set it out as in the case of frame B. For dimensions of c, see Figure 10. Note that one end of frame C is semicircular, and that at the center of this portion a ⁵⁄₁₆ inch hole must be bored through each piece (see D D, Figure 10), so as to enable the coach-screw (the head of which is shown in Figure 11) to pass through for the purpose of hanging frame C to frame A, and allowing it to hinge easily. The other ends of the frame c must be cut at an angle of 45°, to obtain which it is not necessary to have a bevel, for if the thickness of the stuff is set off from the end and a line joined diagonally from this point to the top edge, the angle will be 45°. The cross-rail is set out similarly to frame b; it is 9 ½ inches between the shoulders, and the tenons are 1 inch longer. The setting-out is not complete.

Making a Bed Rest

Begin the actual work by mortising the sides for frame a (Figure 7). The mortises are 1 ⅝ inches, and are widest at bottom for fox-wedging (see Figure 8); note also that the haunching (see Figure 8) tapers from nothing at the end to ⅜ inch at the tenon. Great care must be used in mortising, because if the hole is not perpendicular it will cause the frames to twist, as the tenons will follow the holes, and so tend to take the stuff with it. After mortising, cut the tenons, ripping down first, then sawing the shoulders, afterwards marking width of tenons 1⅝ inches from face edges, and the haunched part of tenon ⅜ inch to nothing (see Figure 8). Next make two cuts in each tenon about ¼ inch from the side and 1 inch long, and prepare and insert in each cut a small wedge of the same thickness as the tenon, ¾ inch long, and tapered ³⁄₁₆ inch to nothing, leaving the wedge projecting at the tenon end. On driving or cramping the frame together the wedges are driven in, and the end of

each tenon is widened so that, with the glue, it is impossible to pull it out. A hole should now be bored in the center of each face edge of stiles 9 inches from the bottom, the hole to be a little less than ¼-inch diameter for the coach-screw to fasten in tight when hanging. It is best done before gluing up, for when together it would be found awkward to bore except with a rather large gimlet. Smooth the edges of the frame so as to leave it clean, and the frame is ready for gluing. The frames b and c can be served in a similar manner. In gluing up the frames, be sure the glue is hot. Get the cramp or cramps together before you begin, and make sure the fox-wedges are right. When the frames are glued up, test them with a large square while applying the cramp; a slight movement of the cramp will correct any error. In frame A, if the diagonals are the same length it is square. If an iron cramp is not to hand, a wooden one can be constructed by screwing cleats of wood on a piece of board, 2 inches or 3 inches wide, and allowing for a pair of wood between the frame and the hammer to avoid dents. When the glue is set, the frames can be smoothed over and the fixing together can begin. A pair of 1¼-inch back flaps and ¾-inch screws will be required to hinge frame B to a (see Figures 12 and 13); about half the knuckle of the hinge projects. The drawings will fully explain the position of the hinges. Two coach-screws, 3 inches by ¼ inch, are needed to hang frame c in position in frame a (see Figure 7); these can be turned in with a pair of pincers. The front should now be covered with webbing. The lengths are cut to the dimensions of frame A, about ¾ inch at each end being turned under and fastened by two ¾-inch clout nails at each end. The spaces between webs should be no more than the width of the webbing, as shown in Figure 14, where it will be noticed that the web-

Figure 14—Webbing on Bed Rest Frame

bing is interlaced; 5 yd. will suffice, though a little more would certainly make the whole a little stiffer. Finish with a light stain and a coat of varnish.

Linen Chest

Deal is the best material probably with which to make the linen chest illustrated by Figure 15. The elevations and inside view of the chest, which is 2 feet, 3 inches high, 4 feet long, and 1 foot 10 inches wide, are represented by Figures 15 to 17. The two uprights at the front corners are strips of 1½-inch material, 2 feet, 3 inches long by 4 inches wide. Figure 18 shows a front and Figure 19 a side view of that to the right hand. In both figures, a denotes where the upper 4 inches is cut away from the front to a depth of ½ inch to receive the end of the upper front rail; while b is a similar cut to take the end of the lower front rail. At c and d are other cuts for the end rails, all being ½ inch deep. A side view of the right-hand back upright is shown by Figure 24. Its length

Figures 15 and 16—Elevations of Linen Chest

and breadth are those of its fellow at the front, but it is of ¾-inch board only; 1½-inch wood may be used, to make the job stronger, but one half will have to be sawn away as far as e to receive the ends of the back-boards. Below e, where the double thickness is needed to form the leg, a second piece of ¾-inches material should be screwed on. At f and g this upright is cut away ½ inch to receive the ends of the rails. The rails are all of ¾-inches board and 4 inches wide. The front ones are 4 feet long. The ends of these are cut away behind for a distance of 4 inches and to the depth of ¼ inch to fit the corresponding cuts in the uprights. To these they are fastened with round-headed screws, as shown. The stiles are also of ¾-inches material, 1 foot, 4 inches long and 3 inches wide, with the exception of the middle one of the front, which is 4 inches. Their tops and bottoms are halved in front for 1½ inches (See H H, Figure 21), corresponding cuts being made in the rails above and below to receive them. Their front edges, as well as those of the rails against the panels, are chamfered off. The paneling of the front is done with ½-inches board, cut to 1-foot, 7-inch lengths. These reach from the bottom of the lower rail to the middle of the upper one, above which is a longitudinal strip of the same board. This may be seen in Figure 17, which shows the inner side of the chest front. This arrangement will make a neater finish

than could be gained by carrying the upright boards to the top. In Figure 17 the position of the stiles is shown by dotted lines, and the joints of the paneling should be set up so that it is covered by them.

The panels which have been thus made are 1 foot, 1 inch by 7½ inches at sight; and, before leaving them, it will be well to finish them by adding the small ornamental spandrils, one of which is illustrated by Figure 22. These are of ¼-inch board, and the curved edge of each is worked to a hollow molding with the gorge. The length of the end rails is 1 foot, 9½ inches. The backs of their ends are cut away ¼ inch deep to fit the openings in the uprights—the front end for 1 inch and the back end for 1½ inches. The end of the chest (Figure 16) has a 3-inches stile, and is paneled like the front. The back is of ¾-inches boarding, the pieces being 4 feet long and (together) 1 foot, 9 inches wide. At the corners, openings will have to be cut, 4 inches by ½ inch, to admit the ends of the rails. The back-boards are screwed to the uprights, and should also be doweled together. For the bottom ½-inches matchboarding will be best, the pieces being 4 feet long and (together) 1 foot, 10 inches wide. Openings are cut at the corners, 4 inches by 1½ inches, to admit the uprights. These boards are screwed to the lower end rails, also to the lower front rail and to the back-board. For their further support, three ledgers (L L, L, Figure 17), 2 inches by 1½ inches, are placed beneath them and screwed to the back

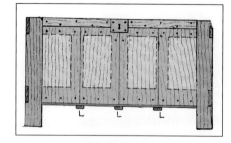

Figure 17—Inside View of Chest Front

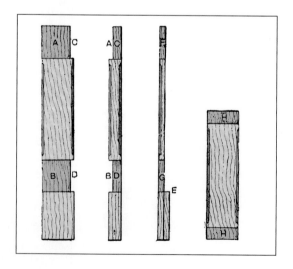

Figures 18 and 19—Front Corner Upright of Linen Chest
Figure 20—Back Corner Upright
Figure 21—Front Middle Stile

Figure 22—Spandril for Linen Chest Panels

and front rail. The ledgers are placed in line with the middle of the stiles.

Linen Chest Base-boards, and Lids

You may prefer to add to the chest the ornamental base-boards and moldings shown in Figures 15 and 16, running along the front and ends. These boards are ¾ inch thick and 4 inches wide. The front one is 4 feet, 1½ inches long; those at the ends are 1 foot, 10¾ inches long. They are mitered where they meet at the corners, and their lower edges are so shaped as to hide the ends of the ledgers Their tops overlap the lower rails 1 inch, and they are screwed to the uprights. Place a strip of molding on these, ¾ inches square, with its front edges simply rounded off. These strips, like the boards below, are mitered at the corners; the front strip is 4 feet, 3 inches long, the end strips 1 foot, 11½ inches. The lid is made of ¾-inch boards, 4 feet, 6 inches long and (together) 2 feet wide. These boards are screwed down to four ledgers, two of which appear in Figures 15 and 16; the others are within the chest, and are placed at equal distances from the outer ones and from each other. As a prop for the lid, have a lath about 1 foot, 4 inches long, working on a screw driven into the inner side of one of the end rails near the front. The chest may be stained to a dark oak color.

Oak Bedstead

Figure 23 shows a beautiful oak bedstead. Leading dimensions are given on Figures 24 and 25, but these dimensions can be varied to your own requirements. Alternative methods of connecting the sides to the foot and head as shown, Figures 26 and 27 illustrating the old bed-screw method of connecting by means of nuts and bolts. In this case a hole A (Figure 26) is bored from the end of the tenon longitudinally in the rail; a mortise is made from the inside of the rail, and a nut B (Figure 27) is inserted, the bolt-head being hidden by the turned wooden button C. A method that is not being generally used is illustrated in Figure 29, the principle being that used on iron bedsteads. A dovetail piece with flanges is screwed to the post, the flanges being let in, and a corresponding dovetail socket piece is fitted and

Figure 23—Oak Bedstead

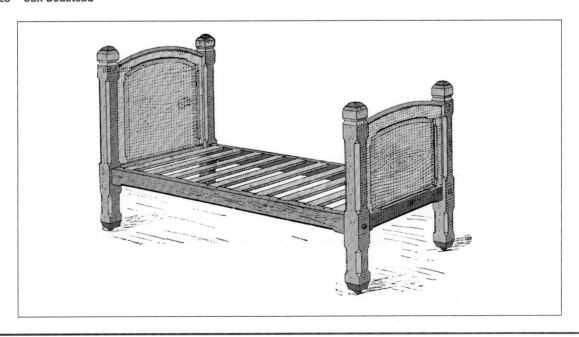

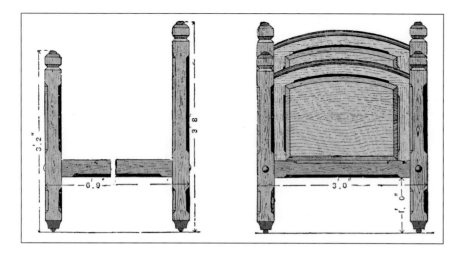

Figures 24 and 25—Side and End Elevations of Oak Bedstead

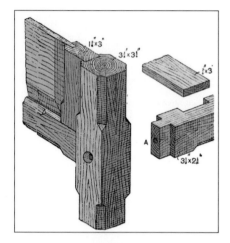

Figure 26—Part of Bedstead Post and Panel

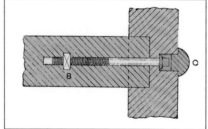

Figure 27—Bed-screw Joint

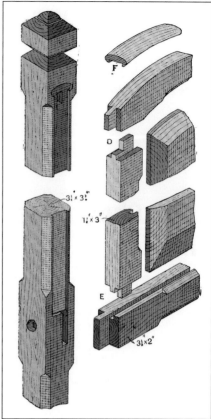

28—Jointing Bedstead Framework

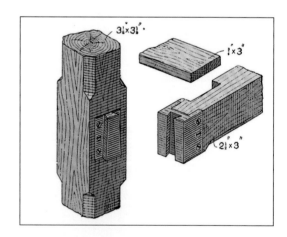

Figures 29—Metal Dovetail for Bed Rails

screwed to each end of the rails. The illustration shows the construction fully. The pieces of timber having been cut to their several sizes, each should be planed to finished dimensions. Then the posts and rails should be set out and the mortising and tenoning done. The forms of joints for connecting the stiles and rails of the head and foot are clearly shown at D and E (Figure 28), and the top rails of the head and foot are curved for a better appearance. The curved top rail has a stub-tenon at each end to enter corresponding mortises in each post. The stiles and rails are ploughed to receive a panel, which is shown in Figures 26 and 28, with splayed margins. Cut an oval-sectioned piece F (Figure 28) to proper sweep and sink out the under-side to fit on to the curved rail, the posts being recessed so as to receive the ends of the capping. After you fit the joints properly, the posts, rails, and stiles need to be stop-chamfered and the tops and bottoms of the posts worked to shape. The joints of the head and foot framings can then be glued together, and when the glue is dry the stiles and rails can be smoothed off. Then the top and bottom rails and stiles are ready to be fit into the posts, which can also be glued together. Four castors should be fixed on the bottom of each post to complete the bedstead.

Framework of Chest of Drawers

The chest of drawers illustrated by Figure 30 is 4 feet wide, 4 feet, 6 inches high, and 2 feet deep outside. The two sides a, 1¼ inches thick, 1 foot, 10¾ inches wide, and 3 feet, 11 ⅛ inches high, support the top B, which is 1½ inches thick, 2 feet wide, and 4 feet long, the grooves being ½ inch deep. Two inner pieces D, 1 foot, 10 ¼ inches wide and 3 feet, 5 ¼ inches long, rest on piece C (see Figures 31 and 32). The top piece F and the two lower partitions G are 5 ½ inches wide and 3 feet, 5 ¼ inches long, the shorter partitions H being 5 ½ inches wide and 10½ inches long. The partition J is 1 foot, 4 ¾ inches wide and 3 feet, 5 ¼ inches long, and the vertical partitions K are 1 foot, ¾ inch deep by 1 foot, 4 ¾ inches long, cut to the shape shown by Figure 33, to rest in ⅜-inch grooves in the pieces B, F, and J. Cut all the partitions, including F and bottom E, as shown in Figure 34, and rest, as shown in Figure 31, in ⅜-inch grooves in the partitions D and K, are put in at each side of the drawer spaces for the drawers to run on (see Figures 32 and 35). A

¼-inch boarding separates the drawers, as seen by the black lines in Figure 32. The plinth is 2½ inches thick and 5 inches deep, and should be cut out to receive the bottom and sides, as shown. The whole is raised slightly from the ground by means of buttons, 1 inch thick, screwed to the plinth. The framework should be strengthened where necessary by triangular blocks.

Figure 30—Chest of Drawers

Figures 31 and 32—Longitudinal and Cross Sections of Chest of Drawers

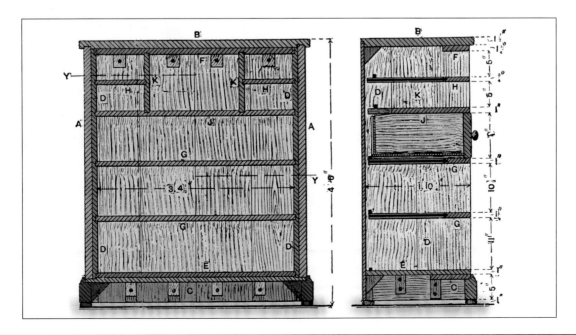

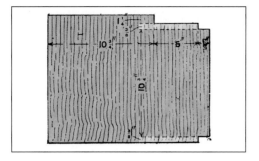

Figures 33—Vertical Partition for Chest of Drawers

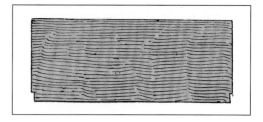

Figures 34—Horizontal Partition for Chest of Drawers

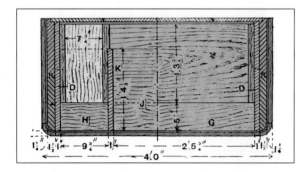

Figure 35—Horizontal Section of Chest of Drawers

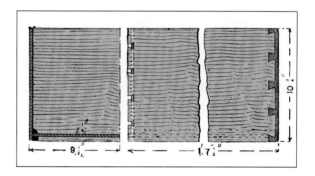

Figures 36 and 37—Drawer Details

The Drawers

Fit the drawers together as shown in Figures 36 and 37, but the bottoms of the long drawers are strengthened with two battens. Make the drawer fronts 1 inch thick, the sides ⅜ inch thick, and the bottom and back ¼ inch thick; the smaller drawers can be made of thinner material. The clear spaces for the drawers are as follows:

1. Bottom drawer, 11 inches deep, 3 feet, 4 ½ inches wide, by 1 foot, 9¾ inches.

Figures 38 and 39—Section and Elevation of Chest of Drawers and Dressing Glass

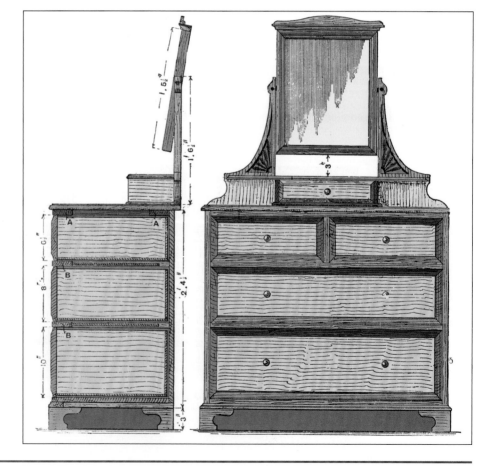

2. Bonnet drawer, 1 foot, 10¾ inches deep, 1 foot, 7 inches wide, by 1 foot, 7½ inches;

3. Four short drawers, 5 inches deep, 9¾ inches wide by 1 foot, 7½ inches

Drawer stops should be fixed on the runners. You can make the handles of either wood or brass. The back should be ½ inch thick, and is strengthened vertically by two battens, and rests in a groove in the top B (see Figure 32). You can make the bottom long drawer deeper than the rest, and, instead of the four small drawers and center bonnet drawer, two short drawers can be put in if you desire. Figure 35 is a section on Y Y (Figure 31).

Chest of Drawers and Dressing Glass

Figure 38 shows the front elevation of a chest of drawers, with a dressing glass and jewel drawer at the top. This chest contains four drawers, one 10 inches deep, one 8 inches deep, and two short drawers 6½ inches deep. The chest measures 3 feet over the gables. Plane and square up the two gables 2 feet, 3½ inches in length by 1 foot, 6 inches in width by ⅞ inch thick. The back edges have to be rebated to receive the back, which consists of ⅝-inch matchboarding. Make the bottom from pine ¾ inch thick and lap-dovetail it to the gable ends. Two pieces A (Figure 39), 3 inches wide by ¾ inch thick, are fixed to the gable ends in the same way as the bottom, the object of one being to allow the back to overlap. The front piece may be of pine, with a thin slip of walnut glued to the edge. The two fore-edges B are also fixed to the gables by grooving the latter to form a dovetail (Figure 40). The fore-edges are slid in from the back. Stop

the grooves at the front, and rebate the fore-edges to bring them flush there.

Bedroom Bookshelves

Figure 41 shows the front view and Figure 42 the side elevation of a bookshelf which may stand on a table or be hung on the wall. The drawers are of the usual pattern and make. A wooden back would strengthen the bookshelf, but for

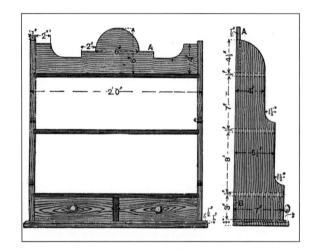

Figures 41 and 42

Figure 43—Ornamented Bedroom Bookshelf

Figure 40—Dovetail Joint in Gable

Figures 44 and 45—Gesso Ornaments for Bookshelf

ordinary use the ornamental top A and the piece of backing B behind the drawers is enough. Figure 43 shows a larger and highly ornamented set of small shelves with lockers beneath. The shelves and sides can be made of ⅝-inch deal, the outer frame being 3 feet, 4 inches by 2 feet, 4 inches by 6 inches deep, and the pieces may be screwed together for simplicity of construction. The back and top should be of well-seasoned ½-inch boards, glued and clamped up together. The outlines of the ornament at the top (Figure 44) and at the lower ends of the side pieces (Figure 45) should be roughly shaped out, and the ornament applied in gesso. The whole of the ornament may be applied this way or it can be carved. The doors of the lockers should be paneled and hinged at the bottom to open downwards. Add a coat of white enamel to finish the work.

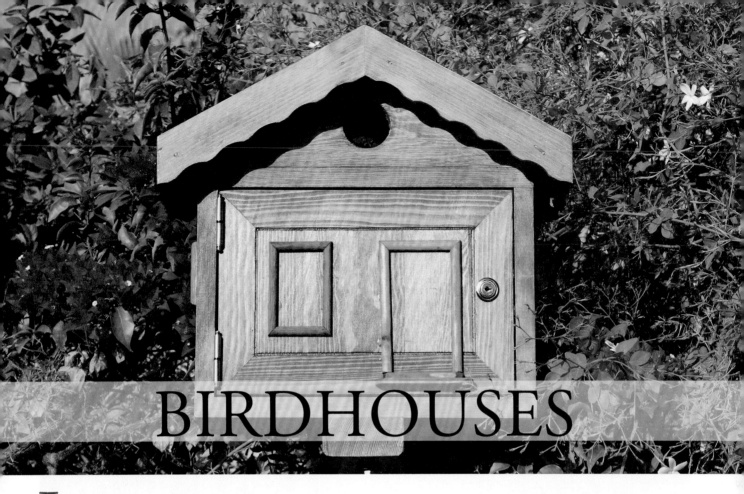

BIRDHOUSES

If you are interested in attracting birds to your yard during the spring, summer, and fall months, in particular, you'll need to provide shelter for the birds. Birdhouses do not need to be very elaborate and they should be rather inconspicuous so birds can easily come and go without attracting predators to their house and nest. All that is really required of a birdhouse is a good hiding place, with an opening just large enough for the bird to fit through, and a strong roof that keeps out the rain.

Birdhouses can be made from a variety of materials— even an old hat tacked to the side of a shed with a hole cut in the top can suffice for a birdhouse. Other usable materials include tin cans, barrels, flowerpots, wooden buckets, and small boxes (preferably wooden or metal).

Most standard birdhouses are made of wood pieces nailed together to look like a miniature house. If you plan to have many birds nesting in your yard, you may want to build a few birdhouses during the winter so they can be ready for springtime, when birds are beginning to nest. To attract a particular kind of bird to your birdhouse, you must make the size of the hole appropriate for the type of bird. For wrens, make the hole about 1 inch; for bluebirds and tree-swallows, the hole should be 1½ inches; for martins, it should be 2½

inches. Below are a few examples of birdhouses you can easily make to attract beautiful birds to your yard.

Bird Ark Birdhouse

This birdhouse is constructed of three tin cans joined together. Both ends of the center can are removed, but the bottom is left on both end cans. To make the bird ark, simply:

1. Cut a hole into the side of the center can and another through the bottom of each can. Do not remove the pieces of tin but bend them out to serve as perches.

A bird ark birdhouse.

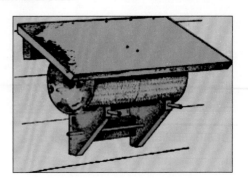

Be creative with your birdhouse designs, and experiment with different shapes and materials.

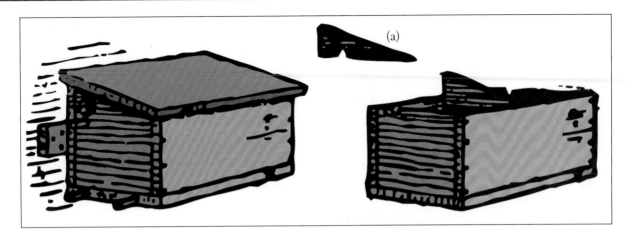

(a)

Lean-to birdhouse.

2. Cut the roof boards of the correct size to project over the ends and sides about 1 inch, nail them together, and fasten them in place by nailing the boards to the connecting blocks between the cans.

3. Fasten the ark between the blocks on a platform or board and then mount the platform on post supports and brace it with brackets, as seen in the picture above. Attach several sticks for perches.

Lean-to Birdhouse

This birdhouse is made out of an empty wooden box (or, if you want to put together your own box, just nail small pieces of wood together of any size you like). If you do make your own box, make sure that the top edges of the end pieces are cut at a slant to allow for a slanted roof. If you use an empty box, add triangular pieces to the edges to create a slant (a).

Cut the center partition, dividing the box into two compartments of equal size as the end pieces. The doorway can be cut with a jackknife—this is easily done if the ends are in two separate pieces because then half of the hole can be cut out of the edge of each piece. After the ends have been pieced out, nail a strip to the back, and then cut the roof board large enough so it will project out from the box about 1 inch.

The lean-to birdhouse can now be mounted on the side of a shed or on a tree and sticks can be fitted into the holes for perches. To mount the house, cut out and nail a wooden bracket to the shed wall and then nail the lean-to birdhouse into the wooden bracket.

Log Cabin Birdhouse

This birdhouse can be made out of any-sized box. Nail pieces together to form the roof, and then thatch the roofing itself to blend into the surrounding environment. The more sloped the roof, the easier the rain can fall off and not

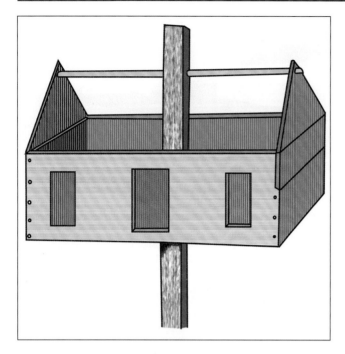

Cross section of a log cabin birdhouse.

penetrate into the house. This house is slightly more elaborate in the sense that the support pole passes through the house to form a "chimney." The windows can be cut out and fake doors painted on for aesthetic purposes. Small branches should be cut to the proper lengths, split, and then nailed all over the exterior of the house to produce a "log cabin"

Decorate your birdhouse with twigs and bark.

look—this also helps the birdhouse to better blend into the surrounding trees and foliage in your yard.

Temple Birdhouse

This is a small birdhouse, perfect for wrens. This birdhouse hangs from a tree branch.

Materials

- Large tin can
- Wooden board about 7 inches square
- Carpet or upholstery tacks
- Earthen flowerpot
- Small cork to plug up the flowerpot hole
- Eye screw
- Short stick
- Wire
- Small nails

Directions

1. Mark the doorway on the side of the can and cut the opening with a can opener.
2. Fasten the can to the square baseboard (A) by driving large carpet tacks through the bottom of the can into the board.

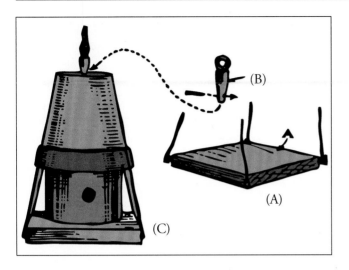

Temple birdhouse.

3. Invert the flowerpot to make the roof. Plug up the drain hole to make the house waterproof (use a cork or other means of stopping up the hole) (B).

4. Screw the eye screw into the top of the plug to attach the suspending wire. Drill a small hole through the lower end of the plug so that a short nail can be pushed through after the plug has been inserted to keep it from coming out.

5. Fasten the flowerpot over the can with wire, passing the loop of wire entirely around the pot and then running short wires from this wire down to small nails driven into the four corners of the base, (C).

6. Now the bird temple can be painted and hung on a tree.

Birdhouses for Specific Bird Species

Species	Floor of cavity (inches)	Depth of cavity (inches)	Entrance above floor (inches)	Diam. of entrance (inches)	Height above ground (feet)
Bluebird	5 x 5	8	6	1½	5 to 10
Robin	6 x 8	8	(a)	(a)	6 to 15
Chickadee	4 x 4	8 to 10	8	1⅛	6 to 15
Tufted titmouse	4 x 4	8 to 10	8	1¼	6 to 15
White-breasted nuthatch	4 x 4	8 to 10	8	1¼	12 to 20
House wren	4 x 4	6 to 8	1 to 6	⅞	6 to 10
Bewick wren	4 x 4	6 to 8	1 to 6	⅞	6 to 10
Carolina wren	4 x 4	6 to 8	1 to 6	1⅛	6 to 10
Dipper	6 x 6	6	1	3	1 to 3
Violet-green swallow	5 x 5	6	1 to 6	1½	10 to 15
Tree swallow	5 x 5	6	1 to 6	1½	10 to 15
Barn swallow	6 x 6	6	(a)	(a)	8 to 12
Martin	6 x 6	6	1	2½	15 to 20
Song sparrow	6 x 6	6	(b)	(b)	1 to 3
House finch	6 x 6	6	4	2	8 to 12
Phoebe	6 x 6	6	(a)	(a)	8 to 12
Crested flycatcher	6 x 6	8 to 10	8	2	8 to 20
Flicker	7 x 7	16 to 18	16	2½	6 to 20
Red-headed woodpecker	6 x 6	12 to I5	12	2	12 to 20
Golden-fronted woodpecker	6 x 6	12 to I5	12	2	12 to 20
Hairy woodpecker	6 x 6	12 to I5	12	2	12 to 20

Species	Floor of cavity (inches)	Depth of cavity (inches)	Entrance above floor (inches)	Diam. of entrance (inches)	Height above ground (feet)
Downy woodpecker	4 x 4	8 to 10	8	1¼	6 to 20
Screech owl	8 x 8	12 to I5	12	3	10 to 30
Sparrow hawk	8 x 8	12 to 15	12	3	10 to 30
Saw-whet owl	6 x 6	10 to 12	10	2½	12 to 20
Barn owl	10 x 18	15 to 18	4	6	12 to 18
Wood duck	10 x 18	10 to 15	3	6	4 to 20

Simple Bird Shelters

Birds seek out dry places to nest and to escape from rain and wet leaves. Simple bird shelters can be constructed out of regular materials, some of which you may have lying about your house or garage.

The bird shelter in illustration (a) has a canvas or heavy muslin roof, supported on two uprights. Five perches are arranged from side to side, and this allows a few birds to rest and retreat from inclement weather. The uprights are 1½ x 3 inches and the strips forming the Y braces are 2 inches wide and ⅞ inch thick. The perches are ¾-inch dow-

If desired, paint your birdhouse or shelter after assembling.

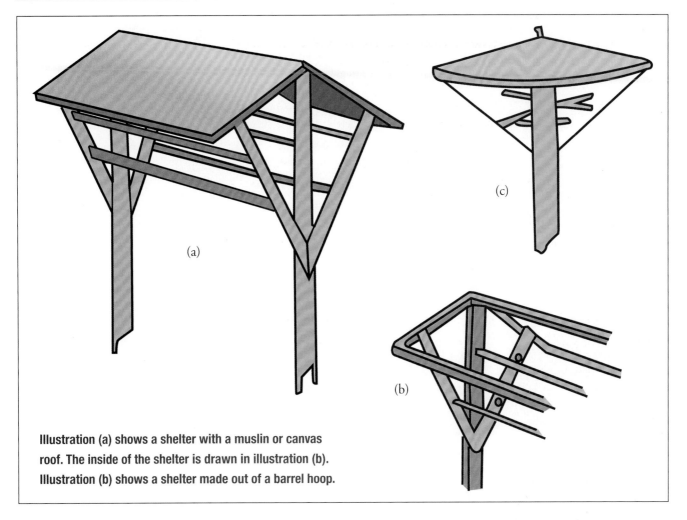

Illustration (a) shows a shelter with a muslin or canvas roof. The inside of the shelter is drawn in illustration (b). Illustration (b) shows a shelter made out of a barrel hoop.

els that are 3 feet long. You can find these at your local hardware store.

Make holes in the uprights to insert and secure the perches. Illustration (b) shows the inside of such a shelter. Make sure the canvas roof is waterproof and is tacked securely around all the edges of the house.

Bird shelters can also be made out of a barrel hoop, as shown in illustration (c). To make this house:

1. Cover a flat barrel hoop loosely with canvas or heavy muslin tacked around all the edges.
2. Drive a wooden peg into the top of a post. Cut a hole in the canvas and slip it over the post.
3. Attach four wires to the hoop at equal distances apart. Pass the lower ends through staples or eye-screws driven into the post about a foot or two from the top.
4. Make two or three holes in the post, and drive round perches through the holes for the birds to sit on.

BRIDGES

If you have a river or brook on your property, you may want to construct a simple bridge. Building these bridges can be fairly easy, especially if you don't plan on transporting very heavy machinery or cars over them. Here are a few different ways to build basic bridges over streams, creeks, or other rather narrow waterways.

Footbridge

This natural-looking footbridge can be built between 8 feet and 12 feet long.

Excavate the banks of the stream or creek to allow for the building of a small, low rubble or stonewall. The sleepers will rest on this wall. The girders are formed of wooden spars (four are used in this plan). The girders should be between 8 and 10 inches in diameter. Lay the girders down and bolt them together in pairs with six ¾-inch-diameter coach bolts. Wedge the posts to fit mortises in the girders.

The posts and top rails should be roughly 4½ to 5½ inches in diameter and the intermediate rails 3 inches in diameter. Finally, join the rails to the posts.

The bridge should be anchored well if it's in a place where flooding is frequent, as you don't want your footbridge floating away in the stream. To do so, drive four short piles into the soil on the inside of the girders, near their ends. Fasten the girders to the piles with coach bolts. The pile tops are hidden by the ends of the floor battens.

Now, if you want to decorate your footbridge, you can use small twigs and nails to make patterns on your bridge.

Small Stream Bridge

If you have a small creek or stream on your property, you may want to construct a simple bridge for easy access to the other side. To build this bridge, you'll need lumber that

A woodern footbridge.

A cross section of the footbridge.

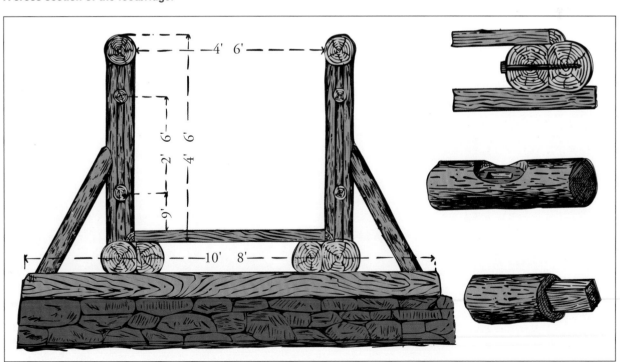

Join the rails of the footbridge to the posts as shown here.

is 6 inches wide and 2 inches thick, and additional lumber for the floor and four side braces.

Directions

1. Saw 11 pieces of wood the length required for the two sides.
2. Bore bolt holes 1½ inches from each end. Use 5⁄8-inch bolts 8½ inches long for where four pieces come together, and use 6½-inch bolts where three pieces meet.
3. Bolt on the A-shaped supports and pieces for the approaches at one time, and then put on the side braces.
4. The sides of the bridge are made of triangles. The first triangle is made of pieces *a*, *b*, and *c*. The second triangle is made of pieces *b*, *d*, and *e*.

Railings can be fashioned like wooden fences (see pages 248–251).

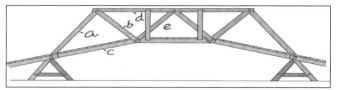

A bridge for a small stream.

5. The piers for this bridge may be made of posts, stone, or even concrete, depending on how permanent you wish your small stream bridge to be.

A Very Simple Bridge

Another very simple way of building a bridge across a creek or stream is to find a narrow part of the waterway and then find two logs that are longer than the creek is wide. These logs should be very sturdy (not rotted out) and thick. Place them across the creek, so they make a narrow beam over the water. Each log should have an extra foot at each end of the creek so they can be securely walked upon with no danger of slipping into the creek bed. Place the logs roughly two feet apart.

If the water comes up close to the bottom of the logs, raise them so the bridge does not get washed away in heavy storms or during the course of the stream rising. Raising the bridge will require a bit more work, as each log will need to

be set into another log on the edge of the streambed or even into stone to make it more permanent.

After you have the two base logs secured, find some sticks that are long enough (and relatively thick) to lay across the tops of the two logs. Or, if you have extra plywood or other boards, those can be used as well. Just make sure to place the sticks or boards fairly close to one another, leaving only little gaps between them. Then, once all the sticks have been laid down, secure these by tying twine or rope to them and the base logs.

If you'd like your bridge to last a little longer, you can pave it with clay or fine cement. Using a shovel, coat the bridge with the clay or cement until it's about 2 inches thick. Then shovel dirt onto the clay mixture, packing it down all over, and make the bridge as thick as you like. However, for just a simple bridge across a narrow stream or creek, the wooden sticks or boards will work just fine and won't require as much time and energy.

For longer bridges, you will need supports underneath.

DOGHOUSES

Doghouses and kennels are easy to construct and are especially useful if you have dogs that primarily live outdoors. A dog kennel needs to protect the dog from harsh winds and heavy rains and should be spacious enough for the dog to move around in comfortably. Doghouses should be located near to your own house so you can have easy access to your pet, and should be situated on a side of your house that creates a natural barrier against the wind and weather. Dogs should not be left outside overnight in very cold weather, even with access to a doghouse. Below are a couple of doghouses and kennels that can be easily constructed for your outdoor pet.

Standard Dog Kennel

This kennel is constructed to be warm and windproof, to direct the rain away from the base by creating large roof overhangs, and to be easily cleaned.

Materials

- Matched boards for the sides, ends, and bottom (standard measurements for the kennel are 30 inches long, 20 inches wide, and 30 inches tall)

Refer to this illustration when making the dog kennel. The kennel raised off the ground is shown by (c); (d) illustrates the parts in contact; (e) is a vertical section of the back end.

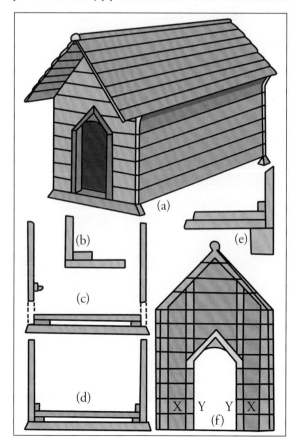

Modify the dog kennel plans here to fit your dog if it is a larger breed.

- Weather boards for the roof
- Strip of sheet metal
- Wooden beading

Directions

1. Make the ends first by nailing lengths of matching boards across uprights of 2 x 1-inch batten (*f*). At the top, halve the battens into the two roof pieces. Set the two outer uprights, X X, in about ¾ inch from the edges to allow the sides to be flush with the outside of the ends. Place these four uprights on the inside.

2. It is advisable to cut out the door—using a pad-saw for the semicircular top—before nailing on Y Y, which should be a little nearer to one another than are the rough edges of the door. Two short verticals on the outside, also projecting beyond the edges, prevent the dog from injuring himself on them.

3. The battens may be omitted from the back end of the kennel, but they ultimately help strengthen the structure and so should be included.

4. When the ends are finished, the horizontal boards for the sides are nailed onto each end (b). Begin at the bottom, arranging the lowest board with its tongue pointing upwards and add the upper boards one by one. The direction in which the tongue points is an important detail—if the boards are put on the wrong way, water will leak more easily into the kennel, rotting the boards and making your dog wet.

This doghouse is wider and has the door set off to one side, which allows for even more protection from the elements.

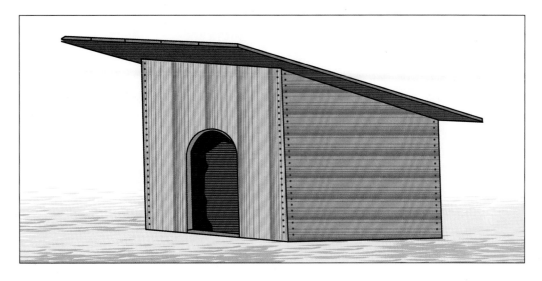

This kennel has a floor that is 2 feet square, is 3 feet 4 inches high in front, and the roof has an overhang of 8 inches.

5. Battens (d) and (e) are nailed inside along the sides of the kennel and a third is nailed across the back, at a distance above the bottom edge equal to the thickness of the bottom boards and of the battens (b) and (c), to which they are attached. At each end a 2 x 2-inch deal, (f), is screwed to (b) and (c) to raise the bottom clear off the ground.

6. The roof weather boards must be long enough to project at least 6 inches beyond the door end, to prevent rain from coming through the entrance. The eaves overhang 3 inches and are supported, as shown in (a), by three brackets cut out of hard wood. Begin laying on the boards, starting at the eaves and finishing at the ridge, which is closed with a 6-inch strip of sheet metal placed on top of a wooden beading.

7. Stain all the exterior surfaces, including the bottom, and fill in the cracks with caulking to keep the water from seeping through.

The inside part of the kennel should be exposed to the sun occasionally by being turned on its end, and the bottom should be cleaned often.

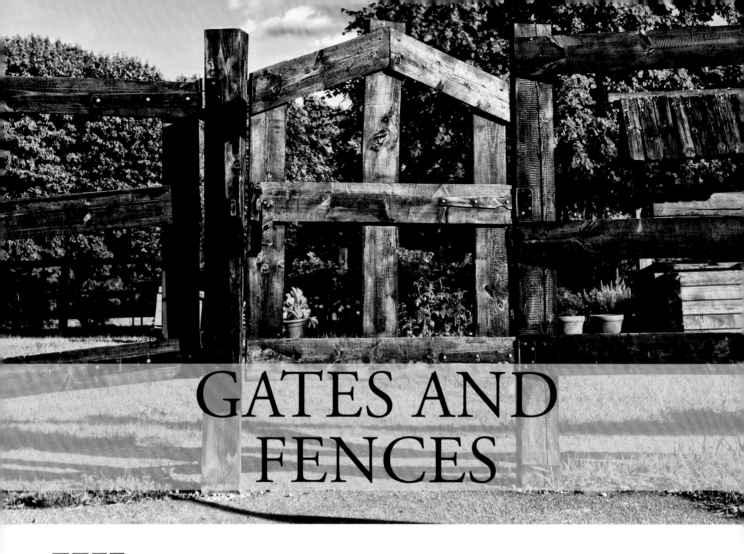

GATES AND FENCES

Whether you are looking to add a lovely fence and gate around your garden plants or you have poultry or other livestock to keep in check, you may need to build a fence, gate, or animal pen. These structures can be attractive if well-built and should be able to stand up to all kinds of weather and animals. Depending on your needs, here are some various fences, gates, and pens you can easily construct in your yard or on your property.

Fences

Fences are perfect for keeping animals or young children in a confined space or for drawing boundaries between yours and your neighbor's property lines—but check with your neighbors before you construct your fence to make sure they don't mind. Also call your local utility companies to make sure that you will not be digging up power or gas lines.

Wooden Fences

Wooden fences allow for good ventilation and an open, airy feel. They can provide protection for young shrubs and plants as well as keep animals and children safe within the yard or fenced-in space.

The most common type of wooden fence consists of horizontal rails nailed to posts or stakes that are placed vertically into the ground. These fences can be constructed with three or four horizontal rails made out of split wood, spruce, or pine wood planks. The posts are usually about 6 feet long and sharpened at the end that will be driven into the ground (to a depth of roughly 8 inches). These posts should be spaced about 6 feet apart.

To keep the pointed, earth-bound ends from rotting, dip them in melted pitch before inserting them into the ground. To do this, boil linseed oil and stir in pulverized coal until it reaches the consistency of paint. Brush a coat of this on the wooden post. Make sure the posts are completely dry before painting them. If properly done, this should keep moisture from seeping into the buried parts of the posts and will keep your fence upright for many, many years.

To drive the posts into the ground, use a very good shovel or a heavy wooden mallet. For longer poles, use a post-hole borer. This saves lots of time and energy and will work with almost all types of soil.

To construct a basic wooden fence, you'll need:

Materials

- Post-hole diggers, a post-hole borer, or a shovel
- 4 x 4 wooden posts (wood that has been treated will last longer but is not necessary)
- 2 x 4 lumber (this too can be treated but does not need to be) or fence boards (which can be purchased at your local home and gardening center)
- Thick, long nails

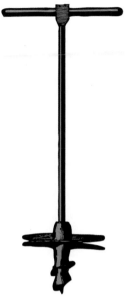

A hole borer lifts the soil from the hole without having to use spades. These borers can be used by hand or electric models can be purchased for the same purpose.

Directions

1. Decide where the fence will be constructed and then lay a line of twine or string to mark out the border.
2. Decide how tall you want your fence to be. Take into consideration what the fence is being used for (if it's for larger animals, such as llamas, you may want a 6-foot-tall fence; if for decoration, a shorter fence may do the trick).
3. Dig holes for your end posts (in all four corners of your fence). Make sure the holes are deep enough to support the end posts. Fill in dirt around the posts and pack in the soil very well.
4. Start digging the remaining holes, keeping them aligned with the end posts.

A simple wooden fence is enough to keep most animals in their pastures.

You can nail wire mesh to the rails of a wooden fence for extra security or to allow vines or other climbing plants to grow up along the posts.

A basic wooden fence.

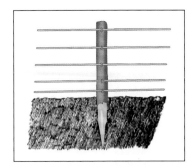

Drive your fence poles far enough into the ground that they stand firmly upright even when moderate pressure is applied to one side.

5. Insert the remaining posts into the holes, piling in the dirt and packing it down as before.
6. Nail on your fence boards, leaving a little space in between. Paint or stain the finished fence if you wish.

   Note: If you want a privacy fence, you can nail thicker boards horizontally or vertically between each post, leaving little space in between.

A G-line wire fence consists of three-ply strands of wire.

Wire Fences

A picket fence is constructed by nailing two or more long boards to posts, and then nailing narrow vertical boards to the horizontal ones.

A very simple fence can be made by driving small tree trunks or branches into the ground in a tight row.

Wire fences are both portable and durable, making them convenient and economical to build. Wire fences usually have a longer staying power than wooden fences since they are less prone to deterioration or rot.

The most common type of wire fence is one that has wire lines strung between wooden posts. The wires are fastened to the posts by galvanized wire staples. The wooden posts should be spaced roughly 6 feet apart and should use five single wires.

A more substantial wire fence can be made with G-line wires. Each line consists of a three-ply strand. Instead of the wires being fastened to the post by staples, holes are bored through the posts and the lines pass through. Straining eye bolts with nuts and washers are attached for tightening up the fence. This type of fence, however, is much more expensive to build and, unless you desire a fence that is incredibly strong, is probably not necessary.

Wire Netting Fence

Galvanized wire netting fences are used for enclosing root gardens and for poultry fences. The standard type of netting used when making this fence is 3-inch mesh netting that is 3 feet x 3 feet, and is rather inexpensive to buy.

A separate strip of 2-inch galvanized wire netting that is 6 inches wide can be laid flat on the ground on the side of

A wire netting fence.

the fence where the poultry are—this way they can not dig underneath, especially once grass and other natural materials hide the wire netting.

To dig in this type of fence, make a trench about 6 inches deep; drop the netting into it; and then fill the trench up with dirt, stones, or even concrete, depending on how permanent you want the fence to be.

Portable Fences

If you need a temporary fence or if you want to easily move your livestock fence to new grazing areas on your property, you may want to consider one of these easily made, movable fences. Below are a few types of portable fences that can be tailored to your specific needs.

Convenient and Portable Fence

Often it is helpful to have a fence that can be quickly erected and disassembled. This fence is very cheap, strong, and convenient to use. It is built out of pine (any other wood can be substituted, but pine is typically lighter and easier to move), 1 x 6 inches for the bottom rail and 1 x 4 inches for the top rails. The braces that hold it upright are 2 x 4 inches and the base (cross piece) is 2 x 6 inches. The base is notched 2 inches and the bottom boards are notched with holes.

The base piece, which is more susceptible to rot, could be made out of a stronger wood, such as oak. Make sure the panels aren't too long or they might warp out of shape. This fence can be put up very quickly and taken down again with ease if you want to move it to another part of your yard or get rid of it for a while.

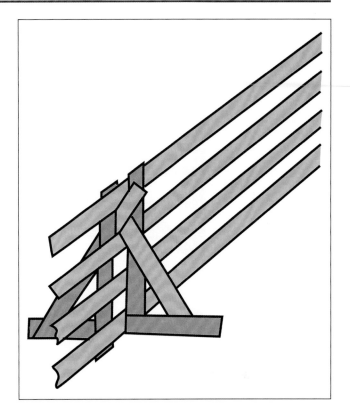

A portable fence.

Scotch Hurdle Fence

This movable fence consists of two posts, each 2 x 3 inches and 4½ feet long. The lower ends are long and pointed which allows them to easily enter the ground and prop up the fence. The brace and two diagonals are made of larch or fir wood. This fence is around 9 feet long and 4 feet high.

The Scotch hurdle fence is easy to set up. The incline should be facing away from any livestock you might have

This Scotch hurdle fence is good for temporary use. If you live in a very windy area, however, this fence may not suit you well, as they do have a tendency to fall over in strong gales.

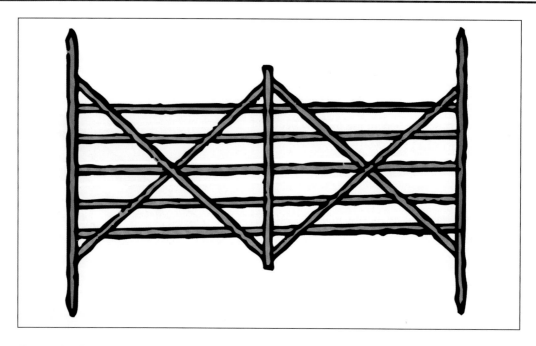

An English hurdle portable fence.

contained inside of it. A stay should be placed between every two hurdles to keep them in position. One wooden peg should be fastened to one end of the hurdle and another peg driven through the other end and into the ground.

English Hurdle Portable Fence

This movable fence is much lighter, cheaper, and more convenient than the Scotch hurdle fence. Usually made of split oak, this fence is tough and impenetrable. It consists of two upright end pieces that are joined by four or five mortised bars 7 to 9 feet long. These are strengthened by an upright bar in the middle and two or more diagonals. The end pieces are long and pointed for setting into the ground. To set these into the earth, use an iron crowbar to avoid splitting the top of the wooden piece.

These fences are set erect and no stay is needed. The two adjoining ends of the fences are connected with a band that is passed over them.

Gates

Gates are a necessary part of any fence or pen and they can be situated in the fence wherever they can be easily accessed. If you have a field, your gate should be roughly 10 feet wide to allow small machinery through.

Most gates are made of either wood or iron (though iron is obviously much more expensive and more complicated to work with). Wooden gates will suffice for most of your homesteading needs. The following are a few simple gates that can be used for your garden, your backyard fences, and your pens housing livestock.

Inexpensive, Simple Gate

A light, useful, and durable gate can be made of sassafras poles (or other tall grass poles) and wire. Dig and place a strong post 4 feet in the ground in the middle of the gateway and balance the gate on it. The lower rail is made of two forked sassafras poles securely nailed together so they can be coiled back over the post.

Easily Opened Gate

To construct this simple gate, take an old wheel (possibly found at an antique store or rummage sale) and fasten it to make a gate that you will be opening frequently. The piece of board (A) drops between the spokes of the wheel and holds the gate either open or closed.

Gates can be useful for entrances to a yard or walkway, as well as for animal pens or pastures.

A basic wire gate can be constructed when you need an opening in your fence.

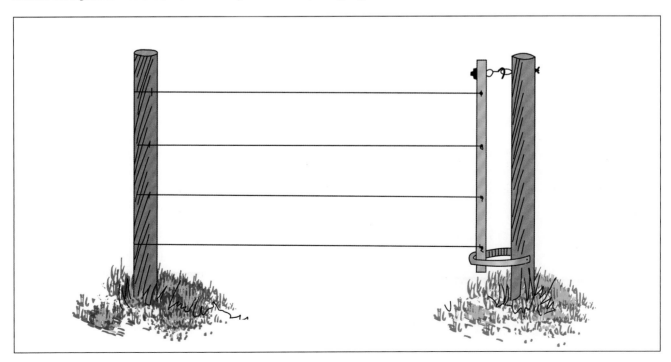

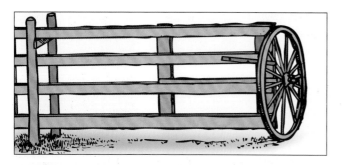

An easily opened gate.

Rustic Garden Gate

If your garden is bordered with a fence, or if you have shrubs or hedges enclosing your backyard, you will want to install a homemade gate through which to enter the garden or yard.

This gate plan is made 6 feet square and 8 feet tall. Any type of timber can be used. To construct the gate:

1. Cut the closing and hinging sides 6 feet long, 6 inches wide, and 2½ inches thick. The three rails are of the same dimensions and can be halved to the stiles or wedged and braced.
2. Separate pieces are fixed on the center to support the gate and to make the frame, on which the boards will be attached.
3. Two gate hinges and hooks can be bolted on or secured from the back with square-headed coach screws.

This rustic garden gate blends in well with natural surroundings.

Closing stile for a garden gate.

Either of these gates will work well with the rustic garden gate.

4. Begin fixing the debarked twigs—they should be as straight as possible.
5. To make the joint ends, start by fixing the outside square then the two inner squares and finally the diagonal filling.
6. The posts should be 9 inches in diameter by 9 feet long (3 feet should be buried underground). Cut three mortises in the posts to insert the rails for the side fencing. These rails are nailed flush to the secondary posts.
7. Dig holes for the posts and pack in the soil tightly (or fill in the holes with cement for a permanent fixture).
8. After a week or so (let the cement dry fully), the gate can be hung on the hinges and the latch positioned correctly.

Simple Gate

This is a simple and appealing gate, especially for fences leading into pastures. The materials required to make this gate vary depending on what purpose the gate will serve.

For a paddock or pasture gate, make it out of seasoned boards, 1 x 6 inches and 12 to 14 feet long. The posts support-

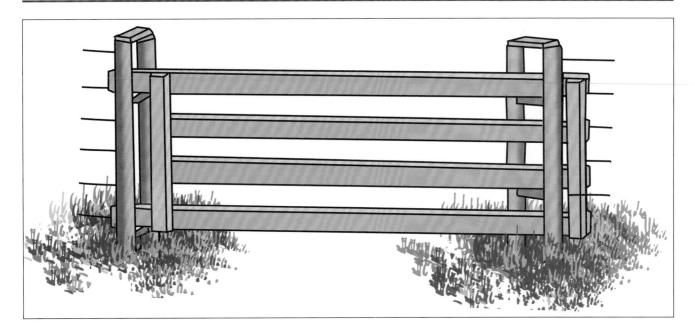

A simple sliding gate can be made for any modern or wire fence.

A bamboo fence or gate can be constructed by lashing the bamboo together with strong rope.

ing the gate should be placed about 5 inches apart, the one on the inside being about 8 inches ahead of the other. These are joined together by cleats or rollers that support the gate and allow it to be pushed back and swing open. If rollers are not obtainable, cleats made of any hard wood are acceptable.

Pens

If you have built a simple stable to house your llamas, sheep, or other animals, it will be beneficial to build a small fence around it as an outdoor pen. A basic wooden fence or a simple wire fence will enclose most of your livestock in an area around the stable or shelter. If you have a llama or two, it is best to have at least a 4-foot fence so they cannot escape. If you have ample space, having a pathway into a larger grazing field or pasture from your pen will allow your animals to come and go as they please. Or, if you want to keep them confined in the pen, a simple gate will suffice for when the animals need to be removed or relocated.

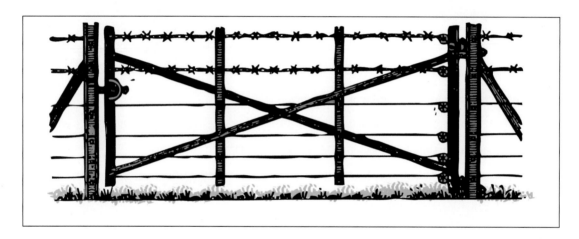

This wire gate is hung on ordinary iron posts. The heel of the gate, made of angle iron, is fitted with winding brackets for tightening the wire bars.

A simple wire swinging gate.

KITCHEN FURNITURE

Kitchen Corner Cupboard

A small corner cupboard as illustrated in Figure 1 will be found very useful, and can be quickly built. The body, outer frame, and door frame can be made from ¾-inch prepared material, to save planning up. You won't need any mortises, grooves, or rebates. Figure 1 gives a general view of the body. To make the body, cut from some 8-inch by ¾-inch material four 2-feet 6-inches lengths. Nail small cleats or battens across them (See c, Figure 2); these will make the two parts that form the back of the cupboard. Figure 3 shows a plan of top and bottom; these should be cut from ¾-inch stuff. Two shelves must be cut from ½-inch or ⅝-inch prepared material to the shape of the top and bottom, but ⁵⁄₁₆ inch less from front to back. This will allow the door frame to come flush with the outer frame when the door is closed. The four pieces referred to being cleated, and the top, bottom, and shelves cut to their proper shape and size, nail the two parts that form the back to the top and bottom; Then place the shelves on the cleats c c, nail with small wire nails, and the body is finished, except for planning the front edges of the back a little, so that the outer frame may lie flat and even against it.

Outer Frame of Corner Cupboard

The outer frame is made from ¾-inch material like so: Cut off two 2-feet, 6-inches lengths, and two 2-feet, ½-inch

Figure 1—Kitchen Corner Cupboard

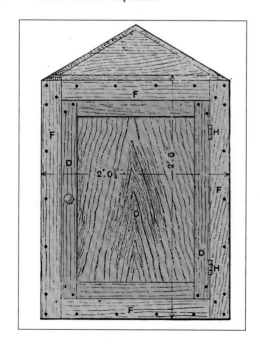

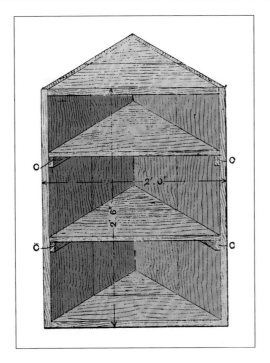

Figure 2—Shelves of Corner Cupboard

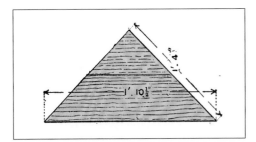

Figure 3—Top or Bottom of Corner Cupboard

Figure 4—Corner Cupboard Door

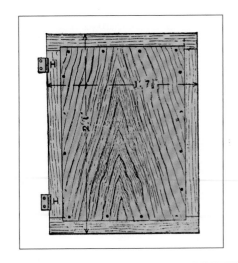

lengths; these should be 2½ inches wide. Halve the ends, and nail these together, and the frame f (Figure 1) is made. A small portion of the frame is cut away with a chisel to receive two butt hinges. This frame is nailed against the front of the body, and the outer edges of the frame are planed to correspond with the angle of the back of the cupboard. The door frame is made from material 2 inches wide, and the ends are halved and put together as the outer frame, which should thus fit nicely within the outer frame. On the inside of the door frame a piece of pine of suitable length and width, and about ⁵⁄₁₆ inch thick, nicely planed up, is nailed to the frame, seen in Figure 4; This piece forms the door-panel and at the same time tends to strengthen the door frame, and looks as well from the outside as if the frame had been mortised and tenoned together, and the panel let into a rebate in the frame. Secure two 1½-inch butt hinges, as shown. Hang the door d (Figure 1) to the outer frame by securing the hinges h h. Fix a knob and fastener as shown in Figure 1, then fill in the nail-holes with putty, and rub up the outer frame, door frame, and panel with glasspaper; then the cupboard may be stained, sized, and varnished, or painted.

Household Tidy

A small cabinet should prove most useful. Figure 5 is part front elevation, part section. In Figure 6, which is a section on n n (Figure 5), a is one of a pair of vertical side pieces, into each of which a shelf b is housed. A top c is joined as shown in Figure 5. A back may be fitted in by grooving the inner edges of the sides, or by rebating and screwing on a strip as in Figure 7. Two small doors hinged to the sides are framed up and paneled flush on the inside. Figure 8 shows

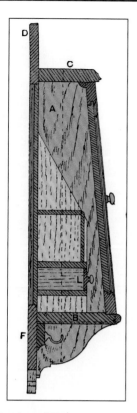

Figure 6—Cross Section of Tidy

a section of them at the meeting stiles. The piece d (Figure 6) should be screwed from behind. e (Figure 5) is a strip to carry hooks for keys, and f (Figure 6), to which this strip is fastened, is screwed to the shelf. A space g (Figure 5) is divided centrally, and is useful for holding small books, etc. The spaces h (Figure 5) may receive small drawers, which may be made of tobacco boxes with the lids removed, and with wooden fronts l (Figure 6) screwed through holes punched in the front of the boxes. Small compartments j (Figure 5) are closed by sliding fronts pierced by center-bit holes, as indicated by the dotted circle. Each space may contain a ball of string, the hole being used help remove the front and admit one end of the string. The vertical portion k is carried well up so that a gum or paste-pot with a protruding brush may be

Figure 5—Part Elevation and Section of Tide

Figure 7—Back of Tidy Jointed to Side

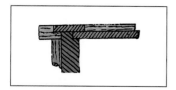

Figure 8—Section through Door Meeting Stiles

Figure 9—Looped Leather Strip

protected if placed in the recess at the side. The inside of each door has a strip of leather near the top fastened transversely (Figure 9), and in the loops formed some such article as a hammer, screwdriver, or sardine-tin opener may be placed. You will find this very handy for many household tools. The cabinet may be hung to the wall by eye-plates, attached one on each side at about the level of m (Figure 5). The total width of the article is 17 inches, the height is 23 ¼ inches, and the depth is 6 inches.

Spice Box with Drawers

To make a spice box like the one in Figure 10, yellow pine or deal is the best material, but, in any case the wood must be thoroughly dry. The following pieces will be required:

1. Back of case, one piece 7¾ inches by 6½ inches by ¼ inch
2. Sides, two 7 ¾ inches by 3 inches by ¼ inch
3. Top and bottom, two 7 ¾ inches by 3¼ inches by ¼ inch
4. Shelves, four 6 ½ inches by 2 ⅝ inches by ³⁄₁₆ inch
5. Partition, one 6 ⅛ inches by 1 inch by ¼ inch
6. Ornament, one 7 ¼ inches by 1¼ inches by ¼ inch
7. Sides of drawers, eighteen 2 ⅜ inches by 1 ⅛ inches by ³⁄₁₆ inch
8. Back of small drawers, eight 2⅝ inches by 1 ⅛ inches by ³⁄₁₆ inch
9. Front of small drawers, eight 3 inches by 1⅜ inches by ¼ inch

Figure 10—Spice Box with Drawers

10. Bottom of small drawers, eight 3 inches by 2⅜ inches by ³⁄₁₆ inch
11. Back of bottom drawer, one 5⅞ inches by 1⅛ inches by ³⁄₁₆ inch
12. Front of bottom drawer, one 6¼ inches by 1⅜ inches by ¼ inch
13. Bottom of drawer, one 6¼ inches by 2⅜ inches by ³⁄₁₆ inch

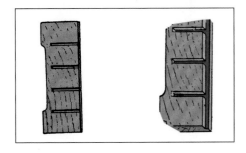

Figures 11 and 12—Spice Box Sides

Figure 13—Spice Box Foot

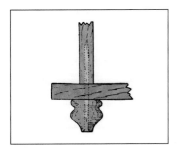

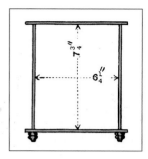

Figure 14—Vertical Outline of Spice Box

These are all finished sizes. The other materials required are nine small brass knobs, two bracket eyes, four wooden feet, and a handful of small nails. To make the case, cut the two side pieces to the shape shown by Figure 11. Cut four grooves (Figures 11 and 12) ⅛ inch deep and 2½ inches long, to fit in the shelves, and treat the back edge of these pieces in the same manner. The width of the grooves is just sufficient to allow the shelves to fit tightly into them. At equal distances from each other carefully mark out the positions for the grooves by dividing the side piece into five equal parts. The places for the grooves being determined, draw lines across representing the widths of the grooves; then cut these lines down to a uniform depth with a chisel, cutting downwards, or guided by a straightedge, draw it along to that it cuts into the wood. The bottoms of the grooves need not be absolutely smooth, as they are not seen when the parts are fitted together. Nail the top and bottom

Figure 15—Spice Box Shelf

Figure 16—Spice Box Case

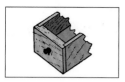

Figure 17—Part of Spice Box Drawer

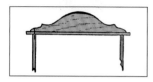

Figure 18—Top of Spice Box

onto the sides. The feet can be put on at the same time by driving the nail first through the foot (see Figure 13). When you complete the case, it should measure inside 7¾ inches by 6 ¼ inches. Next take three of the four shelves and cut them as shown by Figure 15. The piece cut from the middle is 1 inch long and ¼ inch wide. The pieces cut from the sides are ⅜ inch long, and ⅛ inch wide. The fourth and bottom shelf is cut similarly, only the middle piece is left in. Fit the shelves and partition into their respective places, the partition being nailed to the top of the case and to the bottom shelf. If the back is now fixed in its place, the case (Figure 16) may be considered complete. All that's left to

Figure 19—Plate Rack

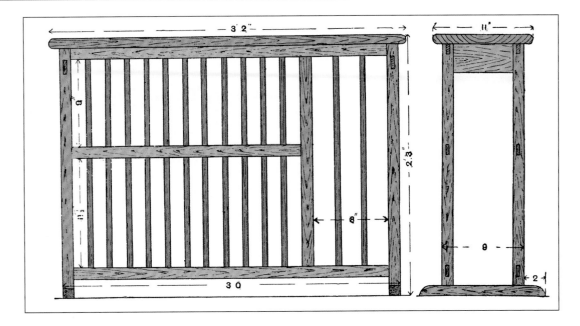

Figures 20 and 21—Front and End Elevations of Plate Rack

be done is to make the drawers. The front of the drawers should be cut so that the sides will fit into them as in Figure 17. After making the drawers, fix a knob on each to serve as a handle for pulling them out. If the remaining piece is cut to the shape shown in Figure 18, a passable ornament will be the result. It is fixed to the top by nails driven down into the two side pieces. By fixing two bracket eyes to the back, the box can either be hung against the wall or stood in any convenient place. To add a finish to the box it can either be stained or polished, or painted—according to your taste.

Figure 22 and 23—Front and End Elevations of Pantry Safe

Pantry Safe or Cupboard Pantry

Figure 22 is a front elevation of a pantry safe which is rectangular in plan, and Figure 23 shows the end elevation. For the three pieces of framing, six stiles, 2 feet, 8 inches by 2 inches by 1 inch will be required. These pieces are mortised and tenoned together, the six top and bottom rails having haunched tenons, as shown in Figure 24. The other three

Figure 24—Top and Bottom Rail Joint
Figure 25—Middle Rail Joint

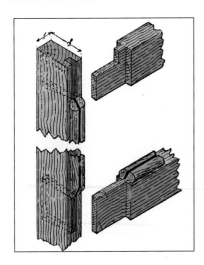

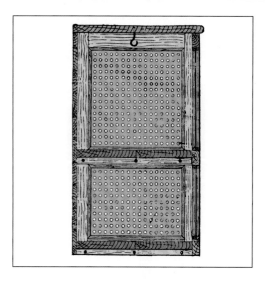

Figure 26—Cross Section of Pantry Safe

rails have tenons and the pieces of framing are glued and wedged together. The two framings for the ends are rebated ½ inch each way on the back edges to receive the back, which is formed of 3-inch by ½-inch tongue and groove-jointed matchboarding. The front edge of the ends has a ¾-inch chamfer on the outside corners. The shelves and top should be got out of a wide board of pine or whitewood. The top, when finished, is 1 foot, 10½ inches long and 1 foot, 6 inches wide, which allows it to project ¾ inch over the front and ends. The projecting should either have a nosing worked on, as shown in Figures 22 and 23, or a chamfer. The shelves are 1 foot, 3¾ inches wide by 1 inch thick, supported by four fillets, 1 foot, 3¾ inches by ¾ inch by ½ inch, screwed to the sides of the safe as shown in Figures 26 and 27. In fixing together, the top may be secured to the ends with 1½-inch screws three through each top rail. The bottom shelf is fixed

Figure 27—Bottom shelf of Pantry Safe

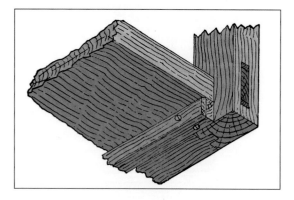

with screws to the fillets, thus securing the lower ends of the sides as shown in Figure 27.

Fitting Together Pantry Safe

Before fixing the back, the framing should be squared, and a temporary lath fastened diagonally across the front to hold it square until the back is completed. Fit in the boards for the back and fasten with 1½-inch oval wire nails. Figure 28 shows the end rebated to receive the back. A 1½-inch screw through the top into the edge of the back will stiffen it considerably. Two or three hooks should be screwed into the top for meat to hang from (see Figure 26), and the middle shelf should be left loose, so that it can be removed when the hooks are used. The door may next be fitted in, and a ⅜-inch bead, worked down each side, will prevent the butt hinges looking unsightly on the hanging side; 2-inch butt hinges are used, and are fixed 3 inch from each end, the whole of the hinge being let into the door stile. The spaces are covered with perforated zinc, which can be fastened on the inside of the safe with tacks, or may be secured with beads which is a neater and better method. About 60 feet of ½-inch beading will be required for the latter method, the beads being mitered at the corners, and fastened with 1-inch brads. The door may be fastened either with an ordinary lock, or with a special catch.

Figure 28—Back Corner of Pastry Safe

Cupboard Pantry Door

The door is ⅞ inch thick, with stiles and rails mortised, tenoned, and wedged together. You will need to mold and groove the edges of the framing to create a panel. You can

Figure 29—Door with Vertical Paneling

substitute a square-edged framing and afterwards pin a molding to the opening, but this is not always the best option.

The panel should be made up of two or three widths of timber, butt-jointed and glued, or you can fill the framework as shown in Figure 29. Notice the grooved, tongued, and beaded boards 3 inches by 3/8 inch placed vertically in the frame. Another method is shown in Figure 30, where the boards are placed diagonally. The other panel consists of perforated zinc or fine gauze, and fits in the rebate of the door framing, a molded slip s keeping the zinc z in position, seen in Figure 31. The rebate in this part of the door stile is made by cutting away the inside piece left after the panel grooving has been run. The front corner stiles are grooved to the sides

Figure 30—Door with Diagonal Paneling

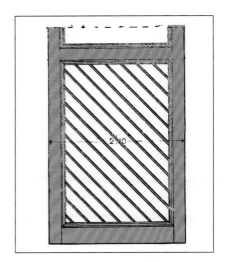

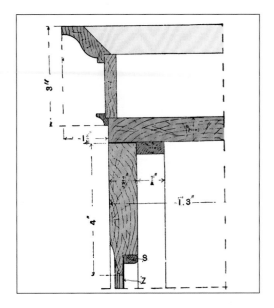

Figure 31—Section of Cornice and Door Rail

and mortised into the top and bottom. The folded edges are beaded to relieve the joint. Hang the door with strong brass butts, and furnish it with a knob, lock, and key.

Cupboard Pantry Interior

The interior of the cupboard is provided with four shelves, 1 foot, 1 inch wide and 1/2 inch thick. The top shelf should be fixed near the center of the perforated zinc, so that the two top spaces can be used for fresh meat. The lower shelves

Figures 32 and 33—Back and Side Elevations of a Basic Kitchen Chair

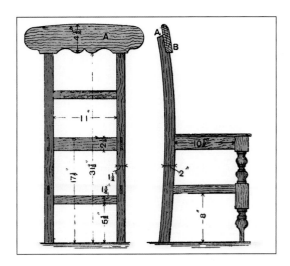

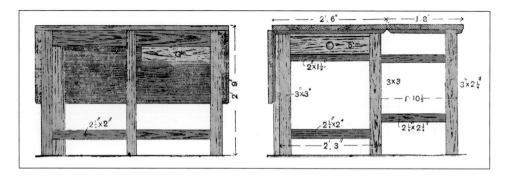

Figures 34 and 35—Side and Front Elevations of Gate-legged Table

can be fitted as needed. The cornice has a small molding to cover the joint with the top (see Figure 31), and is rebated for a frieze panel ⅜ inches thick, surmounted by a cornice molding. The moldings and panel are mitered and keyed at the corners, and strengthened with blocks glued at the angles. When finished, the cornice forms a separate piece from the carcass, so that it can be removed when not in use. The four feet are turned in the lathe, and have dowel ends fitting into the carcass bottom and fixed with wedges. A plinth finish can be adapted if you prefer. This cupboard pantry can be easily converted to a wardrobe if desired by removing the zinc panel and inserting a bevel panel corresponding with the lower one, so that it is desirable to finish the interior neatly.

Gate-Legged Kitchen Table

Figures 34 to 37 illustrate a gate-legged kitchen table which, when closed, is 2 feet, 6 inches wide by 4 feet, 4 inches

Figure 36—Underneath Plan of Gate-legged Table

long. The leading points in the making are as follow: The legs and rails should be planed up square to sizes, and the legs set out for mortising. The mortises for the lower rails are of a simple character, the tenons being stubbed inches. For the long top rails the mortises and tenons at one end should be

Figure 37—Gate-legged Table Framing

Figure 38—Joint of Top Rails and Leg

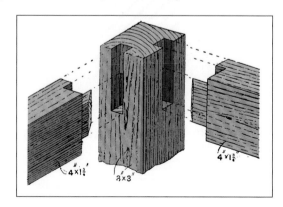

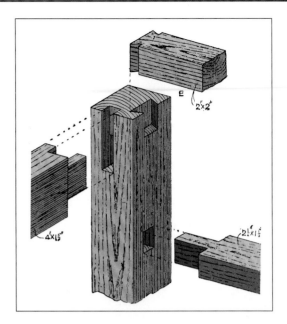

Figure 39—Joint of Rails and Leg at Drawer End

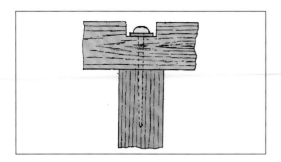

Figure 40—Jointing Gate Leg Stile and Rail

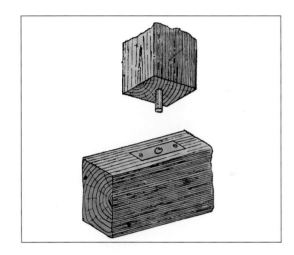

Figure 41—Pin and Socket of Gate Leg Stile and Rail

as shown at A (Figure 38) and at the other end as shown at B (Figure 39). Each end of the top rail C (Figure 37) is of the form shown at C (Figure 38). The two rails F (Figure 37) for the drawer have the joints illustrated at D and E (Figure 39), the upper rail being dovetailed in the top of the legs. The two gate legs and their rails are stub-mortised and tenoned together. Figures 39 and 41 show how the gate-leg stile is jointed to rails. After the joints are made the whole of the framing should be carefully fitted, and the joints numbered; then the framing should be separated, and the internal parts smoothed off. After you glue and fit the joints, cramp them

in position until the glue is dry; odd strips of wood, with a block nailed on each end and a wedge inserted, may be used for this purpose. The top, including the flaps, should be formed of 1½-inch boards, ploughed, tongued, grooved, and glued together. The top and flaps should be planed off true,

Figures 42 and 43—Side and End Elevations of Table with Turned Legs

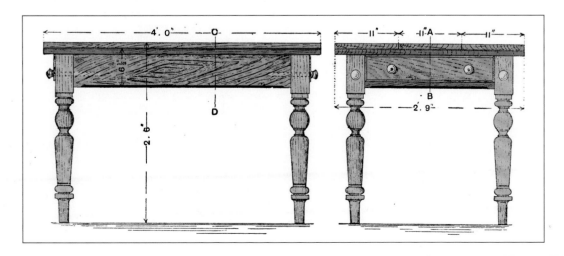

and the top secured to the rails of the framing by 2½-inch screws driven obliquely from the inside of the rails. The flaps may be attached to the top by 2½-inch wrought-iron back-flap hinges as shown in Figure 36. The top and flaps should be strengthened by 2-inch by 1-inch thick fillets, which are screwed on as indicated at Figures 35 and 36. The stiles of the gate legs should be fixed at the bottom end by a pin working in a socket (Figure 41), the upper end being secured by a screw sunk through the top rail as in Figure 40. As the depth of the side rails will not be enough to fix the runners of the drawer, pieces G (Figure 34 and 37) should be added, and to these two runners can be fixed, and also a cross rail; see h

and k (Figure 37). The drawer front should be carefully fitted between the rails and legs, and the sides and back prepared, the back being made wide enough to extend only as far as the plough groove to receive the bottom. The dovetailing can then be set out and made. After this the plough grooves for the bottom should be made. Next some ½-inch boards should be glued up for the bottom, the edges being chamfered to fit into the plough grooves. To secure the bottom it should be nailed into the lower edge of the back, and have strips underneath fixed to the bottom and the sides, these being secured with glue and planed off flush. A knob or handle should be provided and fixed to the front of the drawer.

POULTRY HOUSES

Poultry houses should be warm, dry, well-lighted, and ventilated shelters with convenient arrangements for roosts, feeding space, and nest boxes. In winter, if you're living in a cool climate, light and warmth are of the upmost importance. Fowl will stop laying eggs and their health will suffer when confined in cold, wet, and dark conditions. Windows facing the south or southeast, large enough to admit the sun freely, should be provided and made to slide open to increase circulation during the summer.

Beyond these few requirements, houses for your poultry can be made in a variety of ways and are, generally, relatively easy to construct. Below are many different types of poultry houses that can be used to keep your fowl warm, dry, and healthy.

Simplest Poultry House

While poultry can survive in this type of cheaply and simply built coop, it is best used in warmer climates, where the winter months do not become incredibly cold and not much, if any, snow falls. Also, this type of coop is best suited for only one or two chickens or ducks.

Materials

- Four pieces of 1 x 2-inch boards for the studs and rafters
- Strong nails
- Wire netting
- Tarred paper

1. Take two of the boards and nail them together in a T shape. Repeat with the other two. Set these apart from

each other about 2 feet 10 inches on the centers, and cover them with tightly drawn wire netting (cut to size).

2. Cover the wire netting with tarred paper, creating a barrier between the outside winds and weather and the fowl inside.

Young Poultry Coops

Chicks need extra warmth and protection from predators. This coop, if it houses small chicks, should not hold the other fowl, as they may bully or even harm the young chicks.

This pitched roof chicken coop consists of a pitched roof mounted on three boards, 6 feet high. This coop is 3 feet wide and 2 feet deep. Nail slats across the front to prevent the hen from getting out but to allow the chicks to enter and exit freely into a small, fenced-in area surrounding the coop.

The coop pictured above is similar to the pitched roof chicken coop except that there is a canopy that keeps the rain out and shades the interior of the coop so it does not become too warm. This coop is 3 feet long, 2 feet wide, and 30 inches high at the front and 24 inches high in the back. The coop

A pitched roof chicken coop.

Chicken coop with canopy.

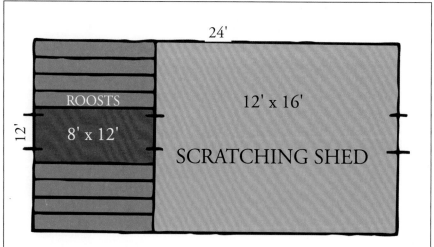

Plan for a henhouse with a scratching shed.

can be constructed from boards with matched edges and should be raised an inch or two above the ground to ensure the floor remains dry. Tack a piece of light canvas or muslin to the roof to serve as the awning.

Practical Henhouse

This simple and efficient henhouse has a shed roof and, as most poultry houses should, faces toward the south. This house can be up to 10 feet wide and as long as you need to accommodate your chickens.

A scratching shed is in the center of the building and has windows that let sunlight in. The sleeping quarters should be kept warm. An open, wire-enclosed front for the scratching

This henhouse has a scratching shed, which allows the chickens access to the open air while still being protected from the elements.

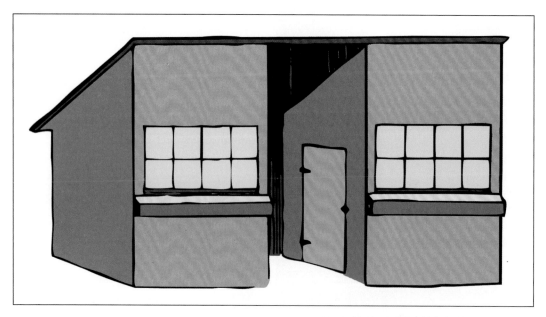

You can build a simple ramp to give your chickens easy access to the coop.

shed should be included, too. The roosts should be made movable, and fresh bedding should be kept on the floor of the henhouse.

The roof of the henhouse should project out 1 foot over the south, east, and west sides. It should also be 5 inches higher than the siding, allowing for free ventilation. Two large windows will admit light and warmth into the henhouse. A laying box should extend the entire length of the room and must be divided into compartments and covered with a hinged lid. This allows the eggs to be gathered simply by raising the lid from the outside. Make sure the floor is cleaned weekly to keep out disease. The inside of the walls should be whitewashed often to keep out moisture and pests.

Two-Room Henhouse

This two-room henhouse has a south-facing front to allow ample sunlight and warmth into the house. It can be made as large as 10 x 12 feet and should be constructed of wood or timber planks. It is divided into two rooms by a partition made out of wire netting. This henhouse can serve two separate yards. A fence constructed in the middle of the house yard should join the center of the front of the building (and at the back as well if you so desire). In this house, both hens and roosters can be kept and are easily separated while allowing each enough space and exercise.

A two-room henhouse with a south-facing front.

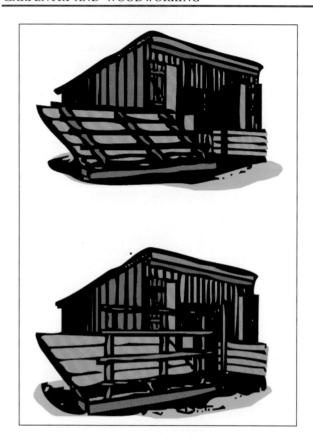

These pictures show how the perches can be moved to allow for easy cleaning.

The platform and perches should be constructed inside of each room. When the perches are in need of cleaning, they are raised up against the wall in the house, in a perpendicular position. To clean the trough, the perches and platform are raised perpendicular to the floor.

Duck Houses

Ducks, while they can survive rather well in any type of poultry house, are happiest when they have either a stream or pond in which to swim, bathe, and gather food. If you have a stream or pond on your property, situating a duck house nearby will help ensure that the duck eggs are safe and secure.

If you are raising a good number of ducks, your duck house should be about 30 feet long and 12 feet high. Doors should be situated in the front of the house and the house should have a few small windows that can be slid open to

A simple duck house.

allow fresh air to circulate within the duck house. The rear of the house should hold the nests (boxes open at the front). A small door should be situated behind each nest so the eggs can be easily removed.

You can use a strip of wire netting to enclose a small, narrow yard in the front of the house. Do not use twine netting, however, as the ducks could get their heads twisted in it and strangle themselves.

Easy, Creative Coops

If you don't have much space in your yard and only have a few chickens to keep, very good coops can be made at a very small cost from items found around your house, yard, or at rummage sales:

1. Barrel Coop
   a. First, drive shingle nails through the hoops on both sides of each stave and clinch them down on the inside.
   b. Divide the barrel in half, if it is big enough, by cutting through the hoops and the bottom.
   c. Drive sticks into the ground to hold the coop in place, and drive a long stick at each side of the opened end just far enough from the coop to allow the front door to be slipped in and out.
   d. The night door can be made from the head of the barrel or any solid board, and the slatted door, used to confine the hen, can be made by nailing upright strips of lath to a cross-lath at top and bottom.
2. Box Coop
   a. Find a box that is roughly 2 to 2½ feet long, 16 inches deep, and 2 feet high and saw a hole, *d*, in one end.

Floor plan for the duck house.

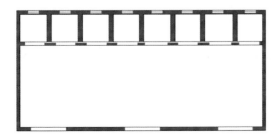

A barrel chicken coop.

   b. Strengthen the box with narrow strips of wood, *b*, *c*, on each side of the hole. This acts as a groove for the door, *a*, to slide in. By doing so, you will have a sliding door that opens and shuts easily.
   c. The front of the coop is enclosed with lath, or narrow strips, placed 2½ to 3 inches apart. The top should be covered with a good grade of roofing paper to make it completely waterproof.
3. Portable Coop—This type of coop will allow you to have a fresh yard for your chickens and other poultry

A simple box coop.

A portable chicken coop.

to scrounge in and is easily transported to any place on your property.

a. The coop is built of ordinary material on a base frame and with a V-shaped roof and side frames. The preferred length of the coop is about 2 feet and the yard should be around 3 to 4 feet.

b. The ridge pole is extended, as shown at each end, to form a handle.

c. If desired, the hen may be allowed to freely roam the yard or can be contained within the coop by slats, as is pictured in the drawing.

Poultry House Aids and Other Considerations

Folding Chicken Roost

This roost is made of 3-inch boards cut to any desired length that will fit within your poultry house. A small bolt fastens the upright pieces at their top ends and the horizontal pieces are fastened on with nails. This roost can be kept at any angle and may be quickly taken out of the house when it is time to clean. This sort of roost will accommodate more fowl in the same space than the flat kind.

Keeping Rats and Mice Out of the Poultry House

If you are building a permanent poultry house, you should try to make it as rodent-proof as possible. If rats and mice can easily enter your poultry house, they will not only steal eggs and spread diseases, but they could scare or even

A folding chicken roost.

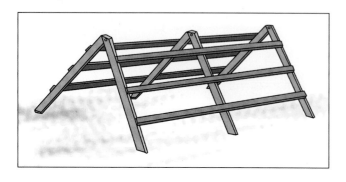

harm the fowl. Cheap and efficient walls can be made of small fieldstones in this way:

1. Dig trenches for the walls below the frost line.

2. Drive two rows of stakes into the trenches, one row at each side of the trench.

3. Set up boards in between the stakes. The boards will hold the stones and cement in place until the cement hardens. The top boards should have a straight upper edge and should be placed level to determine the top of the wall.

4. Place two or three layers of stone in the bottom of the trench, pour in thinly mixed cement, and pound it in. Repeat this until the desired height is reached.

5. The top of the wall should be smoothed off with a trowel and left until the cement completely hardens. The side boards can now be removed and the poultry house built.

Winter Care of Fowl

If chickens and other fowl are not kept warm in the winter, they will stop growing, cease laying eggs, and can become ill. There are several ways you can winterize chicken coops to ensure your birds' comfort and well-being.

Especially if you live in colder climates, having a house with hollow or double side walls will help keep your fowl warm during the winter season. Buildings with hollow side walls are warmer in the winter and are also cooler in the summer. They do not collect as much severe frost and result in less moisture seeping into the henhouse once the frost melts.

The outside walls of chicken coops can be plastered or lined with matched boards and the spaces between the boards filled with wood shavings, sawdust, or hay. The floor should be covered with several inches of dry sand, wood shavings, or straw, and the ventilating holes near the roof should be partly stopped up or shutters arranged to close most of them in very cold weather. You don't want to seal the place up completely, though. Nothing is more important to the health of fowl than pure air. Birds breathe with great rapidity and maintain a relatively high body temperature, so they need plenty of oxygen.

Constructing a solid, insulated roof for your poultry house for the winter is very important. A roof can be built either by sealing the inside with material to exclude draughts or by placing roof boards close together and covering them thoroughly with tarred paper before shingling. An ordinary shingled roof allows too much wind to come into the house and could cause your fowl to get frosted combs or wattles. If this happens, there will not be much, if any, egg production in the winter months.

A drinking fountain for your chickens can be made with a can or bucket and a tray. Cut out one end of the can and poke holes along the edge as shown. Fill with water, cover with a shallow tray, and turn the whole thing over quickly. Chicks will be able to drink water easily without risk of drowning.

Hanging curtains in front of the perches is also a great way to keep your fowl warm during the winter months. Make these curtains of burlap and hang them from the roof in such a way that the perches are enclosed in a little room. Make sure the curtains are long enough to touch the floor all around, and sew the edges of the burlap together, except at the corners. At night, the corners can be pinned together to keep the birds from leaving their sheltered perches. This pseudo–sleeping room allows air to move in without creating drafts and it also helps retain the birds' body heat. This maintains a comfortable temperature for the birds during cold, winter nights.

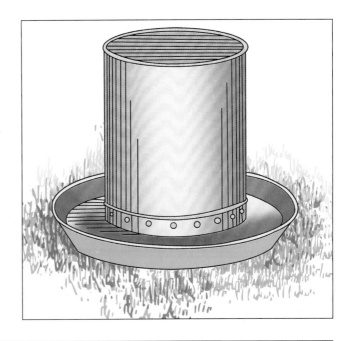

ROOT CELLARS

While most modern houses have basements or crawl spaces in which to keep fresh vegetables and preserves cool and dry, you may want to construct an additional root cellar if you'll be storing significant amounts of these items.

A root cellar is basically an outdoor storage cellar, preferably at least half-way or almost entirely underground. The cellar's primary use is to store fruits, vegetables, nuts, and other perishable foods at a relatively consistent temperature and humidity.

A cellar at least partway underground is ideal because the ground will help keep the cellar at a more uniform temperature, regardless of the time of year. The depth of your root cellar can depend on where you live and how much the temperature fluctuates. If the cellar has some built walls make sure that they are heavily insulated. The floors should be made out of dirt if possible. Near the top of the cellar will naturally be a few degrees warmer than the floor. Keep two thermometers to continually check the temperature—put one in the coolest part of your cellar and the second right outside the door. Open the door or use ventilation to adjust the temperature inside the cellar if necessary.

A root cellar should be dry to prevent the fruits and vegetables from molding. Make sure your root cellar is not damp by building it above the water table, roofing it with waterproof material, and creating a sloped floor when digging it out, to help drainage. Keep humidity as constant as possible. If more humidity is needed you can sprinkle the dirt floor with water, or place damp things such as straw or sawdust inside. Placing a pan or two of water under an open vent will also work. A cellar is too humid if droplets begin to form on the surface of the produce.

Proper ventilation is also important. Make sure your root cellar has some airflow, usually through air-vents in the roof that can be opened or closed. Some produce may rot if there is not adequate air flow around them.

A sloped area in your yard is the perfect place to make a root cellar. To construct the root cellar, follow these simple steps:

1. Make an excavation in the side of the hill, determining how large you'd like your root cellar to be.

2. In the excavation, erect a sturdy frame of timber and planks, or even of logs. Put up planks to stand as side walls, and build a strong roof over the frame.

3. Throw the excavated earth over the structure until it is completely covered by at least 2 feet of soil.

4. On the exposed end, make a door that is large enough for you to enter without ducking. Or, if you like, you can make a "manhole" through which you can enter—this will actually protect your root cellar from frost much better than a full-sized door.

If the soil in the hill is composed of stiff clay, you may not even need to construct any side walls, and the roof can be fitted directly into the clay. Then build up the front of the cellar with planks, bricks, or stone, and create a door.

Root House

If you do not have a large hill on your property and would still like to construct a root cellar, find a knoll or other dry place and remove the soil over a space that is slightly larger than the size of the cellar (or root house if the structure is not built into a hill) and about 2 feet deep. To construct this root house:

1. Select poles or logs of two different sizes. The wider ones should be shorter than the other two.

2. Cut the ends of the logs very flat so they will fit closely together and make a very tight, pen-like structure.

3. Cut two logs in each layer long enough to pass through and fit into the outer pen. This will help fasten the two walls together.

4. Build the doorway up with short logs passing from one layer of poles to the other. These serve as supports to the ends of the wall poles.

5. Fill in the space between these two walls with soil. It is important that these are filled in fully (sod

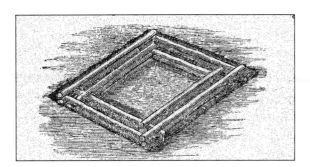

The base of the root house.

The finished root house.

may also be used to pack in spaces between the logs) to protect the inside storage items from frost and to keep the whole structure cool. Pack up the soil as you construct the walls so you can more easily compact it as you build up.

6. When the walls are about 5 or 6 feet on one side and 2 or 3 feet on the other, put the roof on. The roof is made of poles placed close together, secured to the logs, and covered with sod, then 18 inches of soil. It is then finished off with sod once again.

Storing in a Root Cellar or House

Keep your root cellar clean, neat, and organized. Entirely empty all contents once a year, cleaning out storage shelves and any storage baskets and containers. If you use wood baskets and boxes check for rot or decay.

You do not need to wash produce that is going to be to be stored, but all of it should be dry and inspected before you

Most fruits and vegetables should be stored in baskets or boxes, made out of either plastic or wood. This will allow for air flow and help preserve them. Airtight, plastic containers should be used for dried beans, grains, and flours. Some vegetables, such as onions and squashes, require dry storage; do not put them in airtight containers or baskets that allow for only a little airflow, as they will spoil. Potatoes, cabbage, and cauliflower need a cool and moist environment, so wooden boxes or baskets are fine. Most other root vegetables, such as carrots, turnips, radishes, rutabagas, and beets need to be kept very cool (32-40°F) so store them in a wood box under moist sand or in wood boxes with sawdust. Potatoes and most other root vegetables need 90-95% humidity, while squashes, sweet potatoes, garlic, and onions do better in humidity of only 60-75%. If you need to keep your cellar at a higher level humidity, consider storing your squash, onions, etc. somewhere else. Do not store fruits and vegetables together—fruit like apples give off a gas that causes potatoes to sprout and shortens the lives of other root vegetables.

Make sure to go down in your root cellar about once a week to check the produce for any problems. Remove any rotten produce immediately. Check preserves, jams, and pickles for mold. If you see any ripened fruit (such as apples or tomatoes) take it to your kitchen right away and use it. If a squash or pumpkin develops a bad spot cut it out right away and use the rest of the vegetable as soon as possible.

put it in the cellar. Wet produce invites mold. Do not put any produce with bruises or damaged skin in the cellar; find another way to preserve them (canning, etc.) or use them right away. Any decay on one piece of fruit or vegetable will quickly spread to neighboring ones.

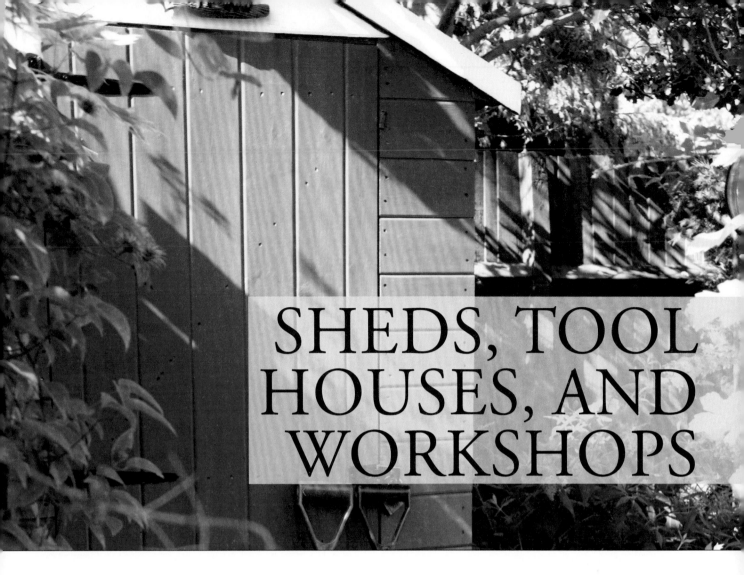

SHEDS, TOOL HOUSES, AND WORKSHOPS

Before building a tool shed, think about what you want to house in it. If you just need it for small tools, such as shovels, buckets, and a wheelbarrow, a smaller shed will be fine. However, if you plan to house machinery, such as a tractor, lawnmower, chainsaw, or rototiller, you'll need a larger shed and you may want to plan for a sliding garage door–style entrance. Will you want a workbench, space to pot plants, shelving, and drawers? Do you want electrical outlets for power tools and lights? Also consider location: It may be more convenient to have it close to your house, or you may prefer to have it nearer the garden. Here are a couple of examples of tool sheds that you can modify to meet your needs.

Medium-sized Tool Shed and Workshop

This shed is large enough to easily store your basic farm machinery. The shed is basically a giant umbrella with posts 30 feet apart in one direction and 12 or 16 feet apart in the other. There are no sides to this shed at all (though you could modify this if you want to store other tools here). If you park your main machinery (tractor, lawnmower, and so on) in the innermost part of the shed, you should still have an overhang of 10 feet. This shed would be most beneficial if it were 10 feet high—that way, most any kind of machine you want to house under it will fit well. Boarding up one, two, or three sides will help prevent snow from drifting in during the winter and rain from rusting your equipment. Making walls will also allow you to hang tools on the inside of the shed, such as clippers, weed whackers, or hoses.

The workshop described here will hold a lot of smaller tools and is a good place to mend harnesses, make repairs, and store grain. The workshop gives about a 30-foot clearance space for the shed below. The entire building is built together using the following materials:

- 2 x 8-inch posts
- Three pieces of 2 x 12-inch wood materials (space these 2 inches apart)
- 2 x 10-inch box plates
- 6 x 6-inch bridge truss
- 2 x 4-inch or 2 x 6-inch beams for the rafters (depending on how much weight they must hold)

The floor of your shed should be either hard dirt or cement and the posts should be anchored firmly into the ground or on stone pillars. A shingle roof will ensure your smaller workshop tools are kept safe from the rain and snow.

Small, Rustic Tool Shed

This small, rustic tool shed is made from "slabs" or "rough planks." If you are using trees from your own property to build the shed, you won't have to bother peeling the bark from the logs or cutting them as exactly. Slabs are cheap to buy (they can be found at saw mills and sometimes at home centers), and create an attractive, "woodsy" look. Although the boards are typically not uniform in size, you can position them in such a way to minimize the number of large cracks in your shed.

These boards may need to be straightened (especially the edges) with a saw or axe, and the interior of the tool shed should be lined with thin boards to cover up cracks and to keep out insects and animals.

When beginning construction on this type of shed, search for boards that lend themselves better to being end posts and those that are better-suited for the walls. The corners of the four main posts (4 inches square) construct a building roughly 7 x 5 feet. Dig holes 2 feet into the ground and fit in the end posts.

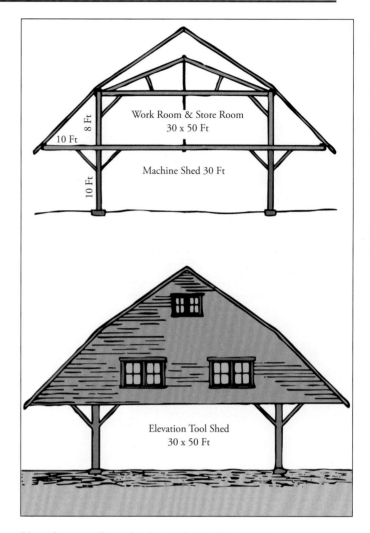

Plans for a medium-sized tool shed and workshop.

On the tops of these posts, rest the wall plates—these should be 3 inches deep. These boards will be at the back and sides of the shed only. The sides will also need cross rails that are around 2 to 3 inches thick with ends flush to the corner posts. Nail the side and back boards to these cross boards to secure them.

Place two door posts in the front of the shed. They should stand 2 feet 8 inches apart and should be about 3 inches square. They should rise about 6 feet or so to attach to the rafters. Fill in the space between the door and corner posts with extra boards.

The roof for this tool shed can be thatched or made of boards and shingles, whichever you prefer. Make raft-

A finished small, rustic tool shed.

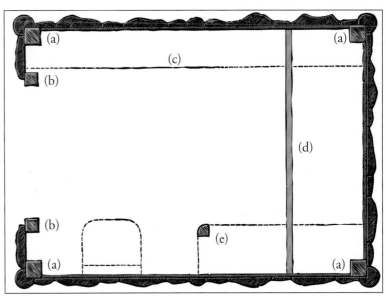

Plans for the inside of a rustic tool shed.

ers and laths out of regular boards, arranging them about 1 foot apart, and the laths should be placed 6 inches apart for thatching. The shed can also be cheaply roofed with galvanized iron or tin roofing.

The door of the tool shed has the slabs nailed to it on the outside only, to make it aesthetically consistent. Attach hinges and the door should be ready. Inside the shed, sets of shelves may be hung in which tools and other items can be stored (c). A wheelbarrow can be stored upright at the back (d) and tools hung from hooks coming down from the rafters. Gardening tools and rakes can be stored on the right-hand side (e) and a chair can sit near the front of the door (f).

Sheds can be customized to meet your wants and needs.

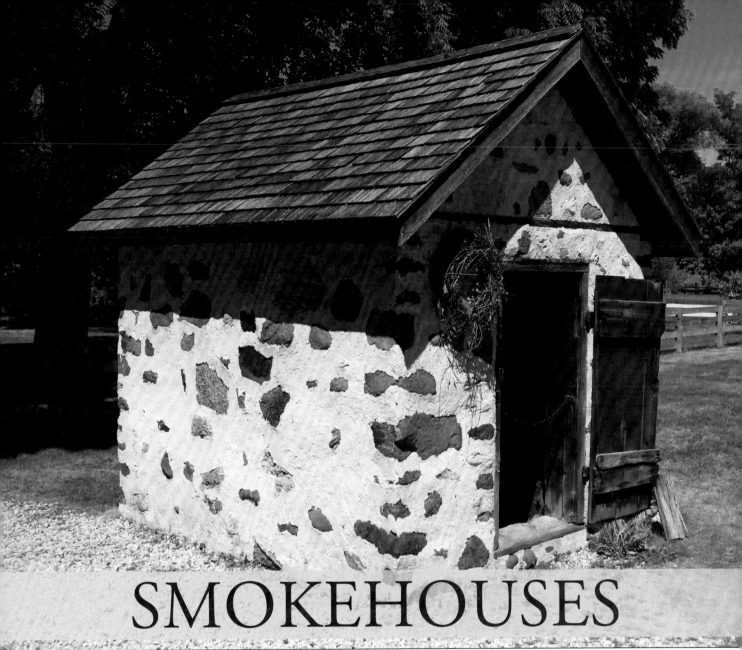

SMOKEHOUSES

If you are slaughtering your own poultry or other livestock, or if you just like the taste of smoked meat, try making your own smokehouse. Smokehouses help expose meats to the action of creosote and empyreumatic vapors resulting from the imperfect combustion of wood. The peculiar taste of smoked meat is from the creosote—this also helps preserve the meat. Other flavors are also imparted onto the meat by the choice of wood that is burned in the smokehouse, such as hickory.

To make a smokehouse you'll need a space (anything from the size of a barrel to a barn-sized area will work) that can be filled with smoke and closed up tightly. You'll also need a way to hang the meat that needs to be cured. In common smokehouses, a fire is made on a stone slab in the middle of the floor. In other instances, a pit is dug about a foot deep into the ground and the fire is built within it. Sometimes a stone slab covers the fire like a standard table. The possibilities are many, depending on your space and needs. Below are a few examples of smokehouses that can be built and used for smoking your own meats.

Standard Smokehouse

This smokehouse diffuses the rising smoke and prevents the direct heat of the fire from affecting the meats that are hung directly above it. In the picture, a section of the smokehouse is shown.

This standard smokehouse is 8 feet square and built of bricks—making it a somewhat permanent structure in your yard. If you want to make it out of wood, be sure to plaster is completely on the inside. The chimney, (c), has an 8-inch flue and the fireplace, (b), is outside, below the level of the floor. From this point, a flue, (f), is carried underneath the chimney into the middle of the floor where it opens up under a stone table, (e).

To kindle the fire, a valve is drawn to directly draft up through the chimney. The woodchips are thrown onto the fire and the valve is then placed to direct the smoke into the brick smokehouse. There are openings, (g, g), in both the upper and lower parts of the chimney that are closed by valves (these can be manipulated from outside the smokehouse). The door of the smokehouse should be made to shut tightly and, when building the smokehouse, be sure that there are no cracks in the brick or mortar through which smoke can easily escape.

This type of smokehouse is convenient because the smoke cools before it is pumped into the chamber and no ashes rise with the smoke. Meat may be kept in this smokehouse all year without tasting too smoky.

Another Brick Smokehouse

A smokehouse of this kind, built 7 x 9 feet, will be sufficient for private use. The bottom of this smokehouse has a brick arch with bricks left out sporadically. This is to allow the extraction of smoke from the house.

Located above the arch are two series of iron rods that have hooks with grooved wheels. You can find these at most local hardware stores. The open archway is for housing the fire and there is a door with steps leading up to it. A series of ventilating holes are situated above the lower bar and below the upper bar. These holes are meant to allow the smoke to escape from the house. By reinserting bricks into these

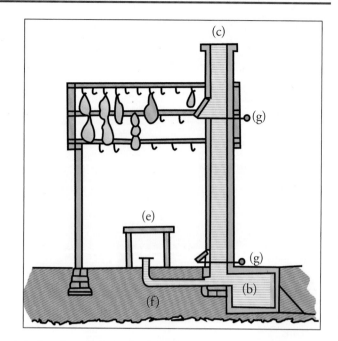

Interior view of a standard smokehouse.

holes, the smoke will stay mostly confined to the inside of the smokehouse.

The arch confines the fire and ashes, preventing any meat that might fall from being ruined or burned. The arch is made over a wooden frame of a few pieces of regular wood board, cut into an oval arch shape. Strips of wood are then nailed to this. When the brickwork is dry, the center is knocked out and removed. A small door can be fashioned to close up the arch when the fire is being kilned.

A Simple Way to Smoke Meats

If you don't want to commit to building a permanent smokehouse in your yard but you would like to smoke meats occasionally, you can use a large cask or barrel as a smokehouse substitute.

To make the barrel into an effective smokehouse, just follow these steps:

1. Dig a small pit and place a flat stone or a brick across it. This is where the edge of the cask will rest.
2. Making sure that half of the pit is beneath the barrel and half is outside, remove the head and bottom of the bar-

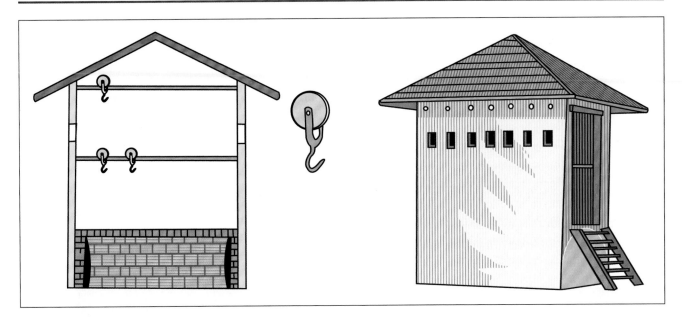

This is a sectional view of a brick smokehouse that can be built to any size.

Fish that are hung and ready to be smoked.

The drawing shows a common smokehouse that is built on a brick wall and over a brick arch. There are a number of holes left in it for smoke to escape. The ash pit is located beneath the arch, and there is also a door that opens to this pit. To reach the meat room door, use a sturdy ladder.

rel (or cut a hole into the bottom slightly larger than the portion of the pit beneath it).

3. Remove the top of the barrel and then hang the meat on cross-sticks. Rest these cross-sticks on crossbars that are made to fit into holes bored into the sides of the barrel, close to the top.

4. Put the lid on top of the barrel and cover it with a sack to confine the smoke inside.

5. Put coals into the pit outside of the cask, and then feed the fire with damp corncobs or a fine brush.

6. Cover the pit with a flat stone that will help regulate the fire and can be removed when more fuel is needed.

A smoke barrel is a simple method for smoking meats.

STABLES

I f you are raising larger livestock—sheep, goats, horses, or llamas, for example—you will need a small stable where they can go for protection during inclement weather and especially during the winter months in cooler regions. Building stables can be relatively easy and inexpensive, and doing it yourself means that you can customize the design to fit your and your animals' needs.

Stables should be built on relatively flat ground that does not become excessively wet or flooded during heavy rains. Laying down a thick bed of gravel or sand below the stable floor will help keep surface water drained. Also consider the positioning of the stable; try to find an area protected from strong winds but also near your own home so you don't have to go too far to tend the animals during bad weather. Facing the stable toward the south or west will help keep a nice breeze flowing through your stable while protecting it from harsh northerly winds. A place to store feed and hay for your

animals is also a worthwhile addition when planning and building a simple stable.

General Stable Construction

When building a stable for your livestock, make sure that the interior walls are weatherproof and free of dampness. To keep moisture out of the stable, the building should be situated on slightly higher ground than that surrounding it. This will keep the ground from getting too damp, and vapors will not be as likely to rise through the floor and foundation walls. If possible, it is best to make the stable floor out of concrete between 4 and 6 inches thick.

The stable walls should be built solid. Brick and stone are preferable to wood, but wooden stables also do an adequate job of providing shelter and are much more common in the United States, due to the availability of wood. If you decide to build your stable using bricks, building the walls one brick (9 inches) thick should be suitable. Internal walls should be built solid, and the foundation must be deep and wide enough to give the whole structure stability. If one side of your stable gets the brunt of driving rain or moisture, it is a good idea to cover it with an extra layer of concrete or stucco, or hang shingles to protect the wall.

The external angles of all of the doors and windows should be rounded. This can be done by using bull-nosed bricks. This way, horses and other livestock will not be injured by coming into contact with any sharp angles or ledges.

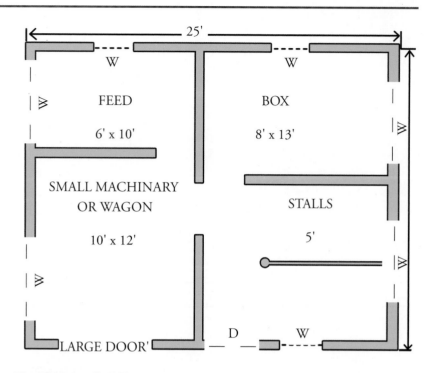

Plans for a small stable.

A Small Stable for Horses, Llamas, or Sheep

This simple stable is inexpensive to build and has plenty of room for two horses, llamas, or sheep, along with feed and tack. Hay and grain can be kept in the loft. Place the windows as high up as possible and hinge them at the bottom so they'll open inwardly to permit the air to pass over the animals without blowing directly on them. Make the stable door a "Dutch door"; that is, a door divided horizontally in the middle so that the upper half may be opened and the lower half remain closed.

TREEHOUSES

Treehouses and brush houses can be used as treetop forts and playhouses, or as relaxing spaces in which you can read and enjoy being surrounded by foliage and other treetop creatures. Building a treehouse requires a lot of work and some good planning. It is imperative that the house is built in trees that are large and sturdy. If you do not have such old trees in your yard, you can always modify these plans to build treetop havens on wooden platforms raised above ground, or construct a simpler brush house.

Low Two-tree Treehouse

This treehouse can be constructed out of ordinary boards and timber. It does not sit up as high in the trees but is still elevated above the ground to give a good view of the yard and surrounding area.

Directions

1. Select a location between two trees that are roughly 6 to 8 feet apart. The trees should have fairly straight trunks and should be at least 15 inches in diameter. Make sure they are healthy and sturdy—not decaying in any way.

2. Using an axe, clear off the brush and small branches up to 20 feet on the tree trunks (or to the height of where the treehouse will be located).

3. Take four or five pieces of spruce (from a lumberyard or home center) that are 2 inches thick, 8 inches wide, and 16 feet long. Saw off and nail two of the pieces to the trunks of the trees 8 feet above the ground. First, cut away some of the bark and wood to make a flat surface on the trunk. You will need 16-inch steel-wire nails to anchor the boards to the trees.

4. Cut two timbers 6 feet long and the other two the length of distance between the tree trunks. In the 6-foot pieces, cut notches on the underside. The ends of the bracket timbers will fit into these notches.

5. Cut the ends of the timbers to form a square frame so that they dovetail. Spike in 6-foot timbers to the tree trunks so that they will rest on the first two timbers that were nailed to the trees (see top right image on page 270).

6. Place the remaining two timbers in position so that the ends fit into those fastened to the trees. Nail them well.

9. Construct a frame 7 feet high at the front and 6 feet high in the back out of 2 x 3-inch spruce. Spike the side timbers, forming the top, to the insides of the tree trunks (see bottom right image on page 270). Mount the bottoms of the uprights on the corners of the floor frame and use four long nails to hold them in place.

10. Now, cut two timbers and arrange them in an upright position at the front, 30 inches apart. The door will be here. Halfway between the floor and the top of the framework, construct timber all around except between the door timbers. This will add strength and will allow the sheathing boards to be nailed. It will also make one more anchoring beam between the tree trunks.

11. Nail the side rails into the tree trunks in a corresponding way to the top (roof) strips.

12. Make the floor from lumber 4, 6, or 10 inches wide. The boards should be planed on both sides.

13. Construct the roof of the same boards. You can lay tarred paper over them and fasten it to the edges with nails. This will help waterproof the roof for at least one year. To make the roof last longer, you can shingle it.

14. Windows can also be made in the side and back walls. These should be about 24 inches square. The door can be constructed out of boards held together with battens. A lock can also be furnished to keep out unwanted visitors.

The treehouse will need a ladder for you to access it. This can either be purchased from a yard sale or can be made out of hickory poles and cross-sticks 20 inches wide. To keep the ladder from slipping while ascending and descending, affix loops to the top of the ladder; these will fit over large, sturdy nails driven into the doorsill, and the ladder will be relatively stable.

A flexible ladder can also be made out of ropes and hung much the same way as the wooden ladder. This type of ladder, though not as sturdy, can be drawn up when people are in the treehouse so no one else can enter.

Inside the treehouse, small chairs and other seats can be constructed and used for relaxing. Narrow shelving can be made and fastened over the windows with brackets. Small things can then be housed on these shelves. A small table may also be housed in the treetop shelter.

Treehouses can be very simple or very elaborate, based on your resources and preferences.

7. Support the first timbers that are spiked to the tree trunks with 15-inch blocks nailed below them. The cross timbers and last ones form the frame. Place the frame into the dovetailed joints at the ends.

8. Cut two more timbers and lay them across the supporting timbers nailed to the trees, so they will fit inside the front and back timbers, and secure them with long nails. The floor frame is now complete.

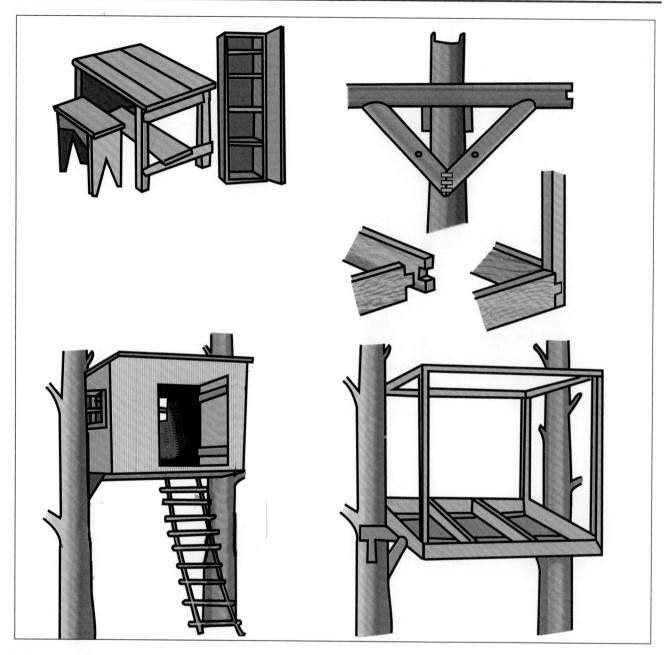

Refer to these illustrations when constructing the low two-tree treehouse.

High Treehouse in One Tree

If you have a tree large and strong enough (oaks are very good for this), a treehouse can be successfully built in its branches. For this plan, the treehouse will be 25 feet above the ground, and below it is a landing from which a rope ladder can be dropped to the ground. A more solid, wooden ladder connects the landing with the deck of the treehouse

and it can be situated through a hole in the deck of the house.

Since every tree is different, it is difficult to give exact dimensions of the frame of this treehouse and how many floorboards should be used. But the construction of a single-tree treehouse is, in many ways, similar to that of the low two-tree treehouse. The trunk of the tree will have to project up through the treehouse and the out-spreading branches

Plans for building a high treehouse in one tree.

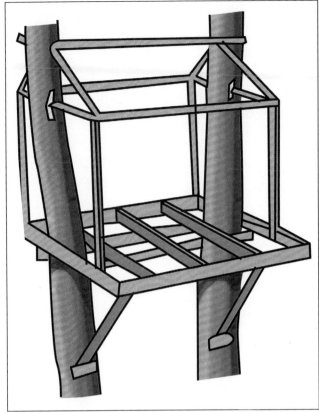

A treehouse can also be built on two trees, as shown here.

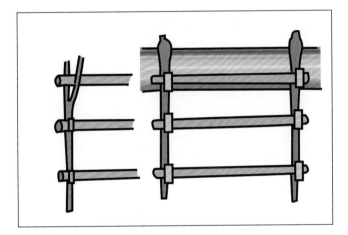

You can construct a wooden ladder or a rope ladder for your treehouse.

will need to support the lower parts of the floor frame. This treehouse can have either a peaked or a flat roof, depending on the structure of the treehouse and the space allowed for a roof within the treetop.

Brace the floorboards well to the main trunk of the tree with long and short brackets or props. These will help make the house secure. Drive large spikes into the tree where the lower ends attach to the trunk. Nailing cleats or blocks under these will help to support and strengthen the structure.

WORKSHOP FURNITURE

Woodworker's Tool Chest

A woodworker's first job, when he has gained enough skill, should be the creation of a good tool chest. This will offer a chance to practice using a range of tools, and the chest itself will keep those tools under lock and key when not in use. An ideal tool box should have a specific place for everything, as the chest illustrated by Figure 1 does. The letter references in Figures 1 and 3 are:

A. bottom plinth

B. top plinth

C. rim round lid

D. compartment for bead-planes, plough, etc.

E. compartment for various tools, planes, etc.

F. compartment for saws

K. bottom till

J. second till

H. top till

L. sliding-board to cover compartment E

M. cleats to hold division between E and F

N. cleats to hold division between D and E

P. runners for sliding-board L

R. runners for tills

S. runners for tills K

The length of the chest must be sufficient to accommodate a rip-saw, so the chest is 33 inches long internally; and if it is made 20 inches wide by 21 inches deep, you will find it convenient for all purposes. For the outside case, use white deal no less than 1 inch thick. In gluing up the front, back, and ends to obtain the necessary width, tongue and dowel the joints, the former being the better method. In dovetailing the framework of the chest, make the pins small, and have them no more than 1½ inches apart; make sure that the joints in the front and back do not come immediately opposite those in the ends, or at some future time the chest may break in two. Figure 2 is a transverse section through the chest, Figures 3 to 6 show the details of construction. The plinths A and B run all round the chest, and are 6 inches and 2½ inches wide, respectively, and 1 inch thick, with the top edge of A and the bottom of B finished with a plain bevel; the tip edge of a ¼-inch bead being worked on it also. The plinths may be mitered at the corners, but it is better

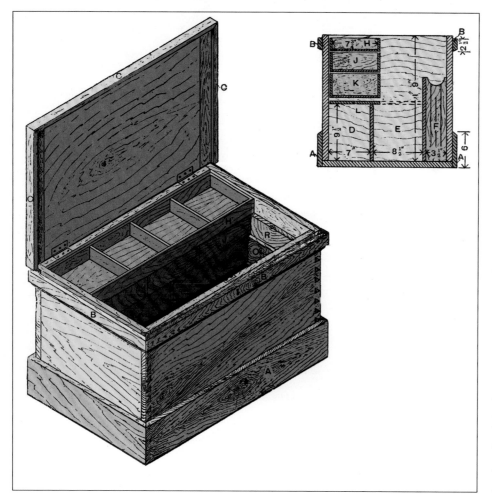

Figure 1—Tool Chest
Figure 2—Cross Section of Tool Chest

Tool Chest Partitions

For the inside fittings of the tool chest use good yellow deal or pine, which you can finish by staining, although sometimes a more fancy wood is used. From Figure 2 it is seen that the chest is divided in its width into three parts: D is for bead-planes, plough, etc.; this is 7 inches wide, and is covered by the sliding tills; E holds miscellaneous tools, best planes, or anything not in everyday use; and F (3½ inches wide inside) is the saw till. These are divided by the two partitions shown, that between D and E are 9 inches high, and between E and F are 14 inches The three tills H, J, and K slide to and fro to give access to compartments beneath, and when in place at the back of the chest form a covering for compartment D; and a sliding-board beneath the tills, when pulled out as shown by dotted lines in Figure 2, covers compartment E. The bench-planes, which are in everyday use, can be packed away on the sliding-board between the tills and the highest partition. Figure 3 shows one end of the chest with the cleats about 1 inch wide by ½ inch thick; between these the partitions fit. The cleats holding the partition between E and F are fixed first, ½ inch apart, and are as shown at M M

to dovetail them, and so obtain extra strength and good appearance. The plinth B is kept down about ¾ inch from the top of the chest to form a rebate for the lid to shut upon. The bottom of the chest is formed with boards 1 inch thick, tongued and grooved, and nailed on crossways—that is, the grain runs from front to back of the chest. The lid also is of 1 inch deal, with the joints tongued and grooved, and the ends clamped. It overhangs the chest all round in about ¹⁄₁₆ inch, and is hung with a pair of strong brass butts, and the self-acting spring lock is put on; then the rim C can be dovetailed together at the corners, and nailed to front and ends. This should result in a good fit where the rim of the lid meets the plinth B. Now you have finished the skeleton of the chest.

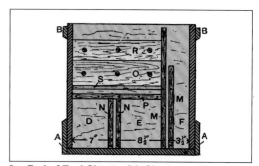

Figure 3—End of Tool Chest with Cleats and Runners

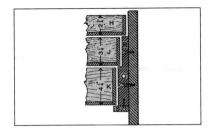

Figure 4—Section of Tool Chest Till Runners

(Figure 3), the one nearer the back of the chest being continued nearly to the top, the other, nearer the front, stopping at the same height as the partition, namely 14 inches. The cleats N must be 8½ inches high from the bottom of the chest, and ¾ inch apart. The back partitions having been placed in position, fix the ledges p, with their top edges 9½ inches from the bottom of the chest; then they run from the back to the long upright cleat m, and on them works the sliding-board L, 9 inches by ¾ inch; this is clamped at the ends for the sake of strength and to make it slide more easily. It must be a good fit endways to avoid jamming against the ends of the chest. The runners for the tills (Figure 4) are made long enough to reach from the back of the chest to the long upright cleat M (Figure 3), and should be made of hard wood. The principle piece R, which forms the runners for the two top tills, is 7½ inches wide by 1 inch thick, rebated to half its thickness at O for a depth of 3¼ inches. A piece of hard wood S, 1½ inches by ½ inch is screwed on to the thick edge of R, and forms the runner for the bottom till. These runners can be fixed in position one on each end of the chest, leaving about ⅛ inch clearance between the bottoms and the top of sliding board L. The partition between compartments E and F can be made and fitted between the cleats M M; along its upper side is a strip of 1½ inches by ½ inch deal, cut to fit between the cleats on each end of the chest, fixed level with the top edge on the side nearest the front of the chest and packed off about 1/16 inch. The slot thus formed can be used as a rack for squares, the stocks resting on top of the partition, and the blades hanging down out of the way inside the saw till.

Racks for Saws and Chisels

The saw racks in the chest, as shown in Figures 5 and 6, are 14 inches long, 3½ inches wide, and 1 inch thick,

shaped at the top ends. Each has three slots, or rather, saw kerfs, and in one rack (Figure 5) the middle kerf runs from the top to within 3 inches of the bottom, the others stopping the same distance from the bottom, and about 1½ inches from the top. In the other racks (Figure 6) the middle slot is stopped at both top and bottom, the others being open at the top end. These two racks are fixed at about 8 inches from each end by screwing through the horn at the top of each to the front of the chest. The partition being then put into its place, screws can be put through it into each saw-rack, which will hold all in place. When placing the saws in the racks, the points are inserted in the closed slots of racks, and the handle ends dropped into the open slots, two saws pointing one way and one the opposite way. To take chisels, a piece of hard wood 2 feet long, 1 inch square, with a series of notches 1 inch apart cut into it wide enough to take the tools, can be screwed to the front of the chest just above the top of the partition; this leaves an equal space at each end to allow the hand to be inserted to remove saws form the rack. The handles of the larger chisels will be just inside the front of the chest, convenient for withdrawal when wanted for use, and the blades with hang out of the way in the saw till.

Tool Chest Tills

The three sliding tills for the inside of the chest are all that remains. They will all be 9 inches wide outside, but varied in depth, as shown by Figure 4, on which dimensions are marked. They should be of ¾ inch stuff, with ½ inch bottoms and divisions, the rims dovetailed together, and the fronts and back rebated to receive the bottoms, the grain of which should run across the width of the tills; and at each end the bottom should be of hard wood. The divisions should be trenched into the sides, forming in k, j, and h two, three, and

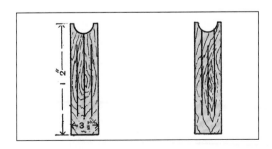

Figures 5 and 6—Saw Racks.

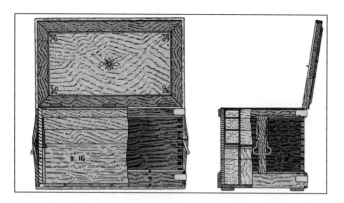

Figure 7 and 8—Elevation and Sections of Tool Chest

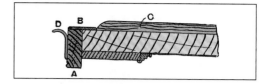

Figure 9—Section through Front of Tool Chest Lid

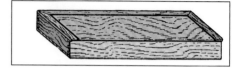

Figure 10—Top Tray of Tool Chest

four compartments respectively. One of the bottom divisions should be fitted up for the brace and bits, with racks for the bits fitted round the brace. Other divisions can be fitted with racks for small chisels, gouges, gimlets, bradawls, and various other tools, the aim throughout being to have a place for all, so that nothing can roll about and get damaged. Turn-buttons to take the tenon and dovetail saws can be screwed on to the under-side of the lid, so that when it is closed they will be in position between the top till and the front of the chest. The purpose of the cleat M, running up higher than its fellow, is to stop the tills from coming into collision with the stocks of squares when in their rack. The sliding-board l can be grasped underneath with the fingers when it is desired to draw it forward, and it should have a couple of thumb-holes cut in its top as a means of pushing it back. Each till should have a pair of flush-rings inserted in the front, so that either can be pulled forward and its contents exposed without the necessity of touching the others. A strong iron handle on each end of chest will now make it complete.

Tool Chest Lid

For the lid a piece of pine 2 feet, 11½ inches by 1 foot, 9 inches by ⅞ inch must be made. In some cases the ends are clamped, but the lid will stand better if properly cross battened. When the lid has been planed, the inside should be roughed with the toothing plane and two or three battens screwed across the grain on the other side to prevent warping. The inside can then be veneered with a center panel and a banding about 2½ inches wide as shown by Figure 7. Use Spanish mahogany veneer for the center panel, and a light or dark fancy wood veneer for the line, which may be about ⅜

inch wide, and for the center and corner inlays. Some workers veneer the center, and have a margin about ¼ inch thick, as shown on the underside of Figure 9, a planted molding being used for covering the edge of the veneer. When the glue is thoroughly dry the battens may be removed from the back, and the mahogany plinth a (Figure 9) screwed to the front and the ends; the parts seen when the lid is open should be polished. Pieces of 1¼ inch by ⅛ inch hoop iron b (Figure 9) can now be screwed round the top of the lid to protect the edges, and the space between filled in with ⅜ inch deal boards c, screwed across the grain of the lid; the ends can be rounded down to the hoop iron to strengthen it and also to prevent warping. The lid can now be hung with three 2¾ inches by ¾ inch brass butt hinges as shown in Figure 7. A strong lock can then be let in the front and a sash lid d (Figure 9) screwed to the plinth at the front. Figure 8 shows that the top plinth at the back is kept above that of the sides and front, to support the lid when open. In addition to the lock, one or two holes should be bored through the lid at both ends and counter-sunk in the hoop iron, so that the lid can be screwed down for traveling, etc. The outside corners of the chest should be protected with angle plates on the plinth as shown on the right-hand side of Figures 7 and 8, and these may be made by bending pieces of 1½ inches No. 16 b.w.g. iron 6 inches long to aright angle, punching the holes and countersinking for No. 10 screws.

Inside Fittings of Tool Chest

For the interior fittings of the chest a small nest of drawers at the back is sometimes used, but some prefer trays, as shown, as the drawers are liable to stick if a tool gets mis-

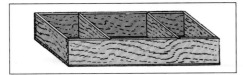

Figure 11—Second Tray of Tool Chest

placed. Also they take a lot of material and labor, and the drawers are most difficult to secure than trays when the chest is packed. The trays should be of ⅜ inch mahogany, dove-tailed together like a drawer, the top trays being fitted with lids, as shown in Figure 10; the total depth over all is 2¼ inches. Figure 11 shows one of the lower trays, and these do not have lids. The cross divisions in the trays may be made to meet requirements, but the following plan of dividing is a good one: Top tray at the back, space 1 foot, 4 inches long at the center with divisions about 8 inches at each end; second tray, the same as the top; and the bottom tray, one division in the center. For the narrow trays, the top one may be divided the same as the top one at the back; the second tray, with a partition 7½ inches from one end; and the bottom tray, with a division 10½ inches from the opposite end to the tray above. The space in the chest below the trays may be divided longitudinally into three compartments. The boards to form the divisions are fixed to an upright piece of wood ⅝ inch thick, B, secured to the ends of the chest. This part of the text is just deep enough to take small planes placed on end. A saw rack may with advantage be fitted in the space under the front trays, and the center space is covered with a board A, which slides back under the back trays. The inside of the chest should be French-polished, and the outside should have three or four coats of good paint.

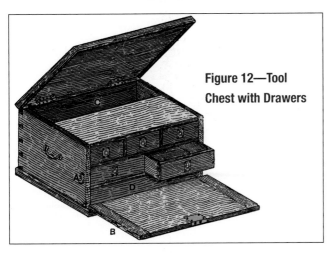

Figure 12—Tool Chest with Drawers

Packing Tool Chest for Transit

In packing the chest for traveling, the bottom divisions should be filled first, and the heaviest tools placed in the center portion. The slide can then be pulled over and fixed with a small screw at one end. The trays can then be filled, and secured either by means of strips fixed across the ends, or by filling the space between them with soft material that will not damage the polish. The lids of the top trays can then be fastened by placing across them two strips at the ends that will just fill up the space between the tops of the trays and the lid, when the latter can be locked and screwed.

Small Tool Chest with Drawers

For the small tool chest with drawers, shown in several views by Figure 12, a handy size is 1 foot, 9 inches by 1 foot, 2 inches by 1 foot deep. The sides, ends, bottom, and top are of red deal finishing about ¾ inch thick. The divisions are of ½ inch material, and the drawer fronts of ⅝ inch material. The sides, backs, and bottoms of the drawers are of ⅜ inch material, but of course these dimensions may be varied to meet your requirements. Figure 12 shows that the front is hinged on the bottom, so as to drop down and to allow of ready access to the drawers. To keep the front from twisting and warping, it must be clamped as shown, and when the front is closed up it is secured to the lid by a lock (see also Figure 13). In addition, a hook A (Figures 12 and 14) and eye B (Figure 14) may be used. The bottom is finished off with a plinth, which is rebated as illustrated in the section (Figure

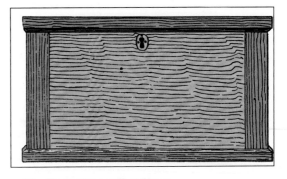

Figure 13—Front View of Tool Chest

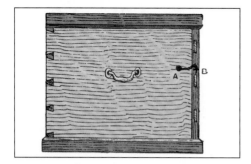

Figure 14—End View of Tool Chest

15) will have extra strength. The lid should be stiffened by a 1¼ inches by ½ inch rim. The well C and the space D under the drawers will be found very useful for large tools.

Utilizing Tool Chest Lids

The insides of tool chest lids are adapted readily to hold hand saws and tenon saws, the ends of which are held in wooden clips. The handle can be fastened by means of a button B, this method being just as suitable for hand saws as for the tenon saw shown. When the button is moved into position, it will allow for the saw to be taken out.

Simple Tool Chest

The simple chest shown in longitudinal section by Figure 21 must be long enough to take the rip saw, and it is as light as possible consistent with strength. Figure 22 is a cross section. The yellow pine is ⅞ inch thick for the body, 1 inch for the lid, and ⅝ inch for the outside plinth and facings; the bottom is of ¾-inch red pine. The plinth has an ovolo molding on it, but you will commonly see an ogee molding. The top facing is in two parts, one being screwed to the edge of the lid and rounded on the top; the other one, upon which are run

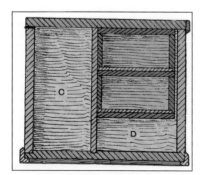

Figure 15—Cross Section of Tool Chest

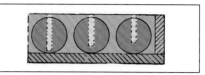

Figure 16—Cross Section of Part of Lid

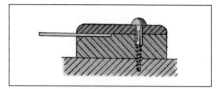

Figure 17—Try-Square Holder

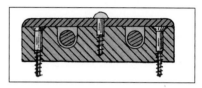

Figure 18—Holder for Pincers

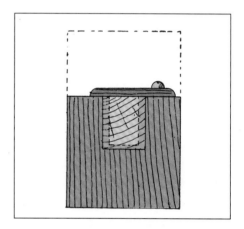

Figure 19—Elevation of Mallet Holder

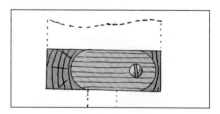

Figure 20—Plan of Mallet Holder

a bead and a chamfer, is merely nailed to the box. The body of the box is dovetailed and glued, but the plinth and top facing are mitered and nailed. The bottom has ploughed and cross-tongued joints waterproofed by painting with white lead. Clamp the top to prevent warping, and screw the facings on after the lid had been fitted to the size of the box. Battens of well painted red or yellow pine should be screwed to the under-side of the bottom, to keep the box clear of wetness. Figure 22 shows the inside arrangement. At the back, a space for smoothing planes, rebate planes, casements, etc., is formed by nailing fillets to the ends of the box, to which the pieces of pine a is screwed. Narrow fillets are nailed to the ends of the box outside of the piece A, and the piece B is screwed to them. A space is thus provided for the tenon saws and hatchet. The method o fixing the hand saws to the inside of the lid is shown by dotted lines in Figure 21, and is on the same principle as that already described. A piece of wood, the thickness of the saw handle, is fitted in the hole, and screwed to the lid. A piece of sheet brass to form a long button then is screwed to the block of wood; this button, when turned round as in Figure 21, prevents the handle of the saw from leaving the lid. The hardwood clip to hold the point of the saw has no recess for the back as there shown. The method of packing the chisels and gouges is seen in Figure 22. A small fillet, with a strip of leather glued to the top edge, is nailed to the bottom, and a thin piece of pine, projecting about 1 inch above the leather, is nailed to the fillet. This receives the points of the chisels and gouges. Another piece, with various sizes of holes cut out, is screwed about 3 inches or 4 inches up, to keep the top part steady. Figure 23 shows the piece with the holes checked out and a thin piece screwed to the front; this is much easier than mortising the holes. The tray C (Figures 21 and 22) is a box the whole length of the inside, lap dovetailed at the back. It is divided into various compartments (two small ones and a large one will be found very handy) by thin pieces of pine, either raggled into the front and back or merely butted and nailed. The bottom, which is ⅝ inches thick, is screwed up. It is very common to have a hardwood flap on the tray, as shown, but this can be dispensed with at will. A back stile is screwed to the back of the tray, and the flap is hinged to it. Fillets D are screwed to the ends of the box, on which the tray slides to and fro.

Tool Cabinet

You can see a useful tool cabinet with paneled doors and three drawers illustrated in Figure 24. It would have a neat appearance if it were made in oak or other hard wood and

polished; or even if made of deal, if stained and varnished. The leading dimensions figured in the vertical section (Figure 25), and in the horizontal section (Figure 26) are only suggestive. The sides, top, and bottom are of ¾-inch material, grooved and tongued together, as indicated in Figure 27. The sides should also be rebated to receive the back, and grooved for shelf as shown in 26 and 27. The two divisions (separating the drawers) should be grooved into the shelf and bottom as shown. The back can be formed of three boards ⅝ inch thick, its upper part being sawn and smoothed to the shape shown in Figure 24. The front edges and ends of the top and bottom may be rounded. Fit all the parts of the case together, and finally secure them by gluing and nailing. Wood about ⅞ inch thick will be required for the stiles and

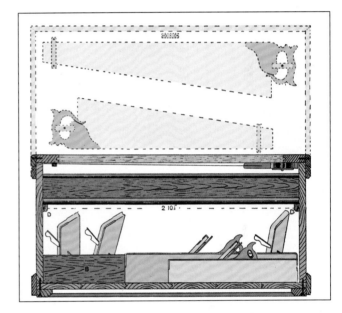

Figure 21—Section of Simple Tool Chest

Figure 22—Cross Section of Simple Tool Chest

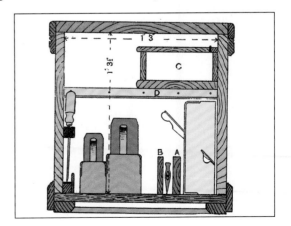

Figure 23—Tool Chest Chisel Rack

Figure 24—Tool Cabinet

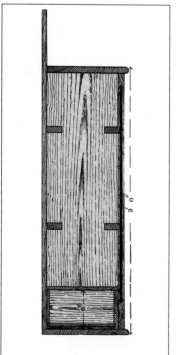

**Figure 25—Vertical
Section of Tool Cabinet**

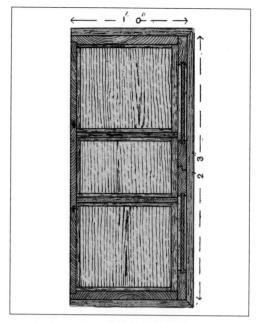

Figure 26—Horizontal Section of Tool Cabinet

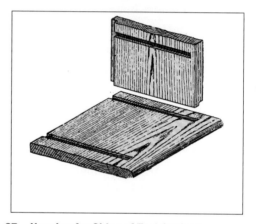

Figure 27—Housing for Sides of Tool Cabinet

rails of the doors; the panels may be about ⅜ inch thick. The doors should be mortised and tenoned together and ploughed to receive panels; they should be finished by being glued, wedged, planed off, fitted to the case, and rebated together (see Figures 24 and 26), after which they can be hung with 3-inch butts. The drawers should be properly dovetailed; ¾-inch wood will do for the fronts, and ½-inch for the sides, backs, and bottoms. Brass flush drop handles will be best for the drawer fronts. Two small bolts secure the door on the left, and there is a 2½-inch cut cupboard lock on the right-hand door. The cabinet could be fixed to a wall with four holdfasts, or it might rest upon a couple of brackets or other similar arrangement. The inside can be fitted with racks, according to requirements.

Strong Stool or Work Bench

A stool or bench is useful for various purposes, both in the household and in the workshop. It can, without much trouble, be taken to pieces, so that it may be conveniently

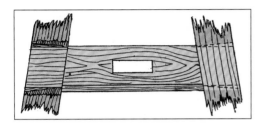

Figure 28—Bottom Rail of Work Bench

stowed away when not in use. When using the stool for domestic use, the sizes of the timber should be about 2 inches by 2 inches for all parts of the frame. The top consists simply of a slab about 1 inch thick, having four holes bored in it to fit over the dowels or pins a. Suitable measurements for the finished article are: Length of top, 2 feet, 9 inches; width

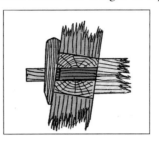

Figure 29—Joint between Rail and Bottom Stretcher

Figure 30—End of Rail in Leg

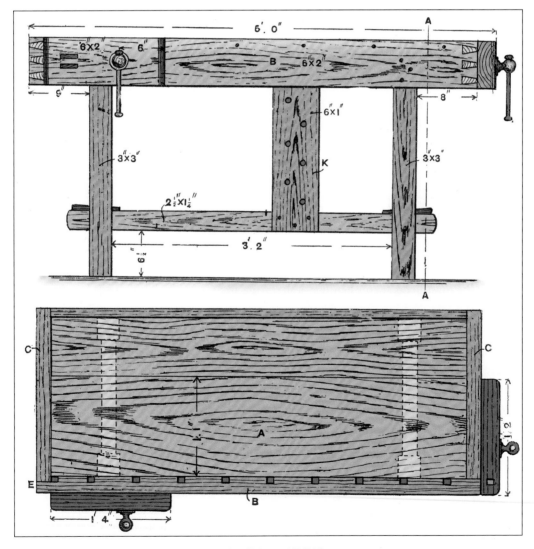

Figures 31 and 32—Side Elevation and Plan of Bench with Side and Tail Vices

of top, 1 foot, 3 inches; and height from ground, about 1 foot, 10 inches. For a work bench, however, you can increase these measurements, being sure to increase the framework proportionately, and the top, instead of fitting over pins or dowels, should be secured with nuts and bolts. The spread of the legs at the bottom should be such that they would occupy the four corners of a rectangle, equal and similar to that of the top; this prevents all tilting, and secures stability for the bench, a point that is often overlooked. Before marking out the framework, make a mold; take a piece of wood, about ⅜ inch thick, and square off one end, as at B, and make the other cut to the desired bevel or splay of the legs, as shown at a; a narrow strip is fastened to the edge, so as to form a fence. With this tool, you should not experience any difficulty in marking out in a proper manner the lines for the necessary joints. Figures 28 to 29 show those in the lower part. These figures are sufficiently explanatory in themselves, and need no further comment. Make all the joints by gluing and wedging, and cut the movable key–wedges from hardwood. A stool or bench made on this principle from good dry wood will stand any amount of rough usage, and should last as long as the timber from which it is made, there being no nails or other source of weakness to lessen its durability.

Bench with Side and Tail Vices

The general view of a bench with side and tail vices is given by Figure 61 and the construction of the bench is dealt with below. Figure 31 is a side elevation, Figure 32 a plan, and Figure 33 a section on A A (Figure 31). Having sawn out the pieces, next plan them true. Then set out the legs and rails, the latter for mortising, and the former for tenons. The mortises go right through, producing a much firmer result than when the tenons are only stubbed in half-way. The haunched mortise and tenons between the top rails and legs, with the tenons of the cross rails through the legs, are shown in Figure 34 and 35; the tenon of the rail is firmly held in position by a wedge, which must be released, and the tenon of the rail lifted up, before it can be withdrawn. The side rails have a bare-faced tenon—that is, have a shoulder on the inside only. When these joints fit suitably, the legs and cross

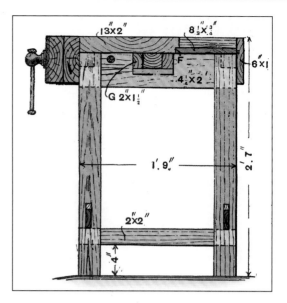

Figure 33—Part elevation and cross Section of Bench

rails should be glued together and cramped up, and the tenons fixed by wedges, which should be glued before insertion. The top should be planed to breadth and thickness, and then the ends cut off and planed square and to length. The front of the back and end cheeks C (Figure 32) should next be carefully set out and worked. At the front end of the side cheek b, the thickness for dovetailing is not the full 2 inches, but is less by ¾ inch than the breadth of the pin hole, as shown at e (Figure 32). After the side cheeks have been dovetailed and fitted together, groove the front cheek on the back for receiving the stop (see Figures 32 and 36). The inside edge of the top should be rebated as shown at F (Figure 33) to receive the well board. This should fit just tight between the end cheeks, the front and side back cheeks being firmly secured to the top plank and well board. Four-inch screws may be used for the front and side cheeks, and 2½-inch screws for the back, the heads being sunk a little below the surface. Glue the side cheeks to the main board of the top. Then mortise and tenor the cheeks and ends of the runners together as in Figure 37, the top of the runner being kept at the same distance from the top of the cheek as the thickness of the top plank; two tenons may be more troublesome to make, but the result will be stronger than when only one tenon is used. Make sure these joints are firmly glued and wedged together, with the runner at right angles to the cheek. In Figures 33 and 36 the construction of the guide boxes for the runners is clearly illustrated, the pieces G, a trifle deeper than the thickness of the runner, being firmly fastened to the top plank with 3½-inch screws. The bottom is formed of ¾ inch boarding screwed to the guides. The box for the tail runner extends from the top rail to the inner surface of the end cheek. You will find wrought-iron bench screws about 18 inches by ⅞ inch, having split collars, to be the most satisfactory, and in fixing them into their places, push in the cheek and runner so they are firmly held in position; then mark the center of the hole for the screw in the cheek, leaving sufficient room for the flange of the box (or nut) for screwing to the side cheek of the bench (see Figure 36). The hole should next be bored through the cheeks of the screw and bench with a bit

slightly larger than the diameter of the screw. Then the collars and boxes can be fixed in position, and the framework of the legs and top fitted together. Notch the top rail of the back legs for the runner, shown by h (Figure 34), and if the work has been done accurately the top will just slide on the upper part of the legs. When the parts are adjusted, the front cheek should be secured to the legs, and the top of the bench to the top rails of the legs with 3½-inch screws. The peg board κ (Figure 31) should be screwed to the front of the bottom rail and to the back of the front cheek. The following are the net sizes of the pieces required (a little in excess of these dimensions should be allowed for waste in working): Top board, 2 inches by 13¼ inches by 4 feet, 8 inches; well board, ¾ inches by 8 inches by 4 feet, 8 inches; peg board, ⅞ inch by 6 inches by 1 foot, 9 inches ; runners, 2 inches by 2½ inches by 2 feet, 3 inches; runner guides, 1½ inches by 2 inches by 3 feet, 2 inches; guide box bottoms, ¾ inch by 5½ inches by 1 foot, 7 inches; screw cheeks, 2½ inches by 6

Figure 34—Joints of Rails and Legs of Bench

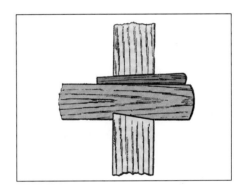

Figure 35—Section showing Rail wedged in Bench Leg

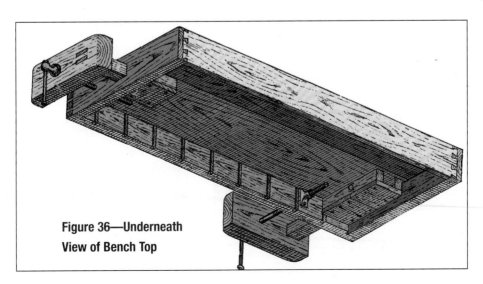

Figure 36—Underneath View of Bench Top

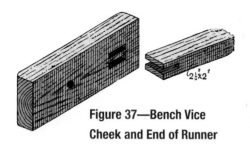

**Figure 37—Bench Vice
Cheek and End of Runner**

inches by 2 feet, 7 inches; front and end cheek, 2 inches by 6 inches by 9 feet; back cheek, 1 inch by 6 inches by 5 feet; legs, 3 inches by 3 inches by 9 feet, 8 inches; top rail (front end), 2 inches by 4 inches by 1 foot, 9 inches; bottom rails (ends), 2 inches by 2 inches by 3 feet, 6 inches ; and bottom rails (front and back), 1¼ inches by 2½ inches by 8 feet, 2 inches.

Portable Folding Bench

The portable folding bench shown ready for use by Figure 63, and with the flap down by Figure 64 is illustrated in side elevation by Figure 38, and in end elevation by Figure 39. An end elevation of the bench when folded is given by Figure 40. Sizes that will meet all ordinary requirements are indicated in the illustrations, which show the construction so clearly that only the leading points need description. The legs and rails are jointed together by plain halving and dovetail-halving. The top is at least 1½ inches thick, and is formed of two boards jointed; to keep it true it should be clamped. The top should be hinged to the rail marked A, and the side of the bench hinged to the top as represented by B

(Figures 39 and 40), 3-inch butt hinges being used for this purpose. The wall-piece C should be firmly screwed to the rail of the top A. The legs should be hinged at the top of this piece, and also at the bottom to the strop marked D, which should be sufficiently thick to project from the wall to the thickness of the wall-piece C. The piece C can be attached to the skirting board with a few screws. The wall-piece C, if against a lath-and-plaster partition, can be firmly and easily fixed to two or three of the studs of the partition with half a dozen screws; if it is against a brick wall, drill a few holes into the wall and drive in hardwood plugs; or, better still, probe the wall with a long fine bradawl until the joints are found (if this is done carefully, little damage will be suffered by the paper), and then with a steel chisel cut some holes about ¾ inch square and about 3 inches or 4 inches deep. These holes may then be fitted with hardwood plugs, into which screws are inserted through the wall-piece. The fitting-up of the screw, cheek, and runner (the last named being of hardwood) is not difficult. The leg to which the screw is attached is larger than the others. The side and top of the bench when folded up can be kept in position by a hook

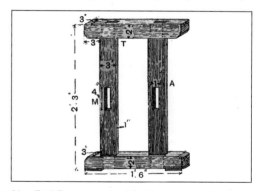

Figure 41—End Framework of Cabinet-worker's Bench

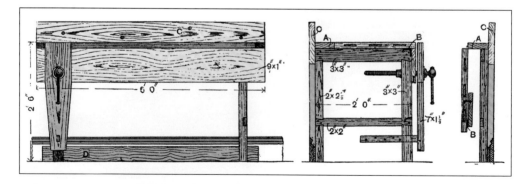

**Figures 38 and 39—Elevations of Folding Bench
Figure 40—Folding Bench with Flap Down**

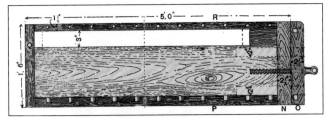

Figure 42—Top of Cabinet-Worker's Bench

and eye as shown. The bench may be made additionally firm by inserting a few screws through the side into the legs, and through the top into the rails. When it is required to remove the bench, all that is necessary is to withdraw these screws.

The Junior Homesteader

Make a Birdhouse!

Teach kids basic woodworking skills with simple, fun projects. This is a small birdhouse that will hang from a tree branch. It's perfect for wrens.

Materials

- Large tin can
- Wooden board about 7 inches square
- Carpet or upholstery tacks
- Earthen flowerpot
- Small cork to plug up the flowerpot hole
- Eye screw
- Short stick
- Wire
- Small nails

Directions

1. Mark the doorway on the side of the can and cut the opening with a can opener.
2. Fasten the can to the square baseboard (A) by driving large carpet tacks through the bottom of the can into the board.
3. Invert the flowerpot to make the roof. Plug up the drain hole to make the house waterproof (use a cork or other means of stopping up the hole) (B).

4. Screw the eye screw into the top of the plug to attach the suspending wire. Drill a small hole through the lower end of the plug so that a short nail can be pushed through after the plug has been inserted to keep it from coming out.
5. Fasten the flowerpot over the can with wire, passing the loop of wire entirely around the pot and then running short wires from this wire down to small nails driven into the four corners of the base (A).
6. Now the bird temple can be painted and hung on a tree.

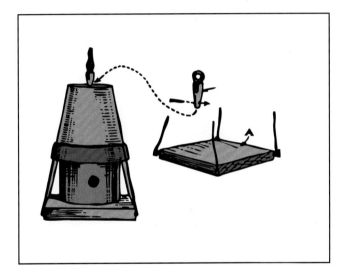

Sawing Stools and Workshop Trestles

The joint most generally used for connecting the legs of the sawing stool to the top beam is shown in Figure 43. These joints may be fastened with nails, but a stronger method is to glue and screw them together. A serviceable trestle for workshop and general use is shown in Figure 44. Quartering of light or heavy scantling will be appropriate, based on the purpose you are using the trestle, and the method of framing it together will change as well. It is mortised and tenoned as follows. The four legs A are mortised into the top B, as shown in Figures 45 and 46. The mortise in B is cut longer than the width of the tenon of A, to allow for driving

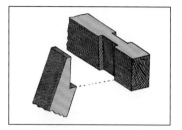

Figure 43—Joint for Sawing Stool

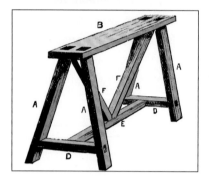

Figure 44—Workshop Trestles

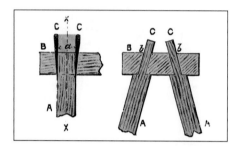

Figures 45 and 46—Legs Mortised and Tenoned into Top of Trestle

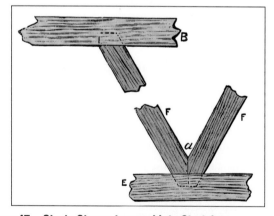

Figure 47—Struts Stump-tenoned into Stretcher

in wedges c. The complete joint is dovetailed, wider at the top than the bottom, and the legs cannot fall out. Drive in the wedges as shown, against the ends of the tenon and the end grain of the top B, and not against the flanks b of the

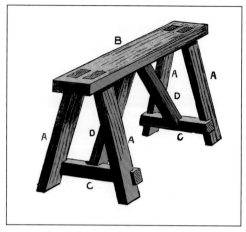

Figure 48—Workshop Trestle

tenons. If they were driven against the flanks b, they would split the top B. The short stretchers D (Figure 44) are mortised into the legs a, and wedges are driven in against the end grain, as in the previous instance. The long stretcher e (Figure 44) between the short ones is mortised in the same fashion. A trestle made thus with close joints will stand much rough usage. If the legs were short and the scantling of large section, as with a sawing stool, strutting would not really be necessary. But it is better to strutt high trestles made of slight scantling, say not exceeding 2 inches by 2 inches, or

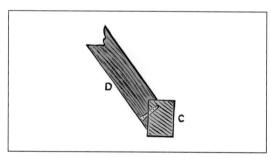

Figure 49—Strut Shouldered upon Stretcher

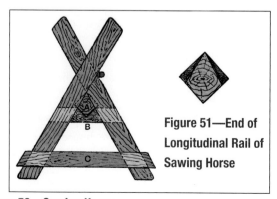

Figure 51—End of Longitudinal Rail of Sawing Horse

Figure 50—Sawing Horse

2 inches by 2½ inches cross section. In Figure 44 the struts at F F are tenoned into *b* and e, but the tenons do not pass through, and are not wedged. It is quite enough to stump-tenon the ends of f f (Figure 47) at top and bottom. The two struts about at *a*, and further steady the framing. A simpler method of framing trestles of this type together is shown in Figure 48. The only members that are mortised are the legs A, into the top B. The cross stretchers c are simply let for about ½ inch into the legs, and screwed or bolted. The struts d are stump-tenoned into the top, but at the other end they are merely shouldered back to fit over the stretchers c (see also Figure 49), and screwed or bolted. For a somewhat heavy trestle this simpler method is quite good enough; but for a lighter trestle, like previously described, the method of framing together with long bottom stretchers and mortised joints throughout makes a firmer job.

Sawing Horse

For sawing wood, and more especially for sawing firewood, a horse is a great help. There are more ways than one of making it; but this is best for ordinary use. This is mainly built of scantling 3 inches square, and is neat, strong, firm, and serviceable. The four pieces forming the legs are about 2½ feet long, and they are so arranged that the upper limbs of the cross have only half the length of the lower ones; They are halved at the intersection, and strongly nailed together, the nails being driven from the outer side and sent well home; for, should they project, they would be likely to catch and blunt the teeth of the saw. All the parts have to be so arranged as to leave nothing which can interfere with the free play of the saw, especially no iron. For this reason the central piece a, which chiefly serves to tie the two pairs of legs together, is kept below the intersections, so that it may fit up closely between the legs. Its two ends are, for a length of 3 inches, cut as shown in section by Figure 51. This piece is of the same scantling as the legs to which it is nailed. For ordinary work, 18 inches will be a good length for it. The upper cross rails B B give support to this piece, and are, as is shown, cut away to receive its lower angle at each end. These

are of 1-inch wood, 2½ inches wide, and about 1 foot long, and are nailed to the inner sides of the legs. A part only of one of these rails is seen in Figure 50, but you can see its extent by the dotted lines. Now nail on the foot-rails c c, 1 inch by 2½ inches to the legs 2 inches from their bottoms. The cross foot-rails are 2 feet long, and those that run lengthwise 20 inches long. Keep the foot-rails as near the ground as indicated, so that your foot can rest comfortably on it when using the saw horse; some part of your weight being thrown on the frame will steady it.

Portable Sawing Horses

A better portable horse for general purposes than the one just mentioned can hardly be desired. If less solid, it would be lacking in firmness, and if too strong, it would be liable to be shaken to pieces by the constant jarring to which it is subjected in use. But sometimes a lighter and more portable horse is desired, and to meet this requirement a shut-up horse may be made as follows: After halving the two pairs of legs as above, cut mortises through them at the intersections, say 1 inches wide by 1½ inches beyond them. In each tenon will be a hole into which a pin, removable at pleasure, can be driven to fasten the frame together for use. In convenience and stability this horse is of course greatly inferior to that described before.

Fixed Sawing Horse

The most simple as well as the firmest of all sawing horses is, however, the primitive fixed one. The legs of this should be about 1 foot or 15 inches longer than those shown in Figure 50, and, pointed at the ends like stakes, they are driven into the ground before being nailed together at their intersections. To connect the pairs, all that is needed is a tie nailed to the legs at the point d (Figure 50), that is, just below the upper limb and on the side upon which the sawyer will stand. Such a horse has the merit of complete immobility; but, of course, it will not serve every purpose.

CRAFTS

BASKETRY

Basketweaving is one of the oldest, most common, and useful crafts. The materials used in making baskets are primarily reed or rattan, raffia, corn husks, splints, and natural grasses. Rattan grows in tropical forests, where it twines about the trees in great lengths. It is numbered according to its thickness, and numbers 2, 3, and 4 are the best sizes for small baskets. For scrap baskets, 3, 5, and 6 are the best sizes. Rattan should be thoroughly soaked before using. Raffia is the outer cuticle of a palm, and comes from Madagascar. Cattail reeds can also be excellent for baskets and may be more readily available, as they frequently grow near ponds or swampy areas. Most basket making materials can also be found at local craft stores.

Small Reed Basket

Most reed baskets have at least sixteen spokes, and for small baskets and where small reeds are used these spokes are often woven in pairs. You can vary the look of your reed basket by combining and interweaving two different colored reeds.

Materials Needed

Sixteen 16-inch spokes, No. 2 reed
Five weavers of No. 2 brown reed

Directions

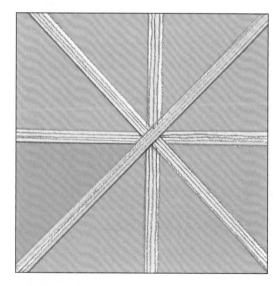

1. Separate the sixteen spokes into groups of four each. Mark the centers and lay the first group on the table in

a vertical position. Across the center of this group place the second group horizontally. Place the third group diagonally across these, having the upper ends at the right of the vertical spokes. Lay the fourth group diagonally with the upper ends at the left of the vertical spokes.

2. Soak the reeds well and then start the basket by laying the weaver's end over the group to the left of the vertical group, just above the center; then bring it under the vertical group, over the horizontal and then under, and so on until it reaches the vertical group again. Repeat this weave three or four times. Then separate the spokes into twos and bring the weaver over the pair at the left of the upper vertical group, and so on, over and under until it comes around again, when it is necessary to pass under two groups of spokes and then continue weaving over and under alternate spokes. At the beginning of each new row the weaver passes under two groups of spokes, always under the last of the two under which it went before and the group at the right of it.

3. Weave the bottom until it is 4 inches in diameter; then wet and turn the spokes gradually up and weave 1 inch. After that, turn the spokes in sharply and draw them in with three rows of weaving. Now weave four rows, going over and under the same spokes, making an ornamental band; then weave three rows of over- and under-weaving, followed by four rows without changing the weave.

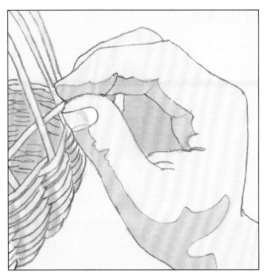

Continue to draw the side in with four rows of over- and under-weaving, and then bind it all off. Finish with the following border:

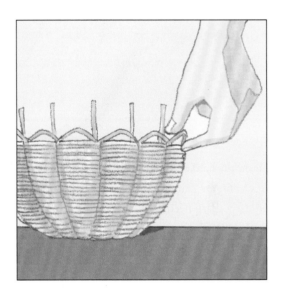

4. Always wet the spokes till they are pliable before starting the border. Bring each group under the first group at the right and over the next and inside the basket. Finally, cut the reeds long enough to allow them to rest on the group ahead.

Note: Leave the first two groups a little loose so that the last ones can be easily woven into them.

Coiled Basket

Sweet grass, corn husks, or any pliable grasses can be used for this type of basket, and with a contrasting color for sewing, the basket can be very attractive.

Materials

A bunch of grasses
A bunch of raffia

Directions

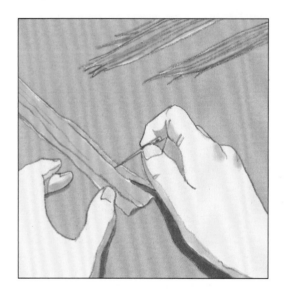

1. Cut off the hard ends of the grasses and take only a small bunch for the center to start. Split the raffia very fine and use a sharp needle for extra help.

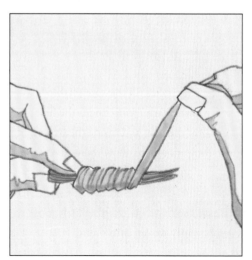

2. Hold the grasses and the end of raffia in your left hand, about 2 inches from the end of the coil, and wind the raffia around the coil to the end of the grasses.

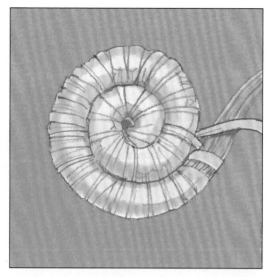

3. Bend the end of the coil into a small round center and sew over and under, binding the first two coils very firmly together. The next time around, leave a very small space between each stitch, and take the stitch only through the upper portion of the coil below. It is necessary that the spaces between the stitches be very small in the first few rows—this will determine the regularity of the spirals to come.

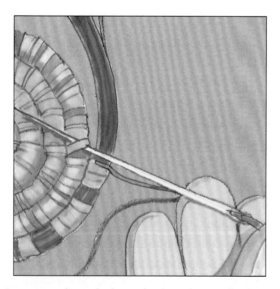

4. In sewing through the coil, place the needle diagonally from the right of the stitch through the coil to the left of the stitch.

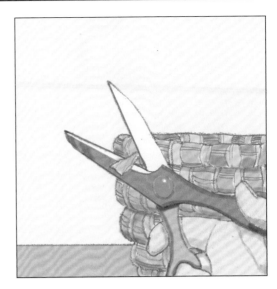

5. When the bottom measures 4 inches across, begin shaping the sides by raising the coil up slightly on the coil below and continue to bind the coils together as before. When the basket measures about 6 inches across, begin shaping the sides by pushing the coil slightly in toward the center.

6. To finish, gradually decrease the size of the coil but do not increase the number of stitches. Fasten the raffia, after the last stitch, by running it through the coil and cut it off close to the border. If necessary, bind the ends more closely by sewing over and over with a thin thread of very fine raffia the same color as the grasses.

BLOWN EGGS

If you want to keep your egg creations to display year after year, blow out the eggs before decorating them. Blown eggs are more fragile than hardboiled eggs, but they won't ever spoil.

Materials

Eggs
Needle
Toothpick
Tiny straw or a syringe

1. Use a needle to poke one hole on each end of the egg. Insert a toothpick and wiggle it around to help widen the holes slightly. Stick the needle back into one hole and move it around inside the egg until the yolk breaks.

2. Insert the straw into one hole and blow through it until the insides of the egg drain out, or use a syringe to draw out the egg. Rinse thoroughly to wash away any remaining egg residue.

3. Bake the eggshell at 400 degrees Fahrenheit for ten minutes.

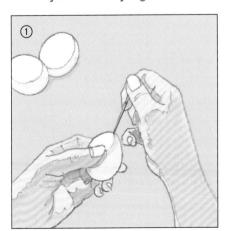

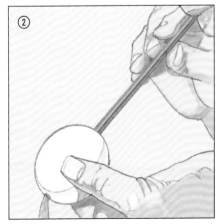

You can also create a stenciled look by painting on your eggs in contrasting colors after they've been dyed.

Leaf- or Flower-Stenciled Eggs

These eggs make a stunning centerpiece when displayed on a plate or in a basket. If you intend to eat the eggs later, use only food coloring to dye them. However, if you are using blown eggs or you do not intend to eat the eggs, fabric dye or other natural dyes and stencil paint will produce richer, more vibrant colors.

Materials

Food coloring, fabric dye, or other natural dye (prepared accordingly)
Large white eggs
White glue
Leaves, ferns, or small flowers
Small paintbrush
Wide, stiff paintbrush

Stencil paint (if you do not intend to eat the egg later)

1. Fill a deep bowl or wide glass half full with the prepared dye solution.
2. Immerse an egg in the dye mixture and allow to sit for a few minutes, or until the egg has reached the desired color. Rinse the egg in cold water and allow it to dry.
3. Using a small paintbrush, paint a thin layer of glue on the back of your leaf or fern. Stick the leaf or fern to the egg.
4. Dip a wide paintbrush in undiluted food coloring (if you intend to eat the eggs later) or stencil paint. Blot the brush on newspaper or paper towel to get rid of excess paint. Dab the ends of the bristles up and down over the leaf, allowing the first layer of color to show through to create a dappled effect.
5. When the egg is completely dry, peel the leaf away from the egg.

Natural Dyes for Easter Eggs

There are many ingredients from nature you can use to dye your Easter eggs. The colors may be more subdued than if you use food coloring or paints, but you can achieve some beautiful pastels with berries, flowers, and other plants and foods. Mix dyestuff or the finished dyes to make more color variations. Note that liquid dyestuffs, such as grape juice, do not need to be simmered with water as described below. Simply add the vinegar and use!

Materials

Dyestuff (see chart below)
Water
White vinegar
Cooking oil or mineral oil

1. Place a handful or two of the dyestuff of your choice (or a couple of tablespoons if using herbs or spices) in a saucepan.
2. Add water until the dyestuff is fully submerged. Simmer on low for about 15 minutes, or until the desired color is reached. Keep in mind that the eggs will turn out paler than the dye appears in the pan.
3. Strain dye into a liquid measuring cup. Add 2 tablespoons of white vinegar for every cup of dye. The dye is now ready for use.
4. After the dyed eggs are dry, rub the eggs with cooking oil or mineral oil to give them a glossier sheen.

Color	Items to Dye With
Blue	Blueberries, red cabbage, purple grape juice
Brown or Beige	Coffee grounds, black walnut shells, black tea leaves
Brown Gold	Dill seeds
Brown	Chili powder
Green	Spinach leaves, liquid chlorophyll
Gray	Purple or red grape juice, beet juice
Lavender	Purple grape juice, violet blossoms plus a little lemon juice, Red Zinger tea
Orange	Yellow onion skins, carrots, paprika
Pink	Beets, crushed cranberries or cranberry juice, crushed raspberries, grape juice
Red	Lots of red onion skins, pomegranate juice, canned cherries, crushed raspberries
Violet or Purple	Violet blossoms, hibiscus tea, small quantity of red onion skins, red wine
Yellow	Orange or lemon peels, carrot tops, chamomile tea, celery seed, green tea, ground cumin, ground turmeric, saffron

BOOKBINDING

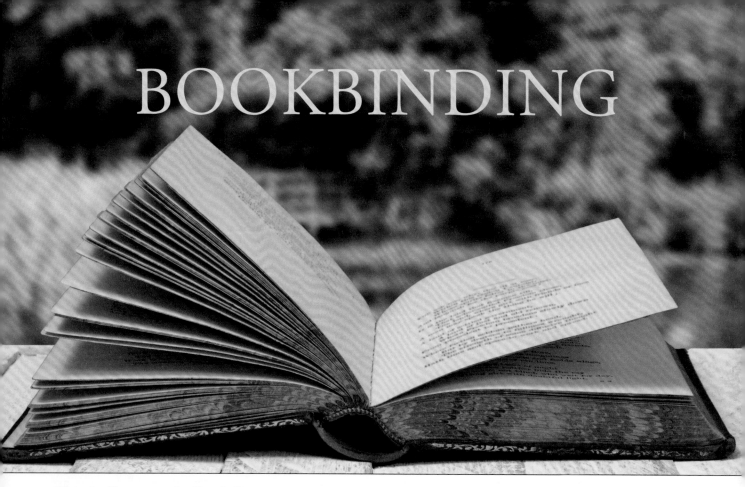

Simple Homemade Book Covers

If you have loose papers that you want bound together, there is an easy way to make a book cover:

1. Take two pieces of heavy cardboard that are slightly larger than the pages of the book you wish to assemble. Make three holes near the edges of each cardboard piece with a hole-punch. Then, punch three holes (at the same distance from each other as in the cardboard pieces) in the papers that are to be in the interior of the book. Be sure that your book is not too thick or the cover won't be as effective.

2. String a narrow ribbon through these holes and tie the ribbons in knots or bows. If the leaves of your book are thin, you can punch more than three holes into the paper and cardboard, and then lace strong string or cord between the holes, like shoelaces.

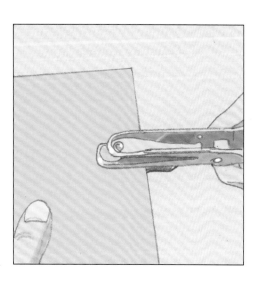

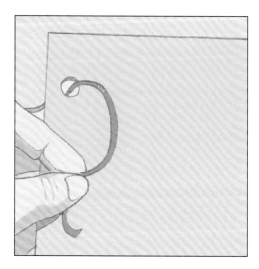

3. To decorate your covers, you can paint them with water-colors or you can simply use colored cardboard. You can also take some fabric, cut it so the fabric just folds over the inside edges of the cardboard pieces, and then hot glue (or use a special paste—see below) the fabric to the cardboard. Or, you can glue on photographs or cutouts from magazines and protect the cover with laminating paper.

To Make Flour Paste for Your Book Covers

Mix ½ cup of flour with enough cold water to make a very thin batter. This must be smooth and free of lumps. Put the batter on top of the stove in a tin saucepan and stir it continually until it boils. Remove the pan from the stove, add three drops of clove oil, and pour the paste into a cup or tumbler and cover.

Bind Your Own Book

Making your own blank book or journal takes careful measuring, folding, and gluing, but the end product will be something unique that you'll treasure for ages. You can also rebind old, worn out books that you'd like to preserve for more years to come.

Before binding anything very special, practice by binding a "dummy" book, or a book full of blank pages—you could use this blank book as a journal if it comes out fairly well, so your efforts will not be wasted. In order to make a blank book, you need to plan out what you'll need—how

thick the book will be, what the dimensions are, the quality of the paper being used (at least a medium-grade paper in white or cream), and so on. Carefully fold and cut the paper to the appropriate size.

Materials

White or cream-colored paper (at least 32 sheets) or the pages you wish to rebind
Stapler or needle and thread
Binder clips or vise
Decorative paper (2 sheets)
Stiff cardboard
Cloth or leather
Silk or cloth cord
Glue
Scissors (or a metal ruler and craft knife)

1. Make four stacks of eight sheets of paper. These stacks, once folded, are called "folios," Four stacks will make a 64-page book. If you wish to make it longer or shorter you can do more or fewer stacks of eight sheets. Carefully fold each stack in half.

2. Unfold the stacks and staple or sew along the crease. If stapling, only use two staples: one at the top of the fold and one at the bottom.

3. Refold all the stacks and pile them on top of each other. Use binder clips or a vise to hold them together. Cut a rectangular piece of fabric that is the same length as the spine

of your book and about five times as wide. So if your stack of folded papers is 8 ½ inches long and one inch high, your fabric should be 8 ½ inches long and five inches wide.

4. Using a hot glue gun or regular white glue, cover the spine with glue and stick on the fabric. The fabric should hang off either side of the spine.

5. For the cover, cut two pieces of sturdy cardboard that are the same size as the pages of your book. Using a metal ruler and a craft knife will help you make the cuts straight and smooth. Place one piece of cardboard at the bottom of your stack of papers and another on the top. Cut another piece of cardboard that is the same height and width as the book's spine, including the pages and both covers.

6. Select a piece of fabric (or leather) to cover your book and lay it flat on a table. Place the three pieces of cardboard on the fabric with the spine between the two cover pieces. Use a ruler to measure and mark a rectangle on the fabric that is 1 inch larger on all sides than the combined pieces of cardboard. Remove the cardboard pieces and cut out the rectangle.

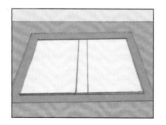

7. Lay the fabric on the table face down. Cover one side of the cardboard pieces with white glue or rubber cement. If using white glue, use a stiff brush, putty knife, or scrap piece of cardboard to spread the glue in an even thin layer so there are no lumps. Place the cardboard

glue-side-down on the fabric so that all three pieces are aligned with the spine between the two covers. Leave a gap of about two thicknesses of the cardboard between the spine and the two covers.

8. Smear glue on the top and bottom edges of the cardboard pieces and fold over the fabric. Then repeat with the outside edges.

9. Smear glue on the inside edges of the cover boards. Don't glue the spine. Place the stack of pages spine-side-down on top of the boards. The extra material that is hanging off of the spine should adhere to the glue on the cover boards. Place two solid bookends, rocks, or jars of food on either side of the papers to hold them upright until they dry thoroughly.

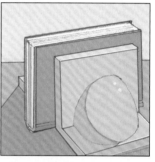

10. Select a decorative piece of paper to use for endpapers. This will cover the inside front and back covers so that you won't see the folded material and cardboard. It can be a solid color or patterned, according to your preference. Cut it to be slightly wider than the pages you started with (before being folded) and not quite as tall.

11. Open your book and cover the inside front cover and first page with glue or rubber cement. Fold your endpaper in half to create a crease, open it back up, and then stick it to the inside  cover and first page, making sure the crease slides into the space between the spine and front cover slightly. Allow to dry and then repeat at the back of the book.

12. Cut two pieces of thin cord for the head and tail band, which will cover the top and bottom of the spine. They should be the same length as the width of the spine. Use a hot glue gun or white glue to adhere them to the  top and bottom of the spine, where the pages are gathered together.

CANDLES

Making candles is a great activity for a fall afternoon. Simple beeswax candles can be completed in a few minutes, but give yourself several hours to make dipped candles. The process is fun, creative, and productive. Give your handmade candles to friends or family or burn them at home to create atmosphere and save on your electricity bill.

TIP

Rather than pouring leftover wax down the drain (which will clog your drain and is bad for the environment), dump it into a jar and set it aside. You can melt it again later for another project.

TIP

When making candles, keep a box of baking soda nearby. If wax lights on fire, it reacts similarly to a grease fire, which is aggravated by water. Douse a wax fire with baking soda and it will extinguish quickly.

Rolled Beeswax Candles

Beeswax candles are cheap, eco-friendly, non-allergenic, dripless, non-toxic, and they burn cleanly and beautifully. They're also very simple and quick to make—perfect for a short afternoon project.

Materials

Sheets of beeswax (you can find these at your local arts and crafts store or from a local beekeeper)

Wick (you can purchase candle wicks at your local arts and crafts store)

Supplies

Scissors
Hair dryer (optional)

Directions

1. Fold one sheet of beeswax in half. Cut along the crease to make two separate pieces.

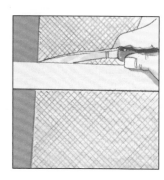

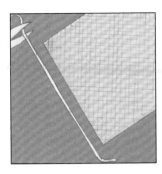

2. Cut your wick to about 2 inches longer than the length of the beeswax sheet.

3. Lay the wick on the edge of the beeswax sheet, closest to you. Make sure the wick hangs off of each end of the sheet.

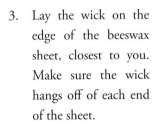

4. Start rolling the beeswax over the wick. Apply slight pressure as you roll to keep the wax tightly bound.

The tighter you roll the beeswax, the sturdier your candle will be and the better it will burn.

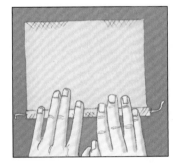

5. When you reach the end, seal off the candle by gently pressing the edge of the sheet into the rolled candle, letting your body heat melt the wax.

6. Trim the wick on the bottom (you may also want to slice off the bottom slightly to make it even so it will stand up straight) and then cut the wick to about ½ inch at the top.

TIP

If you are having trouble using the beeswax and want to facilitate the adhering process, you can use a hair dryer to soften the wax and to help you roll it. Start at the end with the wick and, moving the hair dryer over the wax, heat it up. Keep rolling until you reach a section that is not as warm, heat that up, and continue all the way to the end.

Taper Candles

Taper candles are perfect for candlesticks, and they can be made in a variety of sizes and colors.

TIP

Old crayons can be melted and used instead of paraffin for candle-making.

Materials

Wick (be sure to find a spool of wick that is made specifically for taper candles)
Wax (paraffin is best)
Candle fragrances and dyes (optional)

Supplies

Pencil or chopstick (to wind the wick around to facilitate dipping and drying)
Weight (such as a fishing lure, bolt, or washer)

Dipping container (this should be tall and skinny. You can find these containers at your local arts and craft store, or you can substitute a spaghetti pot)

Materials

Stove
Large pot for boiling water
Small trivet or rack
Glass or candle thermometer
Newspaper
Drying rack

Directions

1. Cut the wick to the desired length of your candle, leaving about 5 additional inches that will be tied onto the pencil or chopstick for dipping and drying purposes. Attach a weight (a fishing lure, bolt, or heavy metal washer) to the dipping end of the wick to help with the first few dips into the wax.

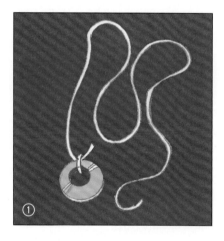

2. Ready your dipping container. Put the wax (preferably in smaller chunks—this will speed up the melting process) into the container and set aside.

3. In a large pot, start to boil water. Before putting the dipping container full of wax into the larger pot, place a small trivet, rack, or other elevating device into the bottom of the larger pot. This will keep the dipping container from touching the bottom of the larger pot and will prevent the wax from burning and possibly combusting.

4. Put the dipping container into the pot and start to melt the wax, keeping a thermometer in the wax at all times. The wax should be heated and melted between 150 and 165°F. Stir frequently in order to keep the chunks of paraffin from burning and to make sure all the wax is thoroughly melted.

(If you want to add fragrance or dye, do so when the wax is completely melted and stir until the additives are dissolved.)

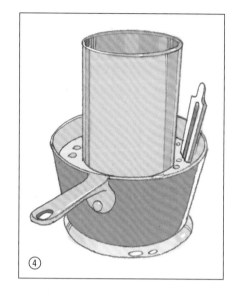

5. Once your wax is completely melted, start dipping. Removing the container from the stove, take your wick that's tied onto a stick and dip it into the wax, leaving it there for a few minutes. Continue to lower the wick in and out of the dipping container, and by the eighth or ninth dip, cut off the weight from the bottom of the wick—the candle should be heavy enough now to dip well on its own.

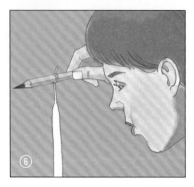

6. To speed up the cooling process—and to help the wax to continue to adhere and build up on the wick—blow on the hot wax each time you lift the candle out of the dipping pot. When the candle is at the desired length and thickness, you may want to lay it down on a very smooth surface (such as a countertop) and gently roll it into shape.

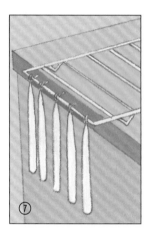
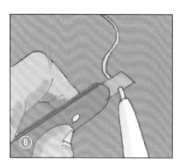

7. On a drying rack, carefully hang your taper candle to dry for a good 24 hours.
8. Once the candle is completely hardened, trim the wick to just above the wax.

Layered Taper Candles

For a more ornate candle, add different shades of food coloring to three or four separate pots of melted wax (or melt down old crayons). Alternate between the different colors of wax as you dip the wick, creating different layers of color. Once the candle is the desired thickness and is mostly cooled, use a paring knife to carefully peel away strips of the wax around the outside of the candle. Allow the wax strips to curl downward as you peel, revealing a rainbow of colors.

Gourd Votives

Small gourds make perfect votive candleholders. Carve a circle out of the top of the gourd, making it the same size as the circumference of the candle you intend to place in it. Gently pry off the top and set the candle in the indentation. If necessary, cut the hole slightly larger, but keep it small enough that the candle fits snugly.

TIP

To make floating candles, pour hot wax into a muffin tin until each muffin cup is about ⅓ full. Allow the wax to cool until a film forms over the tops of the candles. Insert a piece of wick into the center of each candle (use a toothpick to help poke the hole if necessary). Allow candles to finish hardening and then pop them out of the tin. Trim wicks to about ¼ inch.

Jarred Soy Candles

Soy candles are environmentally friendly and easy candles to make. You can find most of the ingredients and materials needed to make soy candles at your local arts and crafts store—or even in your own kitchen!

Materials

1 lb soy wax (either in bars or flakes)

1 ounce essential oil (for fragrance)

Natural dye (try using dried and powdered beets for red, turmeric for yellow, or blueberries for blue)

Supplies

Stove

Pan to heat wax (a double boiler is best)

Spoon

Glass thermometer

Candle wick (you can find this at your local arts and crafts store)

Metal washers

Pencils or chopsticks

Heatproof cup to pour your melted wax into the jar(s)

Jar to hold the candle (jelly jars or other glass jars work well)

Directions

1. Put the wax in a pan or a double boiler and heat it slowly over medium heat. Heat the wax to 130 to 140°F or until it's completely melted.

2. Remove the wax from the heat. Add the essential oil and dye (optional) and stir into the melted wax until completely dissolved.

3. Allow the wax to cool slightly, until it becomes cloudy.

4. While the wax is cooling, prepare your wick in the glass container. It is best to have a wick with a metal disk on the end—this will help stabilize it while the candle is hardening. If your wick does not already have a metal disk at the end, you can easily attach a thin metal washer to the end of the wick. Position the wick in the glass container and wrap the excess wick around the middle of a pen or chopstick. Lay the pencil or chopstick on the rim of the container and position the wick so it falls in the center.

5. Using a heat proof cup or the container from the double boiler, carefully pour the wax into the glass container, being careful not to disturb the wick from the center.

6. Allow the candle to dry for at least 24 hours before cutting off the excess wick and using.

TIP

Add citronella essential oil and a few drops of any of the following other essential oils to make your candle a mosquito repellant:

- Catnip
- Cloves
- Cedarwood
- Lavender
- Lemongrass
- Eucalyptus
- Peppermint
- Rosemary
- Rose geranium
- Thyme

CHRISTMAS WREATHS

How to Make a Christmas Wreath

Supplies:

Small tree branches (from trees such as pine, fir, cedar)
Two wire coat hangers
Wire cutters
Floral wire
Pruning shears
Any decoration you wish to add (ex. Bows, ribbon, pine cones, holly branches, holly berries, etc.)

Cut off the tops of the hangers. Bend the triangular parts of the two coat hangers into circles. Make the cut wire ends into small hooks and loop them around one another to connect the ends of the wires. Overlap the two wire circles and tie them together with floral wire to create one strong circle. You can also purchase a pre-made wire wreath frame from a craft store.

Cut the small branches into eight to ten inch pieces. Make sure you have enough to cover the entire wire circle. Lay them on the wire circle, overlapping them and arranging them so that no wire shows and the wreath looks full. Affix the branches to the circle with floral wire.

If desired, add the holly branches and pine cones. Arrange them decoratively according to taste. Attach them to the wreath with floral wire.

Wrap a ribbon around either the top or the bottom of the wreath and tie a bow, making sure not to crush the branches. If a premade bow is desired instead, attach it to the wreath with floral wire.

Cut another piece of floral wire about six to eight inches long. Slip the piece around the wire frame, looping it around the frame. Bring the two ends up and twist them together. This will form the loop to hang your wreath by. Make sure that the loop will not break or slip off. You can also use thick twine or ribbon, if desired. Hang your wreath.

CORNHUSK DOLLS

This old-fashioned doll makes a wonderful gift for young children and also a unique, decorative, homemade item for your home or for sale at a craft fair. Cornhusk dolls are quite easy to make if you just follow these simple steps:

Materials

Corn husks
Thread or string
Pen or marker
Natural materials for decorations

1. Gather husks from several large ears of corn (you may have these from your garden, if you grow corn, or you might find them at a garden center or a farmers' market in the fall). Select the soft, white husks that grow closest to the ear.

2. Place the stiff ends of two husks together, fold one of the long, soft husks in a strip lengthwise, and wrap it around the ends.

3. Choose the softest and widest husk you can find, fold it across the center, and place a piece of strong thread or string round it and tie it tightly in a knot.

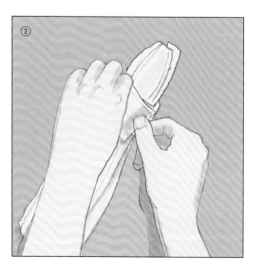

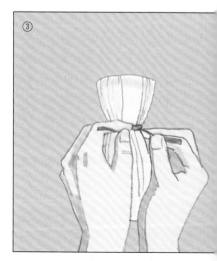

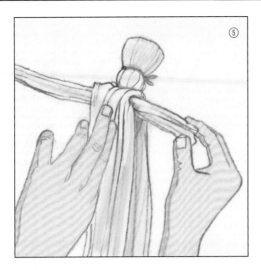

4. Bring this down over the already-wound husks and tie it with a thread underneath. This will form the head and neck of the doll.

5. To make the arms, divide the husk below the neck into two equal parts. Fold two or three husks together and insert them in the space you've made by the division. Hold the arms in place with one hand and use your other hand to fold several layers of husk over each shoulder, allowing them to extend down the back of the figure.

6. When the figure seems substantial enough, use your best husks for the topmost layers and wrap the waist with strong thread, tying it tightly.

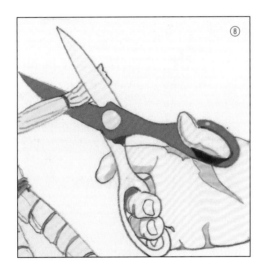

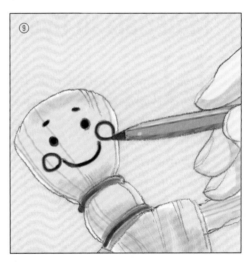

7. Divide the husks below the waist band and make the legs by neatly wrapping each portion with thread. Trim the husks off evenly for the feet.

8. Twist the arms once or twice, tie them, and trim them evenly for the hands.

9. You can draw a face on the doll with a pen or marker or you can glue tiny natural items on the head to make a face.

10. If you want your doll to have clothing or a specific "costume," you can make these from any kind of material and in any way you wish.

DYEING WOOL OR FABRIC

Materials to Dye

Natural materials, such as plant leaves, flowers, barks, roots, and nuts can all be used to dye yarns and cloth. Animal fibers, such as wool and silk, are very easy to dye, and take dye well. Vegetable materials such as cotton and linen, are a bit harder to dye, but with a little knowledge dyeing them can become easy as well.

Supplies

The water you use for dyeing should always be soft water. Most tap water is too hard, and you will have to add a softener to it. The best water to use for dyeing is rain water.

You will need several large pots for dyeing, ones that are able to easily hold four gallons. Make sure that these pots are solely dedicated to dyeing and never use them for cooking. Most people use either stainless steel or enamel pots when working with mordants and dyes.

A larger spoon or paddle will be needed to stir cloths that are being dyed. Wood is best for this. Make sure that it is kept separate and never used for food preparation.

Cheesecloth and string will be needed to make the bundle used to immerse the natural element into water to create the dyebath.

Rubber gloves are essential during the dyeing process. Not only are some of the mordants harmful, but you don't want to end up dyeing your hands!

Mordants

In most cases a material will have to be treated with certain chemicals, called "mordants," in order to set the dye and keep it from fading or washing out over time. Mordants are usually strong and sometimes caustic chemicals.

A yarn or cloth can be treated with a mordant either before or after being dyed, or the mordant can even be added to the dye. Several different colors can also be created using the same dye, if it is combined with different mordants. The overall harshness of the mordant and dye to be used will determine when you should use the mordant; if both are harsh it is best to use mordant either before or after the dyeing to prevent damaging the materials being dyed. The easiest way for a novice dyer to begin is to treat what they would like to dye with the mordant first; that way you will not worry about over-dyeing your fabric or exposing it to too much mordant (if the mordant were mixed in with the dye) and damaging the yarn or cloth.

To create a solution for soaking your yarn or cloth you must first choose the color you want to end up with; this will determine the dyestuff and the mordant you will use. Once you have chosen the mordant, dissolve the correct amount (seen below in "types of mordant") for four gallons of hot water. It is often easiest to dissolve the mordant in a small amount of hot water first, and then add enough water to make four gallons. Four gallons will process one pound of wool. Slowly bring the mixture to a simmer or boil, depending on the directions for the type of mordant used.

To Dye

The first step is always to wash your fabrics—this will get rid of all the natural oils in the material and help the dye to really soak in and adhere. It is easiest to wash wool yarn when it is tied up in a skein.

For one pound of material use three to four gallons of water, enough to cover the material entirely. Add the material to a large pot with the water and a mild detergent or dish soap. Slowly bring the water temperature up and allow it to simmer; silk for 30 minutes, wool for 45 minutes. Allow the water to cool and then rinse it thoroughly. If you are washing cotton or linen add ½ cup sodium carbonate (washing soda) along with the detergent, and then bring the water to a boil. Boil it for an hour and then cool and rinse, the same way as you would with wool or silk.

It is easiest, and best for your material, to complete the dyeing process step-by-step, without allowing your material to dry in-between steps. While the material is being washed, prepare the mordant mixture, then introduce the material to it slowly. Prepare the dyebath while the material is in the mordant mixture. Make sure at each step the water you are taking the material from and the one you are introducing it to is a similar temperature; you do not want to shock the material with a sudden change.

Types of Mordant

For most dyeing jobs, alum (aluminum potassium sulfate) can be used as a mordant. A solution to be used on any wool, silk, or animal fiber should be made of one ounce alum per one gallon water. Simmer any material for one hour in the alum mixture. If you are dyeing plant materials such as cotton or linen, increase the amount of alum to 1 ⅓ ounces per gallon of water and boil the material for one hour.

Chrome (potassium dichromate) can be used to brighten yellows, reds, and greens. Chrome is a highly caustic chemical; gloves should always be worn and it should not be inhaled. A solution can be made from ⅛ ounce of chrome per one gallon of water for animal fiber or ½ ounce for plant fibers. Simmer the animal fibers you want to dye- or boil the plant material- for one hour.

Keep the chrome mordant covered at all times- the fumes are dangerous and any exposure to light will turn your soaking material brown. Either dye chrome soaked materials immediately, or cover and store them in complete darkness. Make sure to dispose of a chrome mixture at a chemical waste facility.

Iron (also called iron sulfate or copperas) can be used to make colors darker, greyer, or duller. If dyeing animal fibers use ⅛ ounce per gallon of water and simmer for one hour. If you are dyeing plant fibers use one ounce of iron for each gallon and boil for one hour.

To dull a color some dyers will add ⅛ ounce of iron to a four gallon dyebath for the last few minutes—you should temporarily remove the fabric from the dyebath and then add the iron, making sure it is dissolved before re-introducing the fabric to be dyed. Use this method only on previously non-mordated material.

Tin (stannous chloride), like chrome, can be used to brighten colors such as reds, oranges, and yellows. It is less caustic than chrome, but can still damage fibers if left too long. For either animal or plant fibers use one ounce of tin per one gallon of water and simmer the materials for one hour. Be sure to thoroughly rinse the materials in clean water after mordanting to prevent damage to the fibers. Alternatively, you can add ⅛ ounce to the dyebath a few minutes before it is finished, making sure to remove and then re-introduce the fabric to make sure the tin properly dissolves into the dyebath. If this method is used the material should have been previously mordated in alum.

Preparing a Dyebath

Choose the color you want; this will determine the raw dyestuff you will need. Natural dyestuff is any plants, leaves, berries, nuts, flowers, roots, or bark. Flowers should be picked just after they bloom. Berries should be ripe and nuts should be fully formed and have just fallen to the ground (make sure not to use any nuts that have been sitting for awhile). All ingredients should be cleaned and checked over for any foreign particles.

Prepare the dyestuff by making sure as much of it as possible will be able to soak. This means shredding any leaves or flowers, cutting up any bark, roots, or branches into small pieces, and crushing any nuts. Wrap the dyestuff in cheesecloth to create a bundle and tie it securely closed with twine.

Cover the bundle in water and bring it to a boil. Cook it until the water is richly colored, adding water as needed to keep the bundle covered. Remove the bundle. You have now created a concentrated dye.

Pour the concentrated dye into a large pot and add enough cold water to make four gallons. Add whatever material you are going to dye and slowly bring the dyebath to a simmer or boil, depending on what material you are dyeing. Cook it until it is the desired shade of color. Cool the material, either inside or out of the dyebath, and then rinse it thoroughly. Alternatively you can cool your material faster and rinse at the same time by dipping it in successively cooler water baths. Gently squeeze out rinsed material and hang it to dry.

Both animal fibers like wool and silk, and plant fibers such as cotton and linen can be dyed in the same way. How-

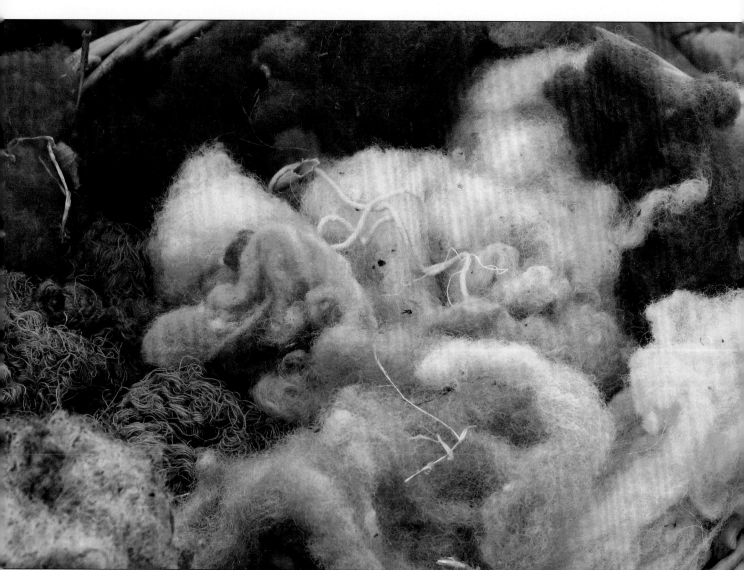

ever, plant fibers do not absorb color as well as animal ones; the times given below are meant for dyeing wool. To dye cotton or linen the material must be boiled from one to two hours in the dyebath.

Types of Dye

Reds—Cochineal, the powder from the crushed cochineal insect, will make rich reds and scarlets. It can be purchased from any dye supplier. Unlike most other plant materials used in dyeing, cochineal can be dissolved directly into the water and does not need to be put into a cheesecloth bundle. It takes about 30 minutes to dissolve completely, after which the water should be boiled for 15 additional minutes to create the dyebath. Simmer any material for ½ hour.

Madder is a root, used since the Middle Ages, to create bright reds and deep oranges. To prepare madder, soak the dried root overnight, then simmer for ½ hour. Any material should first be mordated in alum, and then simmered in the dyebath for ½ hour.

Butterfly weed will produce a lighter reddish-orange. Pick one bushel of blooms and soak them for one hour, then boil them for an additional ½ hour. Simmer material pre-mordated in alum for ½ hour.

Yellows—Many things found in nature can produce yellow dyes.

The flower of the coreopsis will produce a bright yellow. Simmer two bushels of flower heads anywhere from ½ to one hour, until the desired color is reached. Simmer material that has been pre-mordated in alum for ½ hour.

Sophora flower makes a darker yellow. To create the dyebath boil one bushel of flowers for ½ hour. Simmer material pre-mordated in tin and cream of tartar for ½ hour.

Goldenrod, as its name suggests, creates a rich gold-colored dye. Take two pounds of blooming flowers, both the flowers and the stems, and simmer for ½ hour. Add material pre-mordated in alum and simmer for ½ hour more.

The dry, outer layers of onion skins can produce a yellow dye. Take two pounds of the outer skins and simmer them for twenty minutes, making sure not to overcook them. Simmer material that has been pre-mordated in alum for twenty minutes in the dyebath.

Smartweed will produce a dark yellow. To make the dyebath boil two bushels of blooming plants, both flowers and stems, for ½ hour. Simmer material pre-mordated with tin for ½ hour in the dyebath.

The flowers of the marigold will produce a deep yellow, almost verging on tan. Simmer two bushels of flower heads in full bloom for one hour. Add material that has been pre-mordated in alum and simmer for ½ hour.

The berries of the pokeberry create a yellow dye with a red tinge. To prepare the dyebath add 16 quarts and one cup of vinegar to the water and boil for ½ hour. Add material that has been pre-mordated in alum and simmer for ½ hour more.

"Shabby Chic" Dyeing

You don't need to bother with mordants if you're going for a washed out, faded look, and especially if you don't plan to frequently wash whatever fabric you're dyeing. Experiment with beets for a soft pink fabric. Bake the beets in the oven until tender (about an hour) and then peel and chop them into a bowl. Add some water, stir, and add your fabric. The longer you let the fabric soak, the brighter it will be. You can dump the whole mess in a sealable plastic bag and stick it in the fridge if you plan to let it soak for several hours. When you rinse the fabric, most of the dye will come out, but you'll be left with a subtle tint. Onion skins can be boiled in a little water for a yellow dye, and tea or coffee with give an earthy brown. Never wash fabrics dyed this way with other fabrics that you don't want dyed!

Greens—The leaves of the lily of the valley can create two different shades of green, depending on the mordant used. To create the dyebath simmer two pounds of fresh green leaves for one hour. For a soft green add material that has been pre-mordated in alum and simmer for twenty minutes. For a yellow-green color add ⅛ ounce chrome directly into the dyebath and allow it to dissolve, then add un-mordated wool and simmer for twenty minutes.

The privet shrub can create a light green when prepared. Simmer one pound of fresh leaves for ½ to one hour to make the dyebath. Add material pre-mordated in alum and simmer for twenty minutes.

The white queen Anne's lace flower will produce a bright green dye. Gather one bushel of fully bloomed flowers and simmer both the heads and the stems for ½ hour. Add the material that has been pre-mordated in alum, and simmer for ½ hour more.

The leaves of the rhododendron, like those of the lily of the valley, can create two different shades, depending on the mordant used. Prepare three pounds of fresh rhododendron leaves by soaking them overnight and then boiling them for one hour. To create a light green shade, simmer material that has been pre-mordated in alum for ½ hour. To make a very dark green-grey add ½ teaspoon of iron directly to the dyebath, dissolve, and then add un-mordated material and simmer for ½ hour.

Blues—Various shades of blue can be created using indigo. As indigo is not water soluble it needs to be prepared in a special way to produce a dyebath. You will need; sodium corbonate (also called washing soda), hydrosulfite, indigo paste, and water. Mix one ounce washing soda into four ounces water. Add one teaspoon indigo paste. Shake one ounce of hydrosulfite over the mixture and stir gently to combine everything. Add two quarts of warm water and slowly heat it to 350°F, then let it stand for twenty minutes. Shake another one ounce hydrosulfite over the mixture and gently stir it in. The dyebath is now ready; it will have a yellow-green color to it. Immerse your un-mordated material into the dyebath and let it simmer for 20 minutes. As soon as the material is removed and exposed to air it will turn blue.

Browns—Nature is full of things that can dye wool and cloth any shade of tan or brown. The lightest brown can be produced by placing un-mordated material into a dyebath made of tea. Cover ¼ pound dry tea leaves with boiling water and let them steep for 15 minutes. Simmer the material for 20 minutes more.

Another substance that dyes material without any need of mordating is coffee. Boil ½ pounds of unused coffee grounds for 15 minutes. Add your material and simmer for an additional ½ hour.

Tobacco leaves can produce a medium brown color. Add one ounce cream of tartar to your water and boil one pound of cured leaves for ½ hour. Simmer material that has been pre-mordated for ½ hour.

Acorns can create two different shades of brown, depending on the mordant used. To prepare the dyebath gather 7 pounds of just fallen nuts and soak them in water overnight. Boil the nuts for ½ hour and then after adding material simmer for one hour longer. For a tawny brown shade

use material that has been pre-mordated with alum, for a darker brown shade use material pre-mordated with iron.

The logwood, a flowering tree, can also be used to create two different shades of color. The easiest way to obtain logwood is to purchase chips from a dye supplier. Soak four ounces of chips in water overnight, and then boil them for 45 minutes. Add your material and simmer it for ½ hour. Material pre-mordated in iron will become a dark brown color, while material pre-mordated in alum will create a brown so dark it will appear almost black.

To Remember

Always gradually increase or decrease the temperature of water, mordant, or dyebaths when they contain cloths or yarns. It is important never to shock the materials by a sudden change in temperature. Material such as wool will shrink if temperatures change too rapidly.

Always treat materials being dyed gently. Do not twist or wring out cloths or yarn; gently squeeze them of excess water or dye. Do not over-agitate cloths while they are being dyed; gently turn them over in the water to make sure they are dyed evenly.

When dyeing wool it is easiest to create skeins. Tie the skeins loosely, so that the dye can get to all of the fibers. Do not add dry wool to mordant or a dyebath, dampen it first with clean water; this will help the wool absorb the dye evenly.

Natural fabrics such as linen absorb dyes much better than synthetic fabrics.

FLOWER ARRANGING

Decorating your home with beautiful, fragrant flower arrangements doesn't have to mean ordering them from expensive flower shops. You can easily create them yourself, customizing them to your home, your tastes, and your budget. Use your instincts: while there are some rules that will help the novice arranger, there's no shame in ignoring the rules completely and doing entirely your own thing.

Location

First, consider where the arrangement will be placed. A flower arrangement serving as the centerpiece on your dining table during a holiday meal will be very different than the one in your bathroom. It will also help determine the size, generally: a centerpiece cannot be too tall or full if you intend to have conversations at the table.

The color schemes in the room will help you choose the colors you want for your flowers. Choosing between a formal and informal style also helps dictate color choices. Formal styles often opt for a monotone color scheme or just a couple of contrasting colors with fewer types of plants and an elaborate vase. An informal style calls for looser symmetry with more flowers and colors involved and a simple container.

Containers

Possibly the most integral part of the arrangement, containers play a big role in whether the arrangement works or doesn't. It must complement the arrangement and the surrounding in which it's placed. When you pick your container, consider things like color, texture, arrangement size, and line direction (this will be explained in more detail later). You can use a more contemporary container with minimal details, simple shapes, and even lines, or one with more traditional flair, flouting elegance and composed of materials like silver or porcelain.

That being said, don't limit yourself to vases—try using pitchers, baskets, watering cans, terra cotta pots, or anything you can find around the house to put flowers in.

The container should take up ⅓ of the eventual height of the entire display. The remaining ⅔ should be occupied by flowers.

Flower Placement

Generally, the largest and darkest flowers are placed close to the center and base of the design; the smallest and lightest are placed at the outer edges of the arrangements. If you

have a symmetrical arrangement—when the flowers placed in a container can be split exactly down the middle and both sides appear almost as reflections of the other—flowers should be spaced evenly. You can achieve symmetry by placing the same flowers on both sides in the same pattern. If you have an asymmetrical arrangement, the flowers should be placed so that the visual weight is distributed evenly.

A good rule is to cluster the small flowers in groups so they don't get lost among larger blooms. Grouping the flowers makes them easier to handle and allows them to make a distinct impression. To cluster, you can just arrange them close or actually tie them together with rubber bands, floral wire, floral tape, or even ribbon.

Harmony and Contrast

Harmony and contrast are two elements that are used to forge unity within the arrangement and with its surroundings. Harmony is when the arrangement complements its surroundings, being similar to them in a way that makes it seem like it belongs. Elements of the surroundings must exist within the arrangements to achieve harmony, like consistent colors and textures. If the arrangement uses elements varied enough from its surroundings that it cannot blend in completely, then the arrangement has succeeded in finding contrast. The arrangement must stand out to be noticed.

Balance

The finished design must be positioned and secured properly in its container to obtain balance, an essential element in the process of flower arranging. Visible balance is pleasing to the eye. Think about symmetry and asymmetry as you strive for balance.

Some Extra Tips

- Keep the scent of the flowers in mind; the scents of certain flowers might overwhelm a small space, or the scent of one flower may clash with another.

- Keep your flowers in a cool location, away from direct sunlight, heating vents, large lights, etc. Water the plants with very cold water, adding an ice cube to the arrangement every once in a while. Always keep an eye on your water; you don't want a bacteria infestation. If you have one, you must clean the vase and switch out the water.

- Cut your flowers with a pair of scissors or a sharp knife. A clean cut allows water to enter the flower stems, and a ragged one can inhibit water and food absorption allowing your flowers to fade faster.

- Never underestimate the power of a single bloom. Distinctive and elegant, it can be the most eye-catching arrangement in a room full of them.

451

FLOWER PRESERVING

Pressed Flowers and Leaves

Pressed flowers can be used to decorate stationery, handmade boxes, bookmarks, scrapbooks, or picture frames. Kids can glue pressed wildflowers to a blank book and add species names and descriptions to make their own field guides.

Materials

> Large book or newspapers
> Blotting paper
> Weights
> Leaves, ferns, or flowers

1. Have a large book or a quantity of old newspapers and blotting paper, and several weights ready.
2. Use the newspapers for leaves and ferns. Blotting paper is best for the flowers. Both the flowers and leaves should be fresh and without moisture. Place them as nearly as possible in their natural positions in the book or papers, and press, allowing several thicknesses of paper between each layer.
3. Remove the flowers and leaves onto dry papers each day until they are perfectly dried.

Some flowers, like orchids, must be immersed—all but the flower head—in boiling water for a few minutes before pressing, to prevent them from turning black.

In order to preserve your flowers forever, get a blank book or pieces of stiff, white paper on which to mount your preserved flowers and leaves. You can glue them down to the paper with hot glue or regular Elmer's glue. The sooner you mount the specimens, the better. Place them carefully on the paper and, beneath each flower or leaf, write the name of the plant, where it was found, and the date.

GRAPEVINE WREATH

Grapevines make an attractive and natural base for this welcoming wreath. Begin shaping the vines soon after cutting. If you do need to store them for more than a day before using them, remove the leaves and soak the vines for several hours before beginning the wreath to make them more pliable.

Materials

Grape vines
Decoration such as moss, baby's breath, etc.
Florist tape

1. Cut several lengths of vine, keeping them as long as you can manage. Remove the leaves, but don't trim the tendrils.

2. Start with the thicker end of the vine and form a circle. Then begin coiling a second circle, winding and twisting it around the first. Continue until you are almost at the end of the vine and then wind the end around and around, using the vine tendrils or florist tape to secure the end. Add additional vines, winding them around the first circle and tucking the ends into the center.

3. Add moss, baby's breath, yarrow, ferns, feathers, decorative grasses, ribbon, etc. Tuck the ends into the vines and use florist tape as needed for added security.

Plants That Dry Well for Decorative Use

Flowers	Grasses	Herbs
Acrolinium, Baby's Breath, Bachelor's Button, Bells of Ireland, Cockscomb, Coneflower, Delphinium, Foxglove, Globe amaranth, Goldenrod, Heather, Hydrangea, Larkspur, Statice, Strawflower, Yarrow	Bristly foxtail, Cattails, Eulalia grass, Fountain grass, Hare's-tail grass, Indian grass, Northern sea oats, Pampas grass, Plume grass, Quaking grass, Spike grass, Squirrel-tail grass, Switch grass, Wheatgrass	Chamomile, Chives, Dill, Eucalyptus, Fennel, Lavender, Lemongrass, Rosemary, Sage, St. John's wort, Thyme

Dried or silk flowers, pine cones, and fruit can all be used to decorate your wreath.

JEWELRY

Hemp Jewelry

Hemp jewelry is easy to make, the materials required are very affordable, and it is durable and recyclable. Most hemp jewelry is created by making a series of knots and by incorporating beads or charms if you want. By repeating one of two knot patterns, you will make a spiral piece of jewelry; by incorporating two different knot patterns, you will be able to make a flat pattern. It is quite simple to make hemp necklaces and bracelets by following these simple steps.

Materials

A ball of hemp twine (about 1 mm thick and the smoother the better)

Beads or charms (optional—but if you do use these, make sure the stringing holes are large enough for two pieces of hemp to fit through)

Clasp, cord tip, or large bead for securing the jewelry (optional)

Tape (to hold down the hemp while you are knotting)

Scissors

Ruler or measuring tape

Glue (optional—used to attach the securing bead at the end)

1. To start, you will need four lengths of hemp: two for the base strings (these should be a little longer than your overall desired length) and two for knotting (these should be five times the length of the base strings). If you are looking to make a looped closure, you can also cut two lengths of hemp string roughly 3 yards long, fold these strings in half to create four strings, and then simply make a loop at the end where the strings are folded. Whichever method you choose, be sure to tape down the folded end of the string so you have greater ease in knotting your jewelry.

2. Then begin knotting. You may use just one of these knots (spiral pattern) or alternate both (flat pattern) when making your jewelry. Just remember that you are never knotting the two inner (base) strings.

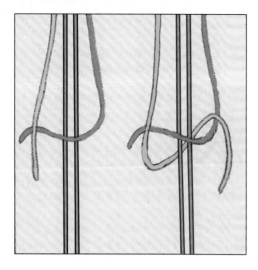

a. First knot: Take the right string, pass it over the base strands, and pull it under the left strand (this should make a shape that looks like a backwards 4). Then, take the left strand, pass it under the base strings, and bring it up and over the right string. Pull both ends and tighten the knot.

b. Second knot: Take the left string, pass it over the base strings, and bring it under the right string (this should make a shape that looks like the number 4). Then, take the right string, pass it under the base strings, and bring it up and over the left string. Pull both ends and tighten the knot.

c. If you want a flat pattern, continue alternating the first and second knots.

3. If you want to add beads or charms to your hemp, simply slide the bead or charm onto the base strings and make a knot around it. It is a good idea to try to space the beads or charms equally on your necklace or bracelet. You can easily do this by counting the number of knots in between each bead or charm and then adding another item after the designated amount of knots.

4. Continue knotting the hemp until the desired length is reached.

5. To finish the jewelry, you may either tie both ends together (the easiest way); knot a larger bead to the end so it will pass through the top loop, securing the jewelry (place a dab of glue onto the bed to help it stay fast); or use metal clasps, which you can buy at your local arts and crafts store.

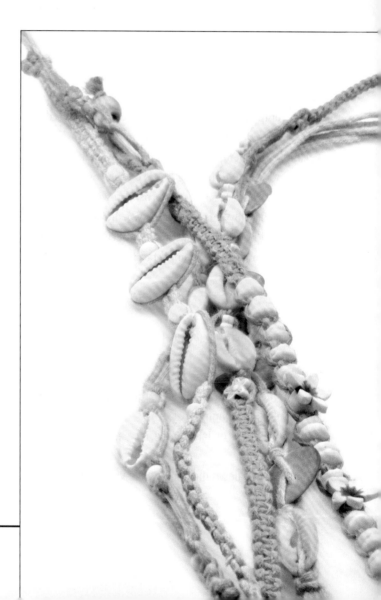

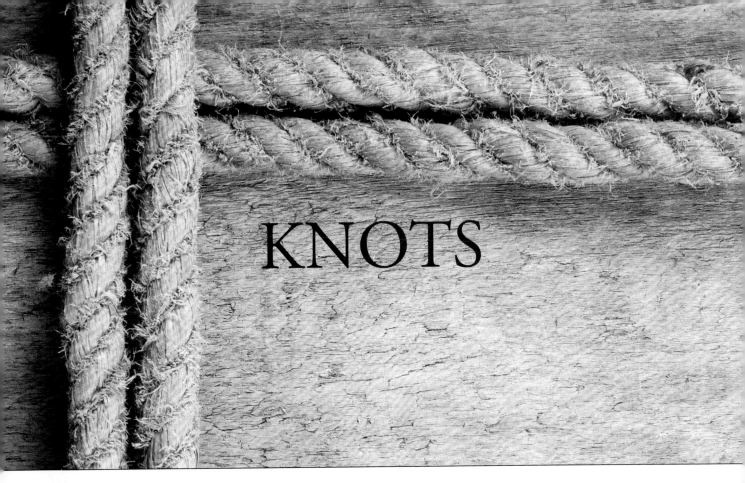

KNOTS

Knowing how to tie a variety of knots is invaluable, especially if you are involved in boating, rock climbing, fishing, or other outdoor activities.

Strong knots are typically those that are neat in appearance and are not bulky. If a knot is tied properly, it will almost never loosen and will still be easy to untie when necessary.

Qualities of a Good Knot

1. It can be tied quickly.
2. It will hold tightly.
3. It can be untied easily.

Three Parts of a Rope

1. **The standing part:** this is the long, unused part of the rope.
2. **The bight:** this is the loop formed whenever the rope is turned back.
3. **The end:** this is the part used in leading.

The best way to learn how to tie knots effectively is to sit down and practice with a piece of cord or rope. Listed below are a few common knots that are useful to know:

- **Bowline knot:** Fasten one end of the line to some object. After the loop is made, hold it in position with your left hand and pass the end of the line up through the loop, behind and over the line above, and through the loop once again. Pull it tightly and the knot is now complete.

- **Clove hitch:** This knot is particularly useful if you need the length of the running end to be adjustable.

- **Halter:** If you need to create a halter to lead a horse or pony, try this knot.

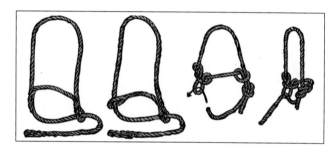

- **Slip knot:** Slip knots are adjustable, so that you can tighten them around an object after they're tied.

- **Square/reef knot:** This is the most common knot for tying two ropes together.

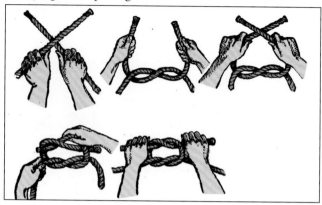

- **Timber hitch:** If you need to secure a rope to a tree, this is the knot to use. It is easy to untie, too.

- **Two half hitches:** Use this knot to secure a rope to a pole, boat mooring, washer, tire, or similar object.

KNITTING

The art of knitting was purportedly invented by the Spanish nobility, as a means of relaxation for noble women in the country. The Scottish also claim to have developed knitting, and King Henry VII was the first to wear knitted stockings in England. Queen Elizabeth also wore knitted silk stockings made by Mistress Montegue.

Whenever and wherever knitting was first "discovered," it is useful, relaxing, and can be done while enjoying a good conversation with a friend.

Knitting Basics

Explanation of Stitches

Knitting can be done in rows of plain or purl stitches or by incorporating a variety of stitches and knitting techniques in one project. However, the simpler stitches are the better when first starting to knit. Just be sure not to pull the thread too tight or to keep it too loose—as you continue to knit, you will learn the proper amount of tension to apply to your string so you create a perfect, knitted item.

General Tip for Beginning Knitters

Hold the needles loosely in your hands and close to the points. In order to knit easily and quickly, your hands should not move too much nor should you make large gestures with the needles.

Tools Needed for Knitting

1. **Gauge**—This measures the knitting needles. Most needles already have their gauge listed on them, but if your needles do not, you should find this measuring tool at your local arts and crafts store.
2. **Knitting needles**—These are made of steel, wood, or plastic and are used to knit your material together.
3. **Knitting shields**—Although these are not a necessary tool for knitting, you may find that you want these so the material does not slip off of your needle.
4. **Material to be knitted**—Beginners should use thicker yarn in their knitted items. When you have become more proficient in knitting, you can experiment with different types of threads and materials to create your various knitted items.

Knitting Terminology

To bring the thread forward—This means to pass the thread between the needles toward the knitter's body.

To cast off—You do this by knitting two stitches, passing the first over the second, and proceeding in this manner until the last stitch, which is secured by passing the thread through it.

To cast on the loops or stitches—Take the material in your right hand and twist it around the little finger, bringing it under the next two fingers, and passing it over the pointer finger. Then, take the end of the material in your left hand (holding the needle with your right), wrap it around the little finger, and then bring it over the thumb and around the second and third fingers. By doing so, you will have formed a loop. Now, you must bring the needle under the lower thread of the material and above the material that is over the right-hand pointer finger under the needle. The thread in the left hand should be pulled tightly, completing this step. You can repeat this process as many times as needed until you've cast the amount of stitches you want.

To cast over—This means to bring the material around the needle (bringing it forward).

To fasten on—This refers to fastening the end of the material when it's needed during the process of knitting. The best way to fasten on is to place the two ends in opposite directions and knit a few stitches with both.

Knitting stitch—In this stitch, the needle must be put through the cast-on stitch and the material should be turned over. This will be taken up and under the loop (or stitch) and then let off. This is also known as a plain stitch and will be continued until an entire round is complete.

A loop stitch—This is made by passing the thread before the needle.

Narrowing—This is to decrease the number of stitches by knitting two together, so you only form one loop.

Purl stitch—This is also known as a seam, ribbed, or turn stitch. It is formed by knitting with the material before the needle and instead of bringing the needle over the upper thread, the material is brought under it. In a sense, this is the opposite of a knitting stitch.

Raising—This is to increase the number of stitches and is made by knitting one stitch in the usual way and then omitting to slip out the left-hand needle. Then, the material is passed forward and a second stitch is formed by pulling the needle under the stitch. The material must be put back to its normal place when the extra stitch is completed.

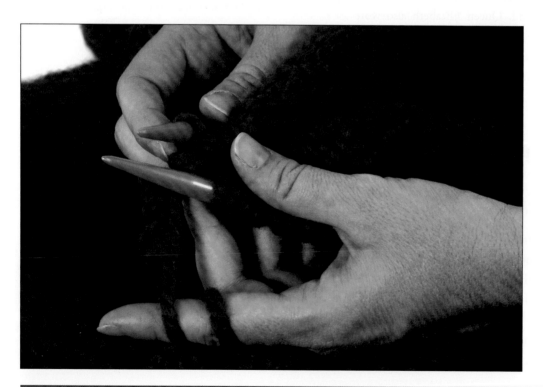

Hold the needles loosely in your hands with the loose yarn wrapped around your pointer finger.

Slip the right needle into the top loop on the left needle, keeping the left needle above the right needle. To do a purl stitch, the right needle would go on top of the right needle.

Wrap the loose strand of yarn over and behind the right needle. For a purl stitch, wrap the yarn behind first and then over the needle. Slip the loop off of the left needle to finish the stitch.

To rib—To alternately knit plain and purled stitches (three plain then three purl, etc.).

A round—This is all of the stitches on two, three, or more needles.

A row—This refers to the stitches from one end of the needle to the other.

To seam—To knit a purl stitch every alternate row.

A slip stitch—This is made by passing the thread from one needle to another without knitting.

To turn—To change the type of stitch.

Welts—These are alternating plain and ribbed stitches and are used for anything that you don't want to twist or curl up.

How to Knit

1. Hold the two needles loosely in your hands. Pass a loop over the left-hand needle near the end of the yarn and hold the right-hand needle loosely. Put the right-hand needle into the loop, passing it from left to right and keeping the right-hand needle under the left needle. Pass the string over this needle—between it and the left-hand needle—and pull the loop up toward the right. Now, bring the right needle up and pass the stitch on it to the left needle by putting the left needle through the left side of the loop, keeping the right needle in the loop. It is ready to begin the next stitch. Repeat.

2. Knitting stitch: After you have made the correct number of stitches, hold the needle that has the stitches on it in your left hand and pass the right needle into the first stitch from left to right. Put the yarn around between the two needles, pull the loop through the other loop on the left needle, and slip that loop off the left needle. Repeat.

3. Purling stitch: Keep the yarn in the front of the work and put the right needle into a stitch from right to left, passing it upwards through the front loop of the stitch. The right needle should be resting on the left. Pass the yarn around the front of the needle and bring it back between the two needles. Pull the right needle slightly back, so as to secure the loop on the right needle and then draw off the loop on the left needle. Repeat. Note: this is basically the knitting stitch, only backwards.

4. Slipping a stitch: This should be done by passing a stitch from one needle to another without knitting it at the beginning of a row. This should always been done when using two needles at the beginning of each row, so the rows remain even.

5. Casting (binding) off: Knit two stitches, passing the first stitch over the second, and then knit a third stitch, passing the second over the third. Continue in this way until all the stitches are off the needle.

Two Simple Knitting Patterns

1. Patent Knitting, or Brioche Knitting
   Cast on any number of stitches divisible by three.

Yarn forward, slip one, knit two together. Work every row in the same way.

2. Cane-work Pattern
   Cast on any number of stitches divisible by four.
   First Row: make one, knit one, make one, knit three. Repeat.
   Second Row: purl.
   Third Row: knit three, make one, slip one, knit two together, pass the slip-stitch over the two knitted together, make one. Repeat.
   Fourth Row: purl.
   Fifth Row: make one, slip one, knit two together, pass the slip-stitch over, make one, knit three. Repeat.
   Sixth Row: purl.
   Seventh Row: repeat the third row.
   Eighth Row: purl.
   Ninth Row: make one, slip one, knit two together, pass the slip-stitch over, make one, knit three. Repeat
   Tenth Row: purl.
   Repeat from the third row until the item is complete.

Simple Scarf

Materials

Mid-weight or 4-ply yarn of any color (use at least one full bundle of yarn)

Knitting needles (size 8 to 10.5 are best for knitting scarves)

Directions

1. Decide how wide you want your scarf to be (26 to 35 stitches are the standard width for a scarf)
2. First row: knit 26 to 36 stitches
3. Second row: knit 26 to 35 stitches (if you want something a little more challenging, purl this row instead)
4. Continuing knitting (or knitting and purling alternately) until you reach the desired length (60 inches is a good length for a scarf)
5. At the end, cast (bind) off the stitches.

Hat

Materials

Yarn of a medium-heavy weight, any color of your choosing

Knitting needles (depending on the head size for the hat, use No. 6 or No. 8 needles)

Directions

1. Cast on 72 stitches.
2. First row: knit 72 stitches.
3. Second row: purl 72 stitches.
4. Continue in this fashion until your hat is about 9 inches tall.

5. To begin to cast (bind) off your hat, follow this pattern:

   a. Knit five stitches, knit two together, and continue to the end of the row.

   b. Purl the next row.

   c. Knit four stitches, knit two together, and continue to the end of the row.

   d. Purl the next row.

   e. Knit three stitches, knit two together, and continue to the end of the row.

f.   Purl the next row.

g.   Knit two stitches, knit two together, and continue to the end of the row.

h.   Purl the next row.

i.   Knit every two stitches together.

j.   Take the excess yarn, pull it through the last stitches, and cut off so only about an inch and a half remains. Sew a seam, put the remaining yarn through the loops, and fold your hat inside out.

Fingerless Mittens

Fingerless mittens are wonderful to use if your hands are cold but you still need to have complete access to things, such as typing on a computer or making a meal. They make wonderful gifts for friends and family members.

Materials

150 yards of worsted-weight yarn or wool/yarn blend
Double-pointed knitting needles, No. 8

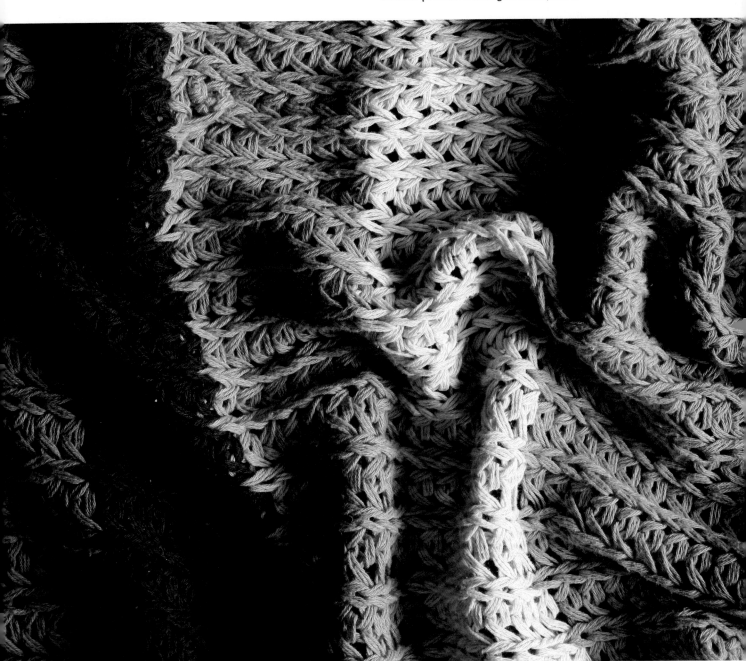

If knitting your blanket for a baby, be sure to use soft yarn that will not irritate a baby's sensitive skin.

Directions

1. To make the cuff, cast off 28 stitches, making sure these stitches are even. Then, begin to knit in the round. Do not twist the stitches. Use the yarn tail to keep track of the round ends. Knit three rounds. Switching to the twisted rib pattern, knit one stitch through the back loop in order to twist it. Purl one, and repeat this pattern until the cuff measures roughly 2½ inches.

2. Now using a stocking stitch, begin the hand and thumb portion of the mitten.

3. First row: knit one, purl one, make one (increase the stitch), knit one, make one, purl one, knit until the end of the round.

4. Second row: knit one, purl one, knit until you reach the next purl stitch in the row above, purl one, knit until the end of the round.

5. Third row: knit one, purl one, make one, knit until the next purl, make one, purl one, knit until the end of the round.

6. Repeat the second and third rows until you have nine stitches between the purls. The glove should now measure about 5 1/2 inches from the edge of the cast off point.

7. Place two purl and nine thumb gore stitches on a piece of scrap yarn. Cast off three stitches and knit four rounds of stocking stitch. Then change to twisted rib stitch and make six rounds. Bind this off very loosely.

8. To make the thumb, put 11 stitches on hold for the thumb onto an extra knitting needle. Pick up three stitches at the base of the thumb and make 14 stitches.

9. Knit one round of only 12 stitches.

10. Using the twisted rib stitch, make six more rounds and bind off loosely.

11. To finish up, weave in the yarn ends and, if necessary, sew closed any holes at the sides of the thumb base.

Knitted Square Blanket

Materials

Thick yarn, any color you like (if you want a multicolored blanket, feel free to use different colored yarn for each individual square)

Knitting needles, No. 6

Directions

1. You'll begin by making smaller squares that will then be sewn together to form a larger blanket.

2. Cast on any number of stitches that can be divided by three. For a square of 6 inches, you'll need 45 stitches.

3. First row: Slip one, knit two. Turn the yarn around the needle and bring it again in front. Then, slip one, knit two together. Purl the last two stitches.

4. Second row: Turn the yarn around the needle, bringing it to the front. Slip one, knit two together. Knit the last two stitches in the row.

5. Continue the pattern in step 4 (alternating purled and knitted last two stitches) until you reach the end and cast off your square.

6. Continue making as many squares as you want to get the desired size of your blanket.

7. When you have all your knitted squares, take a knitting needle and sew each square together.

LAMPSHADES

Making your own lampshades will enable you to match or complement the décor of your room. Choose material that is fire-restistant; a 100 watt light bulb can heat up to more than 200 degrees. Keep in mind that lighter-colored, thin fabrics will let more light through than dark, heavy fabrics.

Materials

Newspaper
Lampshade frame
Fabric
Pins
Scissors
Masking tape
Measuring tape or ruler
Fabric glue

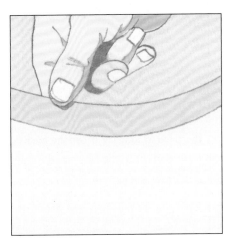

1. Measure the circumference of the top and bottom of your lampshade frame. To facilitate this, you can place a strip of masking tape around one circle and then peel it off and use a ruler or measuring tape to determine the exact length of the tape. Next, measure the distance from the bottom, wide circle of the frame to the top, narrow circle. Add 1 ½ inches to all measurements.

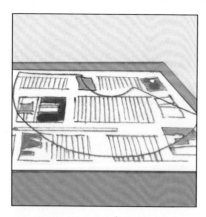

2. Use your measurements to draw a pattern on newspaper. The pattern will be in the shape of a trapezoid. The bottom of the trapezoid is the length of the bottom circle of the frame; the top of the trapezoid is the length of the top circle of the frame; and the distance between them is the height of your frame.

3. Cut out the pattern and lay it on the backside of your fabric. Pin the pattern to the fabric and carefully cut out the fabric. Remove the pattern.

4. Cover the backside of the fabric with fabric glue or spray adhesive. Place the frame onto the fabric and begin roll-

ing it, pressing the fabric to the frame as you go. The fabric should hang over the top and bottom of the frame slightly.

5. Once the fabric is wrapped all the way around the lampshade, tuck in the raw edge of the fabric to create a seam and glue it down. Then glue and tuck the fabric over the top circle and under the bottom circle of the frame. Use clothespins to hold the fabric in place until it dries completely.

6. If desired, add ribbons or tassels to the top or bottom of the shade.

TIP

Fabric with a higher percentage of cotton will adhere to the frame better than synthetic fabrics. Also keep in mind that the darker the fabric, the less light will shine through.

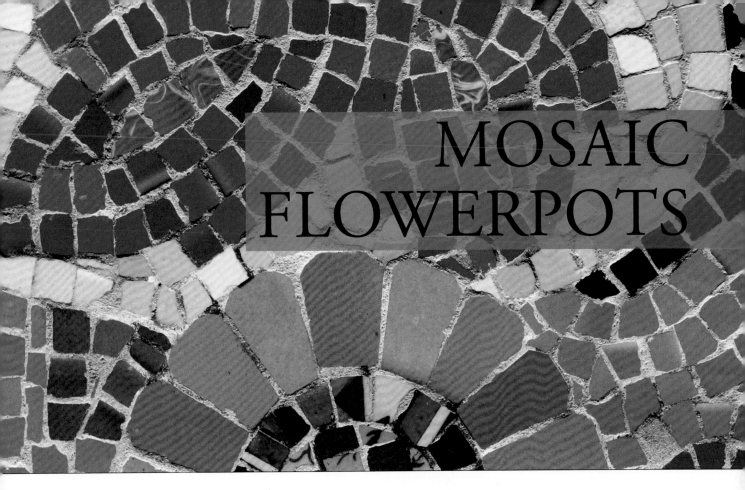

MOSAIC FLOWERPOTS

Spring is the time to start planting seeds so you'll have seedlings to transplant to the garden come summer. Make your pots unique by decorating them yourself with bits of beach glass, pottery, sea shells, or beads. This is a great project to do with kids, but be careful of sharp pieces of glass or pottery.

Materials

Putty knife
Ceramic tile grout
Terra cotta flowerpot
Pieces of beach glass, broken pottery or mirror, tile, beads, charms, etc.

Directions

1. Use the putty knife to spread a thick layer of grout around the outside of the flowerpot (at least ¼ inch thick).
2. Press the pieces into the grout and add more grout around each piece to cover any sharp or rough edges.
3. Allow pot to dry thoroughly. Then wipe away any grout residue from the pieces with a damp sponge.

TIP

Look for nontoxic grout online or at your local hardware store. Alternatively, you can make your own grout by mixing one part Portland cement to two parts sand in a tub. Add water slowly while stirring until the mixture is thick like mud. If desired, add natural iron oxide pigments to the grout for a more colorful background for your mosaic.

For an even simpler "faux grout," mix two parts white sand to one part white glue. Add acrylic paint or concentrated natural dyes as desired. The "faux grout" won't be as strong or as smooth as real grout, but it will work in a pinch.

PAPER MAKING

Instead of throwing away your old newspapers, office paper, or wrapping paper, use it to make your own unique paper! The paper will be much thicker and rougher than regular paper, but it makes great stationery, gift cards, and gift wrap.

Materials

Newspaper (without any color pictures or ads if at all possible), scrap paper, or wrapping paper (non-shiny paper is preferable)

2 cups hot water for every ½ cup shredded paper

2 tsp instant starch (optional)

Supplies

Blender or egg beater

Mixing bowl

Flat dish or pan (a 9 x 13-inch or larger pan will do nicely)

Rolling pin

8 x 12-inch piece of non-rust screen

4 pieces of cloth or felt to use as blotting paper, or at least 1 sheet of Formica

10 pieces of newspaper for blotting

Directions

1. Tear the newspaper, scrap paper, or wrapping paper into small scraps. Add hot water to the scraps in a blender or large mixing bowl.

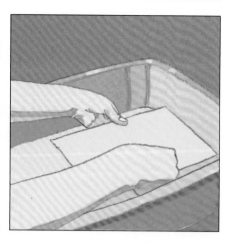

2. Beat the paper and water in a blender or with an egg beater in a large bowl. If you want, mix in the instant starch (this will make the paper ready for ink). You can also add food coloring or natural dye, for colored paper. The paper pulp should be the consistency of a creamy soup when it is complete.

3. Pour the pulp into the flat pan or dish. Slide the screen into the bottom of the pan. Move the screen around in the pulp until it is evenly covered.

4. Carefully lift the screen out of the pan. Hold it level and let the excess water drip out of the pulp for a minute or two.

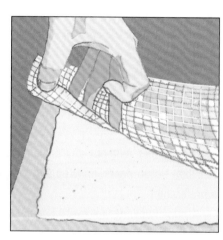

5. With the pulp side up, put the screen on a blotter (felt) that is situated on top of some newspaper. Put another blotter on the top of the pulp and put more newspaper on top of that.

6. Using the rolling pin, gently roll the pin over the blotters to squeeze out the excess water. If you find that the newspaper on the top and bottom is becoming completely saturated, add more (carefully) and keep rolling.

7. Remove the top level of newspaper. Gently flip the blotter and the screen over. Very carefully, pull the screen off of the paper. Leave the paper to dry on the blotter for at least 12 to 24 hours. Once dry, peel the paper off the blotter.

To add variety to your homemade paper:

- To make colored paper, add a little bit of food coloring or natural dye to the pulp while you are mixing in the blender or with the egg beater.

- You can also try adding dried flowers (the smoother and flatter, the better) and leaves or glitter to the pulp.

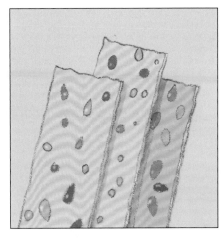

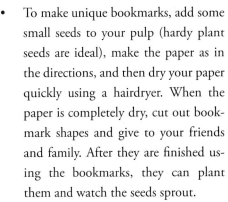

- To make unique bookmarks, add some small seeds to your pulp (hardy plant seeds are ideal), make the paper as in the directions, and then dry your paper quickly using a hairdryer. When the paper is completely dry, cut out bookmark shapes and give to your friends and family. After they are finished using the bookmarks, they can plant them and watch the seeds sprout.

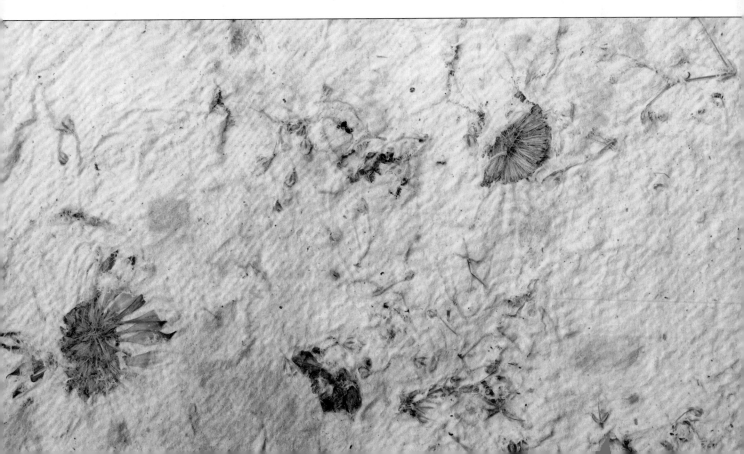

PINE CONE BIRDS

Pine cones lend themselves to all sorts of fanciful creations. With a little string, some construction paper, and a marker you can transform pine cones of varying shapes and sizes into a flock of birds!

Materials:

Pine cones (any size or shape)
Pencil
Scissors
Construction paper
Glue
Colored marker
Strong thread or dental floss

The pine cones will be the bodies of the birds. To add feathers, cut out leaf-shaped pieces of construction paper, draw vein patterns on the pieces, and glue them to the sides or one end of the cone. For a turkey, tilt a round pinecone on its side and glue the feathers upright to the top petals of the cone. Add a neck and head with more construction paper, drawing on the eyes. For an owl, a round cone can be stood upright on its bottom petals—just add big round eyes and an oval beak cut out of construction paper and you have a wise old owl! For a blue bird, use an elongated cone and glue the construction paper wings to either side. Glue a head to the bottom petals of the cone. Don't forget to draw the beak and eyes!

To hang your bird ornaments, secure the strong thread or dental floss around a pine cone petal and tie a knot. Hang them on your Christmas tree or in the window!

PINE CONE BIRD FEEDER

To attract wild birds to your yard, all you need is a pine cone, a bunch of seeds, and a little peanut butter. Keep in mind that wild animals also enjoy backyard treats, so you may end up attracting bears, raccoons, or other unwanted fauna. But in the meantime, the birds will appreciate your generosity!

Materials

> Cord or string (at least two feet long)
> Large pine cone
> Peanut butter
> Birdseed

Directions

1. Loop the cord around the top petals of the pine cone and tie it tightly. Then spoon a little peanut butter between each layer of pine cone petals.

2. Roll the pine cone in the birdseed (you can spread the seeds in a pie dish or on a sheet of waxed paper first) and hang the feeder in a tree.

TIP

Make your own bird seed mix by combining millet, cracked corn, sunflower seeds, safflower seeds, and thistle seeds. Birds also love bits of stale bread, cereals, dried fruit, and nuts.

POTPOURRI

Rose petals and sweet geranium leaves are the primary ingredients in most potpourri. You can also add lavender, sweet verbena leaves, bay leaves, rosemary, dried orange peels, or pine needles. Orrisroot powder will help preserve the scent of the other ingredients.

The best time to collect flowers, seeds, or roots is in the morning, just after the dew has evaporated. Be careful not to bruise flower petals when gathering them, as damaged flowers will lose their scent.

Materials

Flower petals, flower heads, stems, roots, or fruit (see chart below)
Screen
String
Food dehydrator, oven, or solar oven
Spices and orrisroot
Essential oil

1. To dry individual petals or flower heads, lay them on a screen and leave them in a dry place, out of direct sunlight, for about two weeks.
2. For stems or roots, bunch them together, tie with string, and hang upside down for one to two weeks.
3. Flowers or fruit (such as citrus slices or peels) can also be dried in a food dehydrator set at its lowest temperature, or in the oven set to 180 degrees Fahrenheit. Drying in the oven takes several hours. Leave the oven door open slightly to allow moisture to escape. Using a solar oven will take longer but is much more energy efficient!
4. To enhance the fragrance of your potpourri, add spices and orrisroot (available online or from many florists) to your final mixture. Gather violet powder, ground allspice, ground cloves, ground mixed spice, ground mace, whole mace, and/or whole cloves.
5. Once all the components are thoroughly dried, you can mix them together. To make the potpourri smell stronger, you can add a few drops of essential oil and store the mixture in a crock with a tight-fitting lid or a lidded jar for a few weeks or even months. As the mixture sits, the fragrance tends to become much richer.

This chart shows the main types of fragrances and the ingredients associated with them. When experimenting with creating a recipe, first decide whether you want your potpourri to be primarily spicy, sweet, fruity/citrusy, or earthy. Use mostly ingredients from that category and then add smaller amounts from the other categories, if desired.

Fragrance Category

Ingredient	
Spicy	Allspice Berry, Bay, Cardamom, Caraway, Cinnamon, Clove, Clary Sage, Clove, Coriander, Cumin, Fir, Frankincense, Hyssop, Myrrh, Neroli, Nutmeg, Rosewood
Sweet	Anise, Clary sage, Coriander, Frankincense, Geranium, Jasmine, Lavender, Rose, Sandalwood, Vanilla
Fruity/Citrusy	Bergamot, Berries, Lemon, Orange, Lemongrass, Ylang ylang, Wintergreen
Earthy	Cedarwood, Chamomile, Eucalyptus, Fir, Juniper, Myrrh, Patchouli, Pine, Rosemary, Spruce

"You Choose" Potpourri Recipe

Potpourri can be comprised of a mix of flowers, leaves, and herbs—depending on which of these are available to you. The ingredients in this recipe are interchangeable and can all be used together or you can pick and choose which you want to use in making your own potpourri.

Essential Ingredients

1 ounce orrisroot
1 ounce allspice
1 ounce bay salt
1 ounce cloves

Assorted Ingredients (to add as you like)

Rose petals
Lavender
Lemon plant
Verbena
Myrtle
Rosemary
Bay leaf
Violets
Thyme
Mint
Essence of lemon
Essence of lavender

Mix the orrisroot, allspice, bay salt, and cloves together. Combine this with about 12 handfuls of the dried petals and leaves and store in an airtight jar or bowl. A small quantity of essence of lemon and/or lavender may be added but these are not necessary. Let the mixture stand for a few weeks. If it becomes too moist, add additional powdered orrisroot. Once the potpourri is dry and very fragrant, parcel it into bowls to set around the house or give as gifts in jars or sachets.

Rose Potpourri

Ingredients

1 lb rose petals (already pressed and from a jar)
Dried lavender (any proportion you like)
Lemon verbena leaves (any proportion you like)
A dash of orange thyme
A dash of bergamot
1 dozen young bay leaves (dried and broken up)
A pinch of musk
2 ounces orrisroot, crushed
1 ounce cloves, crushed
1 ounce allspice, crushed
½ ounce nutmeg, crushed
½ ounce cinnamon, crushed

Combine all the ingredients together in a large crock, jar, or bowl with a lid. Seal and allow it to sit for a few weeks, until the aroma is to your liking. Parcel into small bowls to fragrance rooms or give as gifts.

A Simple Recipe for Sachet Potpourri

If you are unable to procure large amounts of petals and leaves, here is a simple recipe, using oils as substitutes, that makes fine potpourri sachets.

Ingredients

2 drams alcohol
10 drops bergamot
20 drops eucalyptus oil
4 drops oil of roses
½ tsp cloves
1 ounce orrisroot
¼ tsp cinnamon
½ tsp mace
1 ounce rose sachet powder

Mix these ingredients together in a large stone crock or in a large glass bowl. When the ingredients are thoroughly mixed, store the potpourri in small wooden boxes or sachets and place them around the house. The potpourri gives off a pleasing fragrance to any room or drawer.

Sachet Bags

Sachets are small bags filled with sweet-smelling potpourri. Hang them in a closet or tuck them in a drawer to lend your clothes a gentle fragrance.

Fabric
Scissors
Needle and thread or sewing machine
Ribbon or string

1. Cut two 3 x 5-inch rectangles of fabric. Hem one of the 3-inch sides of both rectangles. Stack the pieces one on top of the other, placing the hemmed edges together and lining up all sides exactly. The right sides of the fabric should be facing toward each other.

2. Sew along three sides of the fabric, about ¼ inch from the edges. Leave one of the 3-inch sides unsewn so that the bag has an opening.

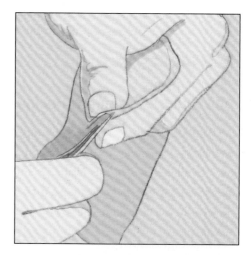

3. Turn the bag inside out so that the right side of the fabric is visible and the seams are hidden. Fill the bag half-full with potpourri and tie a ribbon or string around the top.

POTTERY

Clay is the basic ingredient for making pottery. Clay is decomposed rock containing water (both in liquid and chemical forms). Water in its liquid form can be separated from the clay by heating the mass to a boiling point—a process that restores the clay to its original condition once dried. The water in the clay that is found in chemical forms can also be removed by ignition—a process commonly referred to as "firing." After being fired, clay cannot be restored to any state of plasticity—this is what we term "pottery." Some clay requires greater heat in order to be fired, and these are known as "hard clays." These types of clay must be subjected to a "hard-firing" process. However, in the making of simple pottery, soft clay is generally used and is fired in an over-glaze (soft glaze) kiln.

Pottery clays can either be made by hand (by finding clay in certain soils) or bought from craft stores. If you have clay soil available on your property, the process of separating the clay from the other soil materials is simple. Put the earthen clay into a large bucket of water to wash the soil away. Any rocks or other heavy matter will sink to the bottom of the bucket. The milky fluid that remains—which is essentially water mixed with clay—may then be drawn off and allowed to settle in a separate container, the clear water eventually collecting on the top. Remove the excess water by using a siphon. A repetition of this process will refine the clay and make it ready for use.

You can also purchase clay at your local craft store. Usually, clay sold in these stores will be in a dry form (a grayish or yellowish powder), so you will need to prepare it in order to use it in your pottery. To prepare it for use, you must mix the powder with water. If there are directions on your clay packet, then follow those closely to make your clay. In general, though, you can make your clay by mixing equal parts of clay powder and water in a bowl and allowing the mixture to soak for 10 to 12 hours. After it has soaked, you must knead the mixture thoroughly to disperse the water evenly throughout the clay and pop any air bubbles. Air bubbles, if left in the clay, could be detrimental to your pottery once kilned, as the bubbles would generate steam and possibly crack your creation. However, be careful not to knead your clay mixture too much, or you may increase the chance of air bubbles becoming trapped in the mixture.

If, after kneading, you find that the clay is too wet to work with (test the wetness of the clay on your hands and if it tends to slip around your palm very easily, it is probably too wet), the excess water can be removed by squeezing or blotting out with a dry towel or dry-board.

The main tools needed for making pottery are simply your fingers. There are wooden tools that can be used for adding finer detail or decoration, but typically, all you really need are your own two hands. A loop tool (a piece of fine, curved wire) may also be used for scraping off excess clay where it is too thick. Another tool has ragged edges and this can be used to help regulate the contour of the pottery. Remember that homemade pottery will not always be symmetrical, and that is what makes it so special.

Basic Vase or Urn

Try making this simple vase or urn to get used to working with clay.

1. Take a lump of clay. The clay should be about the size of a small orange and should be rather elastic feeling. Then, begin to mold the base of your object—let's say it is either a bowl or a vase.

2. Continue molding your base. By now, you'll have a rather heavy and thick model, hollowed to look a little like a bird's nest. Now, using

this base as support, start adding pieces of clay in a spiral shape. Press the clay together firmly with your fingers. Make sure that your model has a uniform thickness all around.

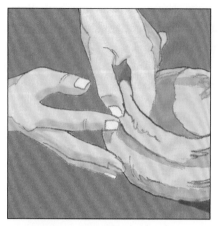

3. Continue molding your clay and making it grow. As you work with the clay, your hands will become more accustomed to its texture and the way it molds, and you will have less difficulty making it do what you want. As you start to elongate and lengthen the model, remember to keep the walls of the piece substantial and not too thin— it is easier to remove extra thickness than it is to add it.

4. Don't become frustrated if your first model fails. Even if you are being extra careful to make your bowl or vase sturdy, there is always the instance when a nearly complete vase will fall over. This usually happens when one side of the structure becomes too thin or the clay is too wet. To keep this from happening, it is sometimes helpful to keep one hand inside the structure and the other outside. If you are building a vase, you can extract one finger at a time as you reach closer and closer to the top of the model.

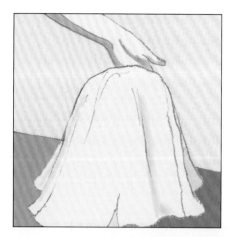

5. Make sure the clay is moist throughout the entire molding process. If you need to stop molding for an extended period of time, cover the item with a moist cloth to keep it from drying out.

6. When your model has reached the size you want, you may turn it upside down and smooth and refine the contours of the object. You can also make the base much more detailed and shaped to a more pleasing design.

7. Allow your model to air dry.

Embellishing Your Clay Models

You may eventually want to make something that requires a handle or a spout, such as a cup or teapot. Adding handles and spouts can be tricky, but not if you remember some simple rules. Spouts can be modeled around a straw or any other material that is stiff enough to support the clay and light enough to burn out in the firing. Keep spouts and handles solid and thick. Also, keeping them closer to the body of your model is more practical, as handles and spouts that are elongated are harder to keep firm and can easily break off.

The simplest way to decorate your pottery is by making line incisions. Line incision designs are best made with wooden, finger-shaped tools. It is completely up to you as to how deep the lines are and into what pattern they are made.

Wheel-working and Firing Pottery

If you want to take your pottery-making one step further, you can experiment with using a potters' wheel and also glazing and firing your model to create beautiful pottery. Look online or at your local craft store for potters' wheels. Firing can leave your pottery looking two different ways, depending on whether you decide to leave the clay natural (so it maintains a dull and porous look) or to give it a color glaze.

Colored glazes come in the form of powder and are generally metallic oxides, such as iron oxides, cobalt oxide, chromium oxide, copper oxide, and copper carbonate. The colors these compounds become will vary depending on the atmosphere and temperature of the kiln. Glazes often come in the form of powder and need to be combined with water in order to be applied to the clay. Only apply glaze to dried pottery, as it won't adhere well to wet clay. Use a brush, sponge, or putty knife to apply the glaze. Your pottery is then ready to be fired.

There are various different kinds of kilns in which to fire your pottery. An over-glaze kiln is sufficient for all processes discussed here, and you can probably find a kiln in your sur-rounding area (check online and in your telephone book for places that have kilns open to the public). Schools that have pottery classes may have over-glaze kilns. It is important, whenever you are using a kiln, that you are with a skilled pottery maker who knows how to properly operate a kiln.

After the pottery has been colored and fired, a simple design may be made on the pottery by scraping off the surface color so as to expose the original or creamy-white tint of the clay.

Unglazed pottery may be worked with after firing by rubbing floor wax on the outer surface. This fills up the pores and gives a more uniform quality to the whole piece.

Pottery offers so many opportunities for personal experimentation and enjoyment; there are no set rules as to how to make a piece of pottery. Keep a journal about the different things you try while making pottery, so you can remember what works best and what should be avoided in the future. Note the kind of clay you used and its consistency, the types of colors that have worked well, and the temperature and positioning within the kiln, if you use firing. Above all, enjoy making unique pieces of pottery!

Making Jars, Candlesticks, Bowls, and Vases

Making pottery at home is simple and easy, and is a great way for you to make personalized, unique gifts for family and friends. Clay can be purchased at local arts and crafts stores. Clay must always be kneaded before you model with it, because it contains air that, if left in the clay, would form air bubbles in your pottery and spoil it. Work out this air by kneading it the same way that you knead bread. Also guard against making the clay too moist, because that causes the pottery to sag, and sagging, of course, spoils the shape.

In order to make your own pottery, you need modeling clay, a board on which you can work, a pie tin on which to build, a knife, a short stick (one side should be pointed), and a ruler.

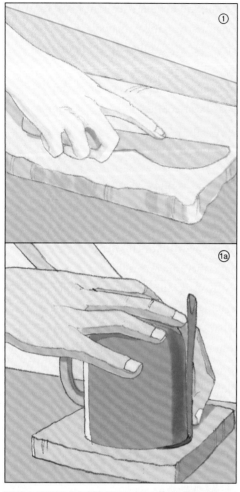

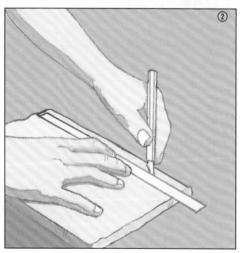

Jars

1. To start a jar, put a handful of clay on the board, pat it out with your hand until it is an inch thick, and smooth off the surface. Then, take

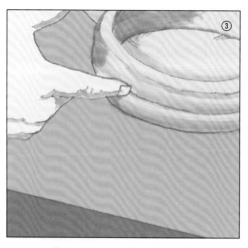

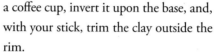

a coffee cup, invert it upon the base, and, with your stick, trim the clay outside the rim.

2. To build up the walls, put a handful of clay on the board and use a knife to smooth it out into a long piece, ¼ inch thick. With the knife and a ruler, trim off one edge of the piece and cut a number of strips ¾ inch wide.

3. Take one strip, stand it on top of the base, and rub its edge into the base on both sides of the strip; then, take another strip and add it to the top of the first one, and continue building in this way, placing one strip on another, joining each to the one beneath it, and smoothing over the joints as you build. Keep doing this until the walls are as high as you want them to be.

4. Remember to keep one hand inside the jar while you build, for extra support.

5. Fill uneven places with bits of clay and smooth out rough spots with your fingers, having moistened your fingers with water first.

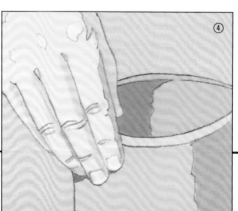

6. When you are finished, add decoration or ornaments to your jar, if desired.

Candlestick

Making a pottery candlestick requires a round base ½ inch thick and 4 inches in diameter. After preparing the base, put a lump of clay in the center, work it into the base, place another lump on top, work it into the piece, and continue in this way until the candlestick has been built as high as you want it. Then, force a candle into the moist clay, twisting it around until it has made a socket deep enough to place a candle into.

A cardboard "templet," with one edge trimmed to the proper shape, will help make it easy to make the walls of the candlestick symmetrical and the projecting cap on the top equal on all sides. Run the edge of the templet around the walls as you work, and it will show you exactly where and how much to fill out, trim, and straighten the clay.

If you want to make a candlestick with a handle, make a base just as stated above. Then cut strips of clay and build up the wall as if building a jar, leaving a center hole just large enough to hold a candle. When the desired height for the wall has been reached, cut a strip of clay ½ inch wide and ½ inch thick, and lay it around the top of the wall with a projection of ¼ inch over the wall. Smooth this piece on top, inside, and outside with your modeling stick and fingers. For the handle, prepare a strip 1 inch wide and ⅜ inch thick, and join one end to the top band and the other end to the base. Use a small lump of clay for filling around where you join the piece, and smooth off the piece on all sides.

When the candlestick is finished, run a round stick the same size as the candle down into the hole, and let it stay put until the clay is dry, to keep the candlestick straight.

Bowls

Bowls are simple to make. Starting with a base, lay strips of clay around the base, building upon each strip as you did when making a jar. Once the bowl is at the desired height and width, allow it to dry.

Vases

Vases can be made as you'd make a jar, only bringing the walls up higher, like a candlestick. Experiment with different shapes and sizes and always remember to keep one hand inside the vase to keep everything even and to prevent your vase from caving in on itself.

Decorating Your Pottery

Pottery may be ornamented by scratching a design upon it with the end of your modeling stick. You can do a simple, straight-line design by using a ruler to guide the stick in drawing the lines. Ornamentation on vases and candlesticks can be done by hand-modeling details and applying them to your item.

Glazing and Firing

Pottery that you buy is generally glazed and then fired in a pottery kiln, but firing is not necessary to make beautiful, sturdy pottery. The clay will dry hard enough, naturally, to keep its shape, and the only thing you must provide for is waterproofing (if the pottery will be holding liquids). To do this, you can take bathtub enamel and apply it to the inside (and outside, if desired) of the pottery to seal off any cracks and to keep the item from leaking.

If you do want to try glazing and firing your own pottery, you will need a kiln. Below are instructions for making your own.

Sawdust Kiln

This small, homemade kiln can be used to bake and fire most small pottery projects. It will only get up to about 1200°F, which is not hot enough to fire porcelain or stoneware. However, it will suffice for clay pinch pots and other decorative pieces.

You will need:

- 20 to 30 red or orange bricks
- Chicken wire
- Sawdust
- Newspaper and kindling
- Sheet metal

1. Choose a spot outdoors that is protected from strong winds. Clear away any dried branches or other flammables from the immediate area. A concrete patio or paved area makes an ideal base, but you can also place bricks or stones on the ground.

2. Stack bricks in a square shape, building each wall up at least four bricks high. Fill the kiln with sawdust.

3. Place the chicken wire on top of the bricks and add another layer or two of bricks. Carefully place your pottery in the center of the mesh, spacing the pieces at least ½ inch apart. Cover the pottery with sawdust.

4. Add another piece of chicken wire, add bricks and pottery, and cover with sawdust. Repeat until your kiln is the desired height.

5. Light the top layer of sawdust on fire, using kindling and newspaper if needed. Cover with the sheet metal, using another layer of bricks to hold it in place.

6. Once the kiln stops smoking, leave it alone until it completely cools down. Then carefully remove the sheet metal lid.

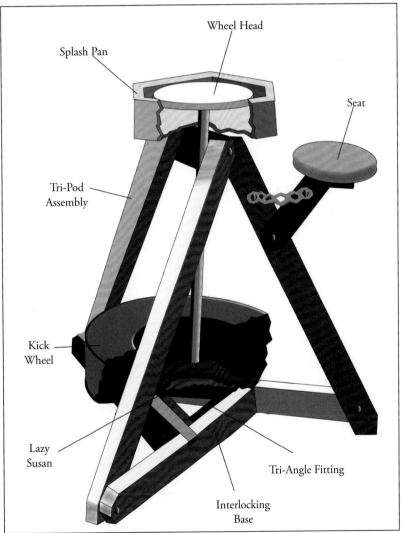

You can make your own pottery wheel using wood, an old lazy susan, and a tire. Place bags of sand inside the tire to give your wheel a sturdier base.

QUILTING

Crazy quilts first became popular in the 1800s and were often hung as decorative pieces or displayed as keepsakes, but they can also be warm and practical. They can be made out of scraps of fabric that are too small for almost any other use, and there is endless room for variation in colors, patterns, and texture.

Quilts are generally made up of many small squares that are sewn together into a large rectangle and layered with batting (a thick layer of fabric to add warmth—usually wool or cotton) and backing (the material that will show on the underside of the quilt).

Non-woven interfacing or lightweight muslin (prewashed)
Fabric pieces in a variety of colors and patterns
Scissors
Ruler
Pencil
Needle or sewing machine
Pins
Thread
Iron

1. Make the foundation squares. If your foundation fabric (interfacing or muslin) is wrinkled, iron it carefully until it is completely flat. Then use a ruler to measure and draw a  13 x 13-inch square in one corner (your final square will be 12 x 12 inches, but it's a good idea to leave yourself a little extra fabric to work with). Repeat until all of the foundation fabric is cut into squares.

2. Cut a small piece of patterned fabric into a shape with three or five straight edges. Pin it right-side up on the center of one foundation square. Cut another small

piece of fabric with straight edges and lay it right-side down on top of the first piece. Sew a ¼-inch seam along the edge where the two fabrics overlap. If the second piece is longer than the first piece, don't sew beyond the edge of the first piece. Turn the second piece of fabric over so that it's facing up and iron it. Trim the second piece to align with the first, so you have one larger shape with straight edges.

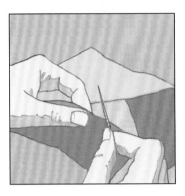

3. Continue with a third piece of fabric, making sure it is large enough to extend the length of the first two patches combined. Sew the seam, flip the fabric upright, iron, trim, and proceed with a fourth piece.

Work clockwise, each piece getting larger as you move toward the edge of the foundation square. Once the square is filled, trim off any overhanging fabric so that you have one neat square. Repeat steps 2 and 3 until all foundation pieces are filled.

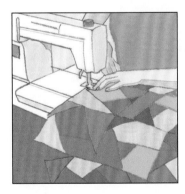

4. Sew all foundation squares together with a ¼- to ½-inch seam.

5. Sandwich the quilt by placing the backing face down, the batting on top of it, and then the foundation on top, with the patterned squares facing up. Baste around the quilt to hold the three layers

together, using long stitches and staying about ¼-inch from the outside edge of the patterned fabric.

6. Binding your quilt covers the rough edges and creates an attractive border around the edges of the quilt. To make the binding, cut strips of fabric 2 ½ inches wide and as long as one side of your quilt plus 2 inches. Fold the fabric in half lengthwise and press.

7. Lay the strip along one edge of the quilt. The raw edges of the quilt and the binding should be stacked together. Leave a ½ inch extra hanging off the first corner. Sew along the length of the quilt, about ¼ inch from the raw edge. Trim the binding, leaving a ½ inch extra. Fold the fabric over the rough edges to the back of the quilt and slip-stitch the binding to the backside. Fold the loose ends of the binding over the edge of the quilt and stitch to the backside. Repeat with all sides of the quilt.

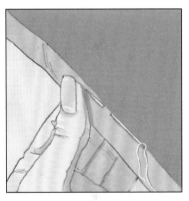

8. Finish your crazy quilt by adding decorative stitching between small pieces of fabric, sewing on buttons, tassels, or ribbons, or using stitching or fabric markers to record important names or dates.

RAG RUGS

The first rag rugs were made by homesteaders over two centuries ago who couldn't afford to waste a scrap of fabric. Torn garments or scraps of leftover material could easily be turned into a sturdy rug to cover dirt floors or stave off the cold of a bare wooden floor in winter. Any material can be used for these rugs, but cotton or wool fabrics are traditional.

Materials

- Rags or strips of fabric
- Darning needle
- Heavy thread

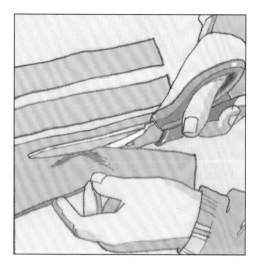

1. Cut long strips of material about 1 inch wide. Sew strips together end to end to make three very long strips (or you can start with shorter strips and sew on more pieces later). To make a clean seam between strips, hold the two pieces together at right angles to form a square corner. Sew diagonally across the square and trim off excess fabric.

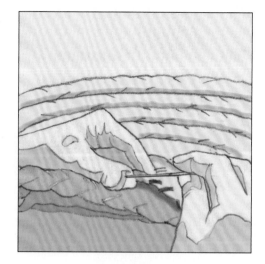

3. Start with one end of the braid and begin coiling it around itself, sewing each coil to the one before it with circular stitches. Keep the coil flat on the floor or on a table to keep it from bunching up.

2. Braid the three strips together tightly, just as you would braid hair.

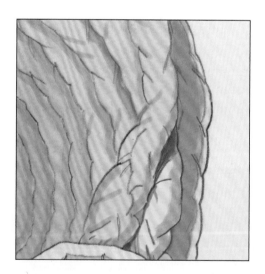

4. When the rug is as large as you want it to be, tack the end of the braid firmly to the edge of the rug.

TIP

Cut strips along the bias to keep them from unraveling.

TIP

Use thinner strips of fabric toward the end of your rug to make it easier to tack to the edge of the rug.

SCRAPBOOKING

Creating a scrapbook is a great way to process and preserve memories. Keep travel brochures from special trips, ticket stubs, event programs, photographs, etc. in a box or file throughout the year. Then, when cold or rainy weather sets in, you can pull out your book and begin arranging your keepsakes in a meaningful way.

Depending on the nature of your keepsakes and your own preferences, you can choose to base your scrapbook on a theme (musical or sporting events, family vacations, your wedding, home renovations, etc.) or to organize it chronologically. Of course, really you can do anything you want, including creating a collage-like book of scattered memories, keepsakes, magazine clippings, or even journal entries. Scrapbooking is a personal project and can be done with as much thoughtful organization or creative chaos as you wish.

There are many ways to create a unified feel to your scrapbook, whether or not the subject matter itself maintains any continuity. Here are a few suggestions:

- **Layout.** Design each page in a similar way, so that (for example) there is always a favorite quote on the top of the page, a photo on the side, text below, and an embellishment across from it. Or create a pattern of layouts, so that every third or fourth page has a similar design.
- **Colors.** Choose two or three colors or a single color palette to use throughout the book. Incorporate these colors in borders, frames for photos, backgrounds, with colored inks, etc.
- **Fonts or Script.** Use the same style of writing for photo captions or other text throughout the book.
- **Backgrounds.** You may want a clean white background if you want to showcase high-quality photographs without any distractions to the eye. Or you can use scrap pieces of wallpaper all in the same style or color, colored card stock, or even thin fabric as the backdrop for your keepsakes.

Once you have a general plan for your scrapbook, you're ready to begin.

1. Create a background. Unless you want a plain white background, select the cardstock or other colored papers

or fabrics, cut them to size, and use rubber cement to glue them to the page. You may choose to use several coordinated pieces of paper to create a patterned background.

2. Matt your photos. Cut squares of paper that are slightly larger than your photos and glue the photos onto them. This will create an attractive border around your photos. Later you can add frames or leave them as they are.

3. Add text or other embellishments. Use neat handwriting or printed text to caption your photos, add favorite quotes, etc. Pieces of ribbon, stickers, or magazine clippings can also be added as accents. Avoid any thick materials that will make your page bumpy and keep the album from closing all the way.

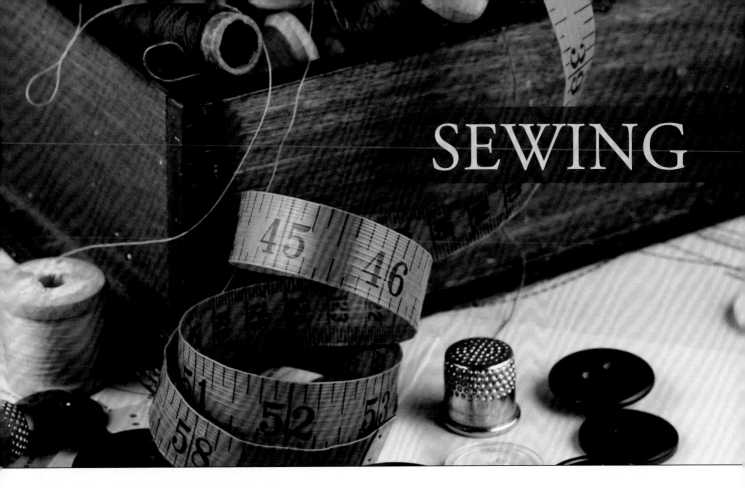

SEWING

Basic Hand Stitching

Supplies:

- Several different sizes of needle
- Straight pins
- Sewing scissors
- Thread in several different colors
- A needle threader (not required, but useful)

Threading a Needle

Estimate the amount of thread a sewing job will need. Cut the end of the thread at a 45 degree angle. Hold up the needle so the eye of the needle is open toward you. Slip the thread into the needle. Knot one end of the thread, leaving one end loose to sew with one thread or knot them together to sew with two threads. Use one strand of thread for places that you do not want the stitching to show, such as hems, and use two strands for extra strength jobs like sewing on buttons. Straighten the threads and begin sewing.

You can also use a needle threader to help you get the end of the thread through the eye of the needle. A needle threader is basically a tool that creates a much larger eye of the needle and makes it easier to thread. Slip the wire loop on the needle threader through the eye of the needle you want threaded. Slip the thread through the metal loop on the threader. Pull the needle up and off the threader loop, taking the thread with it. The needle should now be correctly threaded. Tie a knot at one end of the thread.

Tying a Knot

Place the end of the thread on your pointer finger, holding it in place with your thumb. Loop the thread around the pointer finger and then slide the knot off your finger. Using the end of your pointer finger and thumb, gently slide the knot to the end of the thread.

Pinning a Seam

Correctly pinning a seam is essential in order to have your sewing come out right. To pin a seam take the two pieces of fabric you are going to sew together and lay them on a flat surface. Make sure the front side of the fabric pieces (with the pattern) are on the inside, against one another and the back of the fabric is on the outside, facing you. Brush all of the wrinkles out of the fabric and line up the pieces so their edges match. Look to see where your seam is going to be and pin the fabric near there to keep the two pieces together.

Straight Stitch

This type of stitch is used for very simple hems, sewing two pieces of fabric together, or gathering fabric. Prepare your fabric, if needed, by ironing it flat or by pinning together two pieces. Knot your thread to the fabric, and then simply weave the needle in and out of the fabric in a straight line. Try to keep the stitching as straight as possible. The length of your stitches can be short or long, depending on what you are working on but they should always be a consistent length. If you are gathering your fabric together, you might want to run a second straight stitch next to the first for stability.

Slip Stitch

Insert the needle in a seam allowance or hem edge to anchor the knot on the inside of the garment. Use the tip of the needle to pick up a few threads of the body of the garment, directly under where the thread knot was anchored. Pull the needle through the fabric toward the hem edge. Move the needle over and insert the needle into the hem edge. Repeat the stitch, picking up the threads of the garment fabric in the same direction on each stitch, keeping the stitch spacing as even as possible.

Back Stitch

A back stitch can be used when the stitches will not be seen on the outside of a garment or project. A back stitch is a strong stitch to join two pieces of fabric. Thread a needle a piece of thread that is no longer than a yard long. Longer pieces of thread tend to get knotted. Knot the ends of the threads. Anchor the knot in the inside fabric (usually a seam allowance) near where you need to start sewing. Push the

needle into the fabric where you want to start the seam or joining two pieces of fabric. Bring the needle back through both layers of fabric just in front of the previous stitch for the strongest back stitch. Stitching in this fashion will resemble a machine sewn stitch. The length of the stitch sewn can be adjusted for the look or effect you want. Push the needle back into the fabric in between where the needle came in and out of the fabric to create the first stitch. Bring the needle up through the fabric the same distance you came forward in creating the first stitch. Continue on until you have finished the seam.

Overcast Stitch

This stitch is used to keep a fabric edge from fraying. Use a single knotted thread, and work from right to left. Insert the needle from the underside of your work. Pull the thread through to the knot, and insert the needle from the underside again, one eighth to one quarter inch to the left of the knot. Pull the thread through, but not too tightly or the fabric will curl. The more your fabric frays, the closer the stitches should be. Keep the depth of the stitches uniform, and make them as shallow as possible without pulling the fabric apart.

Sewing on a Button

Thread and knot your needle. Place the needle into the fabric so that the knot will end up on the back of the fabric (out of sight). Make a couple of stitches in the fabric where the button is to be located to anchor the thread. Lay the button on the place you will be attaching it. Bring the needle up through the button. Lay the straight pin, needle or toothpick on top of the button. Take the thread over top of the straight pin, needle or toothpick and bring your needle and thread back down through the button. Repeat to make about six stitches over the straight pin, needle or tooth picks to anchor your button. For a button with four holes, repeat the above steps for the other two holes. Bring the needle and thread to the back of the fabric and knot the thread in the threads that have sewn the button to the garment. Cut the thread. Remove the straight pin, needle or tooth pick and snug the button to the loops of thread by gently tugging the button.

Soap Making

Soap Making

When you make your own soap, you get to choose how you want it to look, feel, and smell. Adding dyes, essential oils, texture (with oatmeal, seeds, etc.), or pouring it into molds will make your soap unique. Making soap requires time, patience, and caution, as you'll be using some caustic and potentially dangerous ingredients—especially lye (sodium hydroxide). Avoid coming into direct contact with the lye; wear goggles, rubber gloves, and long sleeves, and work in a well-ventilated area. Be careful not to breathe in the fumes produced by the lye and water mixture.

Soap is made up of three main ingredients: water, lye, and fats or oils. While lard and tallow were once used exclusively for making soaps, it is perfectly acceptable to use a combination of pure oils for the "fat" needed to make soap. Saponification is the process in which the mixture becomes completely blended and the chemical reactions between the lye and the oils, over time, turn the mixture into a hardened bar of usable soap.

Cold-Pressed Soap

Ingredients

6.9 ounces lye (sodium hydroxide)
2 cups distilled water, cold (from the refrigerator is the best)
2 cups canola oil
2 cups coconut oil
2 cups palm oil

Supplies

Goggles, gloves, and mask (optional) to wear while making the soap
Mold for the soap (a cake or bread loaf pan will work just fine; you can also find flexible plastic molds at your local arts and crafts store)
Plastic wrap or wax paper to line the molds
Glass bowl to mix the lye and water
Wooden spoon for mixing

2 thermometers (one for the lye and water mixture and one for the oil mixture)

Stainless steel or cast iron pot for heating oils and mixing in lye mixture

Handheld stick blender (optional)

Directions

1. Put on the goggles and gloves and make sure you are working in a well-ventilated room.

2. Ready your mold(s) by lining with plastic wrap or wax paper. Set them aside.

3. Slowly add the lye to the cold, distilled water in a glass bowl (*never* add the water to the lye) and stir continually for at least a minute, or until the lye is completely dissolved. Place one thermometer into the glass bowl and allow the mixture to cool to around 110°F (the chemical reaction of the lye mixing with the water will cause it to heat up quickly at first).

4. While the lye is cooling, combine the oils in a pot on medium heat and stir well until they are melted together. Place a thermometer into the pot and allow the mixture to cool to 110°F.

5. Carefully pour the lye mixture into the oil mixture in a small, consistent stream, stirring continuously to make sure the lye and oils mix properly. Continue stirring, either by hand (which can take a very long time) or with a handheld stick blender, until the mixture traces (has the consistency of thin pudding). This may take anywhere from 30 to 60 minutes or more, so be patient. It is well worth the time invested to make sure your mixture traces. If it doesn't trace all the way, it will not saponify correctly and your soap will be ruined.

6. Once your mixture has traced, pour carefully into the mold(s) and let sit for a few hours. Then, when the mixture is still soft but congealed enough not to melt back into itself, cut the soap with a table knife into bars. Let sit for a few days, then take the bars out of the mold(s) and place on brown paper (grocery bags are perfect) in a dark area. Allow the bars to cure for another 4 weeks or so before using.

If you want your soap to be colored, add special soap-coloring dyes (you can find these at the local arts and crafts store) after the mixture has traced, stir them in. Or try making your own dyes using herbs, flowers, or spices.

If you are looking to have a yummy-smelling bar of soap, add a few drops of your favorite essential oils (such as lavender, lemon, or rose) after the tracing of the mixture and stir in. You can also add aloe and vitamin E at this point to make your soap softer and more moisturizing.

To add texture and exfoliating properties to your soap, you can stir some oats into the traced mixture, along with some almond essential oil or a dab of honey. This will not only give your soap a nice, pumice-like quality but it will also smell wonderful. Try adding bits of lavender, rose petals, or citrus peel to your soap for variety.

To make soap in different shapes, pour your mixture into molds instead of making them into bars. If you are looking to have round soaps, you can take a few bars of soap you've just made, place them into a resealable plastic bag, and warm them by putting the bag into hot water (120°F) for 30 minutes. Then, cut the bars up and roll them into balls. These soaps should set in about 1 hour or so.

Soap Oils

Almost any oil can be used to make soap, but different oils have different qualities; some oils create a creamier lather, some create a bubbly lather. Oils that are high in iodine will produce a softer soap, so be sure to mix with oils that are lower in iodine. Online soap calculators are very helpful when creating your own recipes.

Oil	Qualities
Almond Butter	Conditioning, Creamy Lather. Moderate Iodine.
Almond Oil, sweet	Conditioning, Fragrant. High Iiodine.
Apricot Kernel Oil	Conditioning, Fragrant. High Iodine.
Avocado Oil	Conditioning, Creamy Lather. High Iodine.
Babassu Oil	Cleansing, Bubbly. Very Low Iodine.
Canola Oil	Conditioning. Inexpensive. High Iodine.
Cocoa Butter	Creamy Lather. Low Iodine.
Coconut Oil	Bubbly Lather, Cleansing. Low Iodine.
Emu Oil	Conditioning. Creamy Lather. Moderate Iodine,
Evening Primrose Oil	Conditioning. Very High Iodine.
Flax Oil, Linseed	Conditioning. Very High Iodine.
Ghee	Cleansing, Bubbly Lather. Very Low Iodine.
Grapseed Oil	Conditioning. Very High Iodine.
Hemp Oil	Conditioning. Very High Iodine.
Lanolin Liquid Wax	Low Iodine.
Neem Tree Oil	Conditioning, Creamy Lather. High Iodine.
Olive Oil	Conditioning, Creamy Lather. High Iodine.
Palm Oil	Conditioning, Creamy Lather. Moderate Iodine.
Rapeseed Oil	High Iodine.
Safflower Oil	Conditioning. Very High Iodine.

Oil	Qualities
Sesame Oil	Conditioning. High Iodine.
Shea Butter	Conditioning, Creamy Lather. Moderate Iodine.
Ucuuba Butter	Conditioning. Creamy Lather. Low Iodine.

Safflower oil

The Junior Homesteader

How Soap Works

Teach kids how soap cleans with this simple experiment.

Half-fill two Mason jars with water and add a few drops of food coloring. Pour several tablespoons of oil into each jar (corn oil, olive oil, or whatever you have on hand will be fine). You will see that the oil and water form separate layers. This is because the molecules in oil are hydrophobic, meaning that they repel water.

Add a few drops of liquid soap to one of the jars. Close both jars securely and shake for about thirty seconds. The oil and water should be thoroughly mixed.

Let both jars rest undisturbed. The jar with the soap in it will stay mixed, whereas the jar without the soap will separate back into two distinct layers. Why? Soap is made up of long molecules, each with a hydrophobic end and a hydrophilic (water-loving) end. The water bonds with the hydrophilic end and the oil bonds with the hydrophobic end. The soap serves as a glue that sticks the oil and water together. When you rinse off the soap, it sticks to the water, and the oil sticks to the soap, pulling all the oil down the drain.

Natural Dyes for Soap or Candles

Light/Dark Brown	Cinnamon, ground cloves, allspice, nutmeg, coffee
Yellow	Turmeric, saffron, calendula petals
Green	Liquid chlorophyll, alfalfa, cucumber, sage, nettles
Red	Annatto extract, beets, grapeskin extract
Blue	Red cabbage
Purple	Alkanet root

SPINNING WOOL

Equipment Needed:

A spinning wheel or a drop spindle
A drum carder, hand carder, or metal comb for carding
A large tub

A Note on Wool Type

Each different breed of sheep has its own, unique type of wool. You will have to think about what type of wool yarn you want to make and what it will be used for before you get your fleece. Wool comes in many different colors, textures, curl types, and lengths. You can get fine grade, medium, crossbreed, or long wool.

Washing the Fleece

For a beginner it is easier to work with wool that has been washed. Experienced spinners often work with un-washed wool, because the natural grease helps then to spin finer yarn.

Start by inspecting your fleece. Remove any tags, grass, and any larger knots of fleece.

Fill a large tub with hot, steaming water. This first wash is like a pre-soak. Put the fleece in the hot water. Make sure not to agitate, rub, or squeeze your fleece as this will cause it to knot up or "felt." Always treat fleece gently when picking it up or washing it. Let the fleece soak in the water. Dirt will become loosened and should fall to the bottom. If the fleece is very dirty pre-soak it twice.

Dump the dirty water from the tub and fill it with clean, hot water. Add dish soap (any kind you prefer) to the water and agitate it until the water is sudsy. Add the fleece, making sure to dunk it in the water gently so it is all wet, and let it sit for ten minutes. Rinse the fleece, letting water run through it until it comes out clear and without soap suds. If the wash water is very dirty and dark looking you may want to wash the fleece again. Let the fleece air-dry.

Carding the Fleece

The next step is to "card" your fleece (also see page 61). Carding is aligning the fibers of the fleece in a parallel direction, so that the little barbs on the wool fibers can catch easily. Hand cards are combs that are typically square or rectangular paddles. The working face of each paddle can be flat or cylindrically curved. If you are combing the wool, simply lay it out on a flat surface one lock at a time and brush down the locks. If you are drum carding, feed it in a small lock at a time while turning the carder. If you are going to be making a lot of wool yarn a drum carder is a good investment.

You can create several different types of wool bundles as you are carding. A rolag is a roll of fibers created by first carding the fleece, using hand-held combs, and then by gently rolling the fleece off the cards. A sliver is a long bundle of fibers, created by carding and then drawing the wool into long strips where the fibers are parallel. A roving is a long and narrow bundle of fibers, created by carding the fleece and then drawing it into long strips in the same way used to create slivers, except when making a roving draw on the fibers even more while slightly twisting them. A batt is a rectangular layer of carded wool taken off from a drum carder and is usually six to eight inches wide and about two feet long. What type of wool bundle you create is entirely up to you; whichever one would match your equipment, preferences, and the particular project you will use the wool for.

Spinning the Wool

Woolen yarn is made from carded wool. It is soft, light, stretchy, and full of air. Worsted yarn is more complicated to produce, and is created when the fleece is combed and the short fibers are removed. The combed fibers lie parallel to each other and produce a harder, strong yarn. Most spinners make a blend of a woolen and worsted yarn, using techniques from both categories.

To begin spinning your carded wool first set up your drop spindle or spinning wheel with a "lead;" an already spun piece of yarn that you can attach your spinning to. Always spin the wheel or the drop spindle clockwise. Place your non-

dominant hand closest to the wheel or spindle on the fleece, and your dominant further back and closer to you. Your dominant hand will "draw out" the wool, while your non-dominant holds the twist from escaping up the un-spun fiber. Overlap the un-spun and spun fiber about four to five inches, and hold it securely with your non-dominant hand. Turn the wheel clockwise, and your wool will naturally twist. Draw the wool out, and slowly move your hands toward your body (or upward in the case of a drop spindle.) Add more wool as you need it. If working on the drop spindle, after creating a stretch of yarn that is so long that it no longer feels comfortable, wind it onto the spindle. In the case of a wheel, allow the wheel to slowly take it up.

Once you have filled your spindle with new wool yarn, unwind the thread and make a 'skein' by wrapping it around your hand and your elbow- like you would a rope or cord. Tie the skein at intervals with a different yarn (not wool).

You have to set the twist on your yarn, or it will unravel. Thoroughly soak the skein in hot water, being careful not to twist or agitate it. Hang the skein through the top of a hanger to dry. Setting the twist is accomplished by hanging something heavy from the bottom of the skein while it is drying. You want the weight to pull down on the skein so that it is taut but you don't want to break your threads. Once the skein has dried the twist you made should be set and you will have handmade wool yarn!

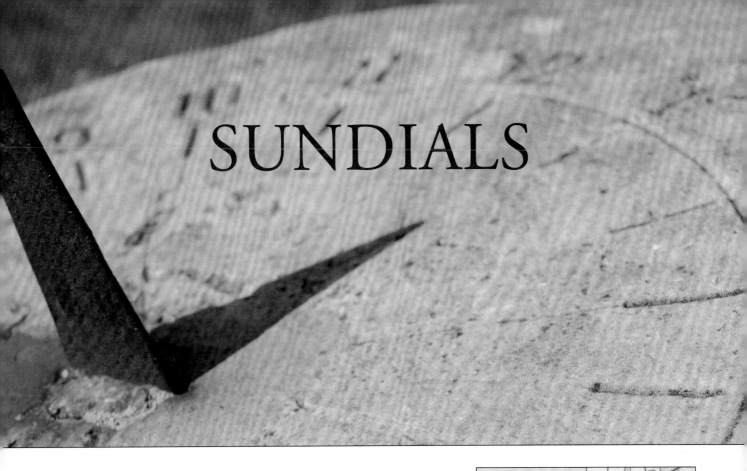

SUNDIALS

Sundials have been used for well over 5,000 years as a means of telling time based on the sun's position. The vertical axis, or gnomon (in this case a chopstick), casts a shadow over the horizontal axis (the wooden disk). The shadow moves as the sun travels across the sky. You can tell what time it is by seeing where the shadow falls.

Materials

One wooden disk, about 12 inches in diameter

One chopstick or wooden dowel

Drill

Air-dry clay

Paint (optional)

Permanent marker

Spray acrylic sealer

1. Drill a hole in the exact center of the wooden disk. The hole should be just large enough for the wider end of the chopstick to fit.

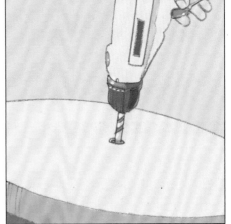

2. Press the clay into the hole and stick the wide end of the chopstick firmly into the clay. Be sure the chopstick is completely vertical—not leaning one way or the other. Allow the clay to harden. If desired, paint the disk and gnomon (chopstick).

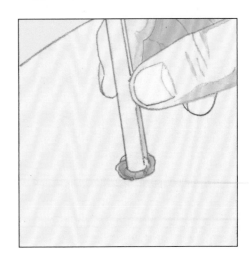

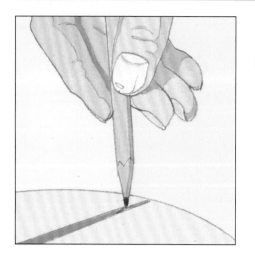

3. Place the sundial in a sunny spot outdoors early in the day. Every hour, on the hour, make a mark where the gnomon's shadow falls and place a number near the mark to indicate the time. You may want to do this in pencil first and then outline it with permanent marker.

4. When all twelve hours are marked, spray the entire sundial with a clear acrylic sealer. Allow to dry and then apply a second coat.

TIP

If you move your sundial from its original location, the time may be off. You'll need to "reset" it by checking a clock and angling the sundial to the correct position to match the real time.

TIP

Adults can make their own non-toxic wood sealer using five parts mineral oil to one part beeswax. Children should not attempt this on their own. Heat the mineral oil in a saucepan over low heat. Add the beeswax, being careful not to splash the oil, and allow it to melt. *Caution: Keep the heat low and stir slowly. Beeswax is very flammable.*

Stir very slowly, being careful not to splash any of the mixture. If it splashes out into the flame, it will ignite. Once all the beeswax is melted, use a funnel to pour the mixture into a mason jar. Once the mixture is cool, use a rag to apply a thin layer to any wood you want to protect from the elements.

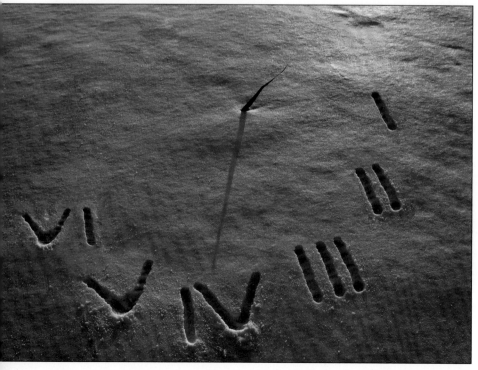

A very temporary sundial can be made in sand or snow to teach children the concept.

TANNING LEATHER

After you have butchered any animal, you have a skin left over. Don't just throw it away! There are simple general tanning methods that will work with any hides you have, whether rabbit, deer, or cow.

Preparing to Tan

The first step, known as "fleshing the hide," is to remove any extra fat or flesh that has stuck to the hide as it was being pulled off the animal. This can be done by first soaking the hide overnight in a salt and water solution to loosen the stuck bits. Rinse the hide in clean water and then apply two layers of salt, waiting for the first to absorb before applying the second. Make sure not to get any salt on the fur side. Fold the hide in half, flesh sides together and then roll the hide up. Place it somewhere to drain and dry overnight.

Put the hide on a smooth surface, flesh side up. A tanners log would be best for this, but if you do not have one you can easily make one by splitting a log in half, smoothing the bark surface, and then placing the log, cut side down, in your workshop. Stretch the hide over the rounded surface. Scrape the flesh side of the hide of any fat and flesh stuck to it using the blunt side of a knife. When all the bits have been removed, wash the hide in soapy water and rinse thoroughly.

If you want to remove the fur from the hide it must be done at this stage, before the tanning begins. Soak the fur in a de-hairing solution, a mixture of one quart hydrated lime to every 5 gallons of water, for five to ten days in a large wooden container. Note that lime is caustic and will burn your skin, so wear gloves and goggles whenever you are working with it! Stir the hide around in the mixture once or twice a day. After five to ten days the hair should be ready to be removed. Rinse the hide, making sure that all the lime is gone; a de-liming agent may be used to make it easier. Put the hide fur-side up on your tanning log. Scrape it in the same way as you did for removing the flesh. At this point, you can simply allow the skin to dry and you'll have rawhide, which is great for moccasins, lacing for snowshoes, cords, and many other uses.

Tanning

Before tanning, the skins must be unhaired, degreased, desalted, and soaked in water over a period of 6 hours to 2 days. Traditionally, tanning was accomplished by using tannic acid—a chemical derived from the bark of many different trees—to process hides. Today chemicals can be used to achieve the same thing.

Tanning begins with the hide being treated with a mixture of common salt and sulfuric acid, in case a chemical tanning is to be done. This is done to bring down the pH of collagen to a very low level so as to facilitate the absorption of mineral tanning agent. This process is known as "pickling." The salt penetrates the hide twice as fast as the acid and checks the ill effect of sudden drop of pH.

For natural tanning, hides are stretched onto frames and immersed for several weeks in vats of increasing concentrations of tannin. This method produces very flexible leather.

For chemical tanning, use basic chromium sulfate. Once the desired level of penetration of chromium into the hide is achieved, the pH of the material is raised again to facilitate the process. This is known as "basification". In the raw state chromium tanned skins are blue. Chemical tanning is faster than vegetable tanning, taking less than one day for this part of the process, and produces a stretchable leather.

Alternately, in large plastic or wooden basin you can mix 5 pounds of salt with 10 gallons of warm water (soft water, such as rain water, is best) and 2 pounds of alum that has been mixed with just enough hot water to dissolve it. Though this solution does not have any toxic vapors, you should still wear gloves to protect your skin. Place the hide in the solution and stir with a big paddle or stick twice a day for 2 days to a week, depending on the type and size of the skin.

Finishing

After the hide has been tanned, remove it from the solution it was in. Rinse

it thoroughly. Hang the hide in a cool ventilated place out of direct sunlight. If the fur is still present be sure to place the hide fur side up. The drying out process should take several days.

When the hide is very slightly damp, fold it in half, flesh side down, roll it up, and let it sit overnight. After this, unfold the hide and work it—pulling, stretching, and twisting it until it is pliable. Rub in finishing oil of choice (slightly warmed) with your hands. If there are rough patches in the fur they can be smoothed down with sandpaper. If the fur is still present, brush and comb it until it is clean and fluffed.

TERRARIUMS

Terrariums are miniature ecosystems that you can create and keep indoors. They're a wonderful way to learn about gardening on a small scale and can add interest to your home décor and oxygen to the air you breathe. Terrariums can contain only plants or can be homes for lizards, turtles, or other small animals. If you do wish to make your terrarium a home for a pet, be sure to include the proper shelter for the animal and a way to provide food and water. Some terrariums even include waterfalls so that animals can have a constant supply of fresh water! The size of the terrarium can be as large as a fish tank or as small as a thimble. Bowls, teapots, jars, and bottles have all been successfully transformed into miniature indoor gardens. Terrariums can be fully enclosed or can have an open top to allow fresh air to circulate. Because one of the benefits of a terrarium is the oxygen that the plants contribute to the air, these directions are for an open top terrarium.

Materials

Container (preferably a clear glass container, so you can easily see your miniature garden)

Coarse sand or pebbles

Sphagnum moss

Soil

Seeds or seedlings

Water

Ornaments (optional)

1. Place a ½-inch to 1-inch layer of coarse sand or pebbles in the bottom of your container. This will help the soil to drain properly.

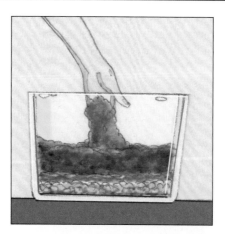

2. Add a layer of moss over the pebbles. The moss acts as a filter, allowing the soil to drain but not to seep down into the pebbles.

3. Pour the soil over the moss and spread evenly. How much soil you use will depend on how big your container is and how large the plants will grow. Pat the dirt down firmly.

4. Plant the seeds or seedlings. Think carefully about how you want the plants to be arranged. You may want taller plants in the center and shorter ones toward the outside so that you can see them all. Add pretty stones, pinecones, figurines, or other ornaments, if desired. Place your terrarium in a sunny spot and water it regularly.

WEDDING CRAFTS

Homemade Weddings

Weddings are wonderful, memorable, and very personal events. Every couple is unique and weddings are an opportunity to celebrate their distinct experiences of love, beauty, and joy. Couples who do some or all of the wedding preparations on their own or with the help of friends and family (rather than relying entirely on professionals) often find their day especially meaningful—and they're less likely to start off their married life in debt.

There are endless ways to add homemade touches to a wedding. The invitations, flowers, centerpieces, favors, food, and even the attire can be made or arranged without the help of professional florists, caterers, and other service providers. Couples should think about what aspects of the wedding matter most to them, what they would most enjoy doing on their own, and how much time they can realistically dedicate to wedding preparations. Delegation is also key—friends and family are often more than happy to be a part of the celebration by assisting with certain tasks.

Bouquets

Once you've chosen the flowers you want in your bouquets (often based on color and season), think about what shape the bouquets will be and how you will arrange them. Smaller bouquets can be secured with a wide ribbon tied around the stems a little below the flower heads. For larger bouquets or for flowers with rough or thorny stems, you may want to bind the stems together with a ribbon that covers the length of the stems.

Materials

Flowers
Floral tape
Ribbon
Corsage pins

1. Arrange the flowers in a bunch, cutting the stems evenly at the bottom. Wrap a piece of floral tape around the stems just below the blossoms. Cut a piece of ribbon about five feet long (or longer for very long stems).

2. Begin wrapping the ribbon around the stems just below the blossoms, leaving a tail of ribbon about two feet long.

3. Wrap the ribbon around and around the stems, working your way down and making sure the ribbon lies flat. When you reach the bottom, wrap over the bottom of the stems and work your way back up to the top.

4. Now you should have two ribbon tails that you can tie into a full bow. Use pearl-headed corsage pins to secure the ribbon in place. Trim the tails if desired.

Hanging Flower Pomander

Flower pomanders should be made as late as possible before the celebration, since the flowers will not be getting adequate water to stay fresh for as long as a bouquet in a vase would. If making the pomanders the day before the wedding, keep them refrigerated as long as possible. Choose

flowers with stiff stems such as roses or carnations, as they'll be easier to insert into the ball.

Ball of floristry foam
2 ½-inch-wide ribbon
Straight pins or a glue gun
Scissors
Flowers

1. Soak floral foam in water until saturated and then set aside on a towel and allow to dry until just damp.

2. Wrap the ribbon once around the outside of the foam ball, pinning it at every inch or securing it to the ball with hot glue. Leave a foot of ribbon loose on each end.

3. Trim your flowers, leaving about an inch of stem for smaller blossoms and two to three inches for larger blossoms. Push the stems all the way into the ball, following the edge of the ribbon. Then do a line of flowers around the circumference so that the ball is divided into corners. Finally, fill in the four sectors, adding ferns or other greenery as desired to close up any spaces. If desired, only cover the top two thirds of the ball with flowers and line the bottom third with large green leaves, secured with pins.

4. Tie the two ends of the ribbon together in a tight knot and hang the pomander in a door or window or on a chandelier, or carry one down the aisle instead of a bouquet.

Wedding Cake

Your cake can be any flavor you like, and any shape, for that matter. But to make a traditional tiered wedding cake, follow the steps below. Be sure to do at least one test run well before the wedding.

Materials

    ½-inch-thick plywood or large cutting board

    Parchment paper

    Tape

    Frosting

    Cake tiers

    Corrugated cardboard

    Tinfoil

    Knife or toothpick

    ¼-inch-thick dowels

    Heavy duty shears

    Cake decorations

1. Prepare the cake board. Cover a ½-inch-thick piece of plywood or a large cutting board with parchment paper, taping it to the underside of the board. The board should be at least 4 inches larger than the largest tier of your cake. This board will stay with your cake until it's eaten, so do a neat job of covering it.

2. Frost and fill the bottom, largest cake tier and place it in the center of the cake board.

3. Cut out two circles of corrugated cardboard that are the same size as the second tier, and two circles the same size as the third tier. Tape the same-sized circles together so that they are double thickness. Cover each cardboard circle with tinfoil, making the foil as smooth as possible.

4. Gently place the medium-sized cardboard circle onto the center of the bottom tier and mark around the circle with a knife or toothpick. Now you can remove the cardboard and still see the circle.

5. Push a ¼-inch-thick dowel into the very center of the cake. Mark the height of the cake on the dowel and remove it. Use heavy-duty shears to cut eight dowels at that length. Insert the dowels into the cake at even intervals around the circle marked on the bottom tier.

6. Place the second cake tier on the medium-sized cardboard circle and fill and frost it. Place it carefully on the center of the bottom tier.

Refer to page 480 for directions for making sachets. Sachets filled with candy, a few homemade cookies, or handmade soap make lovely favors.

7. Repeat steps 5 and 6 with the top layer. To help keep the layers from sliding, sharpen one long dowel with a knife and insert it through the center of the top layer straight down through all the layers. Trim the dowel so that it's even with the top of the cake.

8. Decorate the cake with piping around the bottom of each tier, marzipan fruit, or fresh fruit or flowers.

Edible Flowers for Cake Decorating

Any nontoxic flowers can be used to decorate your cake, as long as you remove the flowers before serving. If you don't want to remove the flowers, choose edible flowers such as the following:

- Apple blossoms
- Orange blossoms
- Cornflowers
- Pansies
- Violets
- Nasturtiums
- Oregano
- Lavender
- Carnations
- Clover
- Calendula
- Jasmine
- Hollyhock
- Impatiens
- Lilacs
- Peonies
- Roses

Programs for the Ceremony

A handmade program will give your guests a unique keepsake by which to remember your ceremony. To make your own paper, refer to page 471.

Paper in various colors
Pen
Paste
Decorations

Cut a piece of sturdy, attractive paper to 7 x 11 inches, or desired size.

Write or print the text on paper of a contrasting color or texture, making sure it fits on a rectangle that is slightly smaller than the original cut rectangle (about 6 x 10 inch).

Cut out the second rectangle and paste it to the first, centering it carefully. Allow to dry and then fold the program in half.

If desired, paste ribbon, pressed flowers, or other accents to the outside of the program.

Lanterns hung from tree branches with wide ribbons create a simple and festive atmosphere.

DISASTER PREPAREDNESS

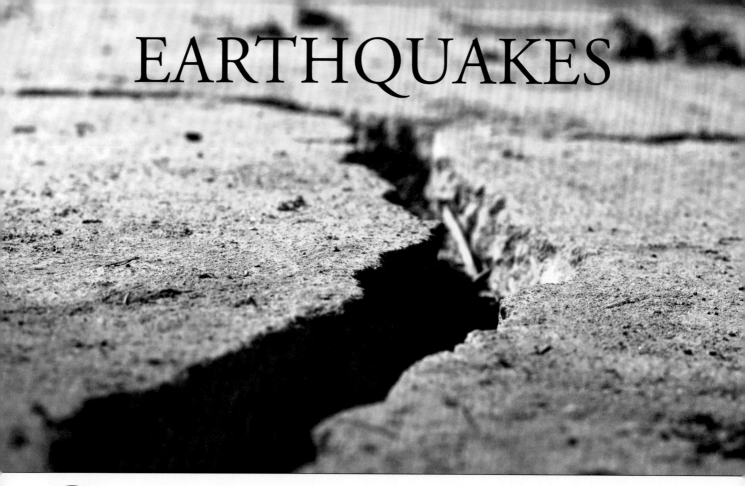

EARTHQUAKES

One of the most frightening and destructive phenomena of nature is a severe earthquake and its terrible aftereffects. Earthquakes strike suddenly, violently, and without warning at any time of the day or night. If an earthquake occurs in a populated area, it may cause many deaths and injuries and extensive property damage.

Although there are no guarantees of safety during an earthquake, identifying potential hazards ahead of time and planning appropriately can save lives and significantly reduce injuries and property damage.

What to Do Before an Earthquake

- Check for hazards in the home. Fasten shelves securely to walls, place large or heavy objects on lower shelves, and store breakable items such as bottled foods, glass, and china in low, closed cabinets with latches. Hang heavy items such as pictures and mirrors away from beds, couches, and anywhere people sit. Brace overhead light fixtures. Repair defective electrical wiring and leaky gas connections and secure a water heater by strapping it to the wall studs and bolting it to the floor. Repair any deep cracks in ceilings or foundations, get-

ting expert advice if there are signs of structural defects. Store weed killers, pesticides, and flammable products securely in closed cabinets with latches and on bottom shelves.

- Identify safe places indoors and outdoors. Safe places include under sturdy furniture such as a heavy desk or table; against an inside wall; away from where glass could shatter around windows, mirrors, pictures, or where heavy bookcases or other heavy furniture could fall over; or in the open, away from buildings, trees, telephone and electrical lines, overpasses, or elevated expressways.

- Educate yourself and family members. Contact your local emergency management office or American Red Cross chapter for more information on earthquakes. Teach children how and when to call 9-1-1, police, or fire department and which radio station to tune to for emergency information. Teach all family members how and when to turn off gas, electricity, and water.

- Have disaster supplies on hand, including a flashlight and extra batteries; portable battery-operated radio and extra batteries; first aid kit and manual; emergency food and water; non-electric can opener; essential medicines; and sturdy shoes.

- Develop an emergency communication plan. In case family members are separated from one another during an earthquake, develop a plan for reuniting after the disaster. Ask an out-of-state relative or friend to serve as the "family contact." After a disaster, it's often easier to call long distance. Make sure everyone in the family knows the name, address, and phone number of the contact person.

What to Do During an Earthquake

Be aware that some earthquakes are actually foreshocks and a larger earthquake might occur. Minimize your movements to a few steps to a nearby safe place and stay indoors until the shaking has stopped and you are sure exiting is safe.

If indoors:

- Drop to the ground; take cover by getting under a sturdy table or other piece of furniture; and hold on until the shaking stops. If there isn't a table or desk near you, cover your face and head with your arms and crouch in an inside corner of the building.

- Stay away from glass, windows, outside doors and walls, and anything that could fall, such as lighting fixtures or furniture.

- Stay in bed if you are there when the earthquake strikes. Hold on and protect your head with a pillow, unless you are under a heavy light fixture that could fall. In that case, move to the nearest safe place.

- Use a doorway for shelter only if it is in close proximity to you and if you know it is a strongly supported, load-bearing doorway.

- Stay inside until shaking stops and it is safe to go outside. Research has shown that most injuries occur when people inside buildings attempt to move to a different location inside the building or try to leave.

- Be aware that the electricity may go out or the sprinkler systems or fire alarms may turn on.

If outdoors:

- Stay outside.
- Move away from buildings, streetlights, and utility wires.
- Once in the open, stay there until the shaking stops. The greatest danger exists directly outside buildings, at exits, and alongside exterior walls. Many of the 120 fatalities from the 1933 Long Beach earthquake occurred when people ran outside of buildings only to be killed by falling debris from collapsing walls. Ground movement

during an earthquake is seldom the direct cause of death or injury. Most earthquake-related casualties result from collapsing walls, flying glass, and falling objects.

If in a moving vehicle:

- Stop as quickly as safety permits and stay in the vehicle. Avoid stopping near or under buildings, trees, overpasses, and utility wires.
- Proceed cautiously once the earthquake has stopped. Avoid roads, bridges, or ramps that might have been damaged by the earthquake.

If trapped under debris:

- Do not light a match.
- Do not move about or kick up dust.
- Cover your mouth with a handkerchief or clothing.
- Tap on a pipe or wall so rescuers can locate you. Use a whistle if one is available. Shout only as a last resort. Shouting can cause you to inhale dangerous amounts of dust.

What to Do After an Earthquake

- Expect aftershocks. These secondary shockwaves are usually less violent than the main quake but can be strong enough to do additional damage to weakened structures and can occur in the first hours, days, weeks, or even months after the quake.
- Listen to a battery-operated radio or television. Listen for the latest emergency information.
- Open cabinets cautiously. Beware of objects that can fall off shelves.
- Stay away from damaged areas unless your assistance has been specifically requested by police, fire, or relief organizations. Return home only when authorities say it is safe.
- Be aware of possible tsunamis if you live in coastal areas. These are also known as seismic sea waves (mistakenly called "tidal waves"). When local authorities issue a tsunami warning, assume that a series of dangerous waves is on the way. Stay away from the beach.
- Help injured or trapped persons. Remember to help your neighbors who may require special assistance, such as infants, the elderly, and people with disabilities. Give

first aid where appropriate. Do not move seriously injured persons unless they are in immediate danger of further injury. Call for help.

- Clean up spilled medicines, bleaches, gasoline, or other flammable liquids immediately. Leave the area if you smell gas or fumes from other chemicals.
- Inspect the entire length of chimneys for damage. Unnoticed damage could lead to a fire.
- Inspect utilities. Check for gas leaks. If you smell gas or hear a blowing or hissing noise, open a window and quickly leave the building. Turn off the gas at the outside main valve if you can and call the gas company from a neighbor's home. If you turn off the gas for any reason, it must be turned back on by a professional. Second, look for electrical system damage. If you see sparks or broken or frayed wires, or if you smell hot insulation, turn off the electricity at the main fuse box or circuit breaker. If you have to step in water to get to the fuse box or circuit breaker, call an electrician first for advice. Finally, check for sewage and water line damage. If you suspect sewage lines are damaged, avoid using the toilets and call a plumber. If water pipes are damaged, contact the water company and avoid using water from the tap.

If water pipes are damaged, you may be able to obtain safe water by melting ice cubes.

FIRST AID

It's impossible to predict when an accident will occur, but the more you educate yourself ahead of time, the better you'll be able to help should the need arise. The first step in an emergency situation should always be to call for help, but there are many things you can do to help the victim while you're waiting for assistance to arrive. The most important procedures are described below.

Drowning

1. As soon as the patient is in a safe place, loosen the clothing, if any.
2. Empty the lungs of water by laying the body breast down and lifting it by the middle, with the head hanging down. Hold for a few seconds until the water drains out.
3. Turn the patient on his breast, face downward.
4. Give artificial respiration: press the lower ribs down and forward toward the head, then release. Repeat about twelve times to the minute.
5. Apply warmth and friction to extremities, rubbing toward the heart.

6. Don't give up! Persons have been saved after hours of steady effort, and after being under water for more than twenty minutes.

Keep a buoy nearby whenever spending time in or near the water.

7. When natural breathing is reestablished, put the patient into a warm bed, with hot-water bottles, warm drinks, fresh air, and quiet.

Sunstroke

1. Move the patient to a cool place, or set up a structure around the patient to produce shade.
2. Loosen or remove any clothing around the neck and upper body.
3. Apply cold water or ice to the head and body, or wrap the patient in cold damp cloths.
4. Encourage the patient to drink lots of water.

Burns and Scalds

1. Cover the burn with a thin paste of baking soda, starch, flour, petroleum jelly, olive oil, linseed oil, castor oil, cream, or cold cream.
2. Cover the burn first with the paste, then with a soft rag soaked in the paste.
3. Shock always accompanies severe burns, and must be treated.

Shock or Nervous Collapse

A person suffering from shock has a pale face, cold skin, feeble breathing, and a rapid, feeble pulse, and will appear listless or half-dead.

1. Place the patient on his back with head low.
2. Give stimulants, such as hot tea or coffee.

Handy Household Hints

Relief from Burns

Raw potato, grated, will give almost instant relief. Another remedy is butter, then baking soda.

3. Cover the patient with blankets.
4. Rub the limbs and place hot-water bottles around the body.

Cuts and Wounds

1. After making sure that no dirt or foreign substance is in the wound, apply a tight bandage to stop the bleeding.
2. Raise the wound above the heart to slow the bleeding.
3. If the blood comes out in spurts, it means an artery has been cut. For this, apply a tourniquet: make a big knot in a handkerchief, tie it around the limb, with the knot just above the wound, and twist it until the flow is stopped.

To bandage a hand, start by wrapping around the knuckles, then go around the thumb and wrap the wrist. Repeat.

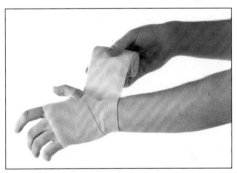

How to Make a Tourniquet

The tourniquet is an appliance used to check severe bleeding. It consists of a bandage twisted more or less lightly around the affected part. The bandage—a cloth, strap, belt, necktie, neckerchief or towel—should be long enough to go around the arm or leg affected. It can then be twisted by inserting the hand, and the blood stopped.

If a stick is used, there is danger of twisting too tightly.

The tourniquet should not be used if bleeding can be stopped without it. When used it should be carefully loosened every 15 to 20 minutes to avoid permanent damage to tissues.

Hemorrhage or Internal Bleeding

Internal bleeding usually comes from the lungs or stomach. If from the lungs, the blood is bright red and frothy, and is coughed up; if from the stomach, it is dark, and is vomited.

1. Help the patient to lie down, with head lower than body.
2. Encourage the patient to swallow small pieces of ice, and apply ice bags, snow, or cold water to the place where the bleeding is coming from.
3. Hot applications may be applied to the hands, arms, feet, and legs, but avoid stimulants, unless the patient is very weak.

Fainting

Fainting is caused by a lack of blood supply to the brain, and is cured by getting the heart to correct the lack.

1. Have the person lie down with the head lower than the body.
2. Loosen the clothing. Give fresh air. Rub the limbs. Use smelling salts.
3. Do not let the person get up until fully recovered.

Snake Bite

1. Put a tight cord or bandage around the limb between the wound and the heart. This should be loose enough to slip a finger under it.
2. Keep the wound lower than the heart. Try to keep the patient calm, as the faster the heart beats, the faster the venom will spread.
3. If you cannot get to a doctor quickly, suck the wound many times with your mouth or use a poison suction kit, if available.

Insect Stings

1. Wash with oil, weak ammonia, or very salty water, or paint with iodine.
2. A paste of baking soda and water also soothes stings.

Poison

1. First, get the victim away from the poison. If the poison is in solid form, such as pills, remove it from the victim's mouth using a clean cloth wrapped around your finger. Don't try this with infants because it could force the poison further down their throat.
2. If the poison is corrosive to the skin, remove the clothing from the affected area and flush with water for 30 minutes.
3. If the poison is in contact with the eyes, flush the victim's eyes for a minimum of 15 minutes with clean water.

How to Put Out Burning Clothing

1. If your clothing should catch fire do not run for help, as this will fan the flames.
2. Lie down and roll up as tightly as possible in an overcoat, blanket, rug, or any woolen article—or lie down and roll over slowly, at the same time beating the fire with your hands. Smother the fire with a coat, blanket, or rug. Remember that woolen material is much less flammable than cotton.

Ice Rescue

1. Always have a rope nearby if you're working or playing on ice. This way, if someone falls through, you can tie one end to yourself and one to a tree or other secure anchor onshore before you attempt to rescue the person.

Handy Household Hints

A Good Eye Wash.

A good eye wash, and one that is safe and harmless, dilute one teaspoon of coarse salt, not the pulverized, in one quart of rainwater. Wash the eyes several times a day, and just before retiring.

2. You could also throw one end to the victim if his head is above water.

3. Do not attempt to walk out to victim. Push out to him or crawl out on a long board or rail or tree trunk.

4. The person in the water should never try to crawl up on the broken ice, but should try merely to support himself and wait for help, if it is at hand.

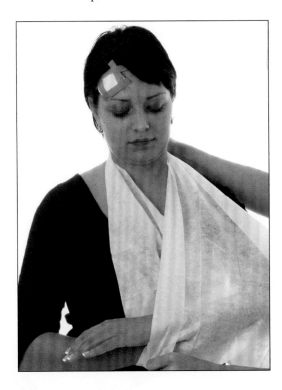

For elbow, arm, or wrist injuries, a simple sling can be made out of any piece of cloth or clothing.

Broken Bone

A simple fracture is one in which the bone is broken but does not break the skin. In a compound fracture the bone is broken and the skin and tissue are punctured or torn. A simple fracture may be converted into a compound fracture by careless handling, as a broken bone usually has sharp, saw-tooth edges, and just a little twist may push it through the skin.

1. Do not move patient without supporting broken member by splints.

2. In a compound fracture, bleeding must be checked—by bandage over compress, if possible or by tourniquet in extreme cases. Then splints may be applied.

3. Where skin is broken, infection is the great danger, so exercise care that compress or dressing is sterile and clean.

Dislocation

A dislocation is an injury where the head of a bone has slipped out of its socket at a joint.

1. Do not attempt to replace the joint. Even thumb and finger dislocations are more serious than usually realized.

2. Cover the joint with cloths wrung out in very hot or very cold water. For shoulder—apply padding and make sling for arm.

3. Seek medical assistance.

For an ankle or foot injury, holding the victim under the knees and behind the back is effective. If neck or spine injury is suspected, do not attempt to move the victim if you can get help to come to the victim instead. If the victim must be moved, the head and neck must first be carefully stabilized.

First Aid Checklist

In order to administer effective first aid, it is important to maintain adequate supplies in each first aid kit. A first aid kit should include:

- Adhesive bandages: These are available in a large range of sizes for minor cuts, abrasions, and puncture wounds
- Butterfly closures: These hold wound edges firmly together.
- Rolled gauze: These allow freedom of movement and are recommended for securing a wound dressing and/ or pads. These are especially good for hard-to-bandage wounds.
- Nonstick sterile pads: These are soft, super absorbent pads that provide a good environment for wound heal-

ing. These are recommended for bleeding and draining wounds, burns, or infections.

- First aid tapes: Various types of tapes should be included in each kit. These include adhesive, which is waterproof and extra strong for times when rigid strapping is needed; clear, which stretches with the body's movement and is good for visible wounds; cloth, recommended for most first aid taping needs, including taping heavy dressings (less irritating than adhesive); and paper, which is recommended for sensitive skin and is used for light and frequently changed dressings.
- Items that also can be included in each kit are tweezers, first aid cream, thermometer, an analgesic or equivalent, and an ice pack.

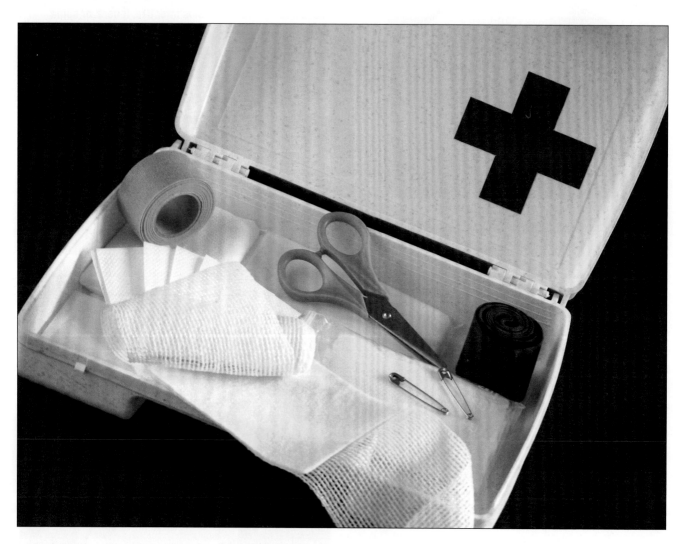

Witch hazel bark can be brewed and used to soothe irritated skin or eyes.

Nature's First Aid

Antiseptic or *wound-wash*: A handful of salt in a quart of hot water.

Balm for wounds: Balsam fir. The gum can be used as healing salve, usually spread on a piece of linen and laid over the wound for a dressing.

Cough remedy: Slippery elm or black cherry inner bark boiled, a pound to the gallon, boiled down to a pint, and given a teaspoonful every hour.

Linseed can be used the same way; add honey if desired. Or boil down the sap of the sweet birch tree and drink it on its own or mixed with the other remedies.

Diuretic: A decoction of the inner bark of elder is a powerful diuretic.

Inflammation of the eyes or skin: Wash with a strong tea made of the bark of witch hazel.

Lung balm: Infusion of black cherry bark and root is a powerful tonic for lungs and bowels. Good also as a skin wash for sores.

Poison ivy: Wash every hour or two with hot soapy water, then with hot salt water.

Linseed oil and seeds.

FLOODS

Floods are one of the most common hazards in the United States. However, not all floods are alike. Some floods develop slowly, sometimes over a period of days. Flash floods can develop quickly, sometimes in just a few minutes and without any visible signs of rain. Flash floods often have a dangerous wall of roaring water that carries rocks, mud, and other debris and can sweep away most things in its path. Overland flooding occurs outside a defined river or stream, such as when a levee is breached, but still can be destructive. Flooding can also occur when a dam breaks, producing effects similar to flash floods.

If your home is by a river, lake, or canal, you are at a higher risk for flooding.

Be aware of flood hazards no matter where you live, but especially if you live in a low-lying area, near water, or downstream from a dam. Even very small streams, gullies, creeks, culverts, dry streambeds, or low-lying ground that appears harmless in dry weather can flood. Every state is at risk from this hazard.

What to Do Before a Flood

To prepare for a flood, you should:

- Avoid building in a flood-prone area unless you elevate and reinforce your home.
- Elevate the furnace, water heater, and electric panel if susceptible to flooding.
- Install "check valves" in sewer traps to prevent floodwater from backing up into the drains of your home.
- Contact community officials to find out if they are planning to construct barriers (levees, beams, or floodwalls) to stop floodwater from entering the homes in your area.
- Seal the walls in your basement with waterproofing compounds to avoid seepage.

If a flood is likely in your area, you should:

- Listen to the radio or television for information.
- Be aware that flash flooding can occur. If there is any possibility of a flash flood, move immediately to higher ground. Do not wait for instructions to move.
- Be aware of streams, drainage channels, canyons, and other areas known to flood suddenly. Flash floods can occur in these areas with or without such typical warnings as rain clouds or heavy rain.

What to Do During a Flood

If you must prepare to evacuate, you should do the following:

- Secure your home. If you have time, bring in outdoor furniture. Move essential items to an upper floor.
- Turn off utilities at the main switches or valves if instructed to do so. Disconnect electrical appliances. Do not touch electrical equipment if you are wet or standing in water.

If you have to leave your home, remember these evacuation tips:

- Do not walk through moving water. Six inches of moving water can make you fall. If you have to walk in water, walk where the water is not moving. Use a stick to check the firmness of the ground in front of you.
- Do not drive into flooded areas. If floodwaters rise around your car, abandon the car and move to higher ground if you can do so safely. You and the vehicle can be quickly swept away.

> **The following are important points to remember when driving in flood conditions:**
>
> Six inches of water will reach the bottom of most passenger cars, causing loss of control and possible stalling.
>
> A foot of water will float many vehicles.
>
> Two feet of rushing water can carry away most vehicles, including sport utility vehicles (SUVs) and pickup trucks.

What to Do After a Flood

After a flood, you should:

- Listen for news reports to learn whether the community's water supply is safe to drink.
- Avoid floodwaters; water may be contaminated by oil, gasoline, or raw sewage. Water may also be electrically charged from underground or downed power lines.
- Avoid moving water.
- Be aware of areas where floodwaters have receded. Roads may have weakened and could collapse under the weight of a car.
- Stay away from downed power lines, and report them to the power company.
- Return home only when authorities indicate it is safe.
- Stay out of any building if it is surrounded by floodwaters.
- Use extreme caution when entering buildings; there may be hidden damage, particularly in foundations.

- Service damaged septic tanks, cesspools, pits, and leaching systems as soon as possible. Damaged sewage systems are serious health hazards.

- Clean and disinfect everything that got wet. Mud left from floodwater can contain sewage and chemicals.

Familiarize yourself with these terms to help identify a flood hazard:

Flood Watch: Flooding is possible. Tune in to NOAA Weather Radio, commercial radio, or television for information.

Flash Flood Watch: Flash flooding is possible. Be prepared to move to higher ground; listen to NOAA Weather Radio, commercial radio, or television for information.

Flood Warning: Flooding is occurring or will occur soon; if advised to evacuate, do so immediately.

Flash Flood Warning: A flash flood is occurring; seek higher ground on foot immediately.

HURRICANES

I f you live in an area particularly prone to hurricanes, taking precautions is especially important. However, hurricanes can form anywhere, so don't assume that just because you're not in the tropics, you're not at risk. Thinking ahead will at the very least give you greater peace of mind.

What to Do Before a Hurricane

If you live in a hurricane-prone area, you may want to take the following precautions:

- Install permanent shutters to protect your windows. Or you can board up windows with ⅝-inch marine plywood. Have the plywood cut to fit and ready to install.
- Install straps or additional clips to fasten your roof securely to the frame structure. This will reduce roof damage.
- Be sure trees and shrubs around your home are well trimmed.
- Clear loose and clogged rain gutters and downspouts.
- Determine how and where to secure your boat, if you have one.

Shutters will help to protect your windows during wind storms.

If you are aware of a hurricane approaching, you should:

- Listen to the radio or TV for information.
- Secure your home, close storm shutters, and secure outdoor objects or bring them indoors.
- Turn off utilities if instructed to do so. If you leave the electricity on, turn the refrigerator thermostat to its coldest setting and keep its doors closed so the food will stay colder longer if the electricity goes out.
- Turn off propane tanks. Avoid using the phone, except for serious emergencies.
- Moor your boat if time permits.
- Draw fresh water in jugs or in your bathtub or sink for use in drinking, bathing, and flushing toilets if the electricity goes out.

What to Do During a Hurricane

You should evacuate if:

- You are directed by local authorities to do so. Be sure to follow their instructions.
- You live in a mobile home or temporary structure—such shelters are particularly hazardous during hurricanes no matter how well fastened to the ground.
- You live in a high-rise building—hurricane winds are stronger at higher elevations.
- You live on the coast, on a floodplain, near a river, or on an inland waterway.
- You feel you are in danger.

If you are unable to evacuate, go to a basement or underground shelter. If you do not have one, follow these guidelines:

- Stay indoors during the hurricane and away from windows and glass doors.
- Close all interior doors—secure and brace external doors.
- Keep curtains and blinds closed. Do not be fooled if there is a lull; it could be the eye of the storm—winds will pick up again.
- Take refuge in a small interior room, closet, or hallway on the lowest level.
- Lie on the floor under a table or another sturdy object.

Watch the sky for warning signs of a serious storm approaching.

WILDFIRES

T he threat of wildland fires for people living near wildland areas or using recreational facilities in wilderness areas is real. Dry conditions at various times of the year and in various parts of the United States greatly increase the potential for wildland fires.

Advance planning and knowing how to protect buildings can lessen the devastation of a wildland fire. There are several safety precautions that you can take to reduce the risk of fire losses. Protecting your home from wildfire is your responsibility. To reduce the risk, you'll need to consider the

Dry grass or other vegetation will encourage fire to spread quickly.

Stucco homes are less vulnerable to fire than homes made of wood.

fire resistance of your home, the topography of your property, and the nature of the vegetation close by.

If you are considering moving to a home or buying land in an area prone to wildfires, consider having a professional inspect the property and offer recommendations for reducing the wildfire risk. Determine the community's ability to respond to wildfire. Are roads leading to your property clearly marked? Are the roads wide enough to allow firefighting equipment to get through?

Learn and Teach Safe Fire Practices

- Build fires away from nearby trees or bushes.
- Always have a way to extinguish the fire quickly and completely.
- Install smoke detectors on every level of your home and near sleeping areas.
- Never leave a fire—even a cigarette—burning unattended.
- Avoid open burning completely, and especially during dry season.

What to Do Before a Wildfire

To prepare your home for a wildfire you should:

- Create a 30-foot safety zone around the house. Keep the volume of vegetation in this zone to a minimum. If you live on a hill, extend the zone on the downhill side. Fire spreads rapidly uphill. The steeper the slope, the more open space you will need to protect your home. Swimming pools and patios can be a safety zone and stonewalls can act as heat shields and deflect flames.

- Remove vines from the walls of the house and move shrubs and other landscaping away from the sides of the house. You should also prune branches and shrubs within 15 feet of chimneys and stovepipes, remove tree limbs within 15 feet of the ground, and thin a 15-foot space between tree crowns.

- Replace highly flammable vegetation such as pine, eucalyptus, junipers, and fir trees with lower growing, less flammable species within the 30-foot safety zone. Check with your local fire department or garden store for suggestions. Also replace vegetation that has living or dead branches from the ground level up (these act as ladder fuels for the approaching fire). Cut the lawn often, keeping the grass at a maximum of 2 inches. Watch grass and other vegetation near the driveway, a source of ignition from automobile exhaust systems. Finally, clear the area of leaves, brush, evergreen cones, dead limbs, and fallen trees.

- Create a second zone at least 100 feet around the house. This zone should begin about 30 feet from the house and extend to at least 100 feet. In this zone, reduce or replace as much of the most flammable vegetation as possible. If you live on a hill, you may need to extend the zone for several hundred feet to provide the desired level of safety.

- Remove debris from under sun decks and porches. Any porch, balcony, or overhang with exposed space underneath is fuel for an approaching fire. Overhangs ignite easily by flying embers and by the heat and fire that get trapped underneath. If vegetation is allowed to grow underneath or if the space is used for storage, the hazard is increased significantly. Clear leaves, trash and other combustible materials away from underneath sun decks and porches. Extend ½-inch mesh screen from all overhangs down to the ground. Enclose wooden stilts with non-combustible material such as concrete, brick, rock, stucco, or metal. Use non-combustible patio furniture and covers. If you're planning a porch or sun deck, use non-combustible or fire-resistant materials. If possible, build the structure close to the ground so that there is no space underneath.

- Enclose eaves and overhangs. Like porches and balconies, eaves trap the heat rising along the exterior siding. Enclose all eaves to reduce the hazard.

- Cover house vents with wire mesh. Any attic vent, soffit vent, louver, or other opening can allow embers and flaming debris to enter a home and ignite it. Cover all openings with ¼ inch or smaller corrosion-resistant wire mesh. If you're designing louvers, place them in the vertical wall rather than the soffit of the overhang.

- Install spark arrestors in chimneys and stovepipes. Chimneys create a hazard when embers escape through the top. To prevent this, install spark arrestors on all chimneys, stovepipes and vents for fuel-burning heaters.

Use spark arrestors made of 12-gauge welded or woven wire mesh screen with openings ½ inch across. Ask your fire department for exact specifications. If you're building a chimney, use non-combustible materials and make sure the top of the chimney is at least 2 feet higher than any obstruction within 10 feet of the chimney. Keep the chimney clean.

- Use fire-resistant siding such as stucco, metal, brick, cement shingles, concrete, or rock. You can treat wood siding with UL-approved fire retardant chemicals, but the treatment and protection are not permanent.

- Choose safety glass for windows and sliding glass doors. Windows allow radiated heat to pass through and ignite combustible materials inside. The larger the pane of glass, the more vulnerable it is to fire. Dual- or triple-pane thermal glass, and fire-resistant shutters or drapes, help reduce the wildfire risk. You can also install non-combustible awnings to shield windows and use shatter-resistant glazing such as tempered or wire glass.

- Prepare for water storage; develop an external water supply such as a small pond, well, or pool.

- Always be ready for an emergency evacuation. Evacuation may be the only way to protect your family in a wildfire. Know where to go and what to bring with you. You should plan several escape routes in case roads are blocked by a wildfire.

Handling Combustibles

- Install electrical lines underground, if possible.
- Ask the power company to clear branches from power lines.
- Avoid using bark and wood chip mulch.
- Stack firewood 100 feet away and uphill from any structure.
- Store combustible or flammable materials in approved safety containers and keep them away from the house.
- Keep the gas grill and propane tank at least 15 feet from any structure. Clear an area 15 feet around the grill. Place a ¼-inch mesh screen over the grill. Always use the grill cautiously but refrain from using it at all during high-risk times.

Safety Measures for New Construction or Remodeling

- Choose locations wisely; canyon and slope locations increase the risk of exposure to wildland fires.
- Use fire-resistant materials when building, renovating, or retrofitting structures.
- Avoid designs that include wooden decks and patios.
- Use non-combustible materials for the roof.
- The roof is especially vulnerable in a wildfire. Embers and flaming debris can travel great distances, land on your roof and start a new fire. Avoid flammable roofing materials such as wood, shake and shingle. Materials that are more fire resistant include single ply membranes, fiberglass shingles, slate, metal, clay and concrete tile. Clear gutters of leaves and debris.

What to Do if a Wildfire is Approaching

- Evacuate your pets and all family members who are notessential to preparing the home. Anyone with medical or physical limitations and the young and the elderly should be evacuated immediately.
- Wear protective clothing.
- Remove combustibles. Clear items that will burn from around the house, including woodpiles, lawn furniture, barbecue grills, tarp coverings, and so on. Move them outside of your defensible space.
- Close outside attic, eave, and basement vents, and windows, doors, and pet doors. Remove flammable drapes and curtains. Close all shutters, blinds, or heavy non-combustible window coverings to reduce radiant heat.
- Close inside doors and open damper. Close all doors inside the house to prevent draft. Open the damper on your fireplace, but close the fireplace screen.
- Shut off any natural gas, propane, or fuel oil supplies at the source.
- Connect garden hoses and fill any pools, hot tubs, garbage cans, tubs, or other large containers with water.

- If you have gas-powered pumps for water, make sure they are fueled and ready.
- Place a ladder against the house in clear view.
- Back your car into the driveway and roll up the windows.
- Disconnect any automatic garage door openers so that doors can still be opened by hand if the power goes out. Close all garage doors.
- Place valuable papers, mementos, and anything you "can't live without" inside the car, ready for quick departure. Any pets still with you should also be put in the car.
- Just before evacuating, turn on outside lights and leave a light on in every room to make the house more visible in heavy smoke.
- Leave doors and windows closed but unlocked. It may be necessary for firefighters to gain quick entry into your home to fight fire. The entire area will be isolated and patrolled by sheriff's deputies or police.

Survival in a Vehicle

- This is dangerous and should only be done in an emergency, but you can survive the firestorm if you stay in your car. It is much less dangerous than trying to run from a fire on foot.
- Roll up windows and close air vents. Drive slowly with headlights on. Watch for other vehicles and pedestrians. Do not drive through heavy smoke.
- If you have to stop, park away from the heaviest trees and brush. Turn headlights on and ignition off. Roll up windows and close air vents.
- Get on the floor and cover up with a blanket or coat.
- Stay in the vehicle until the main fire passes.
- Stay in the car. Do not run! Engine may stall and not restart. Air currents may rock the car. Some smoke and sparks may enter the vehicle. Temperature inside will increase. Metal gas tanks and containers rarely explode.

If Caught in the Open

- The best temporary shelter is in a sparse fuel area. On a steep mountainside, the back side is safer. Avoid canyons, natural "chimneys," and saddles.
- If a road is nearby, lie face down along the road cut or in the ditch on the uphill side. Cover yourself with anything that will shield you from the fire's heat.
- If hiking in the back country, seek a depression with sparse fuel. Clear fuel away from the area while the fire is approaching and then lie face down in the depression and cover yourself. Stay down until after the fire passes!

What to Do After a Wildfire

- Check the roof immediately. Put out any roof fires, sparks, or embers. Check the attic for hidden burning sparks.
- If you have a fire, get your neighbors to help fight it.
- The water you put into your pool or hot tub and other containers will come in handy now. If the power is out, try connecting a hose to the outlet on your water heater.
- For several hours after the fire, maintain a "fire watch." Re-check for smoke and sparks throughout the house.

PART SIX

ENERGY

With the extreme fluctuation in oil prices and ever-growing concerns about the state of our environment, it's no wonder that more and more people are turning to the natural elements for power. Sun, wind, water, and earth have provided for the basic needs of humanity since the beginning of time and it only makes sense to learn how to work with them more efficiently. The term "self-sufficiency," as it is commonly used, is something of a misnomer. We will never be able to meet all of our own needs alone. We don't create the natural world that supplies us with the light, heat, and other resources that we depend on. But we can learn how to make good use of those gifts. In these pages you will find both simple and advanced projects to do so, from fashioning and using solar cookers to building and installing wind turbines to utilizing geothermal systems. There's a lot here, but it's only a sampling of the methods available for harnessing natural energy. Look online or visit your library for more ideas, plans, and tips; you'll also find an extensive list of resources in the back of this book. Remember that the simplest and perhaps most effective way to be energy-efficient is to use less of it. The simple things, like turning off a light when you're not in the room—or even using candlelight in the evenings—can make a big difference. The more you understand about the process of turning the natural elements into usable energy, the more you'll appreciate the value of electricity and want to conserve it in any way you can.

COMPOSTING TOILETS

Toilets come in three common varieties: siphon-jet flush valve toilets (common in most homes), pressurized tank toilets, and gravity flow. These toilets, generally speaking, use up large amounts of water and the waste is flushed into a sewer system and then dumped in a variety of locations. Composting toilets require little to no water, which provides a solution to sanitation and environmental problems in areas that are rural, without sewers, and in the suburbs throughout the world. Although composting toilets are rare in private homes—they are generally found in park facilities and small highway rest stops—these waterless toilets can be utilized by the regular homeowner.

It is astonishing that Americans flush about 4.8 billion gallons of water down toilets every day, according to the U.S. Environmental Protection Agency. Just replacing all existing U.S. toilets with 1.6-gallon-per-flush, ultra-low-flow (ULF) models would save about 5,500 gallons of water per person per year! So, if you are unable to install a composting toilet in your home or on your property, you may choose to install ULF models in your home to help conserve water usage.

The Basics of the Composting Toilet

Composting (or biological) toilet systems contain and process excrement, toilet paper, carbon additive, and, at times, food wastes. These systems rely on unsaturated conditions where aerobic bacteria break down waste—unlike septic systems—much like a compost heap for your gardening necessities. The resulting soil-like material—humus—must be buried or removed. It's a good idea to check state and local regulations regarding proper handling methods.

In many parts of the country, public health officials are realizing that there is a definite need for environmentally sound human waste treatment and recycling methods, and compost toilets are an easy way to work toward these needs. Because they don't require any water, composting toilets are ideal for remote areas and places that have high water tables, shallow soil, and rough terrain. These systems save water and allow for valuable plant nutrients to be recycled in the process.

There are a few key components for establishing a composting toilet:

- Composting reactor that is connected to a micro-flush toilet
- Screened air inlet and exhaust system to remove odors and heat, plus CO_2 and other decomposition byproducts
- Mechanism to provide proper ventilation that will help aerobic organisms in the compost heap
- Process controls
- Access door for the removal of the end product

It is important that the composting toilet separates the solid from the liquid waste and produces a humus-like material with less than 200 MPN per gram of fecal coliform. The compost chamber can be solar or electrically heated to maintain the right temperature for year-round use and bacterial decomposition.

Main Objective of the Composting Toilet

These systems are designed to contain, immobilize, and destroy pathogens. This reduces the risk of human infection and ensures that the toilets do not pollute the environment. If done correctly, the composted material can be handled with little to no risk of harming the individual working with it.

A composting toilet consists of a well-ventilated container that breeds a good environment for unsaturated, moist human excrement that can be decomposed under sanitary conditions. A composting toilet can be large or small, depending on the space and its use. Organic matter is transformed into a humus-like product through the natural breaking down from bacteria and fungi. Most systems like this use the process of continuous composting, which includes a single chamber where the excrement is added to the top and the end product is taken from the bottom.

Advantages of Using a Composting Toilet

Composting toilets can be used practically anywhere a flush toilet can be. They are most likely to be used in homes in

Diagram of a composting toilet.

rural areas, seasonal cabins, recreation areas, and other places where flush toilets are either unnecessary or impractical. They are more cost-effective than establishing a central sewage system and there is no water wasted. These systems—since they aren't using copious amounts of water—also reduce the quantity of wastewater that is disposed of on a daily basis. These toilets can also be used to recycle and compost food wastes, thus reducing the amount of household garbage that is dumped every day. Finally, these toilet systems are beneficial to the environment as they divert nutrient and pathogen-containing effluent from the soil, surface water, and the groundwater.

Disadvantages of Using a Composting Toilet

Composting toilets are a big responsibility; the owner of a composting toilet must be committed to maintaining the system. Removing the compost can be unpleasant if the toilet is not properly set up and they could end up having odor issues.

Composting toilets are being used more frequently in parks around the world.

Successful Management of the Composting Toilet

Composting toilets do not require highly trained people to deal with the sewage as it is relatively harmless to handle. But be sure to maintain your composting toilet so it can be effective and safe. Some composting toilets may need organic bulking agents added to aid the composting process. Adding grass clippings, sawdust, and leaves to your composting toilet reservoir will help aid the process. The end product should be removed every three months for smaller systems and, if composted correctly, should not smell and should not be toxic to humans or animals. Be sure to dispose of the waste materials in accordance with your particular state and local regulations.

Factors That Affect the Rate of Composting

1. Microorganisms—A mix of bacteria and fungi need to be present in order for the excrement to turn into composted material.
2. Moisture—This helps the microorganisms to make simpler compounds before they are metabolized. Moisture should be kept between 40 and 70 percent.
3. pH—The best pH for the composting toilet material should be between 6.5 and 7.5.
4. Carbon to nitrogen ratio—It is important to balance out the nitrogen found in urine with added carbon in your composting toilet.
5. Proper care—Managing your composting toilet well will help keep it efficient and productive.

Making Your Own Composting Toilet

Building your own composting toilet can be inexpensive and takes only a short amount of time to assemble. To construct a composting toilet, you will need the following materials:

- Two or three 5-gallon buckets with lids
- A standard toilet seat (a used one will work just fine) with lid
- ¾ x 3 x 18-inch plywood sheets
- Boards to be cut and used for the sides of the toilet box and for the legs
- Two hinges
- Screws
- Saw and measuring tape
- Bag of sawdust, to be used for soaking up excess moisture in the composting bucket

To begin, cut a hole in one of the pieces of plywood so that it fits the size of the bucket. Then, attach the pieces of plywood together using the hinges. Build a box with the boards and then screw in the solid piece of plywood to the box, allowing for the part with the hole to remain on the top. Attach legs to the box, allowing the bucket to lift just slightly above the hole cut in the top piece of plywood. Then, attach the toilet seat to the plywood top, fitting it securely over the rim of the bucket. Finally, stain or paint the entire composting toilet so it will last longer and match the décor of your bathroom.

Before using your homemade composting toilet, sprinkle 1 to 2 inches of sawdust into the bottom of the bucket. This will help absorb extra moisture and will also add a necessary carbon element that is useful in composting. Sprinkle sawdust into the toilet after each use to facilitate the composting process and to minimize odors. When the first bucket is full, remove and cover (allowing the composting process to continue), insert another bucket, and continue use. When both buckets are full, remove them to your composting pile in your yard. Make a small indent in the center of your composting pile and dump the new compost into the depression, laying old compost and other organic materials on top of the new addition. If used properly, your composting toilet will be odorless and your compost will be rich and ready for use in your garden.

Compost will enrich the soil in your garden to help grow healthier plants.

A geothermal power plant in action.

Geothermal energy (the heat from the earth) is accessible as an alternative source of heat and power. Geothermal energy can be accessed by drilling water or steam wells using a process much like drilling for oil. This resource is enormous but is sadly underused as an energy source. When it is employed, though, it proves to be clean (emitting little or no greenhouse gases), reliable, economical, and domestically found (geothermal energy can be harnessed from almost anywhere and thus makes countries less dependent on foreign oil).

Wells a mile or more deep can be drilled into underground reservoirs to tap steam and very hot water. This can then be brought to the surface and used in a variety of ways—such as to drive turbines and electricity generators. In the United States, most geothermal reservoirs are located in the western states, in Alaska, and in Hawaii. People in more than 120 locations in the United States are using geothermal energy for space and district heating.

Geothermal resources can range from shallow ground water to hot water found in rocks several miles below the surface of the earth. It can even be harnessed, in some cases, from magma (hot molten rock near the earth's core). Geothermal reservoirs of low to moderate temperature (roughly 68 to 302°F) can be used to heat homes, offices, and greenhouses. Curiously, the dehydration of onions and garlic comprises the largest industrial use of geothermal energy in the United States.

Three Main Uses of Geothermal Energy

Some types of geothermal energy usage draw from the earth's temperatures closer to the surface and others require, as noted above, drilling miles into the earth. The three main uses of geothermal energy are:

These greenhouses in Greenland utilize geothermal heating.

1. Direct Use and District Heating Systems—These use hot water from springs and reservoirs near the earth's surface.
2. Electricity Generation—Typically found in power plants, this type of energy requires high-temperature water and steam (generally between 300 and 700°F).

Geothermal power plants are built where reservoirs are positioned only a mile or two from the earth's surface.

3. Geothermal Heat Pumps—These use stable ground or water temperatures near the earth's surface to control building temperatures above the ground.

Additional Resources

The U.S. Department of Energy, in conjunction with the Geo-Heat Center, conducts research, provides technical support, and distributes information on a wide range of geothermal direct-use applications. Some information that is provided revolves around greenhouse informational packages, cost comparisons of heat pumps, low temperature resource assessments, cost analysis for homeowners, and information directed to aquaculture developers.

The greenhouse informational package provides information for people who are looking to develop geothermal greenhouses. This package includes crop market prices for vegetables and flowers, operating costs, heating system specifications, greenhouse heating equipment selection spreadsheets, and vendor information.

Groundwater heat pumps have also been identified as offering substantial savings over other types of pump systems. Informational packets about heat pump systems are provided to answer frequently asked questions concerning the application and usage of geothermal heat pumps.

The Geo-Heat Center examined the costs associated with the installation of district heating systems in single-family residential sectors. They discovered that cost-saving areas included installation in unpaved areas, using non-insulated return lines, and installation in areas that are unencumbered by existing buried utility lines.

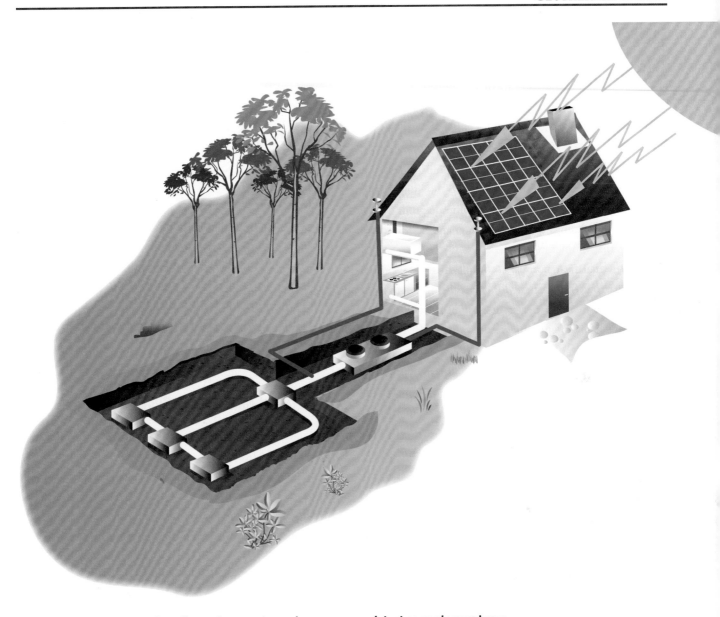

You can combine solar and geothermal energy to produce more consistent power in your home.

Direct Use Geothermal Energy

Since ancient times, people have been directly using hot water as a source of energy. The Chinese, Native Americans, and Romans used hot mineral springs for bathing, cooking, and heating purposes. Currently, a number of hot springs are still used for bathing and many people believe these hot, mineral-rich waters possess natural healing powers.

Besides bathing, the most common direct use of geothermal energy is for heating buildings. This is through district heating systems—these types of systems provide heat for roughly 95 percent of the buildings in Reykjavik, Iceland. District heating systems pipe hot water near the earth's surface directly into buildings to provide adequate heat.

Direct use of geothermal resources is a proven, economic, and clean energy option. Geothermal heat can be piped directly into facilities and used to heat buildings, grow greenhouse plants, heat water for fish farming, and even pasteurize milk. Some northern U.S. cities pipe hot water under roads and sidewalks to melt the snow.

Geothermal Heat Pumps

Even though temperatures above the surface of the earth change daily and seasonally, in general, temperatures in the top 10 feet of the earth's surface stay fairly constant at around 50 to 60°F. This means that, in most places, soil temperatures are typically warmer than air temperatures in the winter and cooler in the summer. Geothermal heat pumps (GHPs) use this constant temperature to heat and cool buildings. These pumps transfer heat from the ground (or underground water sources) into buildings during the winter and do the reverse process in the summer months.

Geothermal heat pumps, according to the U.S. Environmental Protection Agency (EPA), are the most energy-efficient, environmentally clean, and cost-effective systems for maintaining a consistent temperature control. These pumps are becoming more popular, even though most homes still use furnaces and air conditioners. Sometimes referred to as earth-coupled, ground-source, or water-source heat pumps, GHPs use the constant temperature of the earth as the exchange medium (using ground heat exchangers) instead of the outdoor air temperature. In this way, the system can be very efficient on cold winter nights in comparison to air-source heat pumps.

Geothermal heat pumps can heat, cool, and, in some cases, even supply hot water to a house. These pumps are relatively quiet, long-lasting, need little to no maintenance, and do not rely on outside temperatures to function effectively. While geothermal systems are initially more expensive to install, these costs are quickly returned in energy savings

in about five to 10 years. Systems have a life-span of roughly 25 years for inside components and more than 50 years for ground loop systems. Each year, about 50,000 geothermal heat pumps are installed in the United States.

Types of Geothermal Heat Pump Systems

There are four basic types of ground loop heat pump systems: horizontal, vertical, pond/lake, and open-loop systems. The first three are closed-loop systems while the fourth is, as its name suggests, open-loop. The type of system used is generally determined based on the climate, soil conditions, land availability, and local installation costs of the site for the pump. All four types of geothermal heat pump systems can be used for both residential and commercial building applications.

Horizontal Heat Pump System

This closed-loop installation is extremely cost-effective for residential heat pumps and is well suited for new construction where adequate land is available for the system. Horizontal heat pump systems need 4-foot trenches to be installed. These systems are typically laid out using two pipes—one buried 6 feet and the other buried 4 feet below the ground—or by placing two pipes side by side at 5 feet underground in a 2-foot-wide trench.

A geothermal power plant.

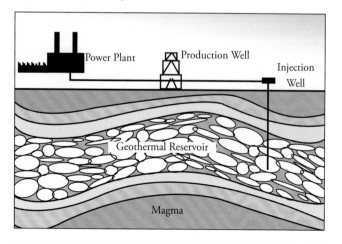

A horizontal closed-loop heat pump system.

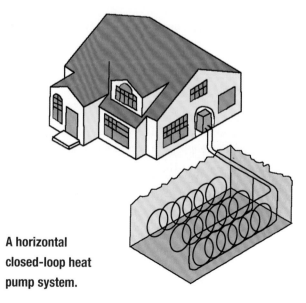

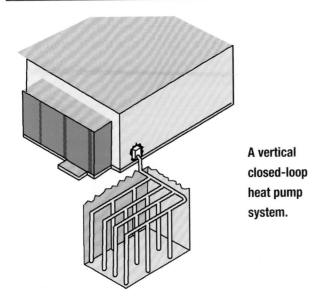

A vertical closed-loop heat pump system.

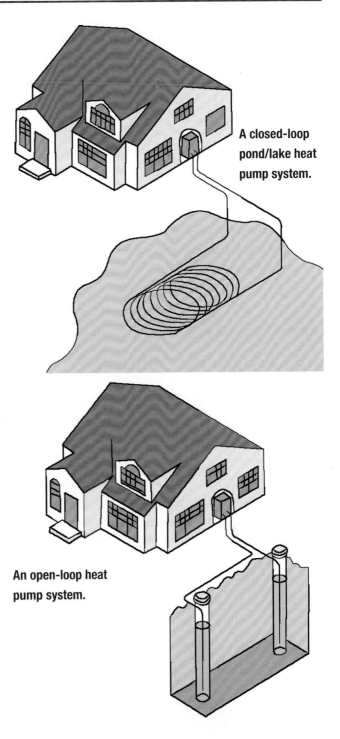

A closed-loop pond/lake heat pump system.

An open-loop heat pump system.

Vertical Heat Pump System

Schools and larger commercial buildings use vertical heat pump systems because they require less land to be effectively used. These systems are best used where the soil is too shallow for trenching. They also minimize any disturbance to established landscaping. To install a vertical system, holes that are roughly 4 inches in diameter are drilled about 20 feet apart and 100 to 400 feet deep. Two pipes are inserted into these holes and are connected at the bottom with a U-bend, forming a loop. The vertical loops are then connected with a horizontal pipe, placed in the trenches, and connected to the heat pump in the building.

Pond/Lake Heat Pump System

Another closed-loop system is the pond/lake heat pump system. If a site has enough water—usually in the form of a pond or even a lake—this system may be the most cost-effective. This heat pump system works by running a supply line pipe underground from a building to the water source. The piping is coiled into circles no less than 8 feet under the surface—this prevents the water in the pipes from freezing. The coils should be placed only in a water source that meets the minimum volume, depth, and quality criteria.

Open-Loop Heat Pump System

An open-loop system uses well or surface body water as the heat exchange fluid that will circulate directly through the geothermal heat pump system. Once this water has circulated through the system, it is returned to the ground through a recharge well or as surface discharge. The system is really only practical where there is a sufficient supply of clean water. Local codes and regulations for proper groundwater discharge must also be met in order for the heat pump system to be utilized.

Selecting and Installing a Geothermal Heat Pump System in Your Home

The heating efficiency of commercial ground-source and water-source heat pumps is indicated by their coefficient of performance (COP)—the ratio of heat provided in Btu per Btu of energy input. The cooling efficiency is measured by the energy efficiency ratio (EER)—the ratio of heat removed to the electricity required (in watts) to run the unit. Many geothermal heat pump systems are approved by the U.S. Department of Energy as being energy efficient products and so, if you are thinking of purchasing and installing this type of system, you may want to check to see if there is any special financing or incentives for purchasing energy efficient systems.

Evaluating Your Site

Before installing a geothermal heat pump, consider the site that will house the system. The presence of hot geothermal fluid containing low mineral and gas content, shallow aquifers for producing the fluid, space availability on your property, proximity to existing transmission lines, and availability of make-up water for evaporative cooling are all factors that will determine if your site is good for geothermal electric development. As a rule of thumb, geothermal fluid temperature should be no less than 300°F.

In the western United States, Alaska, and Hawaii, hydrothermal resources (reservoirs of steam or hot water) are more readily available than the rest of the country. However, this does not mean that geothermal heat cannot be used throughout the country. Shallow ground temperatures are relatively constant throughout the United States and this means that energy can be tapped almost anywhere in the country by using geothermal heat pumps and direct-use systems.

To determine the best type of ground loop systems for your site, you must assess the geological, hydrological, and spatial characteristics of your land to choose the best, most effective heat pump system to heat and cool your home:

1. Geology—This includes the soil and rock composition and properties on your site. These can affect the transfer rates of heat in your particular system. If you have soil with good heat transfer properties, your system will require less piping to obtain a good amount of heat from the soil. Furthermore, the amount of soil that is available also contributes to which system you will choose. For example, areas that have hard rock or shallow soil will most likely benefit from a vertical heat pump system instead of a system requiring large and deep trenches, such as the horizontal heat pump system.

2. Hydrology—This refers to the availability of ground or surface water, which will affect the type of system to be installed. Factors such as depth, volume, and water quality will help determine if surface water bodies can be used as a source of water for an open-loop heat pump system or if they would work best with a pond/lake system. Before installing an open-loop system, however, it is best to determine your site's hydrology so potential problems (such as aquifer depletion or groundwater contamination) can be avoided.

3. Available land—The acreage and layout of your land, as well as your landscaping and the location of underground utilities, also play an important part in the type of heat pump system you choose. If you are building a new home, horizontal ground loops are an economical system to install. If you have an existing home and want to convert your heat and cooling to geothermal energy, vertical heat pump systems are best to minimize the disturbance to your existing landscaping and yard.

Installing the Heat Pumps

Geothermal heat pump systems are somewhat difficult to install on your own—though it can certainly be done. Before

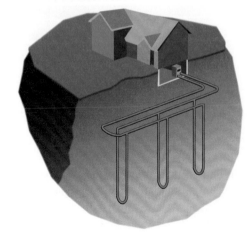

A vertical closed-loop system.

you begin any digging, contact your local utility company to avoid digging into gas pipes or electrical wires.

The ground heat exchanger in a geothermal heat pump system is made up of closed- or open-loop pipe—depending on which type of system you've determined is best suited for your site. Since most systems employed are closed-loop systems, high density polyethylene pipe is used and buried horizontally at 4 to 6 feet deep or vertically at 100 to 400 feet deep. These pipes are filled with an environmentally friendly antifreeze/water solution that acts as a heat exchanger. You can find this at your local home store or contact a contractor to see where it is distributed. This solution works in the winter by extracting heat from the earth and carrying it into the building. In the summertime, the system reverses, taking heat from the building and depositing it into the ground.

Air delivery ductwork will distribute the hot or cold air throughout the house's ductwork like traditional, conventional systems. An air handler—a box that contains the indoor coil and fan—should be installed to move the house air through the heat pump system. The air handler contains a large blower and a filter, just like standard air conditioning units.

One of the advantages of geothermal heating in a home is warm floors, even in the coldest months of the year, due to the piping laid in the floors.

Cost-Efficiency of Geothermal Heat Pump Systems

By installing and using a geothermal heat pump system, you will save on the costs of operating and maintaining your heating and cooling system. While these systems are generally a bit pricier to install, they prove to be more efficient and thus save you money on a monthly and yearly basis. Especially in the colder winter months, geothermal heat pump systems can reduce your heating costs by about half. Annual energy savings by using a geothermal heat pump system range from 30 to 60 percent.

Benefits of Using Geothermal Energy

- It is clean energy. Geothermal energy does not require the burning of fossil fuels (coal, gas, or oil) to produce energy.
- Geothermal fields produce only about ⅙ th of the carbon dioxide that natural gas-fueled power plants do. They also produce little to no sulfur-bearing gases, which reduces the amount of acid rain.
- It is available at any time of day, all year-round.
- Geothermal power is homegrown, which reduces dependence on foreign oil.
- It is a renewable source of energy. Geothermal energy derives its source from an almost unlimited amount of heat generated by the earth. And even if energy is limited in an area, the volume taken out can be reinjected, making it a sustainable source of energy.
- Geothermal heat pump systems use 25 to 50 percent less electricity than conventional heating and cooling systems. They reduce energy consumption and emissions between 44 and 72 percent and improve humidity control by maintaining about 50 percent relative humidity indoors (GHPs are very effective for humid parts of the country).
- Heat pump systems can be "zoned" to allow different parts of your home to be heated and cooled to different temperatures without much added cost or extra space required.

- Geothermal heat pump systems are durable and reliable. Underground piping can last for 25 to 50 years and the heat pumps tend to last at least 20 years.
- Heat pump systems reduce noise pollution since they have no outside condensing unit (like air conditioners).

Alternate "Geothermal" Cooling System

True geothermal energy systems can be very expensive to install and you may not be able to use one in your home at this time. However, here is a fun alternative way to use the concepts of geothermal systems to keep your house

cooler in the summer and your air conditioning bills lower. All you need are a basement, small window fan, and dehumidifier.

Your basement is a wonderful example of how the top layers of earth tend to remain at a stable temperature throughout the year. In the winter, your basement may feel somewhat warm; in the summer, it's nice and refreshingly cool. This is due to the temperature of the soil permeating through the basement walls. And this cool basement air can be used to effectively reduce the temperature in your home by up to five degrees during the summer months. Here are the steps to your alternative "geothermal" cooling system:

1. Run the dehumidifier in your basement during the night, bringing the humidity down to about 60 percent.

2. Keep your blinds and curtains closed in the sunniest rooms in your home.

3. In the morning, when the temperature inside the house reaches about 77°F, open a small window in your basement, just a crack, and open one of the upstairs windows, placing a small fan in it and directing the room air out of the window.

4. With all other windows and outside doors closed, the fan will suck the cool basement air through your home and out the open window. Doing this for about an hour will bring down the temperature inside your home, buying you a couple of hours of reprieve before switching on the AC.

The hot springs at Yellowstone are a natural example of geothermal heating.

GREYWATER

reywater is just wastewater. Greywater, however, does not include toilet wastewater, which is known as blackwater. These two different kinds of water should not be mixed together for basic health reasons. The main differences between greywater and blackwater are:

- Greywater contains less nitrogen than blackwater (and about half of the nitrogen that is found in greywater is organic nitrogen that can be filtered out and used by plants).
- Greywater contains fewer pathogens than blackwater and thus is not as likely to spread organisms that could be potentially harmful to humans.
- Greywater decomposes faster than blackwater and is less likely to cause water pollution because of this factor.

Greywater is not necessarily sewage to begin with, but if left untreated for a couple of days, it will become like blackwater and thus will be unusable. Therefore, it is important to know how best to treat and manage greywater so it can be successfully and safely reused. Much of the material in this section is adapted from Carl R. Lindstrom's excellent site, www.greywater.com.

What is Greywater?

Simply speaking, greywater is wash water—bath, dish, and laundry water that is free from toilet waste and garbage disposal remnants. Greywater, when it is managed properly, can be useful for growing things in your garden or yard. Greywater, in effect, is an excellent source of nutrients for plants when used properly.

Greywater Irrigation Systems

The practice of irrigating with greywater is common in areas where the water supply is short. To have effective greywater irrigation that successfully utilizes the nutrients in the greywater, take precautions before using it in irrigation.

Planning a greywater system requires either an assumption that the system is right for you and your family or an understanding that the system is needed for the house independent of who lives in it.

To assess whether your household could benefit from a greywater system, take inventory of all the sources of greywater in the house. Look at how many gallons of water you use, per person, per day, when doing the laundry, running the dishwasher, and taking a bath or shower, and then add up these numbers. Remember that the typical washing machine uses 30 gallons of water per cycle, a dishwasher uses between 3 and 5 gallons per cycle, and simply washing your hands and brushing your teeth daily wastes about 1 to 5 gallons of water per day. If you were able to recycle and reuse all of that wasted water, you can effectively reduce the amount of water consumption your family has every day and every year.

Once you've decided to use your greywater, check with your local authorities to see if there are any state or local regulations for greywater usage in your area. Once you have the go-ahead to proceed, you can begin reusing your greywater to the benefit of your garden and household.

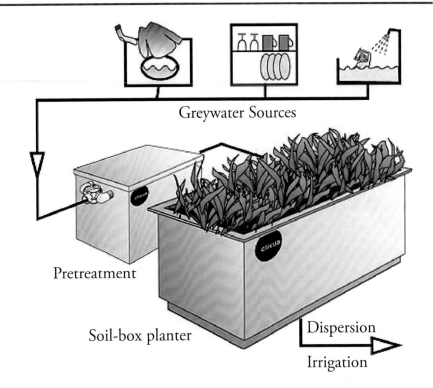

Greywater irrigation.

Aerobic Pretreatment

This type of greywater treatment is suitable for shower, hand-washing, and laundry water. Aerobic pretreatment is a stretch filter technique that removes large particles and fibers to protect the pipes from clogging and transfers the greywater into a biologically active, aerobic soil-zone environment. Here, microorganisms can survive and flourish. Stretch filters retain fibers and large particles and allow the rest of the materials to travel to the next processing stage. The filter is good for sinks and showers at public water facilities.

Anaerobic to Aerobic Pretreatment

If you have food waste entering the water system from dishwashers and kitchen sinks, this is the better option for treating your greywater. This system

Greywater pretreatment.

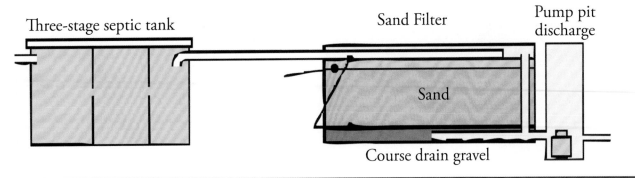

should have a three-stage septic tank to separate the sludge and grease from the water. This waste can then be removed easily. The outgoing water will be anaerobic and will need a sand filter to restore the aerobic conditions to the greywater. The final treatment leads the purified water to be treated in a planter bed. The system, while not inexpensive, is effective and is simple to maintain. A plan for this system can be seen above.

Planter Soil Box

Since 1975, soil boxes have been used to purify greywater. When using a soil box, however, it is vital that the planter bed be well drained to prevent water-logged zones from forming. Therefore, the bottom of the soil box should contain a layer of polyethylene pea gravel to provide for effective drainage. A layer of plastic mosquito netting should be placed over the gravel to prevent the layer of coarse sand from falling through. Atop the coarse sand should be a layer of concrete-mix sand and the top 2 feet should consist of humus-rich topsoil. Clay soils should not be used in soil boxes as they do not effectively allow water to pass through and drain.

Pressure infiltration pipes should be designed to allow for the even distribution of water in both level and uneven terrain. These pipes are easy to clean and should be placed on the soil surface after planting. Then, they should be covered by a 2- to 4-inch layer of wood chip mulch. The pressure infiltration pipes consist of two concentric pipes that expand slightly due to the water pressure when the system is turned

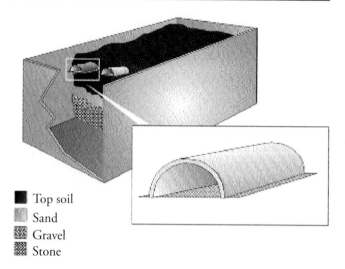

- ■ Top soil
- ▨ Sand
- ▦ Gravel
- ▦ Stone

A leaching chamber.

on. This causes the water to run out along the slot at the bottom of the soil box. When the water pressure is turned off, this causes the sleeve to close and prevents worms, insects, and roots from entering and clogging the pipe.

Gravity/Pressure Leaching Chambers

Leaching chambers can be successful in loading and receiving 2.4 gallons per square foot per day of greywater from a three-bedroom home. Using half of a PVC pipe that is 6 inches in diameter, this leaching chamber can be placed within a trench on a 1- to 2-inch mesh plastic netting to prevent the walls from sinking into the soil. No pre-filtration is used in these chambers. All that is required is a dosing pump chamber to pump every eight hours. The trench should have a minimum surface area of about 100 square feet—this will allow for a loading rate of around 2 to 2½ gallons per square foot per day for an average-sized home.

A planter soil box.

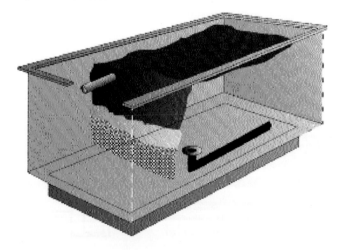

Piping can usually be found in 5-foot sections.

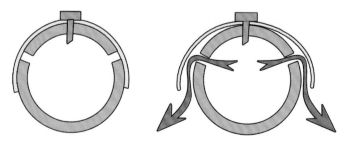

Gravity and Automatic Switch

The illustration below shows an example of an automatic switch system from a shallow leach chamber to one that is below the frost line—an important feature of any greywater system in the northern United States. If a shallow trench freezes and becomes clogged with ice, the water will back up and spill over into the pipe to the deeper, below-the-frost-line trench. It is worth noting that greywater is typically warmer than combined sewage and that the shallow leach zones that are operating in your system tend to stay freer of ice for longer periods of time than in places with combined waste water.

Automatic switching using pump pressure is different from gravity pressure switching. In an automatic switch system, a loop must be arranged indoors where the pressure needed for the shallow infiltration is normally lower than the pressure required to force the water up to the top of the loop. The top of the loop must, then, be no higher than the shut-off head of the pump. About 3 feet of water is a good margin for this system. The system can also be designed to be switched manually by the opening and closing of the valves that feed the different zones and levels of the greywater box.

Options for Using Greywater in Cold Weather

Throughout New England, there are several greywater-irrigated greenhouses that feature a combination of automatically irrigated and fertilized growing beds that provide effective greywater treatment. Since these greenhouses are found in colder, northern states, it is important that these soil beds be deeper to store heat from both the sun and the greywater.

The greenhouse shown here provides enough salad greens for a family of four to six people throughout the long, cold northeastern winters. Growing broccoli, spinach, lettuce, mustard greens, and sorrel in these colder-climate greywater systems can be effective and profitable. To facilitate better distribution of greywater in the soil bed, a pipe-loop system can also be simply constructed to feed the bed from both sides.

An automatic switch system.

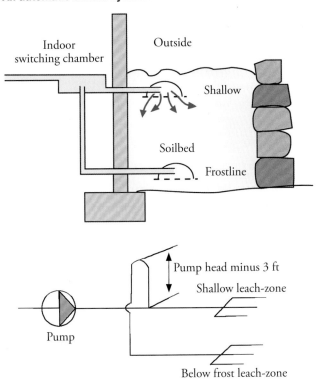

About 3 feet of water is a good margin for this automatic switch system.

An active cooling/passive heating greywater irrigated greenhouse.

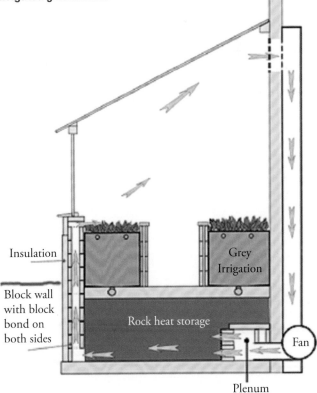

Outdoor Planters

There are many variations of outdoor raised soil beds that are effective in replacing the soil needed for successful leach field treatment of greywater. Houses on ledges or in very sandy soils can be fitted with masonry soil boxes that serve to build up the site's soil profile. Such a strategy has been used in mounds or evapo-transpiration beds (a name derived from the assumption that all of the water will evaporate to the atmosphere even in wet and cold climates).

In parts of the country where construction density makes it very difficult to build a large mound or to locate planters for treating a significant volume of greywater, two adjacent neighbors can agree to build property dividers and plant hedges in their leaching area. This alternative combines privacy, landscaping aesthetics, and good environmental protection. Greywater gardens offer the added benefit of being able to garden at a higher elevation and in a raised garden bed.

Outdoor planters will have a less effective treatment during the winter seasons and during deep freezes. Yet, when relatively warm greywater is injected into the soil, increased biological activity as well as warming of the soil tends to keep the injection area unfrozen for longer periods of time than the surrounding area. Raised beds or planters can also be ideal for compost bins in the fall. The decomposing leaves and grasses act as an insulator as well as a composting fuel source that further insures that the soil beneath does not go into a deep freeze.

Greywater is especially useful in areas that are very dry.

Shallow Subsoil Irrigation

This type of irrigation (2 to 6 inches below the soil level) is preferable to surface irrigation when these factors are in play:

- The water used is "grey" (neither clean nor free of salts)
- The irrigation system is located in a high evaporation locale with water shortages
- It is desired to produce leaf or garden waste compost quickly
- Selective irrigation is needed (for a flower border, shrub, bush, tree, etc.)
- You want to automatically irrigate a drained planter indoors or outdoors

From greywater filter

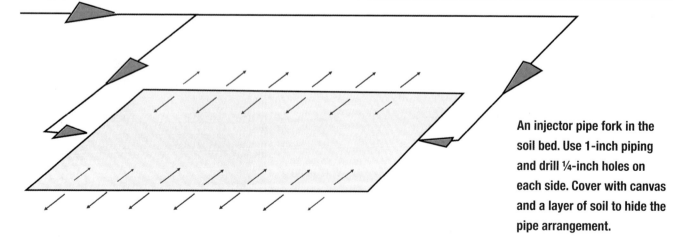

An injector pipe fork in the soil bed. Use 1-inch piping and drill ¼-inch holes on each side. Cover with canvas and a layer of soil to hide the pipe arrangement.

HYDROPOWER

Water is constantly moving through a vast global cycle, evaporating from lakes and oceans, forming clouds, precipitating, and then flowing back into the ocean. The energy of this water cycle, which is mainly driven by the sun, can be tapped to produce electricity or to power machines—a process called hydropower. Hydropower uses water as a type of fuel that is neither reduced nor used up in the process. Since the water cycle is endless and will constantly recharge the system, hydropower is considered a renewable energy.

Hydropower (also known as hydroelectric power) is made when flowing water is captured and turned into elec-tricity. There are many types of hydroelectric facilities that are all powered by the kinetic energy derived from flow-ing water as it moves downstream. Generators and turbines convert this energy into electricity. This is then fed into the electrical grid for use in homes, businesses, and other industries.

A Brief History of Hydropower

Humans have been using water to help them perform work for thousands of years. Water wheels have been employed for grinding grains into flour, to saw wood, and to power textile mills. The technology to use run-ning water to create hydroelectricity has been around for over a hundred years. The modern hydropower turbine was created in the middle of the eighteenth century and developed into direct current technol-ogy. Today, an alternating current is in use and came about when the electric generator was combined with the turbine. The first hydroelectric plant in the United States was built in Appleton, Wisconsin in 1882.

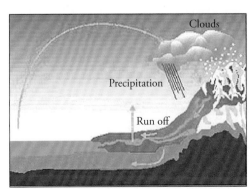

Water's never ending cycle.

Types of Hydropower Plants

There are three types of hydropower plants:

1. Impoundment—Impoundment facilities are the most common type of hydroelectric power plants. This facility, typically a large hydropower system, uses a dam to store river water in a reservoir. Water that is released from the reservoir flows through a turbine, spinning it. This activates a generator to produce electricity. The water may be released either to meet the changing electricity needs or to maintain a constant reservoir level.

2. Diversion—A diversion facility, sometimes referred to as a run-of-river facility, channels a portion of a river through a canal or penstock. This does not always require the use of a dam.

3. Pumped storage—A pumped storage facility stores energy by pumping water from a lower reservoir to an upper reservoir when electricity demands are low. During times when electrical demands are high, water is then released back into the lower reservoir to generate electricity.

Some hydropower plants use dams and others do not. Many dams were originally built for other purposes and then hydropower was added at a later date. In the United States, only 2,400 of the 80,000 dams produce power—the rest are used for recreation, farm ponds, flood control, water supply, and irrigation.

Size of Hydropower Plants

Hydropower plants range in size from small and micro systems, which are operated for individual needs or to sell the power to utilities, to larger projects that produce electricity for utilities, supplying many consumers with electricity.

Micro hydropower plants have a capacity of up to 100 kilowatts. Small hydropower plants have a capacity between 100 kilowatts and 30 megawatts. Large hydropower plants have a capacity of more than 30 megawatts. The small and micro systems can produce enough electricity for a home, farm, or even a small village.

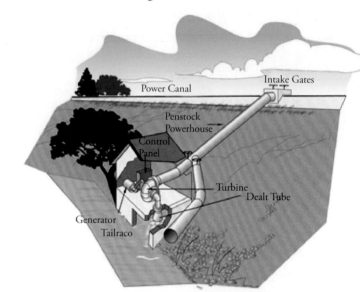

A micro hydropower plant.

Diagram of a hydropower plant.

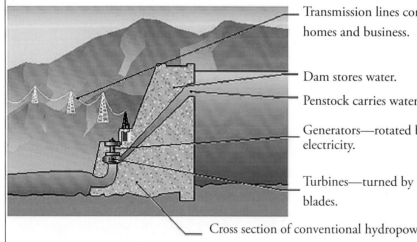

Transmission lines conduct electricity ultimately to homes and business.

Dam stores water.

Penstock carries water to the turbines.

Generators—rotated by the turbines to generate electricity.

Turbines—turned by the force of the water on their blades.

Cross section of conventional hydropower facility that uses an impoundment dam.

Even a small waterfall can provide a lot of power.

Hydropower Turbines

There are two main types of hydropower turbines: impulse and reaction. The type of turbine selected for a project is based on the height of the standing water (the "head") and the flow (volume) of the water at a particular site. It is also determined by how deep the turbine must be set, its efficiency, and its cost.

Impulse Turbine

An impulse turbine typically uses the velocity of water to move the runner and discharges to atmospheric pressure. The water stream then hits each bucket on the runner. The water flows out of the bottom of the turbine after hitting the runner. These turbines are suitable for high head, low flow applications.

Reaction Turbine

A reaction turbine generates power by the combined action of pressure and moving water. The runner is placed in the water stream, which flows over the blades instead of striking each one separately. These turbines are used for sites with lower head and higher flows.

each one separately. These turbines are used for sites with lower head and higher flows.

Diagram of a hydroelectric motor.

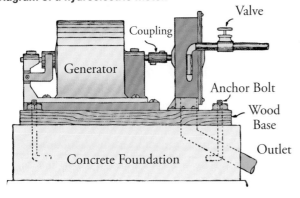

Advantages of Hydropower

- It is fueled by water, making it a clean energy source.
- It does not pollute the air since it does not burn any fossil fuels.
- It is a domestic energy source.
- It relies on the water cycle and is a renewable energy source.
- It is usually available as needed.
- The water flow can be controlled through the turbine to produce energy on demand.
- The plants provide reservoirs for recreation (fishing, swimming, boating), water supply, and food control.

Disadvantages of Hydropower

- It can negatively impact fish populations by hampering fish migration upstream past dams, though there are ways to allow for passage both up- and downstream.
- It can impact the quality and flow of water, causing low dissolved oxygen levels that can negatively impact the riverbank habitats.
- The plants can be impacted by drought, and if they are not receiving adequate water, they cannot produce electricity.
- The plants compete for land use and can cause humans, plants, and animals to lose their natural habitat.

SOLAR ENERGY

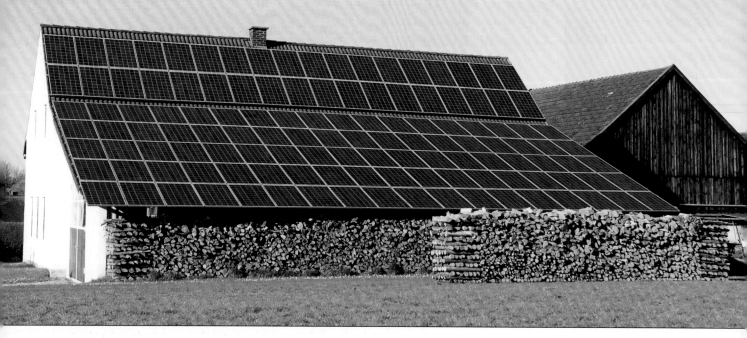

Solar energy is, in its simplest form, the sun's rays that reach the earth (also known as solar radiation). When you step outside on a hot, sunny summer day, you can feel the power of the sun's heat and light. Solar energy can be harnessed to do a variety of things in your home. These include:

- Heating your home through passive solar design or through active solar heating systems
- Generating electricity
- Heating water in your home
- Heating swimming pool water
- Lighting your home both inside and out
- Drying your clothes via a clothesline strung outside in direct sunlight

Solar energy can also be converted into thermal (heat) energy and used to heat water for use in homes, buildings, or swimming pools and also to heat spaces inside homes, greenhouses, and other buildings.

Photovoltaic energy is the conversion of sunlight directly into electricity. A photovoltaic cell, known as a solar or PV cell, is the technology used to convert solar energy into electrical power. A PV cell is a non-mechanical device made from silicon alloys. PV systems are often used in remote locations that are not connected to an electric grid. These systems are also used to power watches, calculators, and lighted road signs.

Advantages of Solar Energy

- It's free.
- Its supplies are unlimited.
- Solar heating systems reduce the amount of air pollution and greenhouse gases that result from using fossil fuels (oil, propane, and natural gas) for heating or generating electricity in your home.
- Solar heating systems reduce heating and fuel bills in the winter.
- It is most cost-effective when used for the entire year.

Disadvantages of Solar Energy

- The amount of sunlight that arrives at the earth's surface is not constant and depends on location, time of day and year, and weather conditions.
- A large surface area is required to collect the sun's energy at a useful rate.

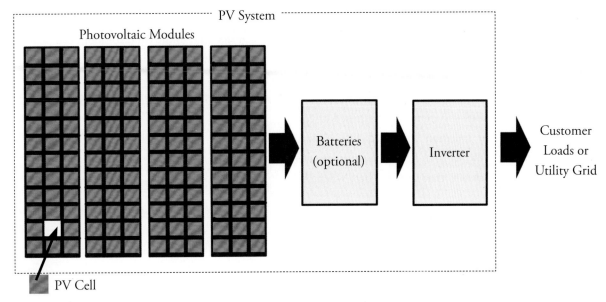

PV system components.

Solar Thermal Energy

Solar thermal (heat) energy is used most often for heating swimming pools, heating water to be used in homes, and heating specific spaces in buildings. Solar space heating systems are either passive or active.

Passive Solar Space Heating

Passive space heating is what happens in a car on a sunny summer day—the car gets hot inside. In buildings, air is circulated past a solar heat surface and through the building by convection—less dense, warm air tends to rise while the denser, cooler air moves downward. No mechanical equipment is needed for passive solar heating.

Passive solar space heating takes advantage of the warmth from the sun through design features, such as large, south-facing windows and materials in the floors and/or walls that absorb warmth during the day and release it at night when the heat is needed most. Sunspaces and greenhouses are good examples of passive systems for solar space heating.

Passive solar systems usually have one of these designs:

1. Direct gain—This is the simplest system. It stores and slowly releases heat energy collected from the sun shining directly into the building and warming up the materials (tile or concrete). It is important that the space does not become overheated.

2. Indirect gain—This is similar to direct gain in that it uses materials to hold, store, and release heat. This material is generally located between the sun and the living space, usually in the wall.

3. Isolated gain—This collects solar energy separately from the primary living area (a sunroom attached to a house can collect warmer air that flows through the rest of the house).

Active Solar Space Heating

Active heating systems require a collector to absorb the solar radiation. Fans or pumps are used to circulate the heated air or the heat-absorbing fluid. These systems often include some type of energy storage system.

There are two basic types of active solar heating systems. These are categorized based on the type of fluid (liquid or air) that is heated in the energy collectors. The collector is the device in which the fluid is heated by the sun. Liquid-based systems heat water or an antifreeze solution in a hydronic collector. Air-based systems heat air in an air collector. Both of these systems collect and absorb solar radiation, transferring solar heat to the interior space or to a storage system, where the heat is then distributed. If the system cannot provide adequate heating, an auxiliary or backup system provides additional heat.

Liquid systems are used more often when storage is included and are well suited for radiant heating systems, boilers with hot water radiators, and absorption heat pumps and coolers. Both liquid and air systems can adequately supplement forced air systems.

Active solar space heating systems are comprised of collectors that absorb solar radiation combined with electric fans or pumps to distribute the solar heat. These systems also have an energy-storage system that provides heat when the sun is not shining.

Another type of active solar space heating system, the medium temperature solar collector, is generally used for solar space heating. These systems operate in much the same way as indirect solar water heating systems but have a larger collector area, larger storage units, and much more complex control systems. They are usually configured to provide solar water heating and can provide between 30 and 70 percent of residential heating requirements. All active solar space heating systems require more sophisticated design, installation, and maintenance techniques than passive systems.

Passive Solar Water Heaters

Passive solar water heaters rely on gravity and on water's natural tendency to circulate as it is heated. Since these heaters contain no electrical components, passive systems are more reliable, easier to maintain, and work longer than active systems. Two popular types of passive systems are:

1. Integral-collector storage systems—These consist of one or more storage tanks that are placed in an insulated box with a glazed side facing the sun. The solar collectors are best suited for areas where temperatures do not often fall below freezing. They work well in households with significant daytime and evening hot-water needs but they do not work as efficiently in households with only morning hot-water draws as they lose most of the collected energy overnight.

2. Thermospyhon systems—These are an economical and reliable choice particularly in newer homes. These systems rely on natural convection of warm water rising to circulate the water through the collectors and into the tank. As water in the collector heats, it becomes lighter and rises to the tank above it and the cooler water flows down the pipes to the bottom of the collector.

In freeze-prone climates, indirect thermosyphons (using glycol fluid in the collector loop) can be installed only if the piping is protected.

Active Solar Water Heaters

Active solar water heaters rely on electric pumps and controllers to circulate the water (or other heat-transfer fluids). Two types of active solar water heating systems are:

1. Direct circulation systems—These use pumps to circulate pressurized potable water directly through the collectors. These systems are most appropriate for areas that do not have long freezes or hard/acidic water.

2. Indirect circulation systems—These pumps heat transfer fluids through the collectors. These heat exchangers then transfer the heat from the fluid to potable water. Some of these indirect circulation systems have overheat protectors so the collector and glycol fluid do not become superheated. Common indirect systems include antifreeze, in which the heat transfer fluid is usually a glycol-water mixture, and drainback, in which pumps circulate the water through the collectors and then the water in the collector loop drains back into a reservoir tank when the pump stops.

A combination of an indirect water heater and a highly efficient boiler can provide a very inexpensive method of water heating.

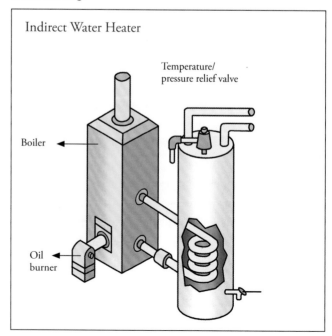

Indirect Water Heater

Temperature/ pressure relief valve

Boiler

Oil burner

Solar panels don't have to be set up on your roof; sometimes it's more efficient to place them in a sunny field near your home.

Installing a Passive Solar Space Heater

A passive solar space heater works when the sun shines through the solar panels to heat the air inside a box. As the air heats up in the box, it rises and moves into the house. Cool air moves into the box and out of the house—in this way, the house is heated without the use of a mechanized heating system. Using a passive solar heater works best if you have a house that faces south and has both basement and first floor windows on that side of the house. If your house meets these requirements (and there aren't too many obstructions that would impede the sun from shining on the heater), then you can begin construction.

The passive solar space heater is made up of a floor and two triangular end walls, all of which can be made simply out of plywood. In between the open space, insulation can be placed. A lid can also be added to cover the heater in the summer.

To build such a solar space heater, first decide where on the southern wall your collector will be located. If you can place the heater in between windows, that is the best option.

You may need to cut through the wall near a window to allow for the proper ventilation but if you don't want to do this, you can also purchase a detachable plywood "chimney" to move the heated air into the house. Next, find the studs that will support the fiberglass panel and find a panel that will be of the appropriate size.

Next, make the base for your solar heating system. The base can be made of ⅜-inch plywood board. Nail the board to a 2 x 4 and level it. Next, add insulation (the kind found on rolls is best), nailing it to the plywood. Then, nail the whole board to the side of the house. Make sloping supports out of 2 x 4s. Make sure the end wall studding is nailed in, and then attach the outside panel to it.

Under the shingles, install flashing or something else that will keep water out of the top of the solar heater. Then, install the fiberglass panels, making sure the edges are caulked so no water can come in. Enclose the edges of the fiberglass with small strips of plywood. Then, install the outer fiberglass panel so that it is flush with the top surface and caulk it. To finish up, paint the inside of the plywood surfaces black to absorb the heat. The inside of the cover panel should be painted white to reflect the light.

A passive solar space heater.

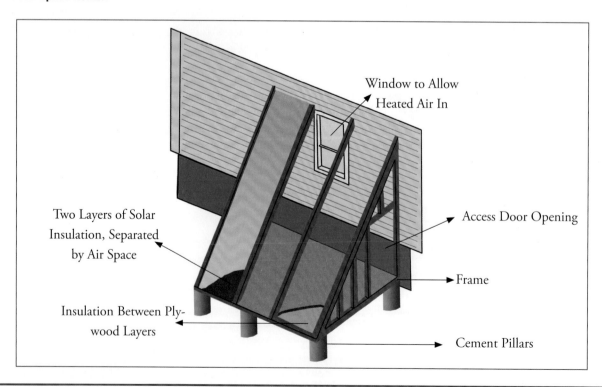

Window to Allow Heated Air In

Two Layers of Solar Insulation, Separated by Air Space

Insulation Between Plywood Layers

Access Door Opening

Frame

Cement Pillars

Building Your Own Solar Water Heater

This very simple and basic solar water heater is a low-pressure system and so should not be combined with your home plumbing system. This type of heater is perfect for camping trips or other smaller water heating uses. Find the supplies online or at a hardware store.

Supplies

- Corrugated, high-density polyethylene draining tube (4 inches is preferred)
- An EPDM rubber cap with clamp (available at hardware stores or online)
- Polyethylene terephthalate bottles (3-liter are preferred—soda bottles are fine)

To construct the water heater, simply stretch the EPDM rubber cap over one end of the draining tube and make certain the clamp is tight. Cut the ends off the bottles and fit them over the other end of the drainage pipe. This will serve as the glazing to heat the water. Each bottle should be able to fit tightly over the other bottle if you cut a small hole in the bottom of each. Fill the tube with water, place it in the sun, and allow the water inside the bottles and drainage tube to heat up. Once it's warm (around 120°F is the maximum it

A solar water heater.

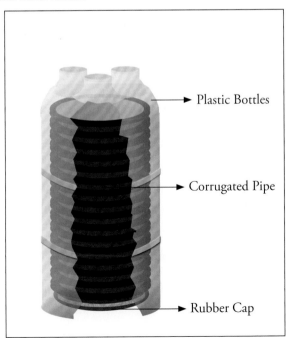

Plastic Bottles

Corrugated Pipe

Rubber Cap

will heat the water), it can be used to wash dishes or clothes, or for a small bath.

Heating a Room Using Collectors

Air collectors can be installed on a roof or an exterior, south-facing wall to facilitate the heating of one or more rooms in a house. Factory-built collectors can be used but you can also make and install your own air collector, though note that this is not always cost-efficient.

The air collector should have an airtight and insulated metal frame and a black metal plate. This will absorb the heat through the glazing on the front. The sun's rays heat the plate, which then heats the air in the collector. A fan or blower can pull the air from the room through to the collector and blow it into the room.

Room Air Heating with Collectors

Air collectors can be installed on a roof or an exterior (south facing) wall for heating one or more rooms. Although factory-built collectors for on-site installation are available, do-it-yourselfers may choose to build and install their own air collectors. A simple window air heat collector can be made for a few hundred dollars. Simple window box collector fans will fit in a window opening. These fans can be active or passive. A passive collector fan allows air to enter the bottom of the collector, rise as it heats, and enter the room. A damper keeps the room air from flowing back into the panel on overcast or cloudy days. Window box systems only provide a small amount of heat as the collectors are rather small.

Solar Collectors

Solar collectors are an essential part of active solar heating systems. These collectors harness the sun's energy and transform it into heat. Then, the heat is transferred to water, solar fluid, or air. Solar collectors can be one of two types:

1. Nonconcentrating collectors—These have a collector area that is the same size as the absorption area. The most common type is flat-plate collectors and these are used when temperatures below 200°F are sufficient for space heating.

Roof Area Needed in Square Feet (Shown in Bold Type)

PV module efficiency (¼)	PV capacity rating (watts)							
	100	250	500	1,000	2,000	4,000	10,000	100,000
4	30	75	150	300	600	1,200	3,000	30,000
8	15	38	75	150	300	600	1,500	15,000
12	10	25	50	100	200	400	1,000	10,000
16	8	20	40	80	160	320	800	8,000

Although the efficiency (percent of sunlight converted to electricity) varies with the different types of PV modules available today, higher-efficiency modules typically cost more. So, a less-efficient system is not necessarily less cost-effective.

Solar collectors on a roof.

2. Concentrating collectors—The area of these collectors gathering the solar radiation is much greater than the absorber area.

Solar thermal energy can be used for solar water heating systems, solar pool heaters, and solar space heating systems. There are many types of solar collectors, such as flat plate collectors, evacuated tube collectors, and integral collector storage systems.

Another Form of Solar Heating: Daylighting

Solar collector panels are not the only way in which the sun's heat can be harnessed for energy purposes. Daylighting uses windows and skylights to bring sunlight into your home.

Using energy-efficient windows, as well as carefully thought-out lighting design, reduces the need for artificial lighting during the daytime. These windows also cut down on heating and cooling problems.

The effectiveness of daylighting in your home will depend on your climate and the design of your house. The sizes and locations of window and skylights should be based on the way in which the sun hits your home and not on the outward aesthetics of your house. Facing windows toward the south is most advantageous for daylighting and for moderating seasonal temperatures. Placing windows that face toward the south will allow more sunlight into your home during the winter months. North-facing windows are also useful for daylighting as they allow a relatively even, natural light into a room, produce little glare, and capture no undesirable summer heat.

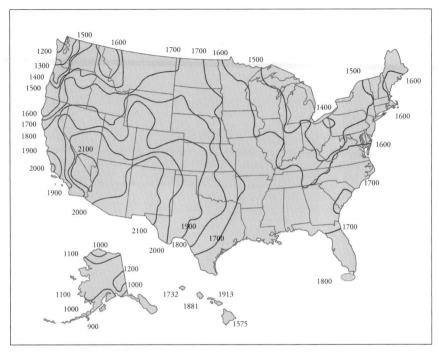

First determine the system's size in kilowatts (kW). A reasonable range is 1 to 5 kW. This value is the "kW of PV" input in the equations. Next, based on your geographic location, select the energy production factor from the map below for the kWh/kW-year input for the equations. Energy from the PV system = (kW of PV) x (kWh/kW-year) = kWh/year. (Divide this number by twelve if you want to determine your monthly energy reduction.) Energy bills savings = (kWh/year) x (Residential Rate)/100 = $/year saved. (Residential Rate in this above equation should be in dollars per kWh; for example, a rate of 10 cents per kWh is input as $0.10/kWh.)

For example, a 2-kW system in Denver, CO, at a residential energy rate of $0.07/kWh will save about $266 per year (1,900 kWh/kW-year x $0.07/kWh x 2kW = $266/year).

Make Your Own Solar Cooking Oven

This type of simple, portable solar oven is perfect for camping trips or if you want to do an outdoor barbeque with additional cooked foods in the summer. This homemade solar oven can reach around 350°F when placed in direct sunlight.

Supplies

A reflective car sunshade or any sturdy but flexible material (such as cardboard) covered with tin foil and cut to the notched shape of a car sunshade

Velcro

A bucket

A cooking pot

A wire grill

A baking bag

Directions

1. Place the car sunshade on the ground. Cut the Velcro into three separate pieces and stick on half of each piece onto the edge near the notch. Then, test the shade to see if the Velcro pieces, when brought together, form a funnel. Place the funnel atop the bucket.

2. Place the cooking pot on the wire grill. Put this all in the baking bag and put it inside the funnel. The rack should now be laying on top of the bucket. Now place the whole cooker in direct sunlight and angle the funnel in the direction of the sun. Adjust the angle as the sun moves.

A simple solar oven.

ILLUSTRATION BY TIMOTHY LAWRENCE

Including plenty of energy-efficient windows in your home will allow sunlight to warm your rooms naturally.

Make Your Own Solar Panels

Making your own solar panels can be tricky and time-consuming, but with the right materials and lots of patience, you can certainly create an effective solar energy panel.

Supplies

Pegboard
Solar cells (quality will be determined on how much power you want to get from your solar panel)
Contact wire
Wire cutters
Solder
Soldering iron
Bolts with washers and wingnuts
Plexiglass
Plywood board
Aluminum framing
Silicone caulking
Screws

Solar ovens can be fashioned in a variety of ways. The goal is to have as much surface area as possible reflecting the sun toward your food.

Directions

1. Apply silicone caulking in vertical strips between the rows of holes on the peg board. Place the solar cells face up along the caulking in straight rows, carefully aligning them so that the wires poke through the holes. The so-

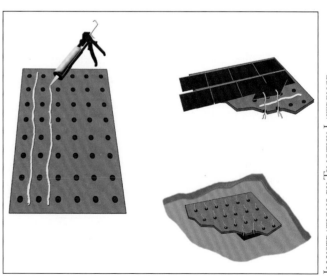

ILLUSTRATIONS BY TIMOTHY LAWRENCE

The more surface area you cover with solar panels, the more power you'll get. It's best to install your panels on the south side of your home.

Installing Your Heat Collector

If possible, install your own solar heat collector on the south side of your house (the side that receives the most sunlight during the day). It can be placed in a window to help minimize your heating costs during the winter months.

A solar heat collector can be made from heavy-duty foam insulation, window glass, sealant, aluminum foil, and heavy-duty tape. Paint the foam panel, or both sides of the aluminum sheets, black and then mount it on cubes that are cemented to the side of your house near a window. This will allow the air to come in on both sides of the heat collector.

All sides of the foam should be covered with aluminum foil and then adhered to the foam board. Then place and seal the glass panels over the foam, sealing it with the sealant and heavy-duty tape if needed. Another piece of foam can be utilized as a cover for the duct at night or during the warm, sunny summer months. Hinge this on with hinge brackets or clasps.

An Alternative Solar Heating Panel

This type of solar panel is different from the expensive, manufactured panels you can purchase and have installed on your roof or the side of your house. It is great for heating air but cannot produce electricity. You can either situate this heater in a south-facing window of your home or place it on the outside, southern wall or on the roof. Heating panels that are on the outside of a house generally create more heat and are much more effective in heating a room or area of your home.

To start, you will need to purchase glass or Plexiglas for your solar heating panel. Either one should be double-paned to keep out moisture. To build the frame for your solar heating panel, use 2 x 4s and create a square or rectangle that will fit your pane of glass. Nail plywood to the back of the frame. Next, take a piece of insulation board and put it at the back of the panel. Heat absorption can be gained through aluminum flashing or copper. After this is inserted, screw down the window frame, if you are using one, and make sure it is caulked well to keep out water.

lar cells should completely cover the board.

2. Place a soft sheet or blanket on the ground or table (to prevent the cells from scratching) and carefully flip the board so that it is face down. Solder together the wires coming out to create one thick wire stemming from each hole. Then use connecting wire or metal strips to connect the wires along horizontal lines. Be sure to connect all positive wires together and all negative wires together, without mixing the two.

3. Drill two holes in the back of your panel and attach a positive and negative bolt, washer, and wingnut. Solder the positive wires to the positive bolt and the negative wires to the negative bolt.

4. Build a watertight frame to size, using aluminum framing for the sides, plywood for the backing, and a plexiglass face to allow the sunlight to shine through. Seal all cracks and edges with silicone sealant.

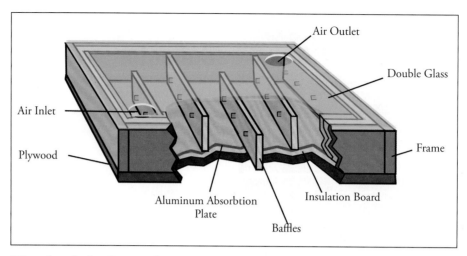

Alternate solar heating panel.

Add the interior boards that line the frame and the baffles to seal the top of the glass. Screw these interior boards to the sides of the panel to keep them secure. Then, cut out the air openings using a jigsaw. One circular opening should be in the lower left and the other in the upper right of your heating panel. Before hanging the panel up, you will need to determine where the studs are in the wall or, if you are installing it on your roof, where the roof rafters are located. It is also important that your openings do not fall on top of a stud or rafter as this will defeat their ability to direct airflow. Screw in boards along the studs or rafters, on which you will then mount the panel.

Once the panel is secured to the wall or roof, begin to install the air delivery system so the hot air can be circulated throughout your home. You may want to add a small fan (one used in a computer will be fine) to your heating panel so you can better circulate the air throughout the system, though this is not necessary to operate your heating panel effectively. If you do choose to use a fan, it must be able to fit inside the wall plate. You will need to drill a hole in your wall where the panel holes are situated on the outside. Cut the hole and add the connector to the ductwork, sliding it through the hole into the room, and seal off the edges of the hole.

Place the fan within the wall plate in the room, and place an electrical box near the fan to turn it on and off. If you aren't familiar with electrical work, you may want to ask an electrician to help you with connecting the electrical wiring. Next, mount the solar panel so it faces to the south,

running a wire into the electrical box inside the room. This will save you money and energy while running your fan. Now turn on the fan and feel the warm air starting to blow through your room.

To finish your outside panel, simply paint the inside black to absorb more heat, add some weather stripping to seal the glass tightly, and screw the glass piece to the panel.

Regulations for Installing and Building Solar Heating Systems

Before you install a solar energy system, learn about the local building codes, zoning, and neighborhood covenants as they apply to these systems. You will most likely need to obtain a building permit to install a solar energy system onto an existing building. Common problems you may encounter as a homeowner in installing a solar energy system are: exceeding roof load, unacceptable heat exchangers, improper wiring, tampering with potable water supplies, obstructing property and yards, and placing the system too close to the street or lot lines. There are also local compliances that must be factored in before installing your system. Contact your local jurisdiction zoning and building enforcement divisions and any homeowner's, neighborhood, or community associations before building and installing any solar heating equipment.

Solar Greenhouse

Greenhouses collect solar energy on sunny days and then store the heat for use in the evening and on days when it is overcast. A solar greenhouse can be situated as a free-standing structure (like a shed or larger enclosure) or in an underground hole. See page 650 for more information on greenhouses

For gardeners who want to grow small amounts of produce, passive solar greenhouses are a good option and help extend the growing season. Active systems take supplemental energy sources to move the solar heated air from its storage facility to other parts of the greenhouse. Solar greenhouses can utilize many of the same features and installation techniques as passive solar heating systems used in homes to stay heated.

While standard greenhouses also rely on the sun's rays to heat their interiors, solar greenhouses are different because they have special glazing that absorbs large amounts of heat during the winter months and also use materials to store the heat. Solar greenhouses have a lot of insulation in areas with little sunlight to keep heat loss at a minimum.

Types of Solar Greenhouses

Two common types of solar greenhouses are the attached solar greenhouse and the freestanding solar greenhouse. Attached solar greenhouses are situated next to a house or shed and are typically lean-to structures. They are limited in the amount of produce they can grow and have passive solar heating systems.

Freestanding solar greenhouses are large structures that are best-suited for producing a large variety and quantity of produce, flowers, and herbs. They can be constructed in the form of either a shed structure or a hoop house. In a shed greenhouse, the south wall is glazed to maximize the heating potential and the north wall is extremely well-insulated. Hoop house greenhouses are rounded instead of shaped like an elongated shed. Solar energy is collected and stored in earth thermal storage and in water. These systems, while common, are not as effective in utilizing solar energy as the shed and lean-to structures.

Sites for Solar Greenhouses

The glazing portion of the solar greenhouse should ideally face directly south to gain the maximum exposure to the sun's heat. Situating the solar greenhouse on a slight slope facing upward will maximize the amount of solar energy it can absorb.

Materials Used in Solar Greenhouse Construction

For a solar greenhouse to collect, circulate, and maintain the greatest amount of heat, it needs to be constructed out of the proper materials. Glazing materials need to allow photosynthetic radiation to get through so it can reach the plants. Clear glass allows direct light into the greenhouse and so should be used as a glazing material. It is also imperative that when the glazing materials are mounted on the greenhouse,

Both flowers and vegetables can thrive in greenhouses year-round.

A solar greenhouse.

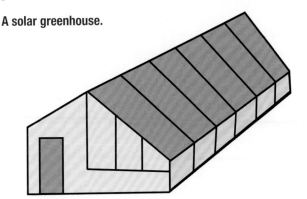

Greenhouses can be made in a range of shapes and sizes and can be attached to or separate from your home.

there are no cracks or holes that can allow for heat to escape. Thus, glazing material should have high heat efficiency and be made of resistant material to hold up in inclement weather and hail.

Solar greenhouses also need to be able to store the heat that is collected for use on cloudy days or at night. The easiest method for storing heat is to situate rocks, concrete, and/or water in the path of the sunlight that is entering the greenhouse. These materials will absorb the heat during the day and release it during the evening hours. Pools of water, rocks, and concrete slabs or small walls should be large enough to absorb and emit enough heat to last for the night or for a few cloudy days.

Phase-change materials may also be used to effectively store heat in your solar greenhouse. These materials consist

of paraffin, fatty acids, and Glauber's salt. These materials store heat as they change into liquid and release it as they turn back into a solid form. They are kept in sealed tubes and many are needed to provide enough heat.

All areas of the greenhouse that are not glazed need to be insulated to keep in the maximum amount of heat. Weather stripping is helpful in sealing doors and vents; foam insulation is helpful for walls. Place a polyethylene film between the insulation and the greenhouse walls to keep these materials dry—if they become too wet or saturated, they will be less effective and may start to mold. The floors of a solar greenhouse can also lose heat so they should be made out of brick or flagstone (with insulation foam underneath) to keep the heat in.

The solar greenhouse needs outdoor insulation as well, which can be attained by placing hay bales along the edges of the greenhouse, or the greenhouse can be situated slightly underground (a pit greenhouse). Of course, if a greenhouse is dug into the soil, it needs to be in an area that is above the water level to minimize leakage.

A solar greenhouse, like any other greenhouse, also needs proper ventilation for the warmer summer months. Vents in the sides of the greenhouse will help create air flow. Ridge vents in the roof will allow the hottest air to escape out of the top of the greenhouse as well. If a greenhouse needs more ventilation, a solar chimney can be hooked up to the passive solar collectors to release extra heat out into the air.

WIND ENERGY

Wind energy is created naturally by circulation patterns in the Earth's atmosphere driven by the heat from the sun. These winds are caused by the uneven heating of the atmosphere by the sun, the irregularities of the earth's surface, and the rotation of the earth. Wind patterns are modified by the earth's terrain, bodies of water, and vegetation. Since the earth's surface is made of very different types of land and water, it absorbs the sun's heat at different rates. During the day, the air above the land heats up very quickly. The warm air over the land expands and rises and the heavier, cooler air rushes in to take its place, creating winds. At night, the winds are reversed as the air cools rapidly over land. This air flow is used for many purposes: sailing, flying kites, and generating electricity.

Small Wind Electric Systems

Small wind electric systems are one of the most cost-effective, home-based renewable energy systems. These systems

A Brief History of Wind Energy

People have been harnessing energy from the wind since ancient times. Wind was used to sail ships and windmills were build to help grind wheat, corn, and other grains. Windmills were also used to pump water and to cut wood at sawmills in the formative years of the American colonies. Even into the early twentieth century, windmills were being used to generate electricity in rural parts of America. The windmill again gained national attention in the early 1980s when wind energy was finally considered a renewable energy source. It continues to be a growing industry throughout the United States.

are nonpolluting and are fairly easy to set up. A small wind electric system can effectively:

- Lower your electricity bills by 50 to 90 percent
- Help you avoid high costs of having utility power lines extended to a remote location
- Help uninterruptible power supplies ride through extended utility outages

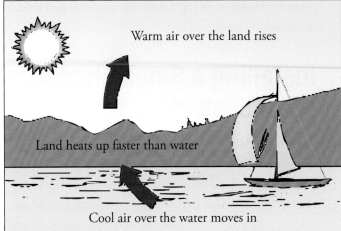

Natural air flow cycle.

1. A wind turbine—This consists of blades attached to a rotor, a generator/alternator mounted on a frame, and a tail
2. A tower
3. Balance-of-system components—i.e., controllers, inverters, and/or batteries

A wind-electric turbine generator, more commonly known as a "wind turbine," converts kinetic energy in the wind into mechanical power. This power can be used directly for specific tasks, like grinding grains or pumping water. A generator can also convert this mechanical power into a high-value, highly flexible, and useful form of energy—electricity.

Wind turbines make electricity by working in the opposite way as a fan. Instead of using electricity to make wind, as a fan does, turbines *use* wind to make electricity. The wind

How Do Small Wind Electric Systems Work?

When the wind spins a wind turbine's blades, a rotor captures the kinetic energy of the wind, converting it into rotary motion to drive the generator. Most turbines have automatic overspeed-governing systems to keep the rotor from spinning out of control on very windy days.

A small wind system can be connected to an electric distribution system (grid-connected) or it can stand alone (off-grid). To capture and convert the wind's kinetic energy into electricity, a home wind energy system must generally be comprised of the following:

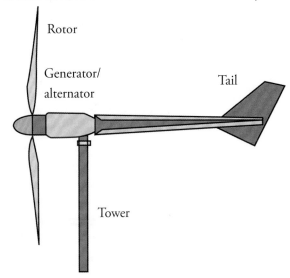

The basic parts of a small wind electric system.

turns the blades, spinning a shaft that connects to a generator, which makes electricity.

Installing a Small Electric Wind System

Small wind electric systems, with the proper installation and maintenance, can last over 20 years. Before installing your system, first find the best site, determine the appropriate size of your wind turbine, decide whether you want a grid-connected or off-grid system, and find out about your local zoning, permitting, and neighborhood covenant requirements.

Many people decide to install these systems on their own (though the manufacturer and/or dealer-should also be able to help you install the small wind electric system). However, before you attempt to install the wind turbine, make sure you can answer these do-it-yourself questions:

1. Can I pour a proper cement foundation?
2. Do I have access to a lift, ladder, or another way to erect the tower safely?
3. Do I know the difference between alternating current (AC) and direct current (DC) wiring?
4. Do I know enough about electricity to safely wire my turbine?
5. Do I know how to safely handle and install batteries?

If the answer to any of these questions is "no," then you should have someone help you install the system (contact the manufacturer or your state energy office).

Evaluating a Potential Site for Your Small Wind Turbine

The site on which you choose to install your system should meet the following criteria:

• Your property has a good wind resource—good annual wind speeds and a prevailing direction for the wind.

• Your home is located on at least one acre of land in a rural area.

• Your local zoning codes and covenants do not prohibit construction of a wind turbine.

• Your average electricity bill is $150 per month or more.

If you live in an area that has complex terrain, be careful when selecting an installation site. If you place your wind turbine on the top of a hill or on an exceptionally windy

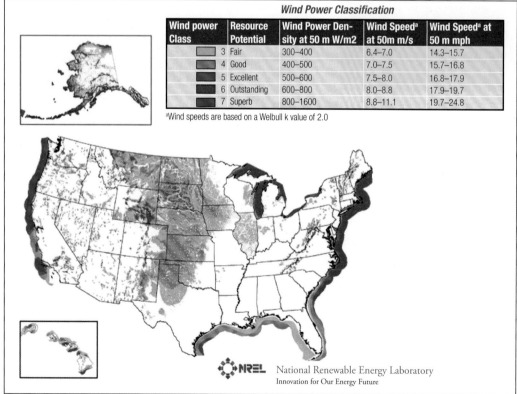

Wind Power Classification				
Wind power Class	Resource Potential	Wind Power Density at 50 m W/m2	Wind Speed[a] at 50m m/s	Wind Speed[a] at 50 m mph
3	Fair	300–400	6.4–7.0	14.3–15.7
4	Good	400–500	7.0–7.5	15.7–16.8
5	Excellent	500–600	7.5–8.0	16.8–17.9
6	Outstanding	600–800	8.0–8.8	17.9–19.7
7	Superb	800–1600	8.8–11.1	19.7–24.8

[a]Wind speeds are based on a Welbull k value of 2.0

This map shows the potential for wind energy in various parts of the United States.

NREL National Renewable Energy Laboratory
Innovation for Our Energy Future

Inside a Wind Turbine

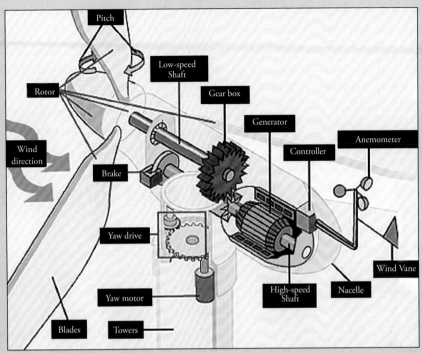

Parts of a wind turbine:

- Anemometer: measures the wind speed and transmits wind speed data to the controller.
- Blades: most turbines have either two or three blades and the wind blows over the blades, causing the blades to lift and rotate.
- Brake: a disc brake, applied mechanically, electrically, or hydraulically, and stops the rotor in emergencies.
- Controller: starts up the machine at wind speeds of about 8 to 16 mph and shuts off the machine at about 55 mph wind speeds. Turbines do not operate at wind speeds above 55 mph because they may be damaged.
- Gear box: gears connect the low-speed shaft to the high-speed shaft and increase the rotational speeds from about 30 to 60 rotations per minute (rpm) to about 1000 to 1800 rpm—the rotational speed required by most generators to produce electricity. The gear box is a costly and heavy part of the wind turbine.
- Generator: usually an off-the-shelf induction generator that produces 60-cycle AC electricity.
- High-speed shaft: drives the generator.
- Low-speed shaft: turned by the rotor at about 30 to 60 rpm.
- Nacelle: sits atop the tower and contains the gear box, low- and high-speed shafts, generator, controller, and brake. Some nacelles are large enough for a helicopter to land on.
- Pitch: Turns the blades out of the wind to control the rotor speed and keep the rotor from turning in winds that are too high or too low to produce electricity.
- Rotor: the blades and hub.
- Tower: made from tubular steel, concrete, or steel lattice. Since wind speed increases with height, taller towers enable turbines to capture more energy and generate more electricity.
- Wind direction: an "upwind" turbine operates facing into the wind while other turbines are designed to face "downwind" or away from the wind.
- Wind vane: measures wind direction and communicates with the yaw drive to orient the turbine properly with respect to the wind.
- Yaw drive: used to keep the rotor facing into the wind as the wind direction changes (not required for downwind turbines).
- Yaw motor: powers the yaw drive.

side, you will have more access to prevailing winds than in a gully or on the sheltered side of a hill. Additionally, consider any existing obstacles—trees, houses, sheds—that may be in the way of the wind's path. You should also plan for future obstructions, such as new buildings or landscaping. Your turbine needs to be positioned upwind of any buildings and trees, and it needs to be 30 feet above anything within 300 feet of its site.

When determining the suitability of your site for a small electric wind system, estimate your site's wind resource. Wind resource can vary significantly over an area of just a few miles because of local terrain's influence on wind flow. Use the following methods to help estimate your wind resource before installing your small electric wind system:

1. Consult a wind resource map. This is used to estimate the wind resource in your area. You can find a specific map for your state at the U.S. Department of Energy's Wind Powering America Program Web site. A general U.S. map is shown in the figure.

2. Obtain wind speed data. The easiest way to quantify the wind resource in your area is by obtaining the average wind speed information from a local airport. Airport wind data are typically measured 20 to 33 feet above ground. Average wind speeds increase with height and may be as much as 15 to 25 percent greater at a usual wind turbine hub (80 feet high) than those measured at airports.

3. Watch vegetation flagging. Flagging is the effect of strong winds on an area of vegetation. For example, if a group of trees on flat ground is leaning significantly in one direction, chances are they've become that way due to strong winds.

4. Use a measurement system. Direct monitoring using a measurement system at a certain site provides the best picture of the available wind resource. These are very expensive, however, and so may not be practical to use.

5. Obtain data from a local small wind system—if there is a small wind turbine near your area, you may be able to obtain information on the annual output of the system, as well as wind speed data.

Small Wind Turbines Used for Homes

Single, small, stand-alone turbines that are sized less than 100 kilowatts are used for homes, telecommunication dishes, and water pumping. Used in residential applications, these small wind turbines can range from 400 watts to 20 kilowatts. In addition to being used for generating electricity and pumping water, they can be used for charging batteries. Most U.S. manufacturers rate their small wind turbines by the amount of power they can safely produce at wind speeds between 24 and 36 mph.

An average home uses about 9,400 kilowatt-hours of electricity per year. Thus, a wind turbine rated in the 5- to 15-kilowatt range would make a significant contribution to this energy demand. Before deciding on a wind turbine you should:

1. Establish an energy budget. Try to reduce the electricity use in your home so you will only need a small turbine.

2. Determine an appropriate height for the wind turbine's tower so it will generate the maximum amount of energy.

3. Remember that a small home-sized wind machine has rotors that are between 8 and 25 feet in diameter and stand around 30 feet tall. If your property does not have enough space to accommodate this, you may not be able to have a powerful enough turbine to help significantly reduce your energy costs.

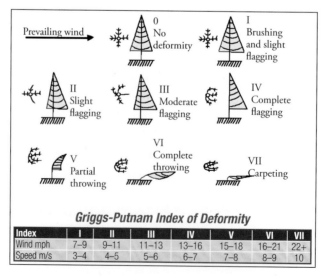

Vegetation flagging is the effect of strong winds on vegetation. It's a good indicator of how strong the winds are in that area.

Index	I	II	III	IV	V	VI	VII
Wind mph	7–9	9–11	11–13	13–16	15–18	16–21	22+
Speed m/s	3–4	4–5	5–6	6–7	7–8	8–9	10

Windmill blades can vary in shape but should always be angled to catch the most wind.

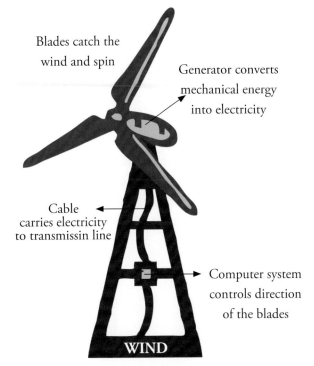

Blades catch the wind and spin

Generator converts mechanical energy into electricity

Cable carries electricity to transmissin line

Computer system controls direction of the blades

WIND

MACHINE

A horizontal-axis wind turbine.

Maintaining Your Small Wind Turbine

To keep your turbine running smoothly and efficiently, do an annual check of the following:

- Check and tighten bolts and electrical connections as necessary.
- Check machines for corrosion.
- Check the guy wires for proper tension.
- Check for and replace any worn leading-edge tape on the turbine blades.
- Replace the turbine blades and/or bearings after 10 years.

Types of Wind Turbines

Modern wind turbines fall into two basic categories: horizontal-axis varieties and vertical-axis designs.

Horizontal-axis Wind Turbines

Most wind machines used today fall into this category. Horizontal-axis wind machines have blades like an airplane propeller. A standard horizontal wind machine stands about 20 stories tall and has three blades spanning 200 feet across. These are the machines most readily found in large fields and on wind farms.

The majority of small wind turbines made today are of the horizontal-axis style. They have two or three blades made of composite material, such as fiberglass. The turbine's frame is a structure to which the rotor, generator, and tail are all attached. The diameter of the rotor will determine the amount of energy the turbine will produce. The tail helps keep the turbine facing into the wind. Mounted on a tower, the wind turbine has better access to stronger winds.

These machines also require balance-of-system components. These parts are required for water pumping systems and other residential uses of your wind turbine. These also vary based on the type of system you are using: either a grid-connected, stand-alone, or hybrid.

For example, if you have a residential grid-connected wind turbine system, your balance-of-systems parts will include:

- A controller
- Storage batteries
- A power conditioning unit (inverter)
- Wiring
- Electrical disconnect switch
- Grounding system
- Foundation for the tower

Vertical-axis Wind Turbines

These machines have blades that go from top to bottom. The most common type looks like a giant two-bladed egg beater. Vertical-axis wind machines are generally 100 feet tall and 50 feet wide. Though these wind turbines have the potential to produce a great deal of energy, they make up only a small percentage of the wind machines that are currently in use due to the cost and effort required to set them up. In addition, they produce a great deal of noise, can be unsightly, hurt the bird population, and require large roads and heavy-duty equipment to get them up and running.

A hybrid wind and solar energy system.

Grid-Connected Small Wind Electric Systems

Small wind energy systems can be connected to the electricity distribution system to become "grid-connected systems." These wind turbines can help reduce your consumption of utility-supplied electricity for appliances, electric heat, and lighting. The utility will make up the difference for any energy that your turbine cannot make. Any excess electricity that is produced by the system, and cannot be used by the household, can often be sent or sold to the utility. One drawback to this system, however, is that during power outages, the wind turbine is required to shut down for safety reasons.

Grid-connected systems are only practical if:

- You live in an area with average annual wind speeds of at least 10 mph.
- Utility-supplied electricity is expensive in your area.
- The utility's requirements for connecting your system to its grid are not exceedingly expensive.
- There are good incentives for the sale of excess electricity.

Mounting Your Small Wind Electric System on a Tower

Since wind speeds increase with height, it is essential that your small wind turbine be mounted on a tower. The higher the tower, the more power the wind system will be able to produce. To determine the best height for your tower, you will need to know the estimated annual energy output and the size of your turbine.

A grid-connected small wind electric system.

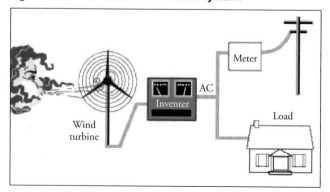

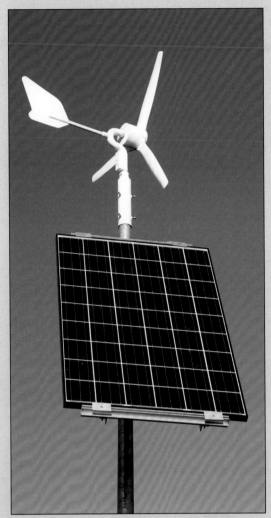

A solar and wind hybrid energy system.

Hybrid power systems combine multiple sources to deliver non-intermittent electric power.

Stand-Alone and Small Hybrid Systems

Wind power can also be used in off-grid systems. These are called stand-alone systems because they are not connected to an electric distribution grid. In these systems, small wind turbines can be used in combination with other components, such as small solar electric systems, to create a hybrid power system. Hybrid power systems provide reliable off-grid power for homes (and even for entire communities in certain instances) that are far from local utility lines.

A hybrid electric system may be a practical system for you if:
- You live in an area with average annual wind speed of at least nine mph.
- A grid connection is not available or can only be made through a very costly extension.
- You would like to become independent from your energy utility company.
- You would like to generate clean power.

Small hybrid systems that combine wind and solar technologies offer several advantages over either single system. In many parts of the United States, wind speeds are low in the summer when the sun shines the brightest and for the longest hours. Conversely, the wind is stronger in the winter when less sunlight is available. These hybrid systems, therefore, are more likely to produce power when you need it.

If there are times when neither the wind nor the solar systems are producing energy, most hybrid systems will then provide power through batteries or an engine generator powered by diesel fuel (which can also recharge the batteries if they run low).

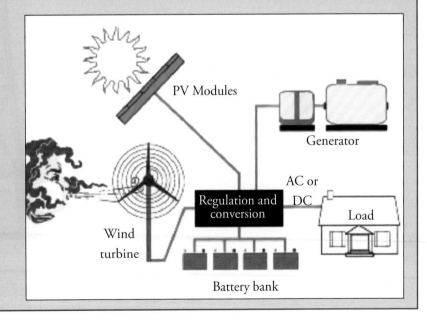

PV Modules

Generator

AC or DC

Regulation and conversion

Load

Wind turbine

Battery bank

There are two types of towers: self-supporting (free-standing) and guyed. Most home wind power systems use a guyed tower as it is the least expensive. Guyed towers consist of these parts:

- Lattice sections
- Pipe
- Tubing (depending on the design)
- Supporting guy wires

These towers are easier to install but they do require lots of space—the radius of the tower must be ½ to ¾ of the tower height.

Tilt-down towers, while more expensive, offer an easy way to maintain smaller, lightweight turbines that are less than 10 kilowatts. These towers can be lowered to the ground during severe weather or unusually high winds.

Generally, it is a good idea to install a small wind turbine on a tower with the bottom of the rotor blades around 30 feet above any obstacle that is within 300 feet from the tower.

Windmills

Windmills are used for pumping water, milling, and operating light machinery all around the world. They are constructed in a variety of shapes and some are very picturesque. When set up properly, windmills cost nothing to operate and if the wheel is made well, it will last for many years without need for major repairs. To make a windmill requires a good understanding of carpentry and workmanship but it is not incredibly difficult or expensive to do.

Constructing a Windmill

Windmills can be of all sizes, though the larger the windmill, the more power it can generate. This windmill and tower can be easily constructed out of wood, an old wheel, and a few iron fittings you may be able to find at a hardware store or home center. Constructing the windmill in sections is the easiest way to create this structure. Simply follow these directions to make your own energy-producing windmill:

The Tower

1. The tower is the first part to be built and should be constructed out of four spruce sticks that are 16 feet long and 4 inches square, in a configuration that measures 30 inches square at the top and 72 inches square at the base.

2. The deck should be 36 inches square and should project 2 inches over the top rails.

3. The rails and cross braces can be spruce or pine strips and should measure 4 inches wide and ⅞ inch thick. Attach these to the corner posts with steel-wire nails.

4. Embed the corner posts 2 feet into the ground, leaving 14 feet above the surface. The rail at the bottom, which is attached to the four posts, should measure 3 feet above the ground. Midway between this and the top rail of the deck, run a middle rail around the post. Make sure that where your wheel will be attached, this point rises at least 2 feet above any obstructions (buildings, trees, etc.) so it can have access to the blowing wind.

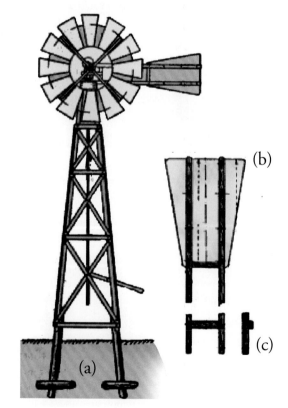

Details of the windmill. Figure (a) shows a general view with the tail turned to "off" position. Figure (b) shows details of the tail, and (c) shows a cross-piece of the tail.

Beveled cross braces fit snugly against the corners.

5. The cross braces should be beveled at the ends so they fit snugly against the corner.

6. The posts, rails, and braces should be planed so they present a nice appearance at the end of the building. A ladder can also be constructed at one side of the tower to allow easy access to the mill.

7. Nail a board across two of the rails halfway up the tower. Secure the lower end of a trunk tightly here if you are constructing a pumping mill. However, if a wooden mill is what you are after, you can use an old wheel from a wagon and six blades of wood.

The Turntable

1. The turntable (d, e, and f) holds the wheel and tail. It should be built of 2½ x 2-inch timber and 2-inch galvanized wrought iron "water" tube and flanges.

2. The upper flange (g and h) supports the timber framing. It should be countersunk, using a half-round file, and screwed tightly onto the tube as far as possible. The end of the tube should project just slightly beyond the face of the flange so that it can be riveted over to fill the countersink.

3. Bolt the two loose flanges to the framework of the tower. Use them with 2-inch pipe with the thread filed away so they may slide freely onto the tube. The upper loose flange should form a footstep bearing and the lower flange a guide for the turntable.

4. Now mount the turntable on the ball bearing to make sure the mill head can turn freely. Screw on two back nuts to guard against any possibility of the turntable being lifted out of place by a strong wind.

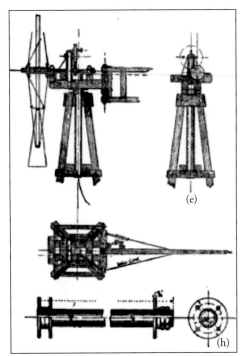

The windmill turntable (d, e, and f) holds the wheel and tail. The upper flange (detailed drawings g and h) forms a support for the timber framing.

The Head

1. This is the part that will carry the wheel spindle.

2. Notch the joints and secure them with 2-inch bolts.

3. The upright, which carries a bolt or pin for the spurwheel to revolve upon, is kept in place in the front and at the sides by a piece of hoop iron.

4. The tail vane swivel is a piece of 5-inch bore tube with back nuts and washers. Pass an iron bolt or other piece of iron through this, screw it to each end, and fit it with four nuts and washers.

The Wheel Shaft

1. Use wrought-iron tubing and flanges to create the wheel shaft. The bore of the tube is at least 5 inches, and the outside diameter should be roughly 1½ inches. Both the tube and the fittings should be of good quality and a thick gauge (steam quality is preferred).

2. If lathe is available, lightly skim it over the tubing. However, if it's not, a careful filing will do just as well to smooth down the edges.

3. Screw the tube higher up on one end to receive the flanges forming the hub. Screw these on and secure

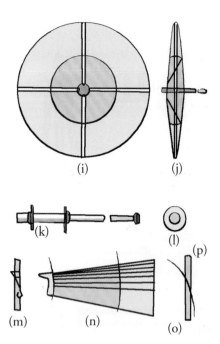

Details of the wheel shaft frame (i, j); front and side views, (k, l); axle of wheel (m); attachment of inner end of vane to inner ring of frame (n); vane on rings (o); attachment of vane to outer brackets by bracket (p).

them on one side with back nuts and on the other with a distance piece made out of a 1½-inch bore tube. Fit a cap to close the open front end of the tube.

4. Grease two plummer blocks with some form of lubrication. These will be the bearings for the shaft.

5. A pinion is needed of at least 2½ inches in diameter at the pitch circle. Bore it to fit the wheel shaft. A spur wheel of 7 inches in diameter should follow that (gear wheels from a lawn mower can be used if available).

The Wheel

1. The wheel should be at least 5 feet in diameter to produce a good amount of energy. The framing consists of an inner and outer ring and four double arms with cross stays and diagonals (a regular wooden wheel will be sufficient, or you can find one made of galvanized steel).

2. Cut each spoke at an angle on one side so that the blades will have the necessary pitch to make the wind turn them.

3. The blades should be 18 inches long, 12 inches wide at the outer ends, and 6 inches wide next to the hub. Each

blade should be only ¾ inch thick. Attach them to the spokes with simple screws.

4. If you desire, you can string a wire between the outer end of each blade to the end of the next spoke. This will help steady the blades.

The Tail

1. Run a fine saw cut up about 2 feet 6 inches from the outer end to receive the vane (optional).

2. Pass a cord over two pulleys and down the turntable tube. It is necessary to attach the end of the cord to a short cylinder of hard wood or metal (about 2 to 3 inches in diameter). This revolves with the turntable but can be slid up or down.

Each spoke should be cut at an angle so that the blades will have the pitch to make the wind turn them.

3. If you plan on using a pump, cut a hole through the axis of the cylinder to fit the pump rod.

4. Cut a groove in the circumference of the cylinder, and bend two pieces of iron into shape and place them into the grooves. Now take the cords from the two bolts, untying the straps. Join these two cords to another cord, which acts as a reel or lever at the base of the tower. In this way, the position of the tail can be regulated from a stationary point.

Total Lift	Gallons per Hour	Bore of Pump	Approximate Stroke
26 ft	100	2 in.	3½ in.
60 ft	50	2 in.	1½ in.
100 ft	25	1½ in.	1½ in.

Adding Pumps to Your Windmill

If you want to use this windmill to pump water, then you may need to do some experimenting with different lengths of pump stroke. Below is a table indicating what should be expected from the pump, and also providing the size of the single-action pump suitable for a given lift (using a ratio of 1 to 3).

Make sure that your pump is not too large; otherwise, it may not start in a light wind or breeze.

The pump is driven by a pin screwed into the side of the spur wheel and is secured with a lock nut. Drill and tap three or four holes at different distances from the center of the wheel so the length of the stroke can be adjusted. If the spokes on the wheel are too thin for drilling, you can use a clamp with a projecting pin instead.

A pump rod—a continuous wooden rod about 1 inch square and thicker at the top end—can be used in connecting the bottom end (by bolting) to the "bow" supplied with the pump. Intermediate joints, if needed, can be fashioned with 1 x ½-inch fish plates roughly 6 inches long. If the pump is no more than 12 feet below the crank pin, one guide will be adequate. The pump rod must be able to revolve with the head and will be need to be thickened up in a circular section where it passes through the guide. Make the guide in two halves and screw or bolt it to a bar running across the tower.

Final Touches

When construction is finished, paint all of the woodwork any color that complements your yard or property and, if desired, lacquer it to protect the wood from rain and snow. A windmill of this size will create at least a one-quarter horsepower in a 15-mph wind.

Building a Small Wind Motor

This small wind motor can easily be made to generate energy for small machines, tool shed lightbulbs, and other small mechanics. The foundation for this wind-wheel can be made out of the front wheel of an old bicycle with the front spindle and cones completely intact.

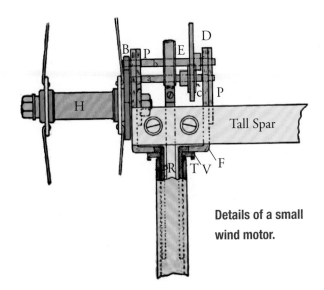

Details of a small wind motor.

Attach eight to 12 vanes of stout sheet tin to the rim. These sheets should be around 8 inches long and 4 to 6 inches wide and should lie at a 30-degree angle to the plane of the rim. The vanes will be much more efficient if they are curved in a circular arc about the same radius as the wheel. The concave side should be positioned to face toward the wind.

On the back of each vane, rivet a rib of strip iron ½ inch thick. This strip should project about ½ inch beyond the tip and 1½ inches at the other end. There, twist and bend it to make a bracket and then bolt the vane to the center line of the rim.

The illustration above shows a side view of the motor with its gearing and supports. *A* is the rim and part of the spokes of a toothed wheel that are attached at several points to the spokes of the bicycle wheel. It is loosely fixed and adjusted until it runs well when the wheel is moved. It should not wobble. *A* drives a smaller cog, *B*, mounted on the same spindle, *a*. This spindle revolves around two plates, *PP*, screwed to *F*. *c* drives a large cog, *D*, and an eccentric, *E*, which moves the eccentric rod, *R*, up and down. This works the small pump at the foot of the mast that supports the windmill. *E* can be quickly made out of a thick disc with two larger discs soldered to it. *R* is a piece of stout brass strip bent around *E* and closed with a screw.

When all of the vanes are in position, connect the tips of the ribs and vanes together with rings of stout wire and solder them on at all the contact points. Screw one of the spindle nuts tightly against its cone. The other end of the spindle should pass through one arm of the stirrup (F) made

out of ½-inch iron 1½ inches wide. This is then secured by a washer and nut on the inside. The stirrup and circular plate (V) are bored to accommodate the end of the iron pipe (T).

Close off the top of the hole (F) and heat the top of the pipe to expand it to fit into the chamber. Clean these parts well and weld them together. It is important that the T is square with the stirrup. Then, cut the pipe off 9 inches below V. Solder a small ring to the underside of V to prevent moisture from working its way along T and ruining your motor.

The tail spar is a wooden bar 1½ x 2½ inches wide and 40 inches long. It is notched to fit the stirrup and tapered off toward the tail. A sheet of sturdy iron, 15 x 12 inches, is then fitted into the saw cut. Two bolts clip the wings of the forked end tightly against the sides of the stirrup. The tail should be able to balance the wheel on the vertical pivot to avoid stressing the joint at the top of *T*.

A wind-wheel this size will spin effectively in a blustery wind but will probably only generate enough energy to power a small pump. This will do nicely to fill a watering can for your garden or for powering other light machinery.

Zoning, Permitting, and Covenant Requirements

Before you invest in or make your own small wind energy system, you should research any zoning and neighborhood covenant issues that may deter your installing a wind turbine system. You can find out about local zoning restrictions by contacting a local building inspector, board of supervisors, or planning board. They will inform you whether or not you'll need a building permit and will provide you with a list of other requirements. Further, your neighbors or homeowners' association may object to a wind machine that will block their view or a system that will be too noisy.

A Pumping Windmill

A pumping windmill can help you pump water from a well or other underground reservoir into a suction-pump. This windmill has a simple wheel with spokes and sails. It consists of a hub, six spokes, a fan tail, and a trunk or pole for attaching the wheel.

The hub is a hexagon 6 x 6 inches. One spoke can be driven into a hole made on either side (Fig. 1). The spokes should be 3 feet long, 3 x 1½ inches at the hub end, and 1 x 1½ inches on the outer end. The spokes are driven into the holes in the hub and pinned to hold them in place.

The hub should be made of hard wood and the holes may be cut with a mortise chisel and mallet. Make sure the holes are spaced evenly so the spokes will light up properly.

Attach triangular pieces of twilled muslin sheeting to the face of each spoke. The loose corner of each can be attached to the next spoke end with a piece of string. This creates an outlet between the leech and the spoke of each space between the spoke so that the wind can pass through. This, in effect, makes the wheel turn.

A pumping windmill.

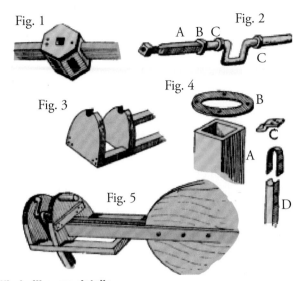

Windmill pump details.

The wheel should be held in place at the top of the supporting post by a shaft passing through the hub and bolted to the front of the wheel with a nut. Fig. 2 is a good example of what this should look like. The shaft should be about 1 inch square where it passes through the hub. At the front end, it should be tightened with a nut and washer. The square part, A, where the end of the hub will be, should be welded at B to hold the hub in the proper place. About an inch beyond the square shoulder, another one, C, should be welded to the shaft. This helps balance the wheel.

Now a crank can be formed, 2 inches wide and 3 inches out from the shaft. Another collar, C, C, should be welded onto the crank and then, beyond this point, the shaft should stick out about 6 inches.

The total length of the shaft is 15 inches, and the whole device can be painted. To attach the fan tail, a head made out of two blocks of wood should be attached and fastened 5 inches apart on the lower rails (Fig. 3). The upper ends of the blocks should be cut so as to allow the shaft to enter them. The collars, C and C, C, are placed at the inside of the blocks. To hold the shaft in place, small iron straps can be screwed tightly over the top of each block.

This head rests on the top of a hollow square post through which the rod passes, connecting the crank with the piston-rod of the pump (Fig. 4 A). A flat iron collar, B, should be screwed tightly at the top. To keep the head properly secured, four iron cleats (Fig. 4 C) should be screwed tightly under the corners of the head to help grip the projecting edge of the collar. This will hold the head rigid while allowing it to move about with the force of the wind.

Apply a little bit of grease or Vaseline to the top of the collar so the head will move easily. The top of the connecting rod should be attached to the crank and bolted to the top of the hard wood rod (Fig. 4 D).

The tail, which is 33 inches long and 24 inches wide at the end, is made of boards that are ¾ inch thick. The tail should be attached to the head (Fig. 5).

To place the windmill over a pump, build a platform that is braced with pieces of wood (see the illustration). Wires can also be run from the upper part of the trunk down to pegs driven into the ground. This will add additional support and steadiness to the upright shaft.

To start the wheel, snap the ends of the sheets to the spoke ends. To stop the wheel, unsnap the ends and furl the sails around the spokes, tying them securely with a piece of yarn or a cotton cord.

The use of large scale windmills is often controversial. They can provide a significant amount of clean energy, but they also clutter ridgelines, produce a lot of noise, and hurt the bird population.

What would happen if we used more wind energy?

According to the American Wind Energy Association, if we increase our nation's wind energy capacity to 20 percent by 2030, it would have the following effects:

- *Reduce Greenhouse Gas Emission:* A cumulative total of 7,600 million tons of CO_2 would be avoided by 2030, and more than 15,000 million tons of CO_2 would be avoided by 2050.
- *Conserve Water:* Reduce cumulative water consumption in the electric sector by 8 percent or 4 trillion gallons from 2007 through 2030.
- *Lower Natural Gas Prices:* Significantly reduce natural gas demand and reduce natural gas prices by 12 percent, saving consumers approximately $130 billion.
- *Expand Manufacturing:* To produce enough turbines and components for the 20 percent wind scenario, the industry would require more than 30,000 direct manufacturing jobs across the nation (assuming that 30 to 80 percent of major turbine components would be manufactured domestically by 2030).
- *Generate Local Revenues:* Lease payments for wind turbines would generate well over $600 million for landowners in rural areas and generate additional local tax revenues exceeding $1.5 billion annually by 2030. From 2007 through 2030, cumulative economic activity would exceed $1 trillion or more than $440 billion in net present value terms.

PART SEVEN

GARDENING

BEES, BIRDS, AND BUTTERFLIES IN YOUR GARDEN

A wonderful part of having a garden is the wildlife it attracts. The types of trees, shrubs, vines, plants, and flowers you choose for your garden and yard affect the types of wildlife that will visit. Whether you are looking to attract birds, butterflies, or bees to your garden, here are some specific types of plants that will bring these creatures to your yard.

Plant Species for Birds

Following is a list of trees, shrubs, and vines that will attract various birds to your yard and garden. Be sure to check with your local nursery to find out which plants are most suitable for your area.

Trees:

- American beech

- American holly
- Balsam fir
- Crabapple
- Flowering dogwood
- Oak

Shrubs:

- Common juniper
- Hollies
- Sumacs
- Viburnums

Vines:

- Strawberry
- Trumpet honeysuckle
- Virginia creeper
- Wild grape

Flowers and Nectar Plants for Hummingbirds, Butterflies, and Bees

If you are looking at attract hummingbirds, butterflies, and bees to your garden, consider planting these nectar-producing shrubs and flowers. Again, check with your local nursery to make sure these plants are suitable for your geographic area. Some common nectar plants are:

- Aster
- Azalea
- Butterfly bush
- Clover and other legumes
- Columbine
- Coneflower
- Honeysuckle
- Lupine
- Milkweeds
- Zinnia

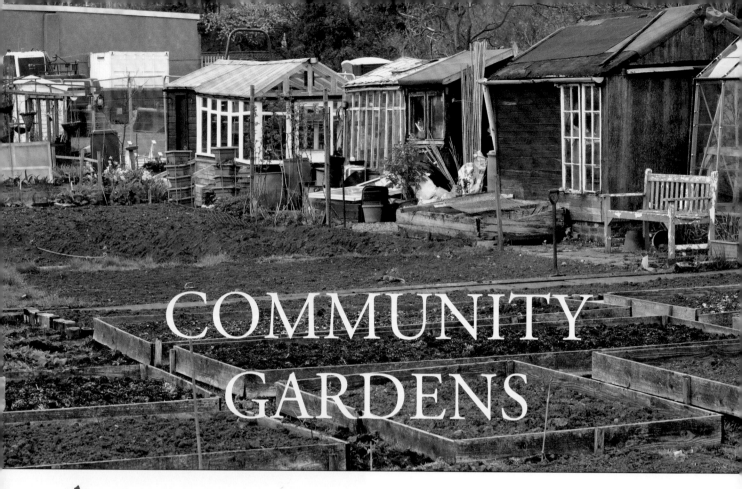

COMMUNITY GARDENS

A community garden is considered any piece of land that many people garden. These gardens can be located in urban areas, out in the country, or even in the suburbs, and they grow everything from flowers to vegetables to herbs—anything that the community members want to produce. Community gardens help promote the growing of fresh food, have a positive impact on neighborhoods by cleaning up vacant lots, educate youth about gardening and working together as a community, and bring whole communities together in one common goal. Whether the garden is just one plot or many individual gardens in a specified area, a community garden is a wonderful way to reach out to fellow neighbors while growing fresh foods wherever you live.

10 Steps to Beginning Your Own Community Garden

The following steps are adapted from the American Community Garden Association:

1. Organize a meeting for all those interested. Determine whether a garden is really needed and wanted, what kind it should be (vegetable, flower, or a combination, and organic or not), whom it will involve, and who benefits. Invite neighbors, tenants, community organizations, gardening and horticultural societies, and building superintendents (if it is at an apartment building)—in other words, anyone who is likely to be interested.

2. Establish a planning committee. This group can be made up of people who feel committed to the creation of the garden and who have the time to devote to it, at least in the initial stages. Choose well-organized people as garden coordinators and form committees to tackle specific tasks such as funding and partnerships, youth activities, construction, and communication.

3. Compile all of your resources. Do a community asset assessment. What skills and resources already exist in the community that can aid in the garden's creation? Contact local municipal planners about possible sites, as well as horticultural societies and other local sources of information and assistance. Look within your community for people with experience in landscaping and gardening.

4. Look for a sponsor. Some gardens "self-support" through membership dues, but for many, a sponsor is essential for donations of tools, seeds, or money. Churches, schools, private businesses, or parks and recreation departments are all possible sponsors.

5. Choose a site for your garden. Consider the amount of daily sunshine (vegetables need at least six hours a day), availability of water, and soil testing for possible pollutants. Find out who owns the land. Can the gardeners get a lease agreement for at least three years? Will public liability insurance be necessary?

6. Prepare and develop the chosen site. In most cases, the land will need considerable preparation for planting. Organize volunteer work crews to clean it, gather materials, and decide on the design and plot arrangement.

7. Organize the layout of the garden. Members must decide how many plots are available and how they will be assigned. Allow space for storing tools and making compost, and allot room for pathways between each plot. Plant flowers or shrubs around the garden's edges to promote goodwill with non-gardening neighbors, pedestrians, and municipal authorities.

8. Plan a garden just for kids. Consider creating a special garden for the children of the community—including them is essential. Children are not as interested in the size of the harvest but rather in the process of garden-

ing. A separate area set aside for them allows them to explore the garden at their own speed and can be a valuable learning tool.

9. Draft rules and put them in writing. The gardeners themselves devise the best ground rules. We are more willing to comply with rules that we have had a hand in creating. Ground rules help gardeners to know what is expected of them. Think of it as a code of behavior. Some examples of issues that are best dealt with by agreed-upon rules are: What kinds of dues will mem-

bers pay? How will the money be used? How are plots assigned? Will gardeners share tools, meet regularly, and handle basic maintenance?

10. Keep members involved with one another. Good communication ensures a strong community garden with active participation by all. Some ways to do this are: form a telephone tree, create an e-mail list, install a rainproof bulletin board in the garden, and have regular community celebrations. Community gardens are all about creating and strengthening communities.

Community gardens provide an opportunity for workers of all ages to connect and learn from each other.

Start a School Garden

The local school might be the perfect place to start a community garden—explain the project to the principal and ask his or her permission to use a portion of the school grounds. You may even want to work with the school food service personnel to use products from the garden in a special school lunch, school-wide salad, or food festival.

Use the following checklist to get your garden up and running:

What to Do

Pick a Spot:

- Make sure the vegetable garden gets at least six hours of sunshine a day—otherwise the seeds produce plants and leaves and not much food. If the plot chosen doesn't have enough sunshine, try growing vegetables that have leaves such as lettuce. Or try vegetables such as carrots, parsley, chives, or zucchini, which need only 4 to 5 hours of daily sun. Keep drainage in mind—a garden needs to drain well, so try to avoid the low spot where puddles form.

How Big Does the Garden Grow?

- A big harvest can be gotten from a garden that is only 4 feet by 4 feet. You probably don't want a garden bigger than 10 feet by 10 feet unless you don't mind the work!

Plan Your Garden:

- Point north. Find the north side of the plot, because that's where the tall plants should go, so they don't shade shorter ones. Stand facing the sunset; north is the direction to the right. Plants are grown in rows, so the garden will be a square or rectangle.

Things You'll Need

- Planting, growing, and harvesting tools
- Seeds, seedlings, and organic material, such as compost, manure, or peat moss
- Long-handled shovels, hoes, rakes, garden spades, and three pronged hand cultivators

- Scissors, knives, and containers (baskets, bowls, or cardboard boxes)

Decide What to Plant:

- Think about what vegetables to grow and decide on the number of plants needed. Involve the children in this part of the planning process.

Design the Site:

- Draw a picture of the garden and plan out what plants will grow in which rows. Figure how far apart the rows should be. Find out how wide the plants will get, add the two widths together, and divide by 2. That's how far apart the rows should be. This will make it easier on planting day.

Once the Garden is Designed—Then What?

Test the Soil:

- If the soil has not been tested, conduct a soil test. Call a local Cooperative Extension Service office, listed under county or state government, for the name and location of a laboratory to do testing for lead content and soil pH or see page 729 to do your own soil testing. A soil test tells two things: (a) lead level of the soil; and (b) whether the soil is *acid* (sour), *alkaline* (sweet), or *neutral* (neither sour nor sweet). Lead is a poison and if it gets into the plants, it will get into the food. Plants will not grow well in soil that is either too acid or too alkaline.

Get the Tools:

- Long-handled shovels, gardening spades, spading forks, hoes, and rakes are all excellent tools for begin-

ning a garden. To care for the garden, use hand tools such as 3-pronged hand cultivators, hose and nozzle, and/or watering cans. If the group doesn't have its own garden tools, find someone who has what is needed and ask to borrow the tools. Or check yard sales to buy used tools.

Prepare the Soil:

- Once the soil is dry enough, dig it and loosen it. Involve children and youth in the process of preparing the soil.

- Remove grass and weeds (roots and all). Take the time to do this well.

- Dig the soil as deep as the blade of the spade and turn the soil. Or, find someone to till the soil with a rototiller.

- If the soil test showed the soil to be too acid, add limestone (lime); if the soil is too alkaline, add ground agricultural sulfur. Sprinkle either lime or sulfur evenly over the garden soil.

- Add organic material such as compost, manure, or peat moss. This helps feed the plants and enriches the soil. Spread evenly on top of the turned soil in a layer no deeper than 3 inches.

- Blend everything using a spading fork, until the soil is so soft that planting can be done with the hands.

- Rake the soil until the soil is smooth and level, with no hills or holes. This will allow the water to seep down to the roots.

Get Ready to Plant:

Children will enjoy accompanying adults on a trip to purchase seeds or seedlings (also called transplants) for the

vegetables you intend to plant. Some plants do better if you start with seedlings rather than seeds. Seedlings are the fastest way to grow plants, and the easiest.

- To identify what you have planted, buy or make stakes, and write the names of the plants on the stakes with a waterproof marker. Save the stakes for later use.

- Here is another place that children and youth can help. If seeds are planted: make a shallow straight line (furrow) in the soil with a finger.

- Put the seeds in the furrow about half an inch apart or as suggested on the seed packaging.

- When the seeds are in the furrow, squeeze the furrow closed with thumb and finger.

- Water the soil right after the seeds are planted—water the plants so that the water comes out like a shower.

- Place the marker stakes in the soil at one end of the row to identify what has been planted.

- If another kind of plant is planted, check the distance between the rows following your design.

- If seedlings are planted: In each row, mark the spot where the plants should go, by poking a hole in the soil using a finger or the end of a pole. Do the entire row at one time.

- Set each plant in the soil so that it isn't too high or too low but just above the root ball. Cover the root ball with soil and press the soil gently so there are no empty spaces near the roots.

- Feed the seedlings with a mixture of fertilizer and water. Water each plant once, let the water soak in, and water a second time. Depending on what plants are grown, this feeding may need to be done every two to three weeks.

Tending the Garden:

Visit the garden daily—check if the garden needs watering, weeding, feeding, and thinning. Make sure to bring the proper tools. Take the children and youth to the garden and have them help care for the plants. Two hours is plenty of time for them to work in the garden at any one time. Children and youth can do weeding, thinning, and harvesting of crops.

Tools Needed to Create and Maintain a Community Garden

When purchasing tools for your community garden, buy high-quality tools. These will last longer and are an investment that will benefit your garden and those working in it for years to come. Every community garden should be equipped with these 10 essential tools:

Gloves: Leather gloves hold up best. If you have roses, get a pair that is thick enough to resist thorn pricks.

Hand fork or cultivator: A hand fork helps cultivate soil, chop up clumps, and work fertilizer and compost into the soil. A hand fork is necessary for cultivating in closely planted beds.

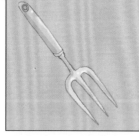

Hand pruners: There are different types and sizes of pruners depending on the type and size of the job. Hand pruners are for cutting small diameters, up to the thickness of your little finger.

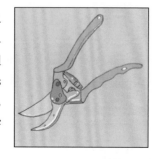

Hoe: A long-handled hoe is key to keeping weeds out of your garden.

Hose: This is the fastest way to transport lots of water to your garden plants. Consider using drip irrigation hoses or tape to apply a steady stream of water to your plants.

Shovels and spades: There are several different types and shapes of shovels and spades, each with its own purpose. There are also different types of handles for either—a D shape, a T shape, or none at all. A shovel is a requisite tool for planting large perennials, shrubs, and trees; breaking ground; and moving soil, leaves, and just about anything else. The sharper the blade, the better.

Trowel: A well-made trowel is your most important tool. From container gardening to large beds, a trowel will help you get your plants into the soil.

Watering can: A watering can creates a fine, even stream of water that delivers with a gentleness that won't wash seedlings or sprouting seeds out of their soil.

Wheelbarrow: Wheelbarrows come in all different sizes (and prices). They are indispensable for hauling soil, compost, plants, mulch, hoses, tools, and everything else you'll need to make your garden a success.

Things to Consider

There are many things that need to be considered when you create your own community garden. Remember that community gardens take a lot of work and foresight in order for them to be successful and beneficial for all those involved.

The Organization of the Garden

Will your garden establish rules and conditions for membership—such as residence, dues, and agreement with any drafted rules and regulations? It is important to know who will be able to use the garden and who will not. Furthermore, deciding how the garden will be parceled out is another key topic to discuss before beginning your community garden. Will the plots be assigned by family size or need? Will some plots be bigger to accommodate larger families? Will your garden incorporate children's plots as well?

Insurance

It is becoming increasingly difficult to obtain leases from landowners without liability insurance. Garden insurance is a new thing for many insurance carriers, and their under-

writers are reluctant to cover community gardens. It helps if you know what you want out of your community garden before you start talking to insurance agents. Two tips: Work with an agent from a firm that deals with many different carriers (so you can get the best policy for your needs); and you will probably have better success with someone local who has already done this type of policy or who works with social service agencies in the area. Shop around until you find a policy that fits the needs of your community garden and its users.

Set Up a Garden Association

Many garden groups are organized very informally and operate successfully. Leaders "rise to the occasion" to propose ideas and carry out tasks. However, as the workload expands, many groups choose a more formal structure for their organization.

A structured program is a conscious, planned effort to create a system so that each person can participate fully and the group can perform effectively. It's vital that the leadership be responsive to the members and their needs.

If your group is new, have several planning meetings to discuss your program and organization. Try out suggestions raised at these meetings, and after a few months of operation, you'll be in a better position to develop bylaws or organizational guidelines. A community garden project should be kept as simple as possible.

Creating Bylaws

Bylaws are rules that govern the internal affairs of an organization. Check out bylaws from other community garden organizations when creating your own. The bylaws of your community garden can be as simple as rules and regulations. This is usually plenty adequate for small community garden affairs.

Bylaws cover these topics:

- The full name of the organization and address
- The organizing members and their addresses
- The purpose, goal, and philosophy of the organization
- Membership eligibility and dues
- Timeline for regular meetings of the committee
- How the bylaws can be rescinded or amended
- Maintenance and cleanup of the community garden
- A hold harmless clause: "We the undersigned members of the [name] garden group hereby agree to hold harmless [name landowner] from and against any damage, loss, liability, claim, demand, suit, cost, and expense directly or indirectly resulting from, arising out of or in connection with the use of the [name] garden by the garden group, its successors, assigns, employees, agents, and invitees."

Sample Guidelines and Rules for Garden Members

Here are some sample guidelines community garden members may need to follow:

- I will pay a fee of $____ to help cover garden expenses.
- I will have something planted in the garden by [date] and keep it planted all summer long.
- If I must abandon my plot for any reason, I will notify the garden leadership.
- I will keep weeds at a minimum and maintain the areas immediately surrounding my plot.
- If my plot becomes unkempt, I understand I will be given one week's notice to clean it up. At that time, it will be reassigned or tilled in.
- I will keep trash and litter out of the plot, as well as adjacent pathways and fences.
- I will participate in the fall cleanup of the garden.
- I will plant tall crops where they will not shade neighboring plots.
- I will pick only my own crops unless given permission by another plot user.

- I will not use fertilizers, insecticides, or weed repellents that will in any way affect other plots.
- I understand that neither the garden group nor owners of the land are responsible for my actions. I therefore agree to hold harmless the garden group and owners of the land for any liability, damage, loss, or claim that occurs in connection with use of the garden by me or any of my guests.

Preventing Vandalism of Your Community Garden

Vandalism is a common fear among community gardeners. Try to deter vandalism by following these simple, preventative methods:

- Make a sign for the garden. Let people know that the garden is a community project.
- Put up fences around your garden. Fences can be of almost any material. They serve as much to mark possession of a property as to prevent entry.
- Invite everyone in the neighborhood to participate in the garden project from the very beginning. If you exclude people, they may become potential vandals.
- Plant raspberries, roses, or other thorny plants along the fence as a barrier to anyone trying to climb the fence.
- Make friends with neighbors whose windows overlook the garden. Trade them flowers and vegetables for a protective eye.
- Harvest all ripe fruit and vegetables on a daily basis to prevent the temptation of outsiders to harvest your crops.
- Plant a "vandal's garden" at the entrance. Mark it with a sign: "If you must take food, please take it from here."

People Problems and Solutions

Angry neighbors and bad gardeners pose problems for a community garden. Neighbors may complain to municipal governments about messy, unkempt gardens or rowdy behavior; most gardens cannot afford poor relations with neighbors, local politicians, or potential sponsors. Therefore, choose bylaws carefully so you have procedures to follow when members fail to keep their plots clean and up to code. A well-organized garden with strong leadership and committed members can overcome almost any obstacle.

10 Benefits of Creating a Community Garden

1. Improves the quality of life for people using the garden.
2. Provides a pathway for neighborhood and community development and promotes intergenerational and cross-cultural connections.
3. Stimulates social interaction and reduces crime.
4. Encourages self-reliance.
5. Beautifies neighborhoods and preserves green space.
6. Reduces family food budgets while providing nutritious foods for families in the community.
7. Conserves resources.
8. Creates an opportunity for recreation, exercise, therapy, and education.
9. Creates income opportunities and economic development.
10. Reduces city heat from streets and parking lots.

School Gardens

A school garden provides children with an ideal outdoor classroom. Within a single visit to a garden, a student can record plant growth, study decomposition while turning a compost pile, and learn more about plants, nature, and the outdoors in general. Gardens also provide students with opportunities to make healthier food choices, learn about nutrient cycles, and develop a deeper appreciation for the environment, community, and each other. While school gardens are typically used for science classes, they can also teach children about the history of their community (what their town was like hundreds of years ago and what people did to farm food), and be incorporated into math curriculum and other school subjects.

How to Start a School Garden

School gardens do not need to start on a grand scale. In fact, individual classrooms can grow their very own container gardens just by planting seeds in small pots, watering them daily, placing them in a sunny corner of the room, and watching the seedlings grow. However, if there is space for a larger, outdoor garden, this is the ideal place to teach children about working as a community, about plants and vegetables, and responsibility. A school garden should eventually become a permanent addition to the school and be maintained year-round.

When starting a school garden, it is important to find someone to coordinate the garden program. This is the perfect way to get parents involved in the school's garden. Establish a volunteer garden committee and assign certain parents particular tasks in the planning, upkeep (even during the summer months when school is generally not in session), and funding for the school's garden.

It is important, so the garden is not neglected, to plan particular classroom activities and lessons that will incorporate the garden and its plants. Assigning students various jobs that relate to the garden will be a wonderful way of introducing them to gardening as well as responsibility and community.

After all the initial planning is done, it is time to choose a spot for the school garden. A place in the lawn that receives plenty of sunlight and that will be close enough to the building for easy access is ideal. It should also be near an outdoor spigot so the plants can be easily watered. If there is enough space, it might be beneficial to have a garden shed, where gardening gloves, tools, buckets, hoses, and other items can be stored for use in the garden. Once you have chosen your spot, it's time to start digging and planting!

Both new and established gardens benefit from the use of compost and mulch. Many schools purchase compost when they initially establish their garden, and then they start making their own compost—which is a wonderful scientific lesson for students as well. You can use grass clippings, yard trimmings, rotten vegetables, and even food scraps from the

cafeteria or students' lunches to build and maintain your compost pile. While some schools choose to make compost piles in the garden, others compost with worm boxes right in the classroom!

Depending on funding and the needs and desires of the school, these gardens can become quite elaborate, with fences, ponds, trellises, trees and shrubs, and other structures. However, all a school garden truly needs is a little bit of dirt and a few plants (preferably an assortment of vegetables, fruit, and wildflowers) that students can study and even eat. Whether the school garden is established only for one class or grade level, or if it is going to be available to everyone at the school, is a factor that will determine the types of plants and the size of the overall garden.

Whether big or small, complex or simple, school gardens provide a wonderful, enriching learning experience for children and their parents alike.

COMPOSTING IN YOUR BACKYARD

Composting is nature's own way of recycling yard and household wastes by converting them into valuable fertilizer, soil organic matter, and a source of plant nutrients. The result of this controlled decomposition of organic matter—a dark, crumbly, earthy-smelling material—works wonders on all kinds of soil by providing vital nutrients, and contributing to good aeration and moisture-holding capacity, to help plants grow and look better.

Composting can be as simple or as involved as you would like, depending on how much yard waste you have, how fast you want results, and the effort you are willing to invest. Since all organic matter eventually decomposes, composting speeds up the process by providing an ideal environment for bacteria and other decomposing microorganisms. The composting season coincides with the growing season, when conditions are favorable for plant growth, so those same conditions work well for biological activity in the compost pile. However, since compost generates heat, the process may continue later into the fall or winter. The final product—called humus or compost—looks and feels like fertile garden soil.

Compost Preparation

While a multitude of organisms, fungi, and bacteria are involved in the overall process, there are four basic ingredients for composting: nitrogen, carbon, water, and air.

A wide range of materials may be composted because anything that was once alive will naturally decompose. The starting materials for composting, commonly referred to as feed stocks, include leaves, grass clippings, straw, vegetable and fruit scraps, coffee grounds, livestock manure, sawdust, and shredded paper. However, some materials that always should be avoided include diseased plants, dead animals, noxious weeds, meat scraps that may attract animals, and dog or cat manure, which can carry disease. Since adding kitchen wastes to compost may attract flies and insects, make a hole in the center of your pile and bury the waste.

For best results, you will want an even ratio of green, or wet, material, which is high in nitrogen, and brown, or dry, material, which is high in carbon. Simply layer or mix landscape trimmings and grass clippings, for example, with dried leaves and twigs in a pile or enclosure. If there is not a good

Junior Homesteader Tip

Compost Lasagna

Watch your produce scraps decompose! And see how some materials don't.

Things You'll Need

- 1 2-liter clear plastic bottle
- 2 cups fruit and vegetable scraps
- 1 cup grass clippings and leaves
- 2 cups soil
- Newspaper clippings or shredded paper
- Styrofoam packing peanuts
- Magic marker

1. Layer all your ingredients, just like you'd make a lasagna. Start with a couple inches of soil, then add the produce scraps, then more dirt, then the grass clippings and leaves, more dirt, the Styrofoam, more dirt, the shredded paper, and top it all off with a little more dirt.

2. Use the magic marker to mark the top of the top layer. Then place the bottle upright in a windowsill or another sunny spot. If there's a lot of condensation in the bottle, open the top to let it air out.

3. Once a week for four weeks check on the bottle and notice how the level of the dirt has changed. Mark it with the marker.

4. At the end of four weeks, dump the bottle out in a garden spot that hasn't been planted, or add it to your compost pile. Notice which items decomposed the most. Remove the items that didn't decompose and discard them in the trash.

Common Composting Materials

Cardboard
Coffee grounds
Corn cobs
Corn stalks
Food scraps
Grass clippings
Hedge trimmings
Livestock manure
Newspapers
Plant stalks
Pine needles
Old potting soil

Sawdust
Seaweed
Shredded paper
Straw
Tea bags
Telephone books
Tree leaves and twigs
Vegetable scraps
Weeds without seed heads
Wood chips
Woody brush

Avoid using:

Bread and grains
Cooking oil
Dairy products
Dead animals
Diseased plant material
Dog or cat manure
Grease or oily foods
Meat or fish scraps
Noxious or invasive weeds
Weeds with seed heads

supply of nitrogen-rich material, a handful of general lawn fertilizer or barnyard manure will help even out the ratio.

Though rain provides the moisture, you may need to water the pile in dry weather or cover it in extremely wet weather. The microorganisms in the compost pile function best when the materials are as damp as a wrung-out sponge—not saturated with water. A moisture content of 40 to 60 percent is preferable. To test for adequate moisture, reach into your compost pile, grab a handful of material, and squeeze it. If a few drops of water come out, it probably has enough moisture. If it doesn't, add water by putting a hose into the pile so that you aren't just wetting the top, or, better yet, water the pile as you turn it.

Air is the only part that cannot be added in excess. For proper aeration, you'll need to punch holes in the pile so it has many air passages. The air in the pile is usually used up faster than the moisture, and extremes of sun or rain can adversely affect this balance, so the materials must be turned or mixed up often with a pitchfork, rake, or other garden tool to add air that will sustain high temperatures, control odor, and yield faster decomposition.

Over time, you'll see that the microorganisms, which are small forms of plant and animal life, will break down the organic material. Bacteria are the first to break down plant tissue and are the most numerous and effective compost-makers in your compost pile. Fungi and protozoans soon join the bacteria and, somewhat later in the cycle, centipedes, millipedes, beetles, sow bugs, nematodes, worms, and numerous others complete the composting process. With the right ingredients and favorable weather conditions, you can have a finished compost pile in a few weeks.

How to Make a Compost Heap

1. Choose a level, well-drained site, preferably near your garden.

2. Decide whether you will be using a bin after checking on any local or state regulations for composting in urban areas, as some communities require rodent-proof bins. There are numerous styles of compost bins

available, depending on your needs, ranging from a moveable bin formed by wire mesh to a more substantial wooden structure consisting of several compartments. You can easily make your own bin using chicken wire or scrap wood. While a bin will help contain the pile, it is not absolutely necessary, as you can build your pile directly on the ground. To help with aeration, you may want to place some woody material on the ground where you will build your pile.

3. Ensure that your pile will have a minimum dimension of 3 feet all around, but is no taller than 5 feet, as not enough air will reach the microorganisms at the center if it is too tall. If you don't have this amount at 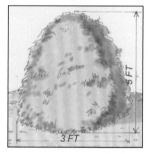 one time, simply stockpile your materials until a sufficient quantity is available for proper mixing. When composting is completed, the total volume of the original materials is usually reduced by 30 to 50 percent.

4. Build your pile by using either alternating equal layers of high-carbon and high-nitrogen material or by mixing equal parts of both together and then heaping it into a pile. If you choose to alternate layers, make each layer 2 to 4 inches thick. Some com-

posters find that mixing the two together is more effective than layering. Adding a few shovels of soil will also help get the pile off to a good start because soil adds commonly found, decomposing organisms to your compost.

5. Keep the pile moist but not wet. Soggy piles encourage the growth of organisms that can live without oxygen and cause unpleasant odors.

6. Punch holes in the sides of the pile for aeration. The pile will heat up and then begin to cool. The most efficient decomposing bacteria thrive in temperatures between 110 and 160 degrees Fahrenheit. You can track this with a compost thermometer, or you can simply reach into the pile to determine if it is uncomfortably hot to the touch. At these temperatures, the pile kills most weed seeds and plant diseases. However, studies have shown that compost produced at these temperatures has less ability to suppress diseases in the soil, since these temperatures may kill some of the beneficial bacteria necessary to suppress disease.

7. Check your bin regularly during the composting season to assure optimum moisture and aeration are present in the material being composted.

8. Move materials from the center to the outside of the pile and vice versa. Turn every day or two and you should get compost in less than four weeks. Turning every other week will make compost in one to three months. Finished compost will smell sweet and be cool and crumbly to the touch.

Other Types of Composting

Cold or Slow Composting

Cold composting allows you to just pile organic material on the ground or in a bin. This method requires no maintenance, but it will take several months to a year or more for the pile to decompose, though the process is faster in warmer climates than in cooler areas. Cold composting works well if you are short on time needed to tend to the compost pile at least every other day, have little yard waste, and are not in a hurry to use the compost.

For this method, add yard waste as it accumulates. To speed up the process, shred or chop the materials by running over small piles of trimmings with your lawn mower, because the more surface area the microorganisms have to feed on, the faster the materials will break down.

Cold composting has been shown to be better at suppressing soil-borne diseases than hot composting and also leaves more non-decomposed bits of material, which can be screened out if desired. However, because of the low temperatures achieved during decomposition, weed seeds and disease-causing organisms may not be destroyed.

Vermicomposting

Vermicomposting uses worms to compost. This takes up very little space and can be done year-round in a basement or garage. It is an excellent way to dispose of kitchen wastes.

Here's how to make your own vermicomposting pile:

1. Obtain a plastic storage bin. One bin measuring 1 foot by 2 feet by 3 ½ feet will be enough to meet the needs of a family of six.

2. Drill 8 to 10 holes about ¼ inch in diameter in the bottom of the bin for drainage.

3. Line the bottom of the bin with a fine nylon mesh to keep the worms from escaping.

4. Put a tray underneath to catch the drainage.

5. Rip shredded newspaper into pieces to use as bedding and pour water over the strips until they are thoroughly moist. Place these shredded bits on one side of your bin. Do not let them dry out.

6. Add worms to your bin. It's best to have about two pounds of worms (roughly 2,000 worms) per one pound of food waste. You may want to start with less food waste and increase the amount as your worm population grows. Redworms are recommended for best composting, but other species can be used. Redworms are the common, small worms found in most gardens and lawns. You can collect them from under a pile of mulch or order them from a garden catalog.

7. Provide worms with food wastes such as vegetable peelings. Do not add fat or meat products. Limit their feed, as too much at once may cause the material to rot.

8. Keep the bin in a dark location away from extreme temperatures.

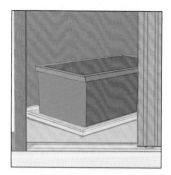

9. Wait about three months and you'll see that the worms have changed the bedding and food wastes into compost. At this time, open your bin in a bright light and the worms will burrow into the bedding. Add fresh bedding and more food to the other side of the bin. The worms should migrate to the new food supply.

10. Scoop out the finished compost and apply to your plants or save to use in the spring.

The Junior Homesteader

Let the kids be in charge of feeding the worms in your compost. They'll be fascinated by the squirmy critters!

Compost Trouble-shooting

Composting is not an exact science. Experience will tell you what works best for you. If you notice that nothing is happening, you may need to add more nitrogen, water, or air; chip or grind the materials; or adjust the size of the pile.

If the pile is too hot, you probably have too much nitrogen and need to add additional carbon materials to reduce the heating.

A bad smell may indicate not enough air or too much moisture. Simply turn the pile or add dry materials to the wet pile to get rid of the odor.

Uses for Compost

Compost contains nutrients, but it is not a substitute for fertilizers. Compost holds nutrients in the soil until plants can use them, loosens and aerates clay soils, and retains water in sandy soils.

To use as a soil amendment, mix 2 to 5 inches of compost into vegetable and flower gardens each year before planting. In a potting mixture, add one part compost to two parts commercial potting soil, or make your own mixture by using equal parts of compost and sand or Perlite.

As a mulch, spread an inch or two of compost around annual flowers and vegetables, and up to 6 inches around trees and shrubs. Studies have shown that compost used as mulch, or mixed with the top 1 inch layer of soil, can help prevent some plant diseases, including some of those that cause damping of seedlings.

As a top dressing, mix finely sifted compost with sand and sprinkle evenly over lawns.

CONTAINER GARDENING

An alternative to growing vegetables, flowers, and herbs in a traditional garden is to grow them in containers. While the amount that can be grown in a container is certainly limited, container gardens works well for tomatoes, peppers, cucumbers, herbs, salad greens, and many flowering annuals. Choose vegetable varieties that have been specifically bred for container growing. You can obtain this information online or at your garden center. Container gardening also brings birds and butterflies right to your doorstep. Hanging baskets of fuchsia or pots of snapdragons are frequently visited by hummingbirds, allowing for up-close observation.

Container gardening is an excellent method of growing vegetables, herbs, and flowers, especially if you do not have adequate outdoor space for a full garden bed. A container garden can be placed anywhere—on the patio, balcony, rooftop, or windowsill. Vegetables such as leaf lettuce, radishes, small tomatoes, and baby carrots can all be grown successfully in pots.

Growing Vegetables in a Container Garden

Here are some simple steps to follow for growing vegetables in containers.

1. Choose a sunny area for your container plants. Your plants will need at least five to six hours of sunlight a day. Some plants, such as cucumbers, may need more.
2. Select plants that are suitable for container growing. Usually their names will contain words such as "patio," "bush," "dwarf," "toy," or "miniature." Peppers, onions, and carrots are also good choices.
3. Choose a planter that is at least five gallons, unless the plant is very small. Poke holes in the bottom if they don't already exist; the soil must be able to drain in order to prevent the roots from rotting. Avoid terracotta or dark colored pots as they tend to dry out quickly.

4. Fill your container with potting soil. Good potting soil will have a mixture of peat moss and vermiculite. You can make your own potting soil using composted soil. Read the directions on the seed packet or label to determine how deep to plant your seeds.

5. Check the moisture of the soil frequently. You don't want the soil to become muddy, but the soil should always feel damp to the touch. Do not wait until the plant is wilting to water it—at that point, it may be too late.

Things to Consider

• Follow normal planting schedules for your climate when determining when to plant your container garden.

• You may wish to line your container with porous materials such as shredded newspaper or rags to keep the soil from washing out. Be sure the water can still drain easily.

Growing Herbs in a Container

Herbs will thrive in containers if cared for properly. And if you keep them near your kitchen, you can easily snip off pieces to use in cooking. Here's how to start your own herb container garden:

1. If your container doesn't already have holes in the bottom, poke several to allow the soil to drain. Pour gravel into the container until it is about a quarter of the way full. This will help the water drain and help to keep the soil from washing out.

2. Fill your container three-quarters of the way with potting soil or a soil-based compost.

3. It's best to use seedlings when planting herbs in containers. Tease the roots slightly, gently spreading them apart with your fingertips. This will encourage them to spread once planted. Place each herb into the pot and cover the root base with soil. Place herbs that will grow taller in the center of your container, and the smaller ones around the edges. Leave about 4 square inches of space between each seedling.

4. As you gently press in soil between the plants, leave an inch or so between the container's top and the soil. You don't want the container to overflow when you water the herbs.

5. Cut the tops off the taller herb plants to encourage them to grow faster and to produce more leaves.

6. Pour water into the container until it begins to leak out the bottom. Most herbs like to dry out between watering, and over-watering can cause some herbs to rot and die, so only water every few days unless the plants are in a very hot place.

Things to Consider

• Growing several kinds of herbs together helps the plants to thrive. A few exceptions to this rule are oregano, lemon balm, and tea balm. These herbs should be planted on their own because they will overtake the other herbs in your container.

• You may wish to choose your herbs according to color to create attractive arrangements for your home. Any of the following herbs will grow well in containers:
  • Silver herbs: artemisias, curry plants, santolinas
  • Golden herbs: lemon thyme, calendula, nasturtium, sage, lemon balm
  • Blue herbs: borage, hyssop, rosemary, catnip
  • Green herbs: basil, mint, marjoram, thyme, parsley, chives, tarragon
  • Pink and purple herbs: oregano (the flowers are pink), lavender

• If you decide to transplant your herbs in the summer months, they will grow quite well outdoors and will give you a larger harvest.

Growing Flowers from Seeds in a Container

1. Cover the drainage hole in the bottom of the pot with a flat stone. This will keep the soil from trickling out when the plant is watered.

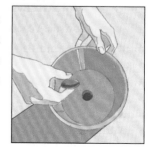

2. Fill the container with soil. The container should be filled almost to the top. For the best results, use potting soil from your local nursery or garden center.

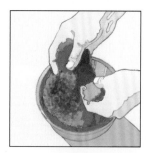

3. Make holes for the seeds. Refer to the seed packet to see how deep to make the holes. Always save the seed packet for future reference—it most likely has helpful directions about thinning young plants.

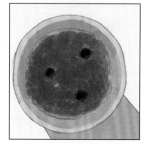

4. Place a seed in each hole. Pat the soil gently on top of each seed.

5. Use a light mist to water your seeds, making sure that the soil is only moist and not soaked.

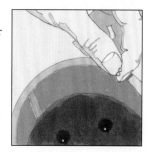

6. Make sure your seeds get the correct amount of sunlight. Refer to the seed packet for the adequate amount of sunlight each seedling needs.

7. Watch your seeds grow. Most seeds take 3 to 17 days to sprout. Once the plants start sprouting, be sure to pull out plants that are too close together so the remaining plants will have enough space to establish good root systems.

8. Remember to water and feed your container plants. Keep the soil moist so your plants can grow. And in no time at all, you should have wonderful flowers growing in your container garden.

Preserving Your Container Plants

As fall approaches, frost will soon descend on your container plants and can ultimately destroy your garden. Container plants are particularly susceptible to frost damage, especially if you are growing tropical plants, perennials, and hardy woody plants in a single container garden. There are many ways that you can preserve and maintain your container garden plants throughout the winter season.

Preservation techniques will vary depending on the plants in your container garden. Tropical plants can be over-wintered using methods replicating a dry season, forcing the plant into dormancy; hardy perennials and woody shrubs need a cold dormancy to grow in the spring, so they must stay outside; cacti and succulents prefer their winters warm and dry and must be brought inside, while many annuals can be propagated by stem cuttings or can just be repotted and maintained inside.

Preserving Tropical Bulbs and Tubers

Many tropical plants, such as cannas, elephant ears, and angel's trumpets can be saved from an untimely death by over-wintering them in a dark corner or sunny window of your home, depending on the type of plant. A lot of bulbous and tuberous tropical plants have a natural dry season (analogous to our winter) when their leafy parts die off, leaving the bulb behind. Don't throw the bulbs away.

1. After heavy frosts turn the aboveground plant parts to mush, cut the damaged foliage off about 4 inches above the thickened bulb.

2. Dig them up and remove all excess soil from the roots.

3. If a bulb has been planted for several years and it's performance is beginning to decline, it may need dividing. Daffodil's, for example, should generally be  divided every three years. If you do divide the bulb, be sure to dust all cut surfaces with a sulfur-based fungicide made for bulbs to prevent the wounds from rotting. Cut the roots back to 1 inch from the bulb and leave to dry out evenly.

4. Rotten bulbs or roots need to be thrown away so infection doesn't spread to the healthy bulbs.

A bulb's or tuber's drying time can last up to two weeks if it is sitting on something absorbent like newspaper and located somewhere shaded and dry, such as a garage or basement. Once clean and dry, bulbs should be stored—preferably at around 50°F—all winter in damp (not soggy) milled peat moss. This prevents the bulbs from drying out any further, which could cause them to die. Many gardeners don't have a perfectly cool basement or garage to keep bulbs dormant. Alternative methods for dry storage include a dark closet with the door cracked for circulation, a cabinet, or underneath a bed in a cardboard box with a few holes punched for airflow. The important thing to keep in mind is that the bulb needs to be kept on the dry side, in the dark, and moderately warm.

If a bulb was grown as a single specimen in its own pot, the entire pot can be placed in a garage that stays above 50°F or a cool basement and allowed to dry out completely. Cut all aboveground plant parts flush with the soil and don't water until the outside temperatures stabilize above 60°F. Often, bulbs break dormancy unexpectedly in this dry pot method. If this happens, pots can be moved to a sunny location near a window and watered sparingly until they can be placed outside. The emerging leaves will be stunted, but once outside, the plant will replace any spindly leaves with lush, new ones.

Preserving Annuals

Many herbaceous annuals can also be saved for the following year. By rooting stem cuttings in water on a sunny windowsill, plants like impatiens, coleus, sweet potato vine cultivars, and purple heart can be held over winter until needed in the spring. Otherwise, the plants can be cut back by half, potted in a peat-based, soilless mix, and placed on a sunny windowsill. With a wide assortment of "annuals" available on the market, some research is required to determine which annuals can be over-wintered successfully. True annuals (such as basils, cockscomb, and zinnias)—regardless of any treatment given—will go to seed and die when brought inside.

Preserving Cacti and Succulents

If you planted a mixed dry container this year and want to retain any of the plants for next year, they should be removed from the main container and repotted into a high-sand-content soil mix for cacti and succulents. Keep them in a sunny window and water when dry. Many succulents and cacti do well indoors, either in a heated garage or a moderately sunny corner of a living room.

As with other tropical plants, succulents also need time to adjust to sunnier conditions in the spring. Move them to a shady spot outside when temperatures have stabilized above 60°F and then gradually introduce them to brighter conditions.

Hardy Perennials, Shrubs, and Vines

Hardy perennials, woody shrubs, and vines needn't be thrown away when it's time to get rid of accent containers. Crack-resistant, four-season containers can house perennials and woody shrubs year-round. Below is a list of specific perennials and woody plants that do well in both hot and cold weather, indoors and out:

- Shade perennials, like coral bells, lenten rose, assorted hardy ferns, and Japanese forest grass are great for all weather containers.

- Sun-loving perennials, such as sedges, some salvias, purple coneflower, daylily, spiderwort, and bee blossom are also very hardy and do well in year-round containers. Interplant them with cool growing plants, like kale, pansies, and Swiss chard, for fall and spring interest.

- Woody shrubs and vines—many of which have great foliage interest with four-season appeal—are ideal for container gardens. Red-twigged dogwood cultivars, clematis vine cultivars, and dwarf crape myrtle cultivars are great container additions that can stay outdoors year-round.

If the container has to be removed, hardy perennials and woody shrubs can be temporarily planted in the ground and mulched. Dig them from the garden in the spring, if you wish, and replant into a container. Or, leave them in their garden spot and start over with fresh ideas and new plant material for your container garden.

Sustainable Plants and Money in Your Pocket

Over-wintering is a great form of sustainable plant conservation achieved simply and effectively by adhering to each plant's cultural and environmental needs. With careful planning and storage techniques, you'll save money as well as plant material. The beauty and interest you've created in this season's well-grown container garden can also provide enjoyment for years to come.

FARMERS' MARKETS

Farmers' markets are an integral part of the urban–farm linkage and have continued to rise in popularity, mostly due to growing consumer interest in obtaining fresh products directly from the farm. Farmers' markets allow consumers to have access to locally grown, farm-fresh produce, enable farmers the opportunity to develop a personal relationship with their customers, and cultivate consumer loyalty with the farmers who grow the produce. Direct marketing of farm products through farmers' markets continues to be an important sales outlet for agricultural producers nationwide. Today, there are more than 4,600 farmers' markets operating throughout the nation.

Who Benefits from Farmers' Markets?

- Small farm operators: Those with less than $250,000 in annual receipts who work and manage their own operations meet this definition (94 percent of all farms).

- Farmers and consumers: Farmers have direct access to markets to supplement farm income. Consumers have access to locally grown, farm-fresh produce and the opportunity to personally interact with the farmer who grows the produce.

- The community: Many urban communities—where fresh, nutritious foods are scarce—gain easy access to quality food. Farmers' markets also help to promote nutrition education, wholesome eating habits, and better food preparation, as well as boosting the community's economy.

Getting Involved in a Farmers' Market

A farmers' market is a great place for new gardeners to learn what sorts of produce customers want, and it also promotes wonderful community relationships between the growers and the buyers. If you have a well-established garden, and know a few other people who also have fruits, vegetables,

and even garden flowers to spare, you may want to consider organizing and implementing your own local farmers' market (if your town or city already has an established farmers' market, you may want to go visit it one day and ask the farmers how you could join, or contact your local Cooperative Extension Service for more information). Joining existing farmers' markets may require that you pay an annual fee, and your produce may also be subject to inspection and other rules established by the market's organization or the local government.

Establishing a farmers' market is not simply setting up a stand in front of your home and selling your vegetables—though you can certainly do this if you prefer. A farmers' market must have a small group of people who are all looking to sell their produce and garden harvests. It is important, before planning even begins, to hold a meeting and discuss the feasibility of your venture. Is there other local competition that might impede on your market's success? Are there enough people and enough produce to make a farmers' market profitable and sustainable? What kind of monetary cost will be incurred by establishing a farmers' market? It is also a good idea to think about how you can sponsor your market—such as though members, nonprofit organizations, or the chamber of commerce. If at all possible, it is best to try to get your entire neighborhood and local government involved and promote the idea of fresh, home-grown fruits and veg-

etables that will be available to the community through the establishment of a farmers' market.

Once you've decided that you want to go ahead with your plan, you should establish rules for your market. Such rules and regulations should determine if there will be a board of directors, who will be responsible for the overall management of the market, who can become a member, where the market will be located (preferably in a place with ample parking, good visibility, and cover in case of bad weather), and when and for how long the market will be open to the public. It is also a good idea to discuss how the produce should be priced and make sure you are following all local and state regulations.

Ideally, it is beneficial for your market's vendors and the community to gather support and involvement from local farmers in your area. That way, your garden vegetable and fruit stand will be supplemented with other locally grown produce and crops from farms, which will draw more people to your market.

Once your farmers' market is up and running, it is important to maintain a good rapport with the community. Be friendly when customers come to your stand, and make sure that you are offering quality products for them to purchase. Price your vegetables and fruits fairly and make your displays pleasing to the eye, luring customers to your stall. Make sure your produce is clearly marked with the name

and price, so your customers will have no doubt as to what they are buying. Make your stall as personal as possible, and always, always interact with the customer. In this way, you'll begin to build relationships with your community members and hopefully continue to draw in business for yourself and the other vendors in the farmers' market.

To find a local farmers' market near you (either to try to join or just to visit), here is a link to a web site that offers extensive listings for farmers' markets by state: http://apps. ams.usda.gov/farmersmarkets.

FLOWER GARDENS

Many flower species will benefit your vegetables when grown alongside them (see page 631). Growing geraniums and marigolds around the outside perimeter of your vegetable garden, for example, will aid your peppers and tomatoes and create an attractive border for the plot. Whether you mix flowers and veggies or start a separate flower plot, these guidelines will get you started.

Choose the Right Size

A flowerbed of around 25 square feet will provide you with room for about 20 to 30 plants—enough room for three types of annuals and two types of perennials. If you want to start even smaller, you can always begin your first flower garden in a container, or create a border from treated wood or bricks and stones around your existing bed. That way, when you are ready to expand your garden, all you need to do is remove the temporary border and you'll be all set. Even a small container filled with a few different types of plants can be a wonderful addition to any yard.

Plan Your Flower Garden

Draw up a plan of how you'd like your garden to look, and then dig a flowerbed to fit that plan. Planning your garden before gathering the seeds or plants and beginning the digging can give you a clearer sense of how your garden will be organized and can facilitate the planting process.

Choose a Good Spot

When choosing where your flower garden will be located, try to find an area that receives at least six hours of direct sunlight per day, as this will be adequate for a large variety of garden plants. Be careful that you will not be digging into utility lines or pipes, and that you place your garden at least a short distance away from fences or other structures.

If you live in a part of the country that is quite hot, it might be beneficial for your flowers to be placed in an area that gets some shade during the hot afternoon sun. Placing your garden on the east side of your home will help your flowers flourish. If your garden will get more than six hours of

sunlight per day, it would be wise to choose flowers that thrive in hot, sunny spaces, and make sure to water them frequently.

It is also important to choose a spot that has good, fertile soil in which your flowers can grow. Try to avoid any areas with rocky, shallow soil or where water collects and pools. Make sure your garden is away from large trees and shrubs, as these plants will compete with your flowers for water and nutrients. If you are concerned that your soil may not contain enough nutrients for your flowers to grow properly, you can have a soil test done, which will tell you the pH of the soil. Depending on the results, you can then adjust the types of nutrients needed in your soil by adding organic materials or certain types of fertilizers.

Start Digging

Now that you have a site picked out, mark out the boundaries with a hose or string. Remove the sod and any weed roots that may re-grow. Use your spade or garden fork to dig up the bed at least 8 to 12 inches deep, removing any rocks or debris you come across.

Once your bed is dug, level it and break up the soil with a rake. Add compost or manure if the soil is not fertile. If your soil is sandy, adding peat moss or grass clippings will help it hold more water. Work any additions into the top 6 inches of soil.

Purchase and Plant Your Seeds or Seedlings

Once you've chosen which types of flowers you'd like to grow in your garden, visit your local garden store or nursery and pick out already-established plants or packaged seeds. Follow the planting instructions on the plant tabs or seed packets. The smaller plants should be situated in the front of the bed. Once your plants or seeds are in their holes, pack in the soil around them. Make sure to leave ample space between your seeds or plants for them to grow and spread out (most labels and packets will alert you to how large your flower should be expected to grow, so you can adjust the spacing as needed).

Water Your Flower Garden

After your plants or seeds are first put into the ground, be sure they get a thorough watering. Then, continue to check your garden to see whether or not the soil is drying out. If so, give your garden a good soaking with the garden hose or watering can. The amount of water your garden needs is dependant on the climate you live in, the exposure to the sun, and how much rain your area has received.

Cutting Your Flowers

Once your flowers begin to bloom, feel free to cut them and display the beautiful blooms in your home. Pruning your flower garden (cutting the dead or dying blooms off the plant) will help certain plants to re-bloom. Also, if you have plants that are becoming top heavy, support them with a stake and some string so you can enjoy their blossoms to the fullest.

Things to Consider

- Annuals are plants that you need to replant every year. They are often inexpensive, and many have brightly colored flowers. Annuals can be rewarding for beginner gardeners, as they take little effort and provide lovely color to your garden. The following season, you'll need to replant or start over from seed.

- Perennials last from one year to the next. They, too, will require annual maintenance but not yearly replanting. Perennials may require division, support, and extra care during winter months. Perennials may also need their old blooms and stems pruned and cut back every so often.

- Healthy, happy plants tend not to be as susceptible to pests and diseases. Here, too, it is easier to practice prevention rather than curing existing problems. Do your best to give your plants good soil, nutrients, and appropriate moisture, and choose plants that are well suited to your climate. This way, your garden will be more likely to grow to its maximum potential and your plants will be strong and healthy.

The Junior Homesteader

Plan a Garden Party

A party in the garden is a great way to celebrate a birthday, last day of school, or other special event. And it will show kids how much fun gardens can be. For invitations, cut large flower shapes out of construction paper and let the kids paste pictures from seed catalogs or gardening magazines onto one side. On the other side, write the date, time, and place of your party. The garden itself will serve as a beautiful backdrop for your celebration, but if you want to go further, choose seasonal decorations such as bouquets of lilacs and apple blossoms in spring, daisy chains in summer, and pumpkins and corn stalks in fall. Once at the party, kids will enjoy decorating their own clay flowerpots by painting them with acrylic paints or, for older kids, using a glue gun to attach stones or beads to the outsides of the pots. Then provide seeds and soil so they can bring a potted plant home and watch it grow. Have a contest to see who can make the most creative faces or sculptures out of fresh produce—and then eat the fun creations! Scavenger hunts are always a hit. Provide a chart showing different types of leaves, grasses, and flowers and see who can find all the items and check them off the chart first. Kids will enjoy coming up with their own games, too. Just spending time together in the garden will create lasting fun memories.

Planters

Barrel Plant Holder

If you have some perennials you want to display in your yard away from your flower garden you can create a planter out of an old barrel. This plant holder is made by sawing an old barrel (wooden or metal) into two pieces and mounting it on short or tall legs—whichever design fits better in your yard. You can choose to either paint it or leave it natural. Filling the planter with good quality soil and compost and planting an array of multi-colored flowers into the barrel planters will brighten up your yard all summer long. If you do not want to mount the barrel on legs, it can be placed on the ground on a smooth and level surface where it won't easily tip over.

Rustic Plant Stand

If you'd like to incorporate a rustic, natural-looking plant stand in your garden or on your patio or deck, one can easily be made from a preexisting wooden box or by nailing boards together. This box should be mounted on legs. To make the legs, saw the piece of wood meant for the leg in half to a length from the top equal to the depth of the box. Then, cross-cut and remove one half. The corner of the box can then be inserted in the middle of the crosscut and the leg nailed to the side of the box.

The plant stand can be decorated to suit your needs and preference. You can nail smaller, alternating twigs or cut branches around the stand to give it a more natural feel or you can simply paint it a soothing, natural color and place it in your yard.

Wooden Window Box

Planting perennial flowers and cascading plants in window boxes is the perfect way to brighten up the front exterior of your home. Making a simple wooden window box to hold your flowers and plants is quite easy. These boxes can be made from preexisting wooden boxes (such as fruit crates) or you can make your own out of simple boards. Whatever method you choose, make sure the boards are stout enough to hold the brads firmly.

The size of your window will ultimately determine the size of your box, but this plan calls for a box roughly 21 x 7 x 7 inches. You can decorate your boxes with waterproof paint or you can nail strips of wood or sticks to the panels. Make sure to cut a few holes in the bottom of the box to allow for water drainage.

FRUIT BUSHES AND TREES

If you take the time to properly plan and care for your fruit bushes and trees, they'll provide you with delicious, nutrient-dense fruit year after year. Some fruit plants, like strawberries, are easy to grow and will reward you with ripe fruit relatively quickly. Fruit trees, like apple or pear, will require more work and time, but with the right maintenance they will bear fruit for generations.

Think carefully about where you choose to plant and then take time to prepare the site. Most fruit plants need at least six hours a day of sun and require well-drained soil. If the soil is not already cultivated and relatively free of pests, spend the first year preparing the site. Planting a cover crop of rye, wheat, or oats will improve the quality of your soil and reduce weeds that could compete with your fruit plants. The cover crops will die in the late fall and add to the organic matter of the soil. Just leave them to decompose on the surface of the soil and then turn them under the soil come spring.

Testing your soil pH the year ahead of planting will give you time to adjust it if necessary to give your plants the best chance of thriving. Fruit trees, grapes, strawberries, blackberries, and raspberries do best if the soil pH is between 6.0 and 6.5. Blueberries require a more acidic soil, around 4.5.

What plants will thrive will depend largely on where you live, your planting zone page, your altitude, and your proximity to large bodies of water (since areas close to water tend to be more temperate). Refer to the chart below for hardiness zones for most fruits, but keep in mind that hardiness varies by variety. Refer to seed catalogs or talk to other local gardeners before settling on a particular variety of fruit to plant.

Fruit	Hardiness Zone	Soil pH	Space between Plants	Space between Rows	Bearing Age (in Years)	Potential Yield (in Pounds)	When to Harvest
Apple	5 to 7	6 to 6.5	7 to 18 (depending on variety)	13 to 24 (depending on variety)	3 to 5	60 to 250	August through October
Apricot	4 to 8	6 to 6.5	15	20	4	100	July to August
Cherry, sweet	5 to 7	6 to 6.5	24	30	7	300	July
Cherry, tart	4 to 7	6 to 6.5	18	24	4	100	July
Peach and nectarine	5 to 8	6 to 6.5	15	20	4 to 5	100	August to September
Pear and quince	4 to 7	6 to 6.5	15 to 20	15 to 20	4	100	August to October
Plum	5 to 7	6 to 6.5	10	15	5	75	July to September
Grapes	5 to 10 (depending on variety)	6 to 6.5	8	9	3	10 to 20	September to October
Blackberry	3 to 9 (depending on variety)	6 to 6.5	2	10	2	2 to 3	July to August
Blueberry	3 to 11 (depending on variety)	4 to 5	4 to 5	10	3 to 6	3 to 10	July to September
Currant	2 to 6	5.5 to 7	4	8	2 to 4	6 to 8	July
Elderberry	3 to 9	6 to 6.5	6	10	2 to 4	4 to 8	August to September
Gooseberry	2 to 6	5.5 to 7	4	10	2 to 4	2 to 4	July to August
Raspberry	3 to 8	5.6 to 6.2	2	8	2	1 to 2	July to September
Strawberry	4 to 9	5.5 to 6.5	12 to 18	12 to 18	1 to 2	1 to 3	May to July

Brambles and Bush Fruits

Most brambles and bush fruits should be planted in the spring after the last frost. Blackberries and raspberries should have any old or damaged canes removed before planting in a 4-inch deep hole or furrow. Do not fertilize for several weeks after planting and even after that use fertilizer sparingly; brambles are easily damaged by over-fertilizing. If the weather is dry, water bushes in the morning, just after the dew has dried, being careful to avoid getting the foliage very wet.

Brambles need to be pruned once a year to keep them healthy and to keep them from spreading out of control. How you prune your brambles will depend on the variety. Fall-bearing brambles (primocane-fruiting brambles) produce fruit the same year they are planted. If planted in spring, they'll produce some berries in late summer or early fall and then again (lower on the canes) in early summer of their second year. For these varieties, there are two pruning choices. The first option produces a smaller yield but higher quality berries in late summer. For this pruning method, mow down the bush all the way to the ground in late fall. The second method will produce berries in summer and fall, and is the same method used for floricane-fruiting brambles. Allow the canes to grow through the first year and prune them gently in the early spring of the following year. Remove any damaged or diseased parts and thin canes to three or four per foot. Trim down the tops of canes to about 4 to 5 feet high so that you can easily pick the berries. Prune similarly every spring.

Blueberries

Blueberries thrive in acidic soil (around pH 4.8). If your soil is naturally over pH 6.5, you're better off planting your blueberries in container gardens or raised beds, where you can more easily manipulate the soil's pH.

Blueberries should be planted in a hole about 6 inches deep and 20 inches in diameter. The crowns should be right at soil level. Surround the stems with about 6 inches of sawdust mulch or leaves and water the bushes in the morning in dryer climates. Blueberries are very sensitive to drought.

Some varieties of blueberries only require pruning if there are dead or damaged branches that need to be removed. Highbush varieties should be pruned every year starting after the third year after planting. When old canes get twiggy, cut them off at soil level to allow new, stronger canes to emerge.

Currants and Gooseberries

These berries thrive in colder regions in well-drained soils. Plant them in holes 12 inches across and 12 inches deep. They should bear fruit in their second year and will give the highest yield in their in the third year. After that the canes will begin to darken, which is a sign that it's time to prune them by mowing or clipping at soil level. By this point, new canes will likely have developed. Currants and gooseberries do not require much pruning, but dead or diseased branches should always be removed.

Fruit Trees

Planting

Once you've decided on which varieties to plant and have planned and prepared the best site, it's time to purchase the trees and plant them. Most young trees come with the roots planted in a container of soil, embedded in a ball of soil and wrapped in burlap (balled-and-burlapped, or B&B), or packed in damp moss or Excelsior. It's best to plant your trees as soon as possible after purchasing, though B&B stock or potted trees can be kept for several weeks in a shady area.

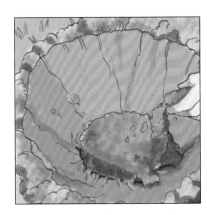

1. To plant, dig a hole that is about 2 feet deep and wide enough to give the roots plenty of room to spread (about 1 ½ feet wide). As you dig, try to keep the sod, topsoil, and subsoil in separate piles.

2. Once the hole is the right size, loosen the dirt at the bottom and then place the sod into the hole, upside down. Then make a small mountain of topsoil in the center of the hole. The roots will sit on top of the mountain and hang over the edges.

3. Next, prune away any broken or damaged roots. Pruning shears or a sharp knife work best. Remove roots that are very tangled or too long to fit in the hole.

4. Place the roots into the hole, on top of the dirt mound. If the tree was grafted, it will have a "graft union," a bulge where the roots meet the trunk. For most trees, this mound should be barely visible above the ground. For dwarf trees, it should be just below the soil level.

5. Most young trees need extra support. Drive a 5- to 6-foot garden stake into the ground a few inches away from the trunk and on the south side. The stake should go about 2 feet deeper than the roots.

6. Begin filling the hole back in with soil, using your fingers to work the soil around the roots and eliminate any air pockets. Add soil until the hole is filled, pat it down until it's slightly lower than the ground level (to help retain water), and then pour a bucket of water over the soil to pack it down further.

7. Prune away all but the three or four strongest branches and then tie the trunk gently to the stake with a soft rag.

8. Spread leaves, mulch, or bark around the base of the tree to protect the roots and help retain the water. Because young tree bark is easily injured, wrap the trunk carefully with burlap from the ground to its lowest branches. This will protect the bark from being scalded by sun (even in winter) or damaged by deer or rodents. Leave the wrap on for the first two or three years.

Pruning

Your fruit trees will benefit from gentle pruning once a year. The goals of pruning are to remove dead or damaged branches, to keep branches from crowding each other, and to keep trees from growing so large that they begin to invade each other's space. Pruning, when done properly, will produce healthier trees and more fruit.

Pruning shears or a pruning saw can be used. When removing a whole branch try to cut flush with the trunk, so that no "stub" is left behind. Stubs soon decay, inviting insects to invade your tree. A cut that is flush with the trunk

will heal over quickly (in one growing season). If you're removing part of a branch, cut slightly above a bud and cut at a slant. Try to choose a bud that slants in the direction you want a new branch to grow.

When removing particularly large branches, care should be taken to ensure that the branch doesn't tear away large pieces of bark from the trunk as it falls. To do this, start below the branch and cut upwards about a third of the way through the branch. Then cut down from the top, starting an inch or two further away from the trunk.

If your pruning leaves a wound that is larger than a silver dollar, use a knife to peel away the bark above and below the wound to create a vertical diamond shape. Then cover the wound with shellac or tree wound paint to protect it from decay and insects.

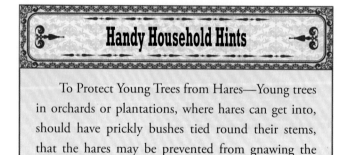

Handy Household Hints

To Protect Young Trees from Hares—Young trees in orchards or plantations, where hares can get into, should have prickly bushes tied round their stems, that the hares may be prevented from gnawing the bark off.

Grapes

Talk to someone at your local nursery or to other growers in your area to determine which grape variety will work best in your location. All varieties fall under the categories of

wine, table, or slipskin. Grapes need full sun to stay healthy and benefit from loose, well-drained, loamy soil.

1. Before planting your vine cutting, soak it in a bucket of water for at least six hours. Cuttings should never dry out.

2. Grape vines need a trellis or a similar support.

3. Dig a hole near the trellis deep enough to accommodate the roots and douse the hole with water.

4. Place the cutting in the hole and fill in the soil around it, adding more water as you go, and then tamping it down firmly.

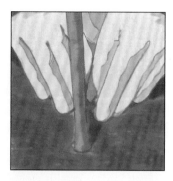

5. Prune away all but the best cane.

6. Tie the cane gently to the support (a stake or the bottom wire of a trellis). Water the vine once a week for at least the first month.

Don't use any fertilizer in the first year as it can actually damage the young vines. If necessary, begin fertilizing the soil in the second year.

Buds will begin to grow after several weeks. After about ten weeks, remove all but the strongest shoots as well as any flower clusters or side shoots. Every year in the late fall, after the last grapes have been picked, remove 90 percent of the new growth from that season. You should be able to harvest fruit in the plant's third year.

Strawberries

Purchase young strawberry plants to plant in the spring after the last frost. Try to find plants that are certified disease-free, since diseases from strawberries can spread through your whole garden. Strawberries thrive with lots of sun and well-drained soil. If you have access to a gentle south-facing slope, this is ideal. Till the top 12 inches of soil. If you planted a cover crop, turn under all the organic matter. If not, be sure to add manure or compost to a reach a rich, slightly acidic soil.

1. Dig a 5- to 7-inch wide hole for each plant. It should be deep enough to accommodate the root system without squishing it.

2. Place the plant in the hole and fill in the soil, tamping it down gently around the plant.

3. Space plants about 12 inches apart on all sides. The roots will shoot out runners that produce more small plants. To allow the plants to focus their energy on fruit production, snip the runners and transplant or discard any new plants.

Matted-row system

An alternate planting method is the matted-row system. This method requires less maintenance but offers a slightly lower quality yield. Space plants about 18 inches apart, allowing the roots to shoot out runners and produce new plants. If planting more than one row, space them three to four feet apart. To aid picking and to keep the plants from competing with each other, prune out the plants on the outer edges of rows by snipping the runners and pulling out the plants.

Strawberry plants should receive at least an inch of water per week. In the first year, snip away or pick blossoms as soon as they develop. You will sacrifice your fruit crop in the first year, but you will have healthier plants and a greater fruit yield for many years afterward.

GRAINS: GROWING AND THRESHING

Grains are a type of grass and they grow almost as easily as the grass in your yard does. There are many reasons for growing your own grains, including supplying feed for your livestock, providing food for you and your family, or to use as a green manure (a crop that will be plowed back into the soil to enrich it). Growing grains requires much less work than growing a vegetable garden, though getting the grains from the field to the table requires a bit more work.

Whether you are growing wheat, oats, barley, or another grain, the process is basically the same:

1. Decide which grain to grow. Most cereal grains have a spring variety and a winter variety. Winter grains are often preferred because they are more nutritious than spring varieties and are less affected by weeds in the spring. However, spring wheat is preferred in cold climates as winter wheat may not survive very harsh winters. If you have trouble finding smaller amounts of seeds to purchase from seed supply houses, try health food stores. They often have bins full of grains you can buy in bulk for eating, and they work just as well for planting, as long as you know what variety of grain you're buying. Winter grains should be planted from late September to mid-October, after most insects have disappeared but before the hard frosts set in. Spring wheat should be planted in early spring.

2. Decide how much grain you want to grow. A 10-foot by 10-foot plot of wheat will provide enough flour for about 20 loaves of bread. An acre of corn will provide feed for a pig, a milk cow, a beef steer, and 30 laying hens for an entire year.

3. Prepare the soil. Rototill or use a shovel to turn over the earth, remove any stones or weeds, and make the plot as even as possible using a garden rake.

4. Sprinkle the seeds over the entire plot. How much seed you use will depend on the grain (refer to the chart). For wheat, use a ratio of around 3 ounces of seed per 100 square feet. Aim to plant about 1 seed per square inch. Rake over the plot to cover all the seeds with earth.

5. Water the seeds immediately after planting and then about once a month throughout the growing season if there's not adequate rainfall.

6. When the grain is golden with a few streaks of green left, it's ready for harvest. For winter grains, harvest is usually ready in June or July. To cut the grains, use a scythe, machete, or other sharp knife, and cut near the base

of the stems. Gather the grains into bundles, tie them with twine, and stand them upright in the plot to finish ripening. Lean three or four bundles together to keep them from falling over. If there is danger of rain, move the sheaves into a barn or other covered area to prevent them from molding. Once all the green has turned to gold, the grains are ready for threshing.

7. The simplest way to thresh is to grasp a bunch of stalks and beat it around the inside of a barrel, heads facing down. The grain will fall right off the stalks. Alternatively, you can lay the stalks down on a hard surface covered by an old sheet and beat the seed heads with a broom or baseball bat. Discard (or compost) the stalks. If there is enough breeze, the chaff will blow away, leaving only the grains. You can also pour the grain and chaff back and forth between two barrels and allow the wind (which can be supplied by a fan if necessary) to blow away the chaff.

8. Store grain in a covered metal trash can or a wooden bin. Be sure it is kept completely dry and that no rodents can get in. To turn the grain berries into flour, see page 104.

An easy way to thresh your grains is to grasp a bunch and beat it against the edge of a barrel. You can also beat the stalks with a broom.

How much grain should you grow to feed your family?

An acre of wheat will supply about 30 bushels of grain, or around 1,800 pounds. The average American consumes about 140 pounds of wheat in a year. The Federal Emergency Management Association (FEMA) recommends the following consumption rates:

- Adult males, pregnant or nursing mothers, active teens ages 14 to 18: 275 lbs./year
- Women, kids ages 7 to 13, seniors: 175 lbs./year
- Children 6 and under: 60 lbs./year

Grain Growing Chart

Type of Grain	Amount of seed per acre (in pounds)	Grain yield per acre (in bushels)	Characteristics and Uses
Amaranth	1	125	Very tolerant of arid environments. High in protein and gluten-free. Use in baking or animal feed.
Barley	100	120 to 140	Tolerates salty and alkaline soils better than most grains. Use in animal feed, soups, as a side dish, and for making beer and malts.
Buckwheat	50	20 to 30	Matures rapidly (60 to 90 days). Rich, nutty flavor perfect for baking and in pancakes.
Field corn	6 to 8	180 to 190	Use in animal feed, corn starch, hominy, and grits.
Grain sorghum	2 to 8	70 to 100	Drought tolerant. Use in animal feed or in baking.
Oats	80	70 to 100	Thrive in cool, moist climates. High in protein. Use in animal feed, baking, or as a breakfast cereal.
Rye	84	25 to 30	Tolerant of cold, dampness, and drought. Use for animal feed, in baking, to make whiskey, or as a cover crop.
Wheat	75 to 90	40 to 70	Hard red winter wheat is used in bread and is highly nutritious. Soft red winter wheat is good for cakes and pastries. Hard red spring wheat is the most common bread wheat. Durum wheat is best for pasta.

GREENHOUSES AND HOOPHOUSES

For an avid gardener, the region in which you live may not be able to support all the types of plants you want to grow. Whether you're interested in reducing, or even eliminating, the produce bill or because gardening is your favorite hobby, a greenhouse can help you garden year-round regardless of where you live by regulating temperature and humidity and diminishing the effects of the weather and the wind.

First, you need to ask yourself a few questions. What type of gardening will you do? What plants are you interested in growing? How much space do you have and where will you place your greenhouse? Will this be permanent or temporary? What is your budget? Once you've answered these, you can narrow down your options.

Types of Greenhouses

The type of the greenhouse depends on the plants you intend to grow.

1. Cold houses and cold frames: These type of greenhouses can protect your plants from the elements, but do not

have any additional heat sources. That means, depending on your climate region, temperatures could drop down below freezing in the winter. Cold houses are great for starting spring crops a few weeks early and for continuing them a little later than usual into the fall.

This is a relatively small, freestanding cold frame house made of steel and plexiglass.

2. Cool houses: The temperature in these greenhouses is regulated between 40-45 degrees. This keeps the temperature above freezing year-round, which is perfect for plants that are frost-sensitive.

3. Warm houses: With the temperature stabilized at about 55 degrees, warm houses can hold a larger variety of plants, perhaps the same amount you'd have in your outdoor garden.

4. Hot houses: Expensive to maintain at a minimum of 65 degrees, hot houses are intended for tropical and exotic plant species. You'll need to install heating and lighting equipment to satisfy these picky plants' needs.

Styles of Greenhouses

The styles have a subcategory of attached greenhouses or freestanding greenhouses. This decision depends on many factors, but particularly the space you have and the money you want to spend. Remember that, regardless of whether your greenhouse is attached or not, you need to consider the location of your greenhouse. Give your plants as much sun as possible by avoiding shadows cast by your house, neighboring houses, or trees.

1. Lean-to: An attached greenhouse that is especially useful when space is lacking. It's also one of the less expensive options. The ridge is attached to the side of the building, which takes care of one of the sides and the door, if one is available. Lean-tos are close to the available electric-

A lean-to-style greenhouse made of steel and plexiglass.

ity, water, and heat that your greenhouse needs. There are disadvantages, of course, namely the limitations of space, sunlight, temperature control, and ventilation. The height of the supporting wall limits the size of the lean-to. It's important to place the lean-to so that it is receiving adequate sun exposure.

2. Even-span: A full-size structure that has one of its ends attached to another building. This structure tends to be the largest and most expensive of the attached greenhouse category, but for good reason. Even-spans have a better shape than lean-tos, allowing for more efficient air circulation that helps maintain the temperature during the winter season. Even-spans can accommodate 2 or 3 benches for growing.

3. Window-mounted: This small greenhouse structure must be attached on the south or east side of a house. It

This lean-to greenhouse is attached to the home and uses post and rafter construction.

This window-mounted greenhouse is perfect for small potted plants.

A Gothic frame freestanding greenhouse.

is a glass enclosure, a special window extending from the house about a foot, that allows a gardener to grow only a few plants. It contains a few shelves for the plants and is relatively cheap.

4. Freestanding: This greenhouse is set apart from other buildings in order to maximize sun exposure. The size can be as large or as small as you desire, but a separate heating system is needed, and water and electricity have to be installed, which can quickly get expensive if you're not careful.

A freestanding greenhouse can be as large or small as space allows. This one has the gothic frame and the siding is plexiglass.

Greenhouse Coverings

The covering of your greenhouse and the frame you choose must be matched correctly. There are a few different coverings to pick from, each having its own life-span. Your decision will depend on how long you plan to have your greenhouse, the money you intend to spend, and how you'd like your greenhouse to look in general.

1. Glass: These coverings last for quite a long time; they're traditional greenhouse walls. Attractive and inexpensive to maintain, glass coverings are a popular choice for those itching to build a greenhouse. When combined with an aluminum frame they are almost maintenance

free, not to mention they're weather-tight, minimizing the cost of heating and preserving humidity. Tempered glass is often used in place of normal glass because it happens to be two-to-three times stronger than the latter. While you can buy and install small pre-fabricated glass greenhouses, most should actually be built by the manufacturer because of how troublesome they are to put together. Glass is, unfortunately, more expensive to construct (at least initially) and more easily broken. It requires a better, sturdier frame and a solid foundation, which means more work for you.

2. Fiberglass: Lightweight, so strong as to be practically hailproof, and with a reputation as good as glass, fiberglass is another excellent option. However this is only true if you purchase a good grade of fiberglass, because the poor ones can discolor and prevent light penetration. Proper light penetration also calls for grades that are clear, transparent, or translucent and they need to be coated in some kind of resin; tedlar-coated greenhouses can last 15-20 years. The resin will eventually wear off and need to be replaced otherwise the exposed fibers of the glass will retain dirt.

3. Double-wall plastic: Two rigid plastic sheets of acrylic or polycarbonate are separated by webs and boast heat retention (and therefore, energy savings) and long-life. You should note that acrylic is a non-yellowing material, and while polycarbonate does yellow faster, it is also protected by a UV-inhibitor coating on the exposed surface and is more flexible than the acrylic.

4. Film-plastic: Film-plastic has many options regarding the actual materials and the grades of each. These covers will have to be replaced more frequently than previous options, but the initial structural costs are very affordable. The frame can be lighter for these coverings, which results in an inexpensive greenhouse. Films can be made of polyethylene (PE), polyvinyl chloride (PVC), copolymers, and other materials. A utility grade of polyethylene that will last about a year can be found at your local hardware store; or, you can opt for the commercial greenhouse polyethylene option which has UV-inhibitors in it to protect against UV rays—these last a little longer, ranging from 12-18 months. Copolymers usually last for 2-3 years. Manufacturers of film-plastic

coatings have upgraded their products so that the film-plastic can block and reflect radiated heat back into the greenhouse which reduces the heating costs. Polyvinyl chloride can cost as much as 2-5 times more than poly-ethylene, but lasts up to five years longer.

Greenhouse Frames

Once you've decided on the covering you'd like to use, you can narrow your framing options.

1. Quonset: If you want plastic sheeting, then this circular frame may be a great choice for you. Its construction is simple and efficient and the frame itself may be made out of galvanized steel pipes or an electrical conduit. The drawbacks of this frame are that, because the sidewalls are on the shorter side, you're left with less storage space and headroom.

Quonset frames work well with plastic coverings.

2. Gothic: Similar to Quonset in shape and the type of covering it calls for, a Gothic frame has higher sidewalls and allows for more headroom. To complete the Gothic look, you may want to opt for wooden arches that join at the ridge.

3. Rigid-frame: Exactly like its name, this frame is rigid and can support heavier coverings. With heavier loads, however, comes a need for a good foundation, so keep that in mind if you decide to build a rigid-frame green-house. The sidewalls are vertical, there aren't any columns or trusses to support the conventional gable roof, and the rafters of the roof are glued to the sidewalls, producing one rigid frame. This will give you a great deal of interior space and air circulation.

4. Post and Rafter: While this is counted as simple construction, with embedded posts and rafters, it does necessitate more wood or metal than other designs. The sidewall posts need to be strong and the posts must be embedded deep into the ground to withstand wind pressures and outward rafter forces. Like the previous design, this frame also offers a good amount of interior space and air circulation.

5. A-Frame: Similar to the Post and Rafter in structure, space, and air circulation, the A-Frame differs simply because a collar-beam ties together the upper parts of the rafters.

This A-frame greenhouse is attached to the home and constructed with ridged steel and glass.

Greenhouses certainly provide you with a lot to think about, and you haven't even started constructing one yet! The question will arise at some point before you start your greenhouse: why shouldn't I use a kit?

And while there's nothing wrong with using a kit, there are a few reasons as to why making your own greenhouse is more beneficial. Kits are not necessarily specialized to be the greenhouse of your dreams. Making your own ensures that you are designing the exact greenhouse for your needs; obviously, it will also be sized to your precise specifications.

Your greenhouse will most likely be stronger than one from a kit, the materials will be easy to locate at a hardware store (rather than specialized ones), and commercially available vents, fans, any accessories you might want to add to your greenhouse will be easier to incorporate because you made your greenhouse yourself; you know what it needs and what it can handle, not to mention there won't be any confusion with whether materials can work together or not.

Lastly, and probably at the forefront of your mind, it can work for your budget! Building your own greenhouse costs roughly half of what a kit does, depending on which greenhouse you decide.

If you aren't averse to putting in a couple of weeks' time and effort, then it's time to build your own greenhouse. In this book, we'll cover the hoop house.

Hoophouses

Hoophouses are small, semi-portable structures that can be used as a small greenhouse structure for starting seedlings outdoors and for growing heat-loving vegetables. A hoophouse provides frost protection, limited insect protection, and season extension. Hoophouse structures are easy to construct and will last many years. You can make them any size, but a structure 4 feet x 10 feet is generally adequate. These dimensions allow easy side access for weeding and adequate hoop arch strength relative to span.

Seeds can be started in flats in the hoophouse. Temperature is regulated by varying the size of the end openings and/or lifting the side wall plastic. After seed trays are removed (to be planted in an outdoor garden), heat-loving plants such as tomatoes, peppers, and melons can be grown directly in the soil in the hoophouse. Plastic can be left in place to keep hoophouse temperatures warm until outside temperatures will support active plant growth or until plant vegetation outgrows the confines of the box. Plants requiring staking, such as tomatoes, can be planted near the edge of the box near a hoophouse support. After the plastic and hoops are removed, a rigid stick or dowel can be inserted into the plastic hoop retainer. Tomato plants can then be tied and supported by the rigid upright stake.

Follow the plans on page 656 to build a hoophouse. First, attach the pipes to the inside of your raised garden bed to form arches.

Next, cover the hoops with plastic sheeting and secure the ends with hand spring clamp.

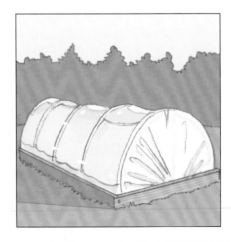

Hoophouses can last for years, but the plastic covering may need periodic changing.

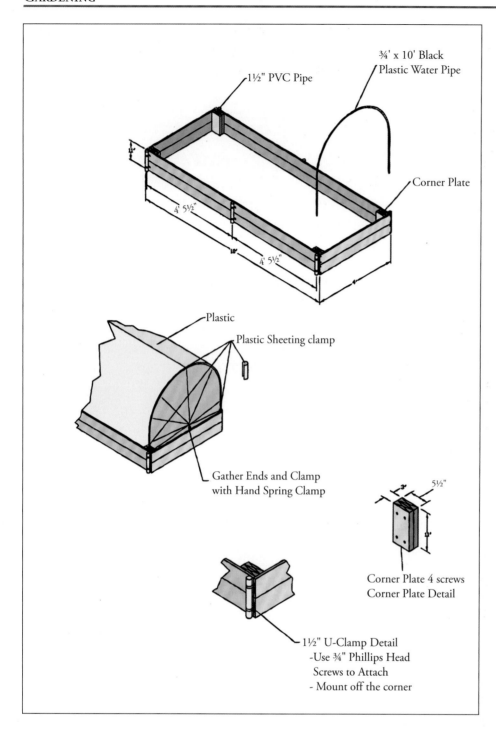

¾' x 10' Black
Plastic Water Pipe

1½" PVC Pipe

Corner Plate

Plastic

Plastic Sheeting clamp

Gather Ends and Clamp
with Hand Spring Clamp

5½"
3"

Corner Plate 4 screws
Corner Plate Detail

1½" U-Clamp Detail
-Use ¾" Phillips Head
Screws to Attach
- Mount off the corner

Materials for Hoophouse
¾' x 5½' x 10' treated wood
(6 each)
1½' PVC pipes
1½" U-Clamp (24 each)
¾" Black Plastic Water Pipe
(35 Lw. Ft.)
Plastic Sheeting (10' x 16')
Hand Spring Clamp (2 ea.)
10 x ¾" Galvanized Phillips Head
Screws (24 ea.)
10 x 2 Torx Head Climatek Plated
Deck Screws (48 ea.)

Hoophouse Pros and Cons

Advantages of Using Hoophouses:

1) Hoophouses allow for earlier soil warming and protects from frost, lengthening the growing season.

2) Small heaters can be used to give additional frost protection.

3) Hoophouses are easily constructed from readily available materials.

4) Hoop/plastic covering can be manipulated and/or removed to control internal temperatures.

Disadvantages of Using Hoophouses:

1) Relatively high cost per square foot of growing space.

2) Internal temperatures can rise quickly on sunny days and kill plants unless the plastic covering is adjusted to allow for adequate ventilation.

3) Hoop covering must be removed at the end of the growing season, as snow load will crush the hoops.

4) Plastic covering will only last 1 to 2 years unless more expensive greenhouse plastic is used.

The quality of the soil is critical to the proper functioning of a hoophouse. The hoophouse may be filled with top-soil that is either purchased or acquired on-site. If the latter is used, be prepared to deal with imported weed seed that often are present in the soil.

Hoophouses will last many years if cared for properly. The plastic covering is the only component that needs periodic replacing. Any clear plastic may be used as a covering, although ultraviolet light will tend to break down plastics not designed for outdoor use after one season. Many types of greenhouse plastics are available and will last for 3 to 10 years.

This freestanding cold frame is made from cement, support bars, and plexiglass windows with frames. It's perfect for starting seedlings in the spring.

HARVESTING YOUR GARDEN

It is essential, in order to get the best freshness, flavor, and nutritional benefits from your garden vegetables and fruits, to harvest them at the appropriate time. The vegetable's stage of maturity and the time of day at which it is harvested are essential for good-tasting and nutritious produce. Overripe vegetables and fruits will be stringy and coarse. When possible, harvest your vegetables during the cool part of the morning. If you are going to can and preserve your vegetables and fruits, do so as soon as possible. Or, if this process must be delayed, make sure to cool the vegetables in ice water or crushed ice and store them in the refrigerator. Here are some brief guidelines for harvesting various types of common garden produce:

Asparagus—Harvest the spears when they are at least 6 to 8 inches tall by snapping or cutting them at ground level. A few spears may be harvested the second year after crowns are set out. A full harvest season will last four to six weeks during the third growing season.

Beans, snap—Harvest before the seeds develop in the pod. Beans are ready to pick if they snap easily when bent in half.

Beans, lima—Harvest when the pods first start to bulge with the enlarged seeds. Pods must still be green, not yellowish.

Broccoli—Harvest the dark green, compact cluster, or head, while the buds are shut tight, before any yellow flowers appear. Smaller side shoots will develop later, providing a continuous harvest.

Brussels sprouts—Harvest the lower sprouts (small heads) when they are about 1 to 1 ½ inches in diameter by twisting them off. Removing the lower leaves along the stem will help to hasten the plant's maturity.

Cabbage—Harvest when the heads feel hard and solid.

Cantaloupe—Harvest when the stem slips easily from the fruit with a gentle tug. Another indicator of ripeness is when the netting on the skin becomes rounded and the flesh between the netting turns from a green to a tan color.

Carrots—Harvest when the roots are ¾ to 1 inch in diameter. The largest roots generally have darker tops.

Cauliflower—When preparing to harvest, exclude sunlight when the curds (heads) are 1 to 2 inches in diameter by loosely tying the outer leaves together above the curd

Squashes should never be kept down cellar when it is possible to prevent it. Dampness injures them. If intense cold makes it necessary to put them there, bring them up as soon as possible, and keep them in some dry, warm place.

with a string or rubber band. This process is known as blanching. Harvest the curds when they are 4 to 6 inches in diameter but still compact, white, and smooth. The head should be ready 10 to 15 days after tying the leaves.

Collards—Harvest older, lower leaves when they reach a length of 8 to 12 inches. New leaves will grow as long as the central growing point remains, providing a continuous harvest. Whole plants may be harvested and cooked if desired.

Corn, sweet—The silks begin to turn brown and dry out as the ears mature. Check a few ears for maturity by opening the top of the ear and pressing a few kernels with your thumbnail. If the exuded liquid is milky rather than clear, the ear is ready for harvesting. Cooking a few ears is also a good way to test for maturity.

Cucumbers—Harvest when the fruits are 6 to 8 inches in length. Harvest when the color is deep green and before yellow color appears. Pick four to five times per week to encourage continuous production. Leaving mature cucumbers on the vine will stop the production of the entire plant.

Eggplant—Harvest when the fruits are 4 to 5 inches in diameter and their color is a glossy, purplish black. The fruit is getting too ripe when the color starts to dull or become bronzed. Because the stem is woody, cut—do not pull—the fruit from the plant. A short stem should remain on each fruit.

Kale—Harvest by twisting off the outer, older leaves when they reach a length of 8 to 10 inches and are medium green in color. Heavy, dark green leaves are overripe and are likely to be tough and bitter. New leaves will grow, providing a continuous harvest.

Lettuce—Harvest the older, outer leaves from leaf lettuce as soon as they are 4 to 6 inches long. Harvest heading types when the heads are moderately firm and before seed stalks form.

Mustard—Harvest the leaves and leaf stems when they are 6 to 8 inches long; new leaves will provide a continuous harvest until they become too strong in flavor and tough in texture due to temperature extremes.

Okra—Harvest young, tender pods when they are 2 to 3 inches long. Pick the okra at least every other day during the peak growing season. Overripe pods become woody and are too tough to eat.

Onions—Harvest when the tops fall over and begin to turn yellow. Dig up the onions and allow them to dry out in the open sun for a few days to toughen the skin. Then remove the dried soil by brushing the onions lightly. Cut the stem, leaving 2 to 3 inches attached, and store in a net-type bag in a cool, dry place.

Peas—Harvest regular peas when the pods are well rounded; edible-pod varieties should be harvested when the seeds are fully developed but still fresh and bright green. Pods are getting too old when they lose their brightness and turn light or yellowish green.

Peppers—Harvest sweet peppers with a sharp knife when the fruits are firm, crisp, and full size. Green peppers will turn red if left on the plant. Allow hot peppers to attain their bright red color and full flavor while attached to the vine; then cut them and hang them to dry.

Potatoes (Irish)—Harvest the tubers when the plants begin to dry and die down. Store the tubers in a cool, high-humidity location with good ventilation, such as the basement or crawl space of your house. Avoid exposing the tubers to light, as greening, which denotes the presence of dangerous alkaloids, will occur even with small amounts of light.

Pumpkins—Harvest pumpkins and winter squash before the first frost. After the vines dry up, the fruit color darkens and the skin surface resists puncture from your thumbnail. Avoid bruising or scratching the fruit while handling it. Leave a 3- to 4-inch portion of the stem attached to the fruit and store it in a cool, dry location with good ventilation.

Radishes—Harvest when the roots are ½ to 1 ½ inches in diameter. The shoulders of radish roots often appear through the soil surface when they are mature. If left in the ground too long, the radishes will become tough and woody.

Rutabagas—Harvest when the roots are about 3 inches in diameter. The roots may be stored in the ground and used as needed, if properly mulched.

Spinach—Harvest by cutting all the leaves off at the base of the plant when they are 4 to 6 inches long. New leaves will grow, providing additional harvests.

Squash, summer—Harvest when the fruit is soft, tender, and 6 to 8 inches long. The skin color often changes to a dark, glossy green or yellow, depending on the variety. Pick every two to three days to encourage continued production.

Sweet potatoes—Harvest the roots when they are large enough for use before the first frost. Avoid bruising or scratching the potatoes during handling. Ideal storage conditions are at a temperature of 55 degrees Fahrenheit and a relative humidity of 85 percent. The basement or crawl space of a house may suffice.

Swiss chard—Harvest by breaking off the developed outer leaves 1 inch above the soil. New leaves will grow, providing a continuous harvest.

Tomatoes—Harvest the fruits at the most appealing stage of ripeness, when they are bright red. The flavor is best at room temperature, but ripe fruit may be held in the refrigerator at 45 to 50 degrees Fahrenheit for 7 to 10 days.

Turnips—Harvest the roots when they are 2 to 3 inches in diameter but before heavy fall frosts occur. The tops may be used as salad greens when the leaves are 3 to 5 inches long.

Watermelons—Harvest when the watermelon produces a dull thud rather than a sharp, metallic sound when thumped—this means the fruit is ripe. Other ripeness indicators are a deep yellow rather than a white color where the melon touches the ground, brown tendrils on the stem near the fruit, and a rough, slightly ridged feel to the skin surface.

Handy Household Hints

Onions should be kept very dry, and never carried into the cellar except in severe weather, when there is danger of their freezing. By no means let them be in the cellar after March; they will sprout and spoil.

HERB GARDENS

Homegrown herbs have flavors and aromas that will transform your cooking. Many herbs also have healing qualities (see page 770–783), and some can even expunge unwanted insects from your home. Best of all, they're easy to grow.

Choosing a Location

First, decide which herbs to grow, and then research how much sun the plant needs, whether it's annual or perennial, how much water it requires, the soil conditions that will ensure the herb will thrive, and whether the herb needs to be controlled in order to avoid over-planting.

If you decide to grow outside, be certain that the spot you choose is sunny (your herb should be in direct sunlight for 4-7 hours), but sheltered from overhanging trees or weeds. The soil should be fertile and drain well. Culinary herbs should be planted at a distance from possible contamination like roadside pollution, agricultural sprays, or pets. If you opt for an indoor garden, a sunny windowsill is essential, preferably south-facing for a lot of natural sunlight.

Preparing the Plot or Pot

There are designs you can abide by, depending on where your garden is and how much time you want to invest. Formal designs rely on geometric patterns and are framed by paved paths and low hedges. Pathways or stones add yet another attractive element to a formal outdoor herb garden, and the soil itself should be a natural shade that contrasts nicely to the plants themselves. Each bed could be planted with one herb to create the maximum impact of color and texture blocks. On the other hand, there's no shame in a simple, practical garden design that is, perhaps, close to your home so you can run out and snip a few sprigs while cooking.

Regardless of the garden design, you must prepare the ground first: remove weeds, fork in the organic matter (compost, for example), and rake the soil bed so it's level. Refrain from adding large amounts of fertilizer because it may produce soft growth. Separate herbs and plant them accordingly, giving each one plenty of space; descriptions of how much space, exactly, can be found on seed packets. After planting, gently firm the soil around the plant and water thoroughly to settle the soil.

If your garden is indoors, you can either use a planter filled with quality soil or opt for a hydroponic kit, which is soil-less and uses special lights and liquid nourishment (see page 668). As with the outdoor herb garden, the indoor one should only have a little fertilizer; herbs thrive in poor to moderate soil.

Caring for Your Herb Garden

Always check that, after watering, the soil is allowing the water to drain. Mulch once a year with some kind of organic bulky material, like shredded bark, which is ideal for drainage. During the growing season, fertilize once a month. Prune your herbs to encourage healthy growth by removing dead leaves or flowers on a daily basis. If you frequently harvest your herbs, the pruning may be done during the process. With an indoor garden, remember to turn the pots so the plant grows evenly on all sides.

Harvesting Your Herbs

Wait for your plants to reach 6-8 inches in height before you harvest any of the leaves. Collect small quantities at a time (a quarter of your plant or less) and handle them as little as possible. Do not cut at random; take advantage of the opportunity to prune and pick at the same time. Always use a sharp knife or scissors to cut the branches, and never bend, break, or tear at them. After snipping, wait for the amount you have snipped to grow back before harvesting again.

Handy Household Hints

Herbs should be gathered while in blossom. If left till they have gone to seed, the strength goes into the seed. Those who have a little patch of ground will do well to raise the most important herbs; and those who have not, will do well to get them in quantities from some friend in the country; for apothecaries make very great profit upon them.

Easy Herb Pesto

Pesto is traditionally made with basil, but mixing the herbs creates a unique spread that's delicious served on crackers or fresh bread, or as a sauce over pasta.

Ingredients

¼ cup almonds or pine nuts
2 cloves garlic
1 cup fresh basil
½ cup fresh cilantro
½ cup fresh parsley
½ cup olive oil
Dash of lemon juice
Salt and pepper to taste

Directions

Combine all the ingredients in a food processor, pulsing to form a paste. Keep leftovers in the refrigerator.

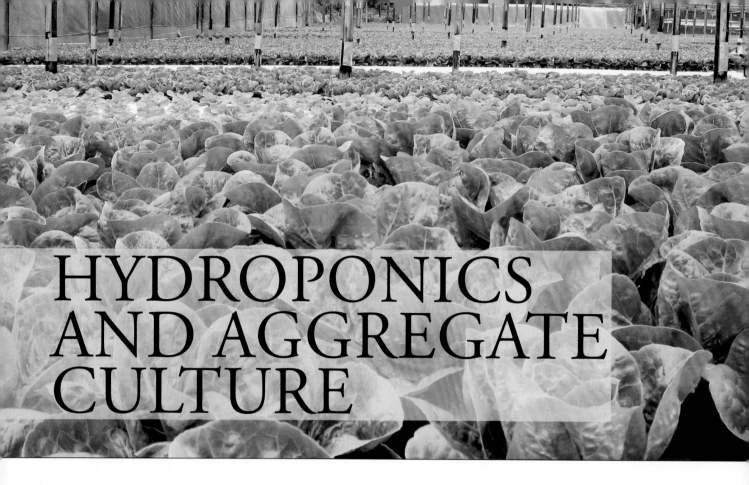

HYDROPONICS AND AGGREGATE CULTURE

Hydroponics is the method of growing plants in a container filled with a nutrient-rich bath (water with special fertilizer) and no soil. Plants grown in soilless cultures still need the basic requirements of plant growth, such as temperature, light (if indoors, use a heat-lamp and set the container near or on a windowsill), water, oxygen (you can produce good airflow by using a small, rotating fan indoors), carbon dioxide, and mineral nutrients (derived from solutions). But instead of planting their roots in soil, hydroponically grown plants have their roots either free-floating in a nutrient-rich solution or bedded in a soil-like medium, such as sand, gravel, brick shards, Perlite, or rockwool. These plants do not have to exert as much energy to gather nutrients from the soil and thus they grow more quickly and, usually, more productively.

The Benefits and Drawbacks of Growing Plants in a Hydroponics System

Benefits:

* Plants can be grown in areas where normal plant agriculture is difficult (such as deserts and other arid places, or cities).
* Most terrestrial plants will grow in a hydroponics system.
* There is minimal weed growth.

- The system takes up less space than a soil system.
- It conserves water.
- No fear of contaminated runoff from garden fertilizers.
- There is less labor and cost involved.
- Certain seasonal plants can be raised during any season.
- The quality of produce is generally consistent.
- Old nutrient solution can be used to water houseplants.

Drawbacks:

- Can cause salmonella to grow due to the wet and confined conditions.
- More difficult to grow root vegetables, such as carrots and potatoes.
- If nutrient solution is not regularly changed, plants can become nutrient deficient and thus not grow or produce.

Types of Hydroponics Systems

There are two main types of soilless cultures that can be used in order to grow plants and vegetables. The first is a water culture, in which plants are supplied with mineral nutrients directly from the water solution. The second, called aggregate culture or "sand culture," uses an aggregate (such as sand, gravel, or Perlite) as soil to provide an anchoring support for the plant roots. Both types of hydroponics are effective in growing soilless plants and in providing essential nutrients for healthy and productive plant growth.

Water Culture

The main advantage of using a water culture system is that a significant part of the nutrient solution is always in contact with the plants' roots. This provides an adequate amount of water and nutrients. The main disadvantages of this system are providing sufficient air supply for the roots and providing the roots with proper support and anchorage.

Water culture systems are not extremely expensive, though the cost does depend on the price of the chemicals and water used in the preparation of the nutrient solutions, the size of your container, and whether or not your are using mechanized

objects, such as pumps and filters. You can decrease the cost by starting small and using readily available materials.

What You'll Need to Make Your Own Water Culture

A large water culture system will need either a wood or concrete tank 6 to 18 inches deep and 2 to 3 feet wide. If you use a wooden container, be sure there are no knots in the wood and seal the tank with non-creosote or tar asphalt.

For small water culture systems, which are recommended for beginners, glass jars, earthenware crocks, or plastic buckets will suffice as your holding tanks. If your container is transparent, be sure to paint the outside of the container with black paint to keep the light out (and to keep algae from growing inside your system). Keep a narrow vertical strip unpainted in order to see the level of the nutrient solution inside your container.

The plant bed should be 3 or more inches deep and large enough to cover the container or tank. In order to support the weight of the litter (where your seeds or seedlings are placed), cover the bottom of the bed with chicken wire and then fill the bed with litter (wood shavings, sphagnum moss, peat, or other organic materials that do not easily decay). If you are starting your plants from seeds, germinate the seeds in a bed of sand and then transplant to the water culture bed, keeping the bed moist until the plants get their roots down into the nutrient solution.

Aeration

A difficulty in using water culture is keeping the solution properly aerated. It is important to try to keep enough space between the seed bed and the nutrient solution so the plant's roots can receive proper oxygen. In order to make sure that air can easily flow into the container, either prop up the seed bed slightly to allow air flow or drill a hole in your container just above the highest solution level.

In order to make sure there is sufficient oxygen reaching the plant roots, you can install an aquarium pump in your water culture system. Just make sure that the water is not agitated too much or the roots may be damaged. You can also use an air stone or perforated pipe to gently introduce air flow into your container.

Water Supply

Your hydroponics system needs an adequate supply of fresh water in order to maintain healthy plant life. Make sure that the natural minerals in your water are not going to adversely affect your hydroponics plants. If there is too much sodium in your water (usually an effect of softened water), it could become toxic to your plants. In general, the minerals in water are typically not harmful to the growth of your plants.

Nutrient Solution

You may add nutrient solution by hand, by a gravity-feed system, or mechanically. In smaller water culture systems, mixing the nutrient solution in a small container and adding it by hand, as needed, is typically adequate.

If you are using a larger setup, a gravity-feed system will work quite well. In this type of system, the nutrient solution is mixed in a vat and then tapped from the vat into your container as needed. You can use a plastic container or larger earthenware jar as the vat.

A pump can also be used to supply your system with adequate nutrient solution. You can insert the pump into the vat and then transfer the solution to your hydroponics system.

When your plants are young, it is important to keep the space between your seed bed and the nutrient solution quite small (that way, the young plant roots can reach the nutrients). As your plants grow, the amount of space between the bed and solution should increase (but do this slowly and keep the level rather consistent).

If the temperature is rather high and there is increased evaporation, it is important to keep the roots at the correct level in the water and change the nutrient solution every day, if needed.

Drain your container every two weeks and then renew the nutrient solution from your vat or by hand. This must be done in a short amount of time, so the roots do not dry out.

Transplanting

When transplanting your seedlings, it's important to make sure that you are careful with the tiny root systems. Gently work the roots through the support netting and down into the nutrient solution. Then fill in the support netting with litter to help the plant remain upright.

How to Build a Simple, Homemade Hydroponics System

Things You'll Need

- External pump
- Air line or tubing
- Air stones
- Waterproof bin, bucket, or fish tank to use as a reserve
- Styrofoam
- Net pots
- Type of growing medium, such as rockwool or grow rocks
- Hydroponics nutrients, such as grow formula, bloom formula, supplements, and pH
- Black spray paint (this is only required if the reservoir is transparent)
- Knife, box cutter, or scissors
- Tape measure

Steps to Building Your Hydroponics System

1. Find a container to use as a reservoir, such as a fish tank, a bin, or a bucket. The reservoir should be painted black if it is not lightproof (or covered with a thick, black trash bag if you want to reuse the tank at some point), and allowed to dry before moving on to the next step. Allowing light to enter the reservoir will promote the growth of algae. Use a reservoir that is the same dimensions (length and width) from top to bottom.

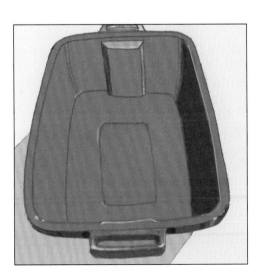

2. Using a knife or sharp object, score a line on the tank (scratch off some paint in a straight line from top to bottom). This will be your water level meter, which will allow you to see how much water is in the reservoir and will give you a more accurate and convenient view of the nutrient solution level in your tank.

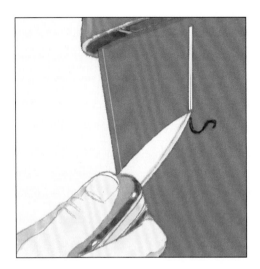

3. Use a tape measure to determine the length and width of your reservoir. Measure the inside of the reservoir from one end to the other. Once you have the dimensions, cut the Styrofoam ¼ inch smaller than the size of the reservoir. For example, if your dimensions are 36 x 20 inches, you should cut the Styrofoam to 35 ¾ x 19 ¾ inches. The Styrofoam should fit nicely in the reservoir, with just enough room to adjust to any water level changes. If the reservoir tapers off at the bottom (the

bottom is smaller in dimension than the top) the floater (Styrofoam) should be 2 to 4 inches smaller than the reservoir, or more if necessary.

4. Do not place the Styrofoam in the reservoir yet. First, you need to cut holes for the net pots. Put the net pots on the Styrofoam where you want to place each plant. Using a pen or pencil, trace around the bottom of each net pot. Use a knife or box cutter to follow the trace lines and cut the holes for pots. On one end of the Styrofoam, cut a small hole for the air line to run into the reservoir.

5. The number of plants you can grow will depend on the size of the garden you build and the types of crops you want to grow. Remember to space plants appropriately so that each receives ample amounts of light.

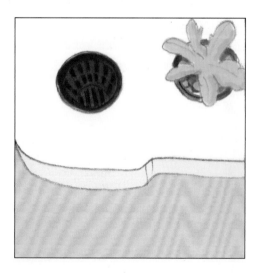

6. The pump you choose must be strong enough to provide enough oxygen to sustain plant life. Ask for advice choosing a pump at your local hydroponics supply store or garden center.

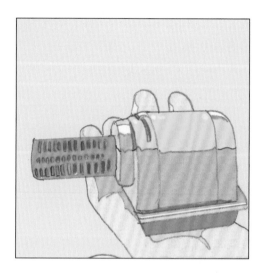

7. Connect the air line to the pump and attach the air stone to the free end. The air line should be long enough to travel from the pump into the bottom of the reservoir, or at least float in the middle of the tank so the oxygen bubbles can get to the plant roots. It also must be the right size for the pump you choose. Most pumps will come with the correct size air line. To determine the tank's capacity, use a one-gallon bucket or bottle and fill the reservoir. Remember to count how many gallons it takes to fill the reservoir and you will know the correct capacity of your tank.

Setting Up Your Hydroponics System

1. Fill the reservoir with the nutrient solution.

2. Place the Styrofoam into the reservoir.

3. Run the air line through the designated hole or notch.

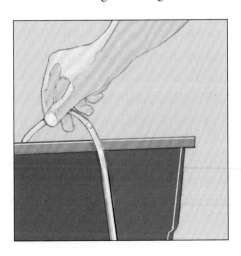

4. Fill the net pots with growing medium and place one plant in each pot.

5. Put the net pots into the designated holes in the Styrofoam.

6. Plug in the pump, turn it on, and start growing with your fully functional, homemade hydroponics system.

Things to Consider

- A homemade hydroponics system like this is not ideal for large-scale production of plants or for commercial usage. This particular system does not offer a way to conveniently change the nutrient solution. An extra container would be required to hold the floater while you change the solution.

- Lettuce, watercress, tomatoes, cucumbers, and herbs grow especially well hydroponically.

The Junior Homesteader

Encourage middle or high school aged kids to come up with their own hydroponic designs. Encourage them to research different methods and find materials around the house. Hydroponic growing systems make great science fair projects!

Aggregate Culture

This type of soil-less growing utilizes different mediums that act in the place of soil to stabilize the plant and its roots. The aggregate in the container is flooded with the nutrient solution. The advantage of this type of system is that there is not as much trouble with aerating the roots. Also, aggregate culture systems allow for the easy transplantation of seedlings into the aggregate medium and it is less expensive.

What You'll Need for an Aggregate Culture System

The container should be watertight to help conserve the nutrient solution. Large tanks can be made of concrete or wood, and smaller operations can effectively be done in glass jars, earthenware containers, or plastic buckets. Make sure to paint transparent containers black.

Aggregate materials may differ greatly, depending on what type you choose to use. Silica sand (well washed) is one of the best materials that can be used. Any other type of coarse-textured sand is also effective, but make sure it does not contain lime. Sand holds moisture quite well and it allows for easy transplantation. A mixture of sand and gravel together is also an effective aggregate. Other materials, such as peat moss, vermiculite, wood shavings, and coco peat, are also good aggregates. You can find aggregate materials at your local garden center, home center, or garden-supply house.

Aeration

Aggregate culture systems allow much easier aeration than water culture systems. Draining and refilling the container with nutrient solution helps the air to move in and out of the aggregate material. This brings a fresh supply of oxygen to the plant roots.

Water Supply

The same water requirements are needed for this type of hydroponics system as for a water culture system. Minerals in the water tend to collect in the aggregate material, so it's a good idea to flush the material with fresh water every few weeks.

Nutrient Solution

The simplest way of adding the nutrient solution to aggregate cultures is to pour it over the aggregates by hand. You may also use a manual gravity-feed system with buckets or vats. Attach the vat to the bottom of the container with a flexible hose, raise the vat to flood the container, and lower it to drain it. Cover the vat to prevent evaporation and replenish it with new nutrient solution once every two weeks.

A gravity drip-feed system also works well and helps reduce the amount of work you do. Place the vat higher than the container, and then control the solution drip so it is just fast enough to keep the aggregate moist.

It is important that the nutrient solution is added and drained or raised and lowered at least once a day. In hotter weather, the aggregate material may need more wetting with the solution. Make sure that the material is not drying out the roots. Drenching the aggregate with solution often will not harm the plants but letting the roots dry out could have detrimental effects.

Always replace your nutrient solution after two weeks. Not replacing the solution will cause salts and harmful fertilizer residues to build up, which may ultimately damage your plants.

Planting

You may use either seedlings or rooted cuttings in an aggregate culture system. The aggregate should be flooded and solution drained before planting to create a moist, compacted seed bed. Seeds may also be planted directly into the aggregate material. Do not plant the seeds too deep, and flood the container frequently with water to keep the aggregate moist. Once the seedlings have germinated, you may start using the nutrient solution.

If you are transplanting seedlings from a germination bed, make sure they have germinated in soilless material, as any soil left on the roots may cause them to rot and may hamper them in obtaining nutrients from the solution.

Making Nutrient Solutions

In order for plants to grow properly, they must receive nitrogen, phosphorous, potassium, calcium, magnesium, sulfur, iron, manganese, boron, zinc, copper, molybdenum, and chlorine. There are a wide range of nutrient solutions that can be used. If your plants are receiving inadequate amounts of nutrients, they will show this in different ways. This means that you must proceed with caution when selecting and adding the minerals that will be present in your nutrient solution.

It is important to have pure nutrient materials when preparing the solution. Using fertilizer-grade chemicals is always the best route to go, as it is cheapest. When housing your nutrient solution, be sure the containers are not transparent (if they are, paint them black to keep out the light). Make sure the containers are closed and not exposed to air. Evaporation from solution concentrates the amount of salt, which could harm your plants.

Pre-mixed Chemicals

Many of the essential nutrients needed for hydroponic plant growth are now available already mixed in their correct proportions. You may find these solutions in catalogs or from garden-supply stores. They are typically inexpensive and only small quantities are needed to help your plants grow strong and healthy. Always follow the directions on the container when using pre-mixed chemicals.

Making Your Own Solution

In the event that you want to make your own nutrient solution, here is a formula for a solution that will provide all the major elements required for your plants to grow.

Salt	Grade	Nutrients	Amt. for 25 gallons of solution
Potassium phosphate	Technical	Potassium, phosphorus	½ ounce (1 Tbsp)
Potassium nitrate	Fertilizer	Potassium, nitrogen	2 ounces (4 Tbsp of powdered salt)
Calcium nitrate	Fertilizer	Calcium, nitrogen	3 ounces (7 Tbsp)
Magnesium sulfate	Fertilizer	Magnesium, sulfur	1 ½ ounces (4 Tbsp)

You can obtain all of these chemicals from garden-supply stores or drugstores.

After all the chemicals have been mixed into the solution, check the pH of the solution. A pH of 7.0 is neutral; anything below 7.0 is acidic and anything above is alkaline. Certain plants grow best in certain pHs. Plants that grow well at a lower pH (between 4.5 and 5.5) are azaleas, buttercups, gardenias, and roses; plants that grow well at a neutral pH are potatoes, zinnias, and pumpkins; most plants grow best in a slightly acidic pH (between 5.5 and 6.5).

To determine the pH of your solution, use a pH indicator (these are usually paper strips). The strip will change color when placed in different levels of pH. If you find your pH level to be above your desired range, you can bring it down by adding dilute sulfuric acid in small quantities using an eyedropper. Keep retesting until you reach your desired pH level.

Plant Nutrient Deficiencies

When plants are lacking nutrients, they typically display these deficiencies outwardly. Following is a list of symptoms that might occur if a plant is lacking a certain type of nutrient. If your plants display any of these symptoms, it is imperative that the level of that particular nutrient be increased.

Deficient Nutrient	Symptoms
Boron	Tip of the shoot dies; stems and petioles are brittle
Calcium	Tip of the shoot dies; tips of the young leaves die; tips of the leaves are hooked
Iron	New upper leaves turn yellow between the veins; edges and tips of leaves may die
Magnesium	Lower leaves are yellow between the veins; leaf margins curl up or down; leaves die
Manganese	New upper leaves have dead spots; leaf might appear netted
Nitrogen	Leaves are small and light green; lower leaves are lighter than upper leaves; weak stalks
Phosphorous	Dark-green foliage; lower leaves are yellow between the veins; purplish color on leaves
Potassium	Lower leaves might be mottled; dead areas near tips of leaves; yellowing at leaf margins and toward the center
Sulfur	Light-green upper leaves; leaf veins are lighter than surrounding area

MACHINERY FOR YOUR GARDEN AND FARM

Though farmers got along for millennia without such luxuries, there are some basic farm tools that will make living off the land much easier and more productive for you. Choose your machinery carefully—you don't need every fancy toy that comes on the market, nor should you spend good money on a poorly-maintained old tractor that's going to fall apart on you first circle around the field. If you're not in desperate need of the machinery, take your time to look around at both new and used options before making any purchases.

Buying a Tractor

First, make sure it is definitely a tractor that you need, and not a similar tool such as a skid steer loader, mini excavator, or backhoe. If you need a machine which has numerous uses and the ability to interact with many kinds of attachments, a tractor is a smart choice. Tractors also offer limited surface impact, so for any work which requires interaction with grass or loose surfaces, a tractor is your best bet. Consider the type of surface the tractor will be used on, the

elevation of the land, the weather conditions of your area, the attachments you will use, and the distance the tractor must be able to travel.

Horsepower is the next important factor in determining the tractor model which will work best for you. A tractor's horsepower is what determines the amount of work it will be able to perform, and also how much you will pay. Typically, tractors with between 25 to 65 horsepower are a comfortable middle ground when it comes to tractor performance. Anything below 25 is for the most basic, least strenuous work, and above 65 is capable of handling heavier agricultural work (tilling fields, baling hay, etc). For most uses, a tractor with a wide front end—instead of a tricycle front end—is safer and more durable.

Depending on your needs, you might want a tractor with a power take-off (PTO) and 3-point hitch. One hydraulic hook-up is sufficient, although more are nice. If the tractor comes with a drawbar with the 3-point hitch, that's great, as those are expensive. Also, if available, you want to get as many weights as possible for the tractor. Weights that fit your tractor are often in high demand. Liquid filled tires are an optional means of adding weight, but the fluid in many of these older tractors is corrosive, so look for possible bad wheel rims. And because of the high cost of new tractor tires, you want very solid tires.

Most new tractors are 4WD, and are generally a better option than 2WD. A 4WD tractor allows for the attachment of a front loader, and also typically have a higher resale value than those with 2WD. On surfaces with poor traction—mud, snow, gravel—the 4WD tractor is definitely the better option. The only situation in which a 2WD tractor may be more beneficial is for basic mowing.

Enclosed cabs do not come standard with all tractors, but are incredibly beneficial for anyone who will be using the tractor in cold weather or under extreme sun. Although tractors with enclosed cabs will cost more, their resale value will also be higher.

Most older tractors will have manual transmissions, whereas hydrostatic transmission has emerged as a popular alternative in newer models for those who are uncomfortable with manual transmission.

The tires that you choose to purchase for your tractor will depend upon the work that your tractor will be doing, Agricultural tires, industrial tires, or turf tires are all tire options which pertain to certain kinds of work.

There are a number of tractor attachments, all of which are geared toward performing a certain task. Box blades, mowers, tillers, plows, backhoes, landscape rakes, spreaders, forks, grapples, and hay bailers are all possible attachments which may be used with your trailer.

Tractor Manufacturers:

Ag-Chem Equipment–www.agchem.com

AGCO–www.agcocorp.com

Ariens–www.ariens.com

B and H Manufacturing–www.bhmfg.com

Batco–www.batcomfg.com

Bobcat–www.bobcat.com

Briggs & Stratton–www.briggsandstratton.com

Brillion–www.brillionfarmeq.com

Bush Hog–www.bushhog.com

Crary–www.crary.com

Cub Cadet–www.mtdproducts.com

Caterpillar–www.cat.com

Fendt–www.fendt.com

Flexi-Coil–www.flexicoil.com

Gehl–www.gehl.com

Hagie Manufacturing–www.hagie.com

JCB–www.jcb.co.uk

John Deere–www.deere.com

Landoll–www.landoll.com

Leon Manufacturing–www.leonsmfg.com

Redrock Engineering–www.redrock-engineering.com

Schulte Manufacturing–www.pima.ca/members/Schulte.html

Shelbourne Reynolds–www.shelbourne.com

Lawn Mowers

There are two basic types of lawn mowers: reel and rotary mowers. Rotary mowers include hand-pushed, self-propelled, and riding mowers. Hand-pushed rotary mowers are suitable for lawns which are relatively small on level ground, but for lawns which are bigger or hilly, a self-propelled rotary mower may be the better choice. Lawns are that are large—an acre or more— often call for a riding rotary mower, which functions much like a tractor. The operator sits on top of the mower to steer and power the device, so the physical strain is significantly less than both other rotary mowers. Riding mowers are more costly than the others, and also take up more storage space.

Reel lawn mowers are comprised of a set of blades on a revolving cylinder. The cylinder moves as the mower is pushed forward, causing the blades to revolve and cut. These mowers require the operator to both push and propel movement, and can also be high-maintenance, as the blades require sharpening. However, they offer the closest cut of all the different kinds of mowers Rotary lawn mowers use a flat, horizontal blade which rotates in a circular motion, which often does not give a close nor as clean a cut as a reel mower, but they're much easier to use.

For a "greener" option, you may want to consider using an electric powered lawn mower instead of a gas powered device. Electric mowers are available in both reel and rotary designs, and are quieter than gas powered alternatives.

When deciding what kind of lawn mower to buy, it is important to consider a number of factors. The size of your

lawn, your budget, and what you need your mower to do. If you have a small lawn with a lot of shrubs, for example, you are not going to want a riding mower, which takes up a lot of room and does not have a great turning radius. Once you determine exactly what kind of mower would address your needs, it's time to shop around and determine which model or brand is your best fit.

Lawn Mower Manufacturers:

Honda–www.hondapowerequipment.com/products/Lawnmowers

Toro—www.toro.com

John Deere–www.deere.com/en_US/homeowners/index.html

Craftsman–www.craftsman.com

Lawn-Boy–www.lawn-boy.com

Cub Cadet–www.cubcadet.com

Troy-Bilt–www.troybilt.com

Husqvarna–www.husqvarna.com/us/homeowner/home

Chain Saws

There are two basic kinds of chain saws: gas or electric powered. Both types have certain advantages and disadvantages, all which must be taken into account when deciding which kind will be best for you. Gasoline-powered chain saws are noisy, require mixing of oil and gas, and must be started with a pull cord. However, they are easily mobile and are capable of a great amount of power. Electric saws, on the other hand, are quieter, lighter, and easier to start than the gas-powered alternative. However, since they require electricity, they are much lest mobile than the cordless gas-powered saw, and do not offer as much power.

The kind of chainsaw that would work best for you depends upon what kind of work needs to be done. If you have area that needs clearing but won't require heavy maintenance, an electric chain saw would probably suit you well. However, a large area with lots of thick branches would need a gas chain saw to really get the job done.

When purchasing your chain saw, you also have to keep in mind the physical exertion that will be required to yield the saw. Although a big, heavy saw will give you a lot of

power with which to cut through heavy branches and shrubbery, if you are not physically capable of maintaining the saw for a long time the work will not get done.

Many modern saws come with a lot of different options and extras that can really help you safely and efficiently operate your saw. Anti-vibration technology, automatic chain oilers, quick-start electronic ignition, mufflers, and quick-adjust chains are just some of the additions that make life a little bit easier for chain saw operators.

If you need a chain saw for a quick backyard fix but do not anticipate on using it again, renting a saw is always a good option. There are a lot of resource on the internet which can help you find a local business that rents out different kinds of tools.

Chain Saw brands:

Husqvarna—www.husqvarna.com
Stihl—www.stihlusa.com
Jonsered—www.jonsered.com
Echo—www.echo-usa.com
Dolmzr—www.dolmarpowerproducts.com

Rototillers

The rotating blades of a rototiller loosen and agitate soil, preparing it for planting and fertilization. There are two standard kinds of rotary blades: counter rotating tines and standard rotating tines. Counter rotating tines are capable of producing deep holes, while standard rotating tines typically create shallower holes.

Rototillers come in all shapes and sizes, ranging from very large tractor attachments to hand-held devices. As with many garden tools, rototillers can be either gas or electric powered. Gas-powered devices typically offer the user more power, but the electric model is generally less expensive and quieter.

The kind of rototiller that you should buy depends on the size of your garden; as with other tools, the cost and strength of a rototiller increases as its horsepower goes up. If you are a homeowner with a small garden, a large horsepower tiller is unnecessary; in that same regard, if your garden is expansive a larger, more powerful model would be better. As with the chain saw, a rototiller is not a necessary purchase if you only plan on using it once or twice, and can be rented instead of bought.

MULCHING: WHY AND HOW TO DO IT

Mulching is one of the simplest and most beneficial practices you can use in your garden. Mulch is simply a protective layer of material that is spread on top of the soil to enrich the soil, prevent weed growth, and help provide a better growing environment for your garden plants and flowers.

Mulches can either be organic—such as grass clippings, bark chips, compost, ground corncobs, chopped cornstalks, leaves, manure, newspaper, peanut shells, peat moss, pine needles, sawdust, straw, and wood shavings—or inorganic—such as stones, brick chips, and plastic. Both organic and inorganic mulches have numerous benefits, including:

1. Protecting the soil from erosion
2. Reducing compaction from the impact of heavy rains
3. Conserving moisture, thus reducing the need for frequent watering
4. Maintaining a more even soil temperature
5. Preventing weed growth
6. Keeping fruits and vegetables clean
7. Keeping feet clean and allowing access to the garden even when it's damp
8. Providing a "finished" look to the garden

Common Organic Mulching Materials

Bark chips
Chopped cornstalks
Compost
Grass clippings
Ground corncobs
Hay
Leaves
Manure
Newspaper
Peanut shells
Peat moss
Pine needles
Sawdust
Straw
Wood shavings

Organic mulches also have the benefit of improving the condition of the soil. As these mulches slowly decompose, they provide organic matter to help keep the soil loose. This improves root growth, increases the infiltration of water, improves the water-holding capacity of the soil, provides a source of plant nutrients, and establishes an ideal environment for earthworms and other beneficial soil organisms.

While inorganic mulches have their place in certain landscapes, they lack the soil-improving properties of organic mulches. Inorganic mulches, because of their permanence, may be difficult to remove if you decide to change your garden plans at a later date.

Mulching Materials

You can find mulch materials right in your own backyard. They include:

1. Lawn clippings. They make an excellent mulch in the vegetable garden if spread immediately to avoid heating and rotting. The fine texture allows them to be spread easily, even around small plants.
2. Newspaper. As a mulch, newspaper works especially well to control weeds. Save your own newspapers and only use the text pages, or those with black ink, as color dyes may be harmful to soil microflora and fauna if composted and used. Use three or four sheets together, anchored with grass clippings or other mulch material to prevent them from blowing away.
3. Leaves. Leaf mold, or the decomposed remains of leaves, gives the forest floor its absorbent, spongy structure. Collect leaves in the fall and chop with a lawnmower or shredder. Compost leaves over winter, as some studies have indicated that freshly chopped leaves may inhibit the growth of certain crops.
4. Compost. The mixture makes wonderful mulch—if you have a large supply—as it not only improves the soil structure but also provides an excellent source of plant nutrients.
5. Bark chips and composted bark mulch. These materials are available at garden centers, and are sometimes used with landscape fabric or plastic that is spread atop the soil and beneath the mulch to provide additional pro-

tection against weeds. However, the barrier between the soil and the mulch also prevents any improvement in the soil condition and makes planting additional plants more difficult. Without the barrier, bark mulch makes a neat finish to the garden bed and will eventually improve the condition of the soil. It may last for one to three years or more, depending on the size of the chips or how well composted the bark mulch is. Smaller chips are easier to spread, especially around small plants.
6. Hay and straw. These work well in the vegetable garden, although they may harbor weed seeds.
7. Seaweed mulch, ground corncobs, and pine needles. Depending on where you live, these materials may be readily available and also can be used as mulch. However, pine needles tend to increase the acidity of the soil, so they work best around acid-loving plants, such as rhododendrons and blueberries.

When choosing a mulch material, think of your primary objective. Newspaper and grass clippings are great for weed control, while bark mulch gives a perfect, finishing touch to a front-yard perennial garden. If you're looking for a cheap solution, consider using materials found in your own yard or see if your community offers chipped wood or compost to its residents.

If you want the mulch to stay in place for several years around shrubs, for example, you might want to consider using inorganic mulches. While they will not provide organic matter to the soil, they will be more or less permanent.

When to Apply Mulch

Time of application depends on what you hope to achieve by mulching. Mulches, by providing an insulating barrier between the soil and the air, moderate the soil temperature. This means that a mulched soil in the summer will be cooler than an adjacent, unmulched soil; while in the winter, the mulched soil may not freeze as deeply. However, since mulch acts as an insulating layer, mulched soils tend to warm up more slowly in the spring and cool down more slowly in the fall than unmulched soils.

If you are using mulches in your vegetable or flower garden, it is best to apply or add additional mulch after the

soil has warmed up in the spring. Organic mulches reduce the soil temperature by 8 to 10 degrees Fahrenheit during the summer, so if they are applied to cold garden soils, the soil will warm up more slowly and plant maturity will be delayed.

Mulches used to help moderate winter temperatures can be applied late in the fall after the ground has frozen, but before the coldest temperatures arrive. Applying mulches before the ground has frozen may attract rodents looking for a warm over-wintering site. Delayed applications of mulch should prevent this problem.

Mulches used to protect plants over the winter should be composed of loose material, such as straw, hay, or pine boughs that will help insulate the plants without compacting under the weight of snow and ice. One of the benefits from winter applications of mulch is the reduction in the freezing and thawing of the soil in the late winter and early spring. These repeated cycles of freezing at night and then thawing in the warmth of the sun cause many small or shallow-rooted plants to be heaved out of the soil. This leaves their root systems exposed and results in injury, or death, of the plant. Mulching helps prevent these rapid fluctuations in soil temperature and reduces the chances of heaving.

General Guidelines

Mulch is measured in cubic feet, so, for example, if you have an area measuring 10 feet by 10 feet, and you wish to apply 3 inches (¼ foot) of mulch, you would need 25 cubic feet to do the job correctly.

While some mulch can come from recycled material in your own yard, it can also be purchased bagged or in bulk from a garden center. Buying in bulk may be cheaper if you need a large volume and have a way to haul it. Bagged mulch is often easier to handle, especially for smaller projects, as most bagged mulch comes in 3-cubic-foot bags.

To start, remove any weeds. Begin mulching by spreading the materials in your garden, being careful not to apply mulch to the plants themselves. Leave an inch or so of space next to the plants to help prevent diseases from flourishing in times of excess humidity.

How Much Do I Apply?

The amount of mulch to apply to your garden depends on the mulching material used. Spread bark mulch and wood chips 2 to 4 inches deep, keeping it an inch or two away from tree trunks.

Scatter chopped and composted leaves 3 to 4 inches deep. If using dry leaves, apply about 6 inches.

Grass clippings, if spread too thick, tend to compact and rot, becoming quite slimy and smelly. They should be applied 2 to 3 inches deep, and additional layers should be added as clippings decompose. Make sure not to use clippings from lawns treated with herbicides.

Sheets of newspaper should only be ¼ inch thick, and covered lightly with grass clippings or other mulch material to anchor them. If other mulch materials are not available, cover the edges of the newspaper with soil.

If using compost, apply 3 to 4 inches deep, as it's an excellent material for enriching the soil.

ORGANIC GARDENING

Organically grown food is food grown and processed using no synthetic fertilizers or pesticides. Pesticides derived from natural sources (such as biological pesticides—compost and manure) may be used. Organic gardening methods are healthier, environmentally friendly, safe for animals and humans, and are typically less expensive for small scale gardening, since you are working with natural materials.

Organic farmers apply techniques first used thousands of years ago, such as crop rotations and the use of composted animal manures and green manure crops, in ways that are economically sustainable in today's world.

Organic farming entails:

- Use of cover crops, green manures, animal manures, and crop rotations to fertilize the soil, maximize biological activity, and maintain long-term soil health.
- Use of biological control, crop rotations, and other techniques to manage weeds, insects, and diseases.
- An emphasis on biodiversity of the agricultural system and the surrounding environment.

- Reduction of external and off-farm inputs and elimination of synthetic pesticides and fertilizers and other materials, such as hormones and antibiotics.
- A focus on renewable resources, soil and water conservation, and management practices that restore, maintain and enhance ecological balance.

Starting an Organic Garden

1. Choose a Site for Your Garden

1. Think small, at least at first. A small garden takes less work and materials than a large one. If done well, a 4 x 4-foot garden will yield enough vegetables and fruit for you and your family to enjoy.
2. Be careful not to over-plant your garden. You do not want to end up with too many vegetables that will end up over-ripening or rotting in your garden.
3. You can even start a garden in a window box if you are unsure of your time and dedication to a larger bed.

The Junior Homesteader

How do microorganisms in the soil affect plants? Take a sample of fertile soil from a field or garden and divide it into two portions. Bake one in an oven at 350 degrees for half an hour (to destroy the microorganisms). Leave the other portion alone as a control. Plant the same number of seeds in each soil sample. Remember to treat both samples the same while the plants are growing. Make sure all the plants receive the same amounts of water and light, and are kept at the same temperature. How do the plants differ as they grow?

Next, discover how some microorganisms and plants form mutually beneficial partnerships. For example, certain bacteria make a natural nitrogen fertilizer for plants in the family called legumes, which includes peas, alfalfa, and soybeans. The nitrogen-fixing bacteria are available from garden supply stores and by mail order. Grow both legumes and non-legume plants with and without the bacteria. Are there differences in how well the plants grow?

2. Make a Compost Pile

Compost is the main ingredient for creating and maintaining rich, fertile soil. You can use most organic materials to make compost that will provide your soil with essential nutrients. To start a compost pile, all you need are fallen leaves, weeds, grass clippings, and other vegetation that is in your yard. (See the Improving Your Soil chapter for more details on how to make compost.)

3. Add Soil

In order to have a thriving organic garden, you must have excellent soil. Adding organic material (such as that in your compost pile) to your existing soil will only make it better. Soil containing copious amounts of organic material is very good for your garden. Organically rich soil:

- Nourishes your plants without any chemicals, keeping them natural
- Is easy to use when planting seeds or seedlings and it also allows for weeds to be more easily picked

- Is softer than chemically treated soil, so the roots of your plants can spread and grow deeper
- Helps water and air find the roots

4. Weed Control

1. Weeds are invasive to your garden plants and thus must be removed in order for your organic garden to grow efficiently. Common weeds that can invade your garden are ivy, mint, and dandelions.
2. Using a sharp hoe, go over each area of exposed soil frequently to keep weeds from sprouting. Also, plucking off the green portions of weeds will deprive them of the nutrients they need to survive.
3. Gently pull out weeds by hand to remove their root systems and to stop continued growth. Be careful when weeding around established plants so you don't uproot them as well.
4. Mulch unplanted areas of your garden so that weeds will be less likely to grow. You can find organic mulches, such as wood chips and grass clippings, at your local garden store. These mulches will not only discourage weed growth but will also eventually break down and help enrich the soil. Mulching also helps regulate soil temperatures and helps in conserving water by decreasing evaporation. (See the Mulching In Your Garden and Yard chapter for more on mulching.)

5. Be Careful of Lawn Fertilizers

If you have a lawn and your organic garden is situated in it, be mindful that any chemicals you place on your lawn may find their way into your organic garden. Therefore, refrain from fertilizing your lawn with chemicals and, if you wish to return nutrients to your grass, simply let your cut grass clippings remain in the yard to decompose naturally and enrich the soil beneath.

Things to Consider

- "Organic" means that you don't use any kinds of chemicals or materials, such as paper or cardboard, that contain chemicals, and especially not fertilizer or pesticides.

Make sure that these products do not find their way into your garden or compost pile.

- If you are adding grass clippings to your compost pile, make sure they don't come from a lawn that has been treated with chemical fertilizer.

- If you don't want to start a compost pile, simply add leaves and grass clippings directly to your garden bed. This will act like a mulch, deter weeds from growing, and will eventually break down to help return nutrients to your soil.

- If you find insects attacking your plants, the best way to control them is by picking them off by hand. Also practice crop rotation (planting different types of plants in a given area from year to year), which will hopefully reduce your pest problem. For some insects, just a strong stream of water is effective in removing them from your plants.

Shy away from using bark mulch. It robs nitrogen from the soil as it decomposes and can also attract termites.

PEST AND DISEASE MANAGEMENT

Pest management can be one of the greatest challenges to the home gardener. Yard pests include weeds, insects, diseases, and some species of wildlife. Weeds are plants that are growing out of place. Insect pests include an enormous number of species from tiny thrips that are nearly invisible to the naked eye, to the large larvae of the tomato hornworm. Plant diseases are caused by fungi, bacteria, viruses, and other organisms—some of which are only now being classified. Poor plant nutrition and misuse of pesticides also can cause injury to plants. Slugs, mites, and many species of wildlife, such as rabbits, deer, and crows, can be extremely destructive as well.

Identify the Problem

Careful identification of the problem is essential before taking measures to control the issue in your garden. Some insect damage may at first appear to be a disease, especially if no visible insects are present. Nutrient problems may also mimic diseases. Herbicide damage, resulting from misapplication of chemicals, can also be mistaken for other problems. Learning about different types of garden pest is the first step in keeping your plants healthy and productive.

Insects and Mites

All insects have six legs, but other than that they are extremely different depending on the species. Some insects include such organisms as beetles, flies, bees, ants, moths, and butterflies. Mites and spiders have eight legs—they are not, in fact, insects but will be treated as such for the purposes of this section.

Insects damage plants in several ways. The most visible damage caused by insects is chewed plant leaves and flowers. Many pests are visible and can be readily identified, including the Japanese beetle, Colorado potato beetle, and numerous species of caterpillars such as tent caterpillars and tomato hornworms. Other chewing insects, however, such as cutworms (which are caterpillars) come out at night to eat, and burrow into the soil during the day. These are much harder to identify but should be considered likely culprits if young plants seem to disappear overnight or are found cut off at ground level.

Sucking insects are extremely common in gardens and can be very damaging to your vegetable plants and flowers. The most known of these insects are leafhoppers, aphids, mealy bugs, thrips, and mites. These insects insert their mouthparts into the plant tissues and suck out the plant juices. They also may carry diseases that they spread from plant to plant as they move about the yard. You may suspect that these insects are present if you notice misshapen plant leaves or flower petals. Often the younger leaves will appear curled or puckered. Flowers developing from the buds may only partially develop if they've been sucked by these bugs. Look on the undersides of the leaves—that is where many insects tend to gather.

Other insects cause damage to plants by boring into stems, fruits, and leaves, possibly disrupting the plant's ability to transport water. They also create opportunities for disease organisms to attack the plants. You may suspect the presence of boring insects if you see small accumulations of sawdust-like material on plant stems or fruits. Common examples of boring insects include squash vine borers and corn borers.

Integrated Pest Management (IPM)

It is difficult, if not impossible, to prevent all pest problems in your garden every year. If your best prevention efforts have not been entirely successful, you may need to use some control methods. Integrated pest management (IPM) relies on several techniques to keep pests at acceptable population levels without excessive use of chemical controls. The basic principles of IPM include monitoring (scouting), determining tolerable injury levels (thresholds), and applying appropriate strategies and tactics to solve the pest issue. Unlike other methods of pest control where pesticides are applied on a rigid schedule, IPM applies only those controls that are needed, when they are needed, to control pests that will cause more than a tolerable level of damage to the plant.

Monitoring

Monitoring is essential for a successful IPM program. Check your plants regularly. Look for signs of damage from insects and diseases as well as indications of adequate fertility and moisture. Early identification of potential problems is essential.

There are thousands of insects in a garden, many of which are harmless or even beneficial to the plants. Proper identification is needed before control strategies can be adopted. It is important to recognize the different stages of insect development for several reasons. The caterpillars eating your plants may be the larvae of the butterflies you were trying to attract. Any small larva with six spots on its back is probably a young ladybug, a very beneficial insect.

Thresholds

It is not necessary to kill every insect, weed, or disease organism invading your garden in order to maintain the plants' health. When dealing with garden pests, an economic threshold comes into play and is the point where the damage caused by the pest exceeds the cost of control. In a home garden, this can be difficult to determine. What you are growing and how you intend to use it will determine how much damage you are willing to tolerate. Remember that larger plants, especially those close to harvest, can tolerate more damage than a tiny seedling. A few flea beetles on a radish seedling may warrant control, whereas numerous Japanese beetles eating the leaves of beans close to harvest may not.

If the threshold level for control has been exceeded, you may need to employ control strategies. Effective and safe strategies can be discussed with your local Cooperative Extension Service, garden centers, or nurseries.

Mechanical/Physical Control Strategies

Many insects can simply be removed by hand. This method is definitely preferable if only a few, large insects are causing the problem. Simply remove the insect from the plant and drop it into a container of soapy water or vegetable oil. Be aware that some insects have prickly spines or excrete oily substances that can cause injury to humans. Use caution when handling unfamiliar insects. Wear gloves or remove insects with tweezers.

Many insects can be removed from plants by spraying water from a hose or sprayer. Small vacuums can also be used to suck up insects. Traps can be used effectively for some insects as well. These come in a variety of styles depending on the insect to be caught. Many traps rely on the use of pheromones—naturally occurring chemicals produced by the insects and used to attract the opposite sex during mating. They are extremely specific for each species and, therefore, will not harm beneficial species. One caution with traps is that they may actually draw more insects into your yard, so don't place them directly into your garden. Other traps (such as yellow and blue sticky cards) are more generic and will attract numerous species. Different insects are attracted to different colors of these traps. Sticky cards also can be used effectively to monitor insect pests.

Other Pest Controls

Diatomaceous earth, a powder-like dust made of tiny marine organisms called diatoms, can be used to reduce damage from soft-bodied insects and slugs. Spread this material on the soil—it is sharp and cuts or irritates these soft organisms. It is harmless to other organisms. In order to trap slugs, put out shallow dishes of beer.

Beneficial Insects that Help Control Pest Populations

Insect	Pest Controlled
Green lacewings	Aphids, mealy bugs, thrips, and spider mites
Ladybugs	Aphids and Colorado potato beetles
Praying mantises	Almost any insect
Ground beetles	Caterpillars that attack trees and shrubs
Seedhead weevils and other beetles	Weeds

Biological Controls

Biological controls are nature's way of regulating pest populations. Biological controls rely on predators and parasites to keep organisms under control. Many of our present pest problems result from the loss of predator species and other biological control factors.

Some biological controls include birds and bats that eat insects. A single bat can eat up to 600 mosquitoes an

hour. Many bird species eat insect pests on trees and in the garden.

Chemical Controls

When using biological controls, be very careful with pesticides. Most common pesticides are broad spectrum, which means that they kill a wide variety of organisms. Spray applications of insecticides are likely to kill numerous beneficial insects as well as the pests. Herbicides applied to weed species may drift in the wind or vaporize in the heat of the day and injure non-targeted plants. Runoff of pesticides can pollute water. Many pesticides are toxic to humans as well as pets and small animals that may enter your yard. Try to avoid using these types of pesticides at all costs—and if you do use them, read the labels carefully and avoid spraying them on windy days.

Some common, non-toxic household substances are as effective as many toxic pesticides. A few drops of dishwashing detergent mixed with water and sprayed on plants is extremely effective in controlling many soft-bodied insects, such as aphids and whiteflies. Crushed garlic mixed with water may control certain insects. A baking soda solution has been shown to help control some fungal diseases on roses.

cutworms

Alternatives to Pesticides and Chemicals

When used incorrectly, pesticides can pollute water. They also kill beneficial as well as harmful insects. Natural alternatives prevent both of these events from occurring and save you money. Consider using natural alternatives to chemical pesticides: Non-detergent insecticidal soaps, garlic, hot pepper spray, 1 teaspoon of liquid soap in a gallon of water, used dishwater, or a forceful stream of water from a hose all work to dislodge insects from your garden plants.

Another solution is to consider using plants that naturally repel insects. These plants have their own chemical defense systems, and when planted among flowers and vegetables, they help keep unwanted insects away.

Natural Pest Repellants

Pest	Repellant
Ant	Mint, tansy, or pennyroyal
Aphids	Mint, garlic, chives, coriander, or anise
Bean leaf beetle	Potato, onion, or turnip
Codling moth	Common oleander
Colorado potato bug	Green beans, coriander, or nasturtium
Cucumber beetle	Radish or tansy
Flea beetle	Garlic, onion, or mint
Imported cabbage worm	Mint, sage, rosemary, or hyssop
Japanese beetle	Garlic, larkspur, tansy, rue, or geranium
Leaf hopper	Geranium or petunia
Mice	Onion
Root knot nematodes	French marigolds
Slugs	Prostrate rosemary or wormwood
Spider mites	Onion, garlic, cloves, or chives
Squash bug	Radish, marigolds, tansy, or nasturtium
Stink bug	Radish
Thrips	Marigolds
Tomato hornworm	Marigolds, sage, or borage
Whitefly	Marigolds or nasturtium

Plant Diseases

Plant disease identification is extremely difficult. In some cases, only laboratory analysis can conclusively identify some diseases. Disease organisms injure plants in several ways: Some attack leaf surfaces and limit the plant's ability to carry on photosynthesis; others produce substances that clog plant tissues that transport water and nutrients; still other disease organisms produce toxins that kill the plant or replace plant tissue with their own.

Symptoms that are associated with plant diseases may include the presence of mushroom-like growths on trunks of trees; leaves with a grayish, mildewed appearance; spots on leaves, flowers, and fruits; sudden wilting or death of a plant or branch; sap exuding from branches or trunks of trees; and stunted growth.

Misapplication of pesticides and nutrients, air pollutants, and other environmental conditions—such as flooding and freezing—can also mimic some disease problems. Yellowing or reddening of leaves and stunted growth may indicate a nutritional problem. Leaf curling or misshapen growth may be a result of herbicide application.

Pest and Disease Management Practices

Preventing pests should be your first goal when growing a garden, although it is unlikely that you will be able to avoid all pest problems because some plant seeds and disease organisms may lay dormant in the soil for years.

Aphids

Diseases need three elements to become established in plants: the disease organism, a susceptible species, and the proper environmental conditions. Some disease organisms can live in the soil for years; other organisms are carried in infected plant material that falls to the ground. Some disease organisms are carried by insects. Good sanitation will help limit some problems with disease. Choosing resistant varieties of plants also prevents many diseases from occurring. Rotating annual plants in a garden can also prevent some diseases.

Plants that have adequate, but not excessive, nutrients are better able to resist attacks from both diseases and insects. Excessive rates of nitrogen often result in extremely succulent vegetative growth and can make plants more susceptible to insect and disease problems, as well as decreasing their winter hardiness. Proper watering and spacing of plants limits the spread of some diseases and provides good aeration around plants, so diseases that fester in standing water cannot multiply. Trickle irrigation, where water is applied to the soil and not the plant leaves, may be helpful.

Removal of diseased material certainly limits the spread of some diseases. It is important to clean up litter dropped from diseased plants. Prune diseased branches on trees and shrubs to allow for more air circulation. When pruning diseased trees and shrubs, disinfect your pruners between cuts with a solution of chlorine bleach to avoid spreading the disease from plant to plant. Also try to control insects that may carry diseases to your plants.

You can make your own natural fungicide by combining 5 teaspoons each of baking soda and hydrogen peroxide with a gallon of water. Spray on your infected plants. Milk diluted with water is also an effective fungicide, due to the potassium phosphate in it, which boosts a plant's immune system. The more diluted the solution, the more frequently you'll need to spray the plant.

PLANNING A GARDEN

A Plant's Basic Needs

Before you start a garden, it's helpful to understand what plants need in order to thrive. Some plants, like dandelions, are tolerant of a wide variety of conditions, while others, such as orchids, have very specific requirements in order to grow successfully. Before spending time, effort, and money attempting to grow a new plant in a garden, learn about the conditions that particular plant needs in order to grow properly.

Environmental factors play a key role in the proper growth of plants. Some of the essential factors that influence this natural process are as follows:

1. Length of Day

The amount of time between sunrise and sunset is the most critical factor in regulating vegetative growth, blooming, flower development, and the initiation of dormancy. Plants utilize increasing day length as a cue to promote their growth in spring, while decreasing day length in fall prompts them to prepare for the impending cold weather. Many plants require specific day length conditions in order to bloom and flower.

2. Light

Light is the energy source for all plants. Cloudy, rainy days or any shade cast by nearby plants and structures can significantly reduce the amount of light available to the plant. In addition, plants adapted to thrive in shady spaces cannot tolerate full sunlight. In general, plants will only be able to survive where adequate sunlight reaches them at levels they are able to tolerate.

3. Temperature

Plants grow best within an optimal range of temperatures. This temperature range may vary drastically depending on the plant species. Some plants thrive in environments where the temperature range is quite wide; others can only survive within a very narrow temperature variance. Plants can only survive where temperatures allow them to carry on life-sustaining chemical reactions.

4. Cold

Plants differ by species in their ability to survive cold temperatures. Temperatures below 60°F injure some tropical plants. Conversely, arctic species can tolerate temperatures well below zero. The ability of a plant to withstand cold is a function of the degree of dormancy present in the plant and its general health. Exposure to wind, bright sunlight, or rapidly changing temperatures can also compromise a plant's tolerance to the cold.

5. Heat

A plant's ability to tolerate heat also varies widely from species to species. Many plants that evolved to grow in arid, tropical regions are naturally very heat tolerant, while sub-arctic and alpine plants show very little tolerance for heat.

6. Water

Different types of plants have different water needs. Some plants can tolerate drought during the summer but need winter rains in order to flourish. Other plants need a consistent supply of moisture to grow well. Careful attention to a plant's need for supplemental water can help you to select plants that need a minimum of irrigation to perform well in your garden. If you have poorly drained, chronically wet soil, you can select lovely garden plants that naturally grow in bogs, marshlands, and other wet places.

7. Soil pH

A plant root's ability to take up certain nutrients depends on the pH—a measure of the acidity or alkalinity—of your soil. Most plants grow best in soils that have

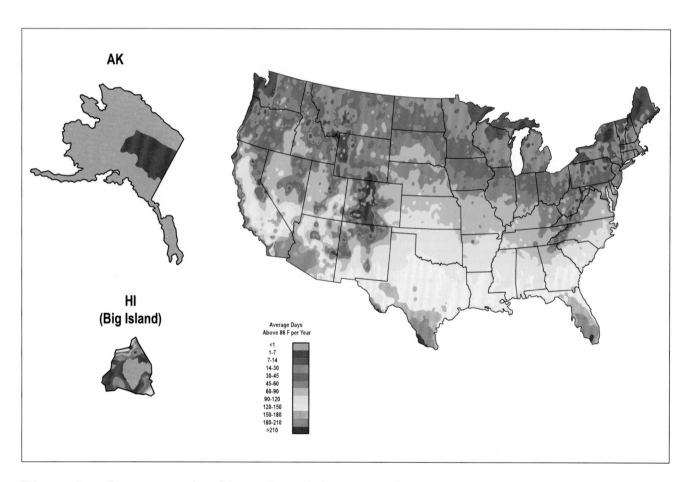

This map shows the average number of days each year that an area experiences temperatures over 86°F ("heat days"). Zone I has less than one heat day and zone 12 has more than 210 heat days, Most plants begin to suffer when it gets hotter than 86°F, though different plants have different levels of tolerance.

a pH between 6.0 and 7.0. Most ericaceous plants, such as azaleas and blueberries, need acidic soils with pH below 6.0 to grow well. Lime can be used to raise the soil's pH, and materials containing sulfates, such as aluminum sulfate and iron sulfate, can be used to lower the pH. The solubility of many trace elements is controlled by pH, and plants can only use the soluble forms of these important micronutrients.

The Junior Homesteader

We eat lots of different plant parts!

Foods We Eat That Are Roots

Beet	Radish
Carrot	Rutabaga
Onion	Sweet potato
Parsnip	Turnip
Potato	Yam

Foods We Eat That Are Stems

Asparagus	Broccoli
Bamboo shoots	Celery
Bok choy	Rhubarb

Foods We Eat That Are Leaves

Brussels sprouts	Lettuce
Cabbage	Mustard greens
Chard	Parsley
Collards	Spinach
Endive	Turnip greens
Kale	Watercress

Foods We Eat That Are Flowers

Broccoli	Cauliflower

Foods We Eat That Are Seeds

Black beans	Lima beans
Butter beans	Peas
Corn	Pinto beans
Dry split peas	Pumpkin seeds
Kidney beans	Sunflower seeds

Foods We Eat That Are Fruits

Apple	Melon
Apricot	Orange
Artichoke	Papaya
Avocado	Peach
Banana	Pear
Bell pepper	Pineapple
Berries	Plum
Cucumber	Pomegranate
Date	Pumpkin
Eggplant	Squash
Grapefruit	Strawberry
Grapes	Tangerine
Kiwifruit	Tomato
Mango	

Plant and Gardening Glossary

Annual—a plant that completes its life cycle in one year or season.

Arboretum—a landscaped space where trees, shrubs, and herbaceous plants are cultivated for scientific study or educational purposes, and to foster appreciation of plants.

Axil—the area between a leaf and the stem from which the leaf arises.

Bract—a leaflike structure that grows below a flower or cluster of flowers and is often colorful. Colored bracts attract pollinators, and are often mistaken for petals. Poinsettia and flowering dogwood are examples of plants with prominent bracts.

Cold hardy—capable of withstanding cold weather conditions.

Conifers—plants that predate true flowering plants in evolution; conifers lack true flowers and produce separate male and female strobili, or cones. Some conifers, such as yews, have fruits enclosed in a fleshy seed covering.

Cultivar—a cultivated variety of a plant selected for a feature that distinguishes it from the species from which it was selected.

Deciduous—having leaves that fall off or are shed seasonally to withstand adverse weather conditions, such as cold or drought.

Herbaceous—having little or no woody tissue. Most plants grown as perennials or annuals are herbaceous.

Hybrid—a plant, or group of plants, that results from the interbreeding of two distinct cultivars, varieties, species, or genera.

Inflorescence—a floral axis that contains many individual flowers in a specific arrangement; also known as a flower cluster.

Native plant—a plant that lives or grows naturally in a particular region without direct or indirect human intervention.

Panicle—a pyramidal, loosely branched flower cluster; a panicle is a type of inflorescence.

Perennial—persisting for several years, usually dying back to a perennial crown during the winter and initiating new growth each spring.

Shrub—a low-growing, woody plant, usually less than 15 feet tall, that often has multiple stems and may have a suckering growth habit (the tendency to sprout from the root system).

Taxonomy—the study of the general principles of scientific classification, especially the orderly classification of plants and animals according to their presumed natural relationships.

Tree—a woody perennial plant having a single, usually elongated main stem, or trunk, with few or no branches on its lower part.

Wildflower—a herbaceous plant that is native to a given area and is representative of unselected forms of its species.

Woody plant—a plant with persistent woody parts that do not die back in adverse conditions. Most woody plants are trees or shrubs.

Choosing a Site for Your Garden

Choosing the best spot for your garden is the first step toward growing the vegetables, fruits, and herbs that you want. You do not need a large space to get started—in fact, often it's wise to start small so that you don't get overwhelmed. A normal garden that is about 25 feet square will provide enough produce for a family of four, and with a little ingenuity (utilizing pots, hanging gardens, trellises, etc.) you can grow more than that in an even smaller space.

Five Factors to Consider When Choosing a Garden Site

1. Sunlight

Sunlight is crucial for the growth of vegetables and other plants. For your garden to grow, your plants will need at

When planning out your garden, first sketch a diagram of what you want your garden to look like. What sorts of plants to you want to grow? Do you want a garden purely for growing vegetables or do you want to mix in some fruits, herbs, and wildflowers? Choosing the appropriate plants to grow next to each other will help your garden grow well and will provide you with ample produce throughout the growing season (see the charts on page 705). If you live in the northern hemisphere, plant taller plants at the north end of your garden so that they won't block sunlight from reaching the smaller plants. If you live in the southern hemisphere, this is reversed.

The Junior Homesteader

You can structure your plants to double as playhouses for the kids. Here are a few possibilities:

- Bean teepees are the best way to support pole bean plants, and they also make great hiding places for little gardeners. Drive five or six poles that are 7 to 8 feet tall into the ground in a circle with a four-foot diameter. Bind the tops of the poles together with baling twine or a similar sturdy string. Plant your beans at the bottoms of the poles so they'll grow up and create a tent of vines.

- Vine tunnels can be made out of poles and any trailing vines—gourds, cucumbers, or morning glories are a few options. Drive several 7- to 8-foot poles (bamboo works well) into the ground in two parallel lines so that they create a pathway. The poles should be at least 3 feet apart from each other. Then lash horizontal poles to the vertical ones at 2-, 4-, and 6-foot heights. You can also lash poles across the top of the tunnel to connect the two sides and create a roof. Plant your trailing vines at the bases of the poles and watch your tunnel fill in as the weeks go by.

- Wigwams and huts are easily fashioned by planting your sunflowers or corn in a circular or square shape. To make a whole house, plant "walls" that are a few rows thick and create several "rooms," leaving gaps for doors.

Making a bean teepee.

least six hours of direct sunlight per day. In order to make sure your garden receives an ample amount of sunlight, don't select a garden site that will be in the shade of trees, shrubs, houses, or other structures. Certain vegetables, such as broccoli and spinach, grow just fine in shadier spots, so if your garden does receive some shade, make sure to plant those types of vegetables in the shadier areas. However, on the whole, if your garden does not receive at least six hours of intense sunlight per day, it will not grow as efficiently or successfully.

2. Proximity

Think about convenience as you plot out your garden space. If your garden is closer to your house and easy to reach, you will be more likely to tend it on a regular basis and to harvest the produce at its peak of ripeness. You'll find it a real boon to be able to run out to the garden in the middle of making dinner to pull up a head of lettuce or snip some fresh herbs.

3. Soil Quality

Your soil does not have to be perfect to grow a productive garden. However, it is best to have soil that is fertile, is

If your garden is not close to your house, you may want to construct a small potting shed in which to keep your tools.

full of organic materials that provide nutrients to the plant roots, and is easy to dig and till. Loose, well-drained soil is ideal. If there is a section of your yard where water does not easily drain after a good, soaking rain, this is not the spot for your garden; the excess water can easily drown your plants. Furthermore, soils that are of a clay or sandy consistency are not as effective in growing plants. To make these types of soils more nutrient-rich and fertile, add in organic materials (such as compost or manure).

4. Water Availability

Water is vital to keeping your garden green, healthy, and productive. A successful garden needs around 1 inch of water per week to thrive. Rain and irrigation systems are effective in maintaining this 1-inch-per-week quota. Situating your garden near a spigot or hose is ideal, allowing you to keeping the soil moist and your plants happy.

5. Elevation

Your garden should not be located in an area where air cannot circulate or where frost quickly forms. Placing your garden in a low-lying area, such as at the base of a slope, should be avoided. Lower areas do not warm as quickly in the spring, and will easily collect frost in the spring and fall. Your garden should, if at all possible, be on a slightly higher elevation. This will help protect your plants from frost and you'll be able to start your garden growing earlier in the spring and harvest well into the fall.

Tools of the Trade

Gardening tools don't need to be high-tech, but having the right ones on hand will make your life much easier. You'll need a spade or digging fork for digging holes for seeds or seedlings (or, if the soil is loose enough, you can just use your hands). Use a trowel, rake, or hoe to smooth over the garden surface. A measuring stick is helpful when spacing your plants or seeds (if you don't have a measuring stick, you can use a precut string to measure). If you are planting seedlings or established plants, you may need stakes and string to tie them up so they don't fall over in inclement weather or when they start producing fruit or vegetables. Finally, if you are interested in installing an irrigation system for your garden, you will need to buy the appropriate materials for this purpose. See page 608 for more on gardening tools.

Companion Planting

Plants have natural substances built into their structures that repel or attract certain insects and can have an effect on the growth rate and even the flavor of the other plants around them. Thus, some plants aid each other's growth when planted in close proximity and others inhibit each other. Smart companion planting will help your garden remain healthy, beautiful, and in harmony, while deterring certain insect pests and other factors that could be potentially detrimental to your garden plants.

Here is a chart that lists various types of garden vegetables, herbs, and flowers and their respective companion and "enemy" plants.

Vegetables

Type	Companion plant(s)	Avoid
Asparagus	Tomatoes, parsley, basil	Onion, garlic, potatoes
Beans	Eggplant	Tomatoes, onions, kale
Beets	Mint	Runner beans
Broccoli	Onions, garlic, leeks	Tomatoes, peppers, mustard
Cabbage	Onions, garlic, leeks	Tomatoes, peppers, beans
Carrots	Leeks, beans	Radish
Celery	Daisies, snapdragons	Corn, aster flower
Corn	Legumes, squash, cucumbers	Tomatoes, celery
Cucumber	Radishes, beets, carrots	Tomatoes
Eggplant	Marigolds, mint	Runner beans
Leeks	Carrots	Legumes
Lettuce	Radish, carrots	Celery, cabbage, parsley
Melon	Pumpkin, squash	None

Type	Companion plant(s)	Avoid
Peppers	Tomatoes	Beans, cabbage, kale
Onion	Carrots	Peas, beans
Peas	Beans, corn	Onions, garlic
Potato	Horseradish	Tomatoes, cucumber
Tomatoes	Carrots, celery, parsley	Corn, peas, potatoes, kale

Herbs

Type	Companion Plant(s)	Avoid
Basil	Chamomile, anise	Sage
Chamomile	Basil, cabbage	Other herbs (it will become oily)
Cilantro	Beans, peas	None
Chives	Carrots	Peas, beans
Dill	Cabbage, cucumbers	Tomatoes, carrots
Fennel	Dill	Everything else
Garlic	Cucumbers, peas, lettuce	None
Oregano	Basil, peppers	None
Peppermint	Broccoli, cabbage	None
Rosemary	Sage, beans, carrots	None
Sage	Rosemary, beans	None
Summer savory	Onion, green beans	None

Flowers

Types	Companion Plant(s)	Avoid
Geraniums	Roses, tomatoes	None
Marigolds	Tomatoes, peppers, most plants	None
Petunia	Squash, asparagus	None
Sunflower	Corn, tomatoes	None
Tansy	Roses, cucumbers, squash	None

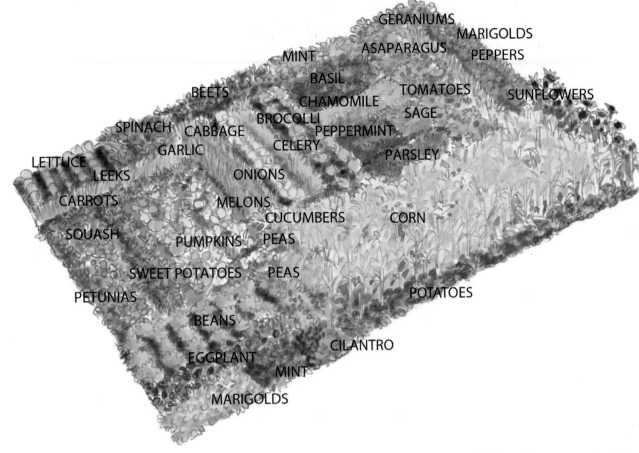

GERANIUMS
MARIGOLDS
ASAPARAGUS
PEPPERS
MINT
BASIL
TOMATOES
BEETS
CHAMOMILE
SUNFLOWERS
BROCOLLI
SAGE
SPINACH CABBAGE
PEPPERMINT
GARLIC
CELERY
LETTUCE
PARSLEY
LEEKS
ONIONS
CARROTS
MELONS
CUCUMBERS
CORN
SQUASH
PUMPKINS PEAS
SWEET POTATOES PEAS
PETUNIAS
POTATOES
BEANS
CILANTRO
EGGPLANT
MINT
MARIGOLDS

Here's a sample plan for a garden that utilizes companion planting. The proportions can be varied depending on your desired quantities of each crop and the shape of your garden plot. Remember to leave space in your garden to walk between rows of plants.

Shade-Loving Plants

Most plants thrive on several hours of direct sunlight every day, but certain plants actually prefer the shade. When buying seedlings from your local nursery or planting your own seeds, read the accompanying label or packet before planting to make sure your plants will thrive in a shadier environment.

Flowering plants that do well in partial and full shade include:

- Bee balm
- Bellflower
- Bleeding heart
- Cardinal flower
- Coleus
- Columbine
- Daylilies
- Dichondra
- Ferns
- Forget-me-not
- Globe daisy
- Golden bleeding heart
- Impatiens
- Leopardbane
- Lily of the valley
- Meadow rue
- Pansy
- Periwinkle
- Persian violet
- Primrose
- Rue anemone
- Snapdragon
- Sweet alyssum
- Thyme

Vegetable plants that can grow in partial shade include:

- Arugula
- Beans
- Beets
- Broccoli
- Brussels sprouts
- Cauliflower
- Endive
- Kale
- Leaf lettuce
- Peas
- Radish
- Spinach
- Swiss chard

Kids as young as toddlers will enjoy being involved in a family garden. Encourage very young children to explore by touching and smelling dirt, leaves, and flowers. Just be careful they don't taste anything non-edible. If space allows, assign a small plot for older children to plant and tend all on their own. An added bonus is that kids are more likely to eat vegetables they've grown themselves!

Daylilies thrive in shady areas.

PLANTING AND TENDING YOUR GARDEN

Once you've chosen a spot for your garden (as well as the size you want to make your garden bed), and prepared the soil with compost or other fertilizer, it's time to start planting. Find seeds at your local garden center, browse through seed catalogs, and order seeds that will do well in your area. Alternatively, you can start with bedding plants (or seedlings) available at nurseries and garden centers.

Read the instructions on the back of the seed package or on the plastic tag in your plant pot. You may have to ask experts when to plant the seeds if this information is not stated on the back of the package. Some seeds (such as tomatoes) should be started indoors in small pots or seed trays before the last frost, and only transplanted outdoors when the weather warms up. For established plants or seedlings, be sure to plant as directed on the plant tag or consult your local nursery about the best planting times.

Seedlings

If you live in a cooler region with a shorter growing period, you will want to start some of your plants indoors. To do this, obtain plug flats (trays separated into many small cups or "cells") or make your own small planters by poking holes in the bottoms of paper cups. Fill the cups two-thirds full with potting soil or composted soil. Bury the seed at the recommended depth, according to the instructions on the package. Tamp down the soil lightly and water. Keep the seedlings in a warm, well-lit place, such as the kitchen, to encourage germination.

Once the weather begins to warm up and you are fairly certain you won't be getting any more frosts (you can contact your local extension office to find out the "frost free" date for your area) you can begin to acclimate your seedlings to the great outdoors. First place them in a partially shady spot outdoors that is protected from strong wind. After a couple of days, move them into direct sunlight, and then finally transplant them to the garden.

Sprouting Seeds for Eating

Seeds can be sprouted and eaten on sandwiches, salads, or stir-fries any time of the year. They are delicious and full of vitamins and proteins. Mung beans, soybeans, alfalfa, wheat, corn, barley, mustard, clover, chickpeas, radishes, and lentils all make good sprouts. Find seeds for sprouting from your local health food store or use dried peas, beans, or lentils

Germination Temperatures of Selected Vegetable Plants

Broccoli 77°F	Eggplant 85°F	Onion 70°F	Summer Squash 80°F
Cabbage 86°F	Herbs 65°F	Pepper 85°F	Tomato 85°F
Cucumber 86°F	Melon 90°F	Pumpkin 85°F	Winter Squash 80°F

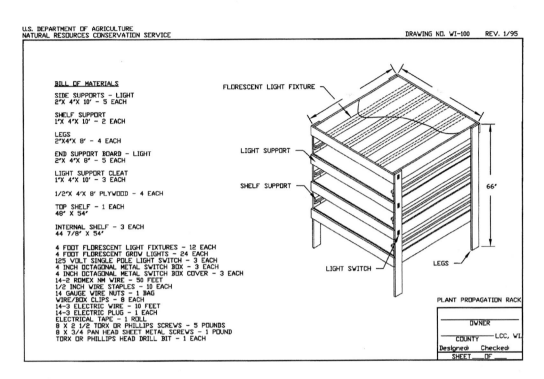

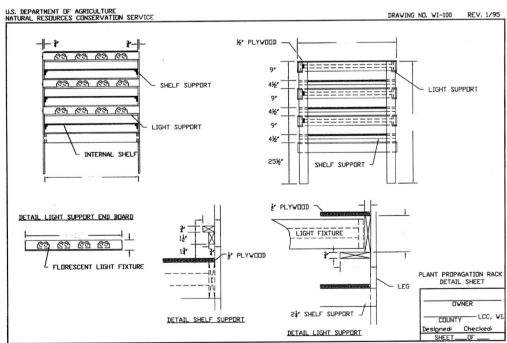

Follow the plans on this page and page 710 to make your own propagation rack for starting seeds indoors. Though a propagation rack is not necessary, it will help to ensure your seedlings receive the light and warmth they need to stay strong and healthy.

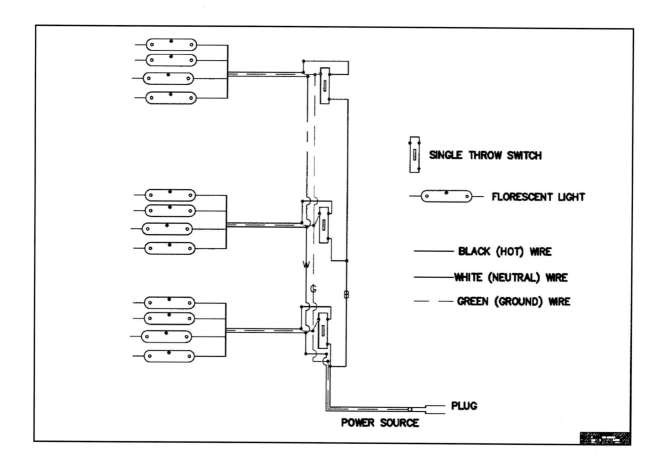

SINGLE THROW SWITCH

FLORESCENT LIGHT

BLACK (HOT) WIRE
WHITE (NEUTRAL) WIRE
GREEN (GROUND) WIRE

PLUG
POWER SOURCE

from the grocery store. Never use seeds intended for planting unless you've harvested the seeds yourself—commercially available planting seeds are often treated with a poisonous chemical fungicide.

To grow sprouts, follow these steps:

1. Thoroughly rinse and strain the seeds.

2. Place in a glass jar, cover with cheesecloth secured with a rubber band, and soak overnight in cool water. You'll need about four times as much water as you have seeds.

3. Drain the seeds by turning the jar upside down and allowing the water to escape through the cheesecloth.

Keep the seeds at 60 to 80ºF and rinse twice a day, draining them thoroughly after every rinse.

4. Once sprouts are 1 to 1 ½ inches long (generally after 3 to 5 days), they are ready to eat.

The Junior Homesteader

How do different treatments change how fast seeds sprout?

Experiment with sprouting seeds under different temperatures, or after being soaked for different times or in different liquids. Or, see how one kind of treatment affects different types of seeds. What makes seeds sprout the fastest? Do some look healthier than others? Write up a lab report to document your findings.

RECOMMENDED PLANTS TO START AS SEEDLINGS

CROP [s] small seed [l] large seed (planting cell size)	WEEKS BEFORE TRANSPLANTING	SEED PLANTING DEPTH (Inches)	TRANSPLANT SPACING	WITHIN ROW / BETWEEN ROW
Broccoli [s]	(1)	4–6	¼–½	8–10" 18–24"
Cabbage [s]	(1)	4–6	¼–½	18–24" 30"
Cucumber [l]	(2)	4–5	½	2' 5–6'
Eggplant [s]	(2)	8	¼	18" 18–24"
Herbs [s]	(1)	4	¼	4–6" 12–18"
Lettuce [s]	(2)	4–5	¼	12" 12"
Melon [l]	(3)	4–5	¼	2–3' 6'
Onion [s]	(*)	8	¼	4" 12"
Pepper [s]	(2)	8	¼	12–18" 2–3'
Pumpkin [l]	(3)	2–4	1	5–6' 5–6'
Summer Squash [l]	(3)	2–4	¾–1	18" 2–3'
Tomato [s]	(3)	8	¼	18"–24" 3'
Watermelon [l]	(3)	4–5	½–¾	3–4' 3–4'
Winter Squash [l]	(3)	2–4	1	3–4' 4–5'

Watering Your Soil

After your seeds or seedlings are planted, the next step is to water your soil. Different soil types have different watering needs. You don't need to be a soil scientist to know how to water your soil properly. Here are some tips that can help to make your soil moist and primed for gardening:

1. Loosen the soil around plants so water and nutrients can be quickly absorbed.
2. Use a 1- to 2-inch protective layer of mulch on the soil surface above the root area. Cultivating and mulching help reduce evaporation and soil erosion.
3. Water your plants at the appropriate time of day. Early morning or night is the best time for watering, as evaporation is less likely to occur at these times.
4. Do not water your plants when it is extremely windy outside. Wind will prevent the water from reaching the soil where you want it to go.

Types of Soil and Their Water Retention

Knowing the type of soil you are planting in will help you best understand how to properly water and grow your garden plants. Three common types of soil and their various abilities to absorb water are listed below:

1. Clay Soil

In order to make this type of soil more loamy, add organic materials, such as compost, peat moss, and well-rotted leaves, in the spring before growing and also in the fall after harvesting your vegetables and fruits. Adding these organic materials allows this type of soil to hold more nutrients for healthy plant growth. Till or spade to help loosen the soil.

Since clay soil absorbs water very slowly, water only as fast as the soil can absorb the water.

2. Sandy Soil

As with clay soil, adding organic materials in the spring and fall will help supplement the sandy soil and promote better plant growth and water absorption.

Left on its own (with no added organic matter) the water will run through sandy soil so quickly that plants won't be able to absorb it through their roots and will fail to grow and thrive.

3. Loam Soil

This is the best kind of soil for gardening. It's a combination of sand, silt, and clay. Loamy soil is fertile, deep, easily crumbles, and contains organic matter. It will help promote the growth of quality fruits and vegetables, as well as flowers and other plants.

Loam absorbs water readily and stores it for plants to use. Water as frequently as the soil needs to maintain its moisture and to promote plant growth.

RAISED BEDS

I f you live in an area where the soil is quite wet (preventing a good vegetable garden from growing the spring), or find it difficult to bend over to plant and cultivate your vegetables or flowers, or if you just want a different look to your backyard garden, you should consider building a raised bed.

A raised bed is an interesting and affordable way to garden. It creates an ideal environment for growing vegetables, since the soil concentration can be closely monitored and, as it is raised above the ground, it reduces the compaction of plants from people walking on the soil.

Raised beds are typically 2 to 6 feet wide and as long as needed. In most cases, a raised bed consists of a "frame" that is filled in with nutrient-rich soil (including compost or organic fertilizers) and is then planted with a variety of vegetables or flowers, depending on the gardener's preference. By controlling the bed's construction and the soil mixture that goes into the bed, a gardener can effectively reduce the amount of weeds that will grow in the garden.

When planting seeds or young sprouts in a raised bed, it is best to space the plants equally from each other on all sides. This will ensure that the leaves will be touching once the plant is mature, thus saving space and reducing the soil's moisture loss.

How to Make a Raised Bed

Things You'll Need

- Forms for your raised bed (consider using 4 x 4-inch posts cut to 24 inches in height for corners, and 2 x 12-inch boards for the sides)
- Nails or screws
- Hammer or screwdriver
- Plastic liner (to act as a weed barrier at the bottom of your bed)
- Shovel
- Compost or composted manure
- Soil (either potting soil or soil from another part of your yard)
- Rake (to smooth out the soil once in the bed)
- Seeds or young plants
- Optional: PVC piping and greenhouse plastic (to convert your raised bed to a greenhouse)

1. Plan Out Your Raised Bed

1. Think about how you'd like your raised bed to look, and then design the shape. A raised bed is not extremely complicated, and all you need to do is build an open-top and open-bottom box (if you are ambitious, you can create a raised bed in the shape of a circle, hexagon, or star). The main purpose of this box is to hold soil.

2. Make a drawing of your raised bed, measure your available garden space, and add those measurements to your drawing. This will allow you to determine how much material is needed. Generally, your bed should be at least 24 inches in height.

3. Decide what kind of material you want to use for your raised bed. You can use lumber, plastic, synthetic wood, railroad ties, bricks, rocks, or a number of other items to hold the dirt. Using lumber is the easiest and most efficient method.

4. Gather your supplies.

2. Build Your Raised Bed

1. Make sure your bed will be situated in a place that gets plenty of sunlight. Carefully assess your placement, as your raised bed will be fairly permanent.

2. Connect the sides of your bed together (with either screws or nails) to form the desired shape of your bed. If you are using lumber, you can use 4 x 4-inch posts to serve as the corners of your bed, and then nail or screw the sides to these corner posts. By doing so, you will in-

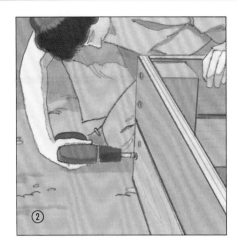

crease strength of the structure and ensure that the dirt will stay inside.

3. Cut a piece of gardening plastic to fit inside your raised bed. This will significantly reduce the amount of weeds growing in your garden. Lay it out in the appropriate location.

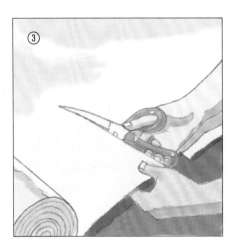

4. Place your frame over the gardening plastic (this might take two people).

3. Start Planting

1. Add some compost into the bottom of the bed and then layer potting soil on top of the compost. If you have soil from other parts of your yard, feel free to use that in addition to the compost and potting soil. Plan on filling at least ⅓ of your raised bed with compost or composted manure (available from nurseries or garden centers in 40-pound bags).

2. Mix in dry organic fertilizers (like wood ash, bone meal, and blood meal) while building your bed. Follow the package instructions for how best to mix it in.

3. Decide what you want to plant. Some people like to grow flowers in their raised beds; others prefer to grow vegetables. If you do want to grow food, raised beds are excellent choices for salad greens, carrots, onions, radishes, beets, and other root crops.

Things to Consider

1. To save money, you can dig up and use soil from your yard. Potting soil can be expensive, and yard soil is just as effective when mixed with compost. However, removing grass and weeds from the soil before filling your raised beds can be time-consuming.

2. Be creative when building your raised planting bed. You can construct a great raised bed out of recycled goods or old lumber.

3. You can convert your raised bed into a greenhouse. Just add hoops to your bed by bending and connecting PVC pipe over the bed. Then clip greenhouse plastic to the PVC pipes, and you have your own greenhouse.

4. Make sure to water your raised bed often. Because it is above ground, your raised bed will not retain water as well as the soil in the ground. If you keep your bed narrow, it will help conserve water.

5. Decorate or illuminate your raised bed to make it a focal point in your yard.

6. If you use lumber to construct your raised bed, keep a watch out for termites.

7. Beware of old pressure-treated lumber, as it may contain arsenic and could potentially leak into the root systems of any vegetables you might grow in your raised bed. Newer pressure-treated lumber should not contain these toxic chemicals.

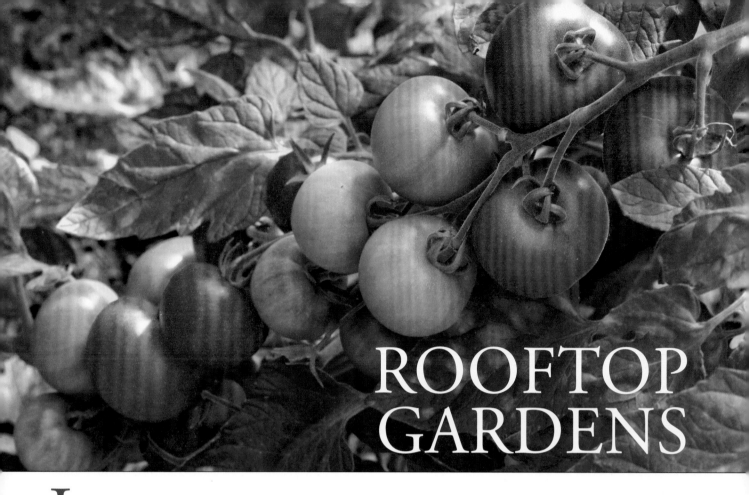

ROOFTOP GARDENS

If you live in an urban area and don't have a lawn, that does not mean that you cannot have a garden. Whether you live in an apartment building or own your own home without yard space, you can grow your very own garden, right on your roof!

Benefits of Rooftop Gardening

- Create more outdoor green space within your urban environment.
- Grow your own fresh vegetables—even in the city.
- Improve air quality and reduce CO_2 emissions.
- Help delay storm water runoff.
- Give additional insulation to building roofs.
- Reduce noise.

Is Your Roof Suitable for a Rooftop Garden?

Theoretically, any roof surface can be greened—even sloped or curved roofs can support a layer of sod or wild-flowers. However, if the angle of your roof is over 30 degrees you should consult with a specialist. Very slanted roofs make it difficult to keep the soil in place until the plants' roots take hold. Certainly, a flat roof, approximating level ground conditions, is the easiest on which to grow a garden, though a slight slant can be helpful in allowing drainage.

Also consider how much weight your roof can bear. A simple, lightweight rooftop garden will weigh between 13 and 30 pounds per square foot. Add to this your own weight—or that of anyone who will be tending or enjoying the garden—gardening tools, and, if you live in a colder climate, the additional weight of snow in the winter.

Will a Rooftop Garden Cause Water Leakage Or Other Damage?

No. In fact, planting beds or surfaces are often used to protect and insulate roofs. However, you should take some precautions to protect your roof:

1. Cover your roof with a layer of waterproof material, such as a heavy-duty pond liner. You may want to place an old rug on top of the waterproof material to help it stay in place and to give additional support to the materials on top.

2. Place a protective drainage layer on top of the waterproof material. Otherwise, shovels, shoe heels, or dropped tools could puncture the roof. Use a coarse material such as gravel, pumice, or expanded shale.

3. Place a filter layer on top of the drainage layer to keep soil in place so that it won't clog up your drainage. A lightweight polyester geotextile (an inexpensive, non-woven fabric found at most home improvement stores) is ideal for this. Note that if your roof has an angle greater than 10 degrees, only install the filter layer around the edges of the roof as it can increase slippage.

4. Using moveable planters or containers, modular walkways and surfacing treatment, and compartmentalized planting beds will make it easier to fix leaks should they appear.

How to Make a Rooftop Garden

Preparation

1. Before you begin, find out if it is possible and legal to create a garden on your roof. You don't want to spend lots of time and money preparing for a garden and then find out that it is prohibited.

2. Make sure that the roof is able to hold the weight of a rooftop garden. If so, figure out how much weight it can hold. Remember this when making the garden and use lighter containers and soil as needed.

Setting Up the Garden

1. Install your waterproof, protective drainage, and filter layers, as described above. If your roof is angled, you may want to place a wooden frame around the edges of the roof to keep the layers from sliding off. Be sure to use rot-resistant wood and cut outlets into the frame to allow excess water to drain away. Layer pebbles around the outlets to aid drainage and to keep vegetation from clogging them.

2. Add soil to your garden. It should be 1 to 4 inches thick and will be best if it's a mix of ¾ inorganic soil (crushed brick or a similar granular material) and ¼ organic compost.

Planting and Maintaining the Garden

1. Start planting. You can plant seeds or seedlings, or transplant mature plants. Choose plants that are wind-resistant and won't need a great deal of maintenance. Sedums make excellent rooftop plants as they require very little attention once planted, are hardy, and are attractive throughout most of the year. Most vegetables can be grown in-season on rooftops, though the wind will make taller vegetables (like corn or beans) difficult

to grow. If your roof is slanted, plant drought-resistant plant varieties near the peak, as they'll get less water.

2. Water your garden immediately after planting, and then regularly throughout the growing season, unless rain does the work for you.

Things to Consider

1. If you live in a very hot area, you may want to build small wooden platforms to elevate your plants above the hot rooftop. This will help increase the ventilation around the plants.

2. When determining whether or not your roof is strong enough to support a garden, remember that large pots full of water and soil will be very heavy, and if the roof is not strong enough, your garden could cause structural damage.

3. You can use pots or other containers on your rooftop rather than making a full garden bed. You should still first find out how much weight your roof can hold and choose lightweight containers.

4. Consider adding a fence or railing around your roof, especially if children will be helping in the garden.

SAVING SEEDS

If you have a unique heirloom variety plant that you want to preserve or if you don't want to buy new seeds every year, you can save seeds from your healthy plants. Saving seeds is relatively easy for dry plants, like beans, where the seeds are easily distinguishable from the vegetable or fruit. In these cases, simply scrape the seeds from the vegetable, place them in a single layer in a glass dish, and leave them near a sunny window to dry for one week.

For some plants, like tomatoes, the seeds are surrounded by a wet pulp. For these plants, remove the seeds from the flesh of the fruit or vegetable with your fingers and then rinse thoroughly in a wire mesh strainer. You may need to soak them for a while to remove all residue. Then dry as described above.

Once seeds are thoroughly dried, store them in labeled envelopes in a cool, dry place.

Keep in mind that plants often naturally cross-pollinate, especially when different types of plants are near each other in a garden, resulting in a hybrid seed. Hybrid seeds are unpredictable and often grow into inferior plants. Also, most seeds that you buy today are already hybrids. If you plan to save your seeds, invest in heirloom variety or open-pollinated seeds.

Hybrid vs. Heirloom Seeds

Plants are like any other living thing in that there are male and female parts and it takes both to create offspring. Some plants contain both the male and female parts within their own flowers (self-pollinators), and others have separate male plants and female plants. With the latter type, bees or birds carry the pollen from male plants to the ovule of female plants. Thus, plants can be bred to have certain characteristics and qualities by ensuring that the desired male and female plants are in close proximity and that undesirable potential "parents" are kept at a distance. Nowadays, seeds can also be artificially cross-bred and genetically altered.

Seed manufacturers frequently breed seeds to be high yield, often at the expense of disease resistance, since the majority of plants now are grown with pesticides anyway. These hybrid seeds are high-maintenance; they require special fertilizers, they're less hardy, and they are more susceptible to disease. However, with the right supports (pesticides, herbicides, and irrigation) they will produce a greater volume of plants. Other seeds are bred for other characteristics, such as size of the fruit or vegetable.

There are several concerns regarding the popularity of hybrid seeds. One is that it creates too much dependence on the major seed producers, as well as suppliers of pesticides and other inorganic gardening products. Since hybrid plants do not produce reliable seeds, farmers must return to the seed supplier every year before they begin planting and then often depend on pesticides to keep their plants healthy. This is an especially serious issue in poorer countries where the people are at the mercy of major seed supply companies.

Heirloom varieties are much more diverse than hybrids. Not only does this mean that by using them you'll be harvesting more interesting (and often more flavorful) produce, but you'll also be helping to prevent a potential food shortage disaster. Because major seed suppliers are breeding seeds for specific purposes, they're narrowing down the varieties of seeds they provide to only those that best meet their needs. This will become a major problem if a disease attacks those plants. If there are many varieties, some will resist the disease. If there are only a few varieties available, they might all be wiped out, as happened with the Irish potato famine of the 1840s.

Heirloom seeds are generally more expensive than hybrids, but you only have to buy them once, since you can save their seeds at the end of every growing season to plant the following spring. Thus, it makes sense to incorporate as many heirloom varieties into your garden as you can.

SOIL AND FERTILIZER

Nutrient-rich, fertile soil is essential for growing the best and healthiest plants—plants that will supply you with quality fruits, vegetables, and flowers. Sometimes soil loses its fertility (or has minimum fertility based on the region in which you live), and so measures must be taken in order to improve your soil and, subsequently, your garden.

Soil Quality Indicators

Soil quality is an assessment of how well soil performs all of its functions now and how those functions are being preserved for future use. The quality of soil cannot just be determined by measuring row or garden yield, water quality, or any other single outcome, nor can it be measured directly. Thus, it is important to look at specific indicators to better understand the properties of soil. Plants can provide us with clues about how well the soil is functioning—whether a plant is growing and producing quality fruits and vegetables or failing to yield such things is a good indicator of the quality of the soil it's growing in.

In short, indicators are measurable properties of soil or plants that provide clues about how well the soil can func-

tion. Indicators can be physical, chemical, and biological properties, processes, or characteristics of soils. They can also be visual features of plants.

Useful indicators of soil quality:

- are easy to measure
- measure changes in soil functions
- encompass chemical, biological, and physical properties
- are accessible to many users
- are sensitive to variations in climate and management

Indicators can be assessed by qualitative or quantitative techniques, such as soil tests. After measurements are collected, they can be evaluated by looking for patterns and comparing results to measurements taken at a different time.

Examples of soil quality indicators:

1. Soil Organic Matter—promotes soil fertility, structure, stability, and nutrient retention and helps combat soil erosion.
2. Physical Indicators—these include soil structure, depth, infiltration and bulk density, and water hold capacity.

Quality soil will retain and transport water and nutrients effectively; it will provide habitat for microbes; it will promote compaction and water movement; and, it will be porous and easy to work with.

3. Chemical Indicators—these include pH, electrical conductivity, and extractable nutrients. Quality soil will be at its threshold for plant, microbial, biological, and chemical activity; it will also have plant nutrients that are readily available.

4. Biological Indicators—these include microbial biomass, mineralizable nitrogen, and soil respiration. Quality soil is a good repository for nitrogen and other basic nutrients for prosperous plant growth; it has a high soil productivity and nitrogen supply; and there is a good amount of microbial activity.

Soil and Plant Nutrients

Nutrient Management

There are 20 nutrients that all plants require. Six of the most important nutrients, called macronutrients, are: calcium, magnesium, nitrogen, phosphorous, potassium, and sulfur. Of these, nitrogen, phosphorus, and potassium are essential to healthy plant growth and so are required in relatively large amounts. Nitrogen is associated with lush vegetative growth, phosphorus is required for flowering and fruiting, and potassium is necessary for durability and disease resistance. Calcium, sulfur, and magnesium are also required in comparatively large quantities and aid in the overall health of plants.

The other nutrients, referred to as micronutrients, are required in very small amounts. These include such elements as copper, zinc, iron, and boron. While both macro- and micronutrients are required for good plant growth, over-application of these nutrients can be as detrimental to the plant as a deficiency of them. Over-application of plant nutrients may not only impair plant growth, but may also contaminate groundwater by penetrating through the soil or may pollute surface waters.

Soil Testing

Testing your soil for nutrients and pH is important in order to provide your plants with the proper balance of nutrients (while avoiding over-application). If you are establishing a new lawn or garden, a soil test is strongly recommended. The cost of soil testing is minor in comparison to the cost of plant materials and labor. Correcting a problem before planting is much simpler and cheaper than afterwards.

Once your garden is established, continue to take periodic soil samples. While many people routinely lime their soil, this can raise the pH of the soil too high. Likewise, since many fertilizers tend to lower the soil's pH, it may drop below desirable levels after several years, depending on fertilization and other soil factors, so occasional testing is strongly encouraged.

Home tests for pH, nitrogen, phosphorus, and potassium are available from most garden centers. While these may give you a general idea of the nutrients in your soil, they are not as reliable as tests performed by the Cooperative Extension Service at land grant universities. University and other commercial testing services will provide more detail, and you can request special tests for micronutrients if you suspect a problem. In addition to the analysis of nutrients in your soil, these services often provide recommendations for the application of nutrients or how best to adjust the pH of your soil.

The test for soil pH is very simple. pH is a measure of how acidic or alkaline your soil is. A pH of 7 is considered neutral. Below 7 is acidic and above 7 is alkaline. Since pH greatly influences plant nutrients, adjusting the pH will often correct a nutrient problem. At a high pH, several of the micronutrients become less available for plant uptake. Iron deficiency is a common problem, even at a neutral pH, for such plants as rhododendrons and blueberries. At a very low soil pH, other micronutrients may be too available to the plant, resulting in toxicity.

Phosphorus and potassium are tested regularly by commercial testing labs. While there are soil tests for nitrogen, these may be less reliable. Nitrogen is present in the soil in several forms that can change rapidly. Therefore, a precise analysis of nitrogen is more difficult to obtain. Most university soil test labs do not routinely test for nitrogen. Home testing kits often contain a test for nitrogen that may give you a general, though not necessarily completely accurate, idea of the presence of nitrogen in your garden soil.

Organic matter is often part of a soil test. Organic matter has a large influence on soil structure and so is highly

desirable for your garden soil. Good soil structure improves aeration, water movement, and retention. This encourages increased microbial activity and root growth, both of which influence the availability of nutrients for plant growth. Soils high in organic matter tend to have a greater supply of plant nutrients compared to many soils low in organic matter. Organic matter tends to bind up some soil pesticides, reducing their effectiveness, and so this should be taken into consideration if you are planning to apply pesticides in your garden.

Tests for micronutrients are usually not performed unless there is reason to suspect a problem. Certain plants have greater requirements for specific micronutrients and may show deficiency symptoms if those nutrients are not readily available.

How to Test Your Soil

1. If you intend to send your sample to the land grant university in your state, contact the local Cooperative Extension Service for information and sample bags. If you intend to send your sample to a private testing lab, contact them for specific details about submitting a sample.

2. Follow the directions carefully for submitting the sample. The following are general guidelines for taking a soil sample:

- Sample when the soil is moist but not wet.
- Obtain a clean pail or similar container.
- Clear away the surface litter or grass.
- With a spade or soil auger, dig a small amount of soil to a depth of 6 inches.
- Place the soil in the clean pail.
- Repeat steps 3 through 5 until the required number of samples has been collected.
- Mix the samples together thoroughly.
- From the mixture, take the sample that will be sent for analysis.
- Send immediately. Do not dry before sending.

3. If you are using a home soil testing kit, follow the above steps for taking your sample. Follow the directions in the test kit carefully so you receive the most accurate reading possible.

Enriching Your Soil
Organic and Commercial Fertilizers

Once you have the results of the soil test, you can add nutrients or soil amendments as needed to alter the pH. If you need to raise the soil's pH, use lime. Lime is most effective when it is mixed into the soil; therefore, it is best to apply before planting (if you apply lime in the fall, it has a better chance of correcting any soil acidity problems for the next growing season). For large areas, rototilling is most effective. For small areas or around plants, working the lime into the soil with a spade or cultivator is preferable. When working around plants, be careful not to dig too deeply or roughly so that you damage plant roots. Depending on the form of lime and the soil conditions, the change in pH may be gradual. It may take several months before a significant change is noted. Soils high in organic matter and clay tend

to take larger amounts of lime to change the pH than do sandy soils.

If you need to lower the pH significantly, especially for plants such as rhododendrons, you can use aluminum sulfate. In all cases, follow the soil test or manufacturer's recommended rates of application. Again, it's best to mix the fertilizer into the soil before planting.

There are numerous choices for providing nitrogen, phosphorus, and potassium, the nutrients your plants need to thrive. Nitrogen (N) is needed for healthy, green growth and regulation of other nutrients. Phosphorus (P) helps roots and seeds properly develop and resist disease. Potassium (K) is also important in root development and disease resistance. If your soil is of adequate fertility, applying compost may be the best method of introducing additional nutrients. While compost is relatively low in nutrients compared to commercial fertilizers, it is nontoxic and wonderful for your soil. By keeping the soil loose, compost allows plant roots to grow well throughout the soil, helping them to extract nutrients from a larger area. A loose soil enriched with compost is also an excellent habitat for earthworms and other beneficial soil microorganisms that are essential for releasing nutrients for plant use. The nutrients from compost are also released slowly, so there is no concern about "burning" the plant with an over-application of synthetic fertilizer.

Manure is also an excellent organic source of plant nutrients. Manure should be composted before applying, as fresh manure may be too strong and can injure plants. Be careful when composting manure. If left in the open, exposed to rain, nutrients may leach out of the manure and the runoff can contaminate nearby waterways. Make sure the manure is stored in a location away from wells and any waterways and that any runoff is confined or slowly released into a vegetated area. Improperly applied manure also can be a source of pollution. If you are not composting your own manure, you can purchase some at your local garden store. For best results, work composted manure into the soil around the plants or in your garden before planting.

If preparing a bed before planting, compost and manure may be worked into the soil to a depth of 8 to 12 inches. If adding to existing plants, work carefully around the plants so as not to harm the existing roots.

Green manures are crops that are grown and then tilled into the soil. As they break down, nitrogen and other plant nutrients become available. These manures may also provide additional benefits of reducing soil erosion. Green manures, such as rye and oats, are often planted in the fall after the crops have been harvested. In the spring, these are tilled under before planting.

With all organic sources of nitrogen, whether compost or manure, the nitrogen must be changed to an inorganic form before the plants can use it. Therefore, it is important to have well-drained, aerated soils that provide the favorable habitat for the soil microorganisms responsible for these conversions.

There are also numerous sources of commercial fertilizers that supply nitrogen, phosphorus, and potassium, though it is preferable to use organic fertilizers, such as compost and manures. However, if you choose to use a commercial fertilizer, you should understand how to read the amount of nutrients contained in each bag. The first number on the fertilizer analysis is the percentage of nitrogen; the second number is phosphorus; and the third number is the potassium content. A fertilizer that has a 10-20-10 analysis contains twice as much of each of the nutrients as a 5-10-5. How much of each nutrient you need depends on your soil test results and the plants you are fertilizing.

As mentioned before, nitrogen stimulates vegetative growth while phosphorus stimulates flowering. Too much nitrogen can inhibit flowering and fruit production. For many flowers and vegetables, a fertilizer higher in phosphorus than nitrogen is preferred, such as a 5-10-5. For lawns, nitrogen is usually required in greater amounts, so a fertilizer with a greater amount of nitrogen is more beneficial.

Fertilizer Application

Commercial fertilizers are normally applied as a dry, granular material or mixed with water and poured onto the garden. If using granular materials, avoid spilling on sidewalks and driveways because these materials are water soluble and can cause pollution problems if rinsed into storm sewers. Granular fertilizers are a type of salt, and if applied too heavily, they have the capability of burning the plants.

Soil Test Reading	What to Do
High pH	Your soil is alkaline. To lower pH, add elemental sulfur, gypsum, or cottonseed meal. Sulfur can take several months to lower your soil's pH, as it must first convert to sulfuric acid with the help of the soil's bacteria.
Low pH	Your soil is too acidic. Add lime or wood ashes.
Low nitrogen	Add manure, horn or hoof meal, cottonseed meal, fish meal, or dried blood.
High nitrogen	Your soil may be over-fertilized. Water the soil frequently and don't add any fertilizer.
Low phosphorus	Add cottonseed meal, bonemeal, fish meal, rock phosphate, dried blood, wood ashes.
High phosphorous	Your soil may be over-fertilized. Avoid adding phosphorous-rich materials and grow lots of plants to use up the excess.
Low potassium	Add potash, wood ashes, manure, dried seaweed, fish meal, cottonseed meal.
High potassium	Continue to fertilize with nitrogen and phosphorous-rich soil additions, but avoid potassium-rich fertilizers for at least two years.
Poor drainage or too much drainage	If your soil is a heavy, clay-like consistency, it won't drain well. If it's too sandy, it won't absorb nutrients as it should. Mix in peat moss or compost to achieve a better texture.

If using a liquid fertilizer, apply directly to or around the base of each plant and try to contain it within the garden only.

In order to decrease the potential for pollution and to gain the greatest benefits from fertilizer, whether it's a commercial variety, compost, or other organic materials, apply it when the plants have the greatest need for the nutrients. Plants that are not actively growing do not have a high requirement for nutrients; thus, nutrients applied to dormant plants, or plants growing slowly due to cool temperatures, are more likely to be wasted. While light applications of nitrogen may be recommended for lawns in the fall, generally, nitrogen fertilizers should not be applied to most plants in the fall in regions of the country that experience cold winters. Since nitrogen encourages vegetative growth, if it is applied in the fall it may reduce the plant's ability to harden properly for winter.

In some gardens, you can reduce fertilizer use by applying it around the individual plants rather than broadcasting it across the entire garden. Much of the phosphorus in fertilizer becomes unavailable to the plants once spread on the soil. For better plant uptake, apply the fertilizer in a band near the plant. Do not apply directly to the plant or in contact with the roots, as it may burn and damage the plant and its root system.

TIP

A Cheap Way to Fertilize

If you are looking to save money while still providing your lawn and garden with extra nutrients, you can do so by simply mowing your lawn on a regular basis and leaving the grass clippings to decompose on the lawn, or spreading them around your garden to decompose into the soil. Annually, this will provide nutrients equivalent to one or two fertilizer applications and it is a completely organic means of boosting a soil's nutrient content.

Fertilizing Tips

It is easiest to apply fertilizer before or at the time of planting. Fertilizers can either be spread over a large area or confined to garden rows, depending on the condition of your soil and the types of plants you will be growing. After spreading, till the fertilizer into the soil about 3 to 4 inches deep. Only spread about one half of the fertilizer this way and then dispatch the rest 3 inches to the sides of each row and also a little below each seed or established plant. This method, minus the spreader, is used when applying fertilizer to specific rows or plants by hand.

How to Properly Apply Fertilizer to Your Garden

- Apply fertilizer when the soil is moist, and then water lightly. This will help the fertilizer move into the root zone where its nutrients are available to the plants, rather than staying on top of the soil where it can be blown or washed away.

- Watch the weather. Avoid applying fertilizer immediately before a heavy rain system is predicted to arrive. Too much rain (or sprinkler water) will take the nutrients away from the lawn's root zone and could move the fertilizer into another water system, contaminating it.

- Use the minimum amount of fertilizer necessary and apply it in small, frequent applications. An application of two pounds of fertilizer, five times per year, is better than five pounds of fertilizer twice a year.

- If you are spreading the fertilizer by hand in your garden, wear gardening gloves and be sure not to damage the plant or roots around which you are fertilizing.

For information on composting, see page "Composting in Your Backyard," page 614.

TERRACING

Terraces can create several mini-gardens in your backyard. On steep slopes, terracing can make planting a garden possible. Terraces also prevent erosion by shortening a long slope into a series of shorter, more level steps. This allows heavy rains to soak into the soil rather than to run off and cause erosion and poor plant growth.

Materials Needed for Terraces

Numerous materials are available for building terraces. Treated wood is often used in terrace building and has several advantages: it is easy to work with, it blends well with plants and the surrounding environment, and it is often less expensive than other materials. There are many types of treated wood available for terracing—railroad ties and landscaping timbers are just two examples. These materials will last for years, which is crucial if you are hoping to keep your terraced garden intact for quite a while. There has been some concern about using these treated materials around plants,

but studies by Texas A&M University and the Southwest Research Institute concluded that these materials are not harmful to gardens or people when used as recommended.

Other materials for terraces include bricks, rocks, concrete blocks, and similar masonry materials. Some masonry materials are made specifically for walls and terraces and can be more easily installed by a homeowner than other materials. These include fieldstone and brick. One drawback is that most stone or masonry products tend to be more expensive than wood, so if you are looking to save money, treated wood will make a sufficient terrace wall.

How High Should the Terrace Walls Be?

The steepness of the slope on which you wish to garden often dictates the appropriate height of the terrace wall. Make the terraces in your yard high enough so the land between them is fairly level. Be sure the terrace material is strong enough and anchored well to stay in place through freezing and thawing, and during heavy rainstorms. Do not underestimate the pressure of waterlogged soil behind a wall—it can be enormous and will cause improperly constructed walls to bulge or collapse.

Many communities have building codes for walls and terraces. Large projects will most likely need the expertise of a professional landscaper to make sure the walls can stand up to water pressure in the soil. Large terraces also need to be built with adequate drainage and to be tied back into the slope properly. Because of the expertise and equipment required to do this correctly, you will probably want to restrict terraces you build on your own to no more than a foot or two high.

Constructing a Terrace

The safest way to build a terrace is by using the cut and fill method. With this method, little soil is disturbed, giving you protection from erosion should a sudden storm occur while the work is in progress. This method will also require little, if any, additional soil. Here are the steps needed to build your own terrace:

1. Contact your utility companies to identify the location of any buried utility lines and pipes before starting to dig.

2. Determine the rise and run of your slope. The rise is the vertical distance from the bottom of the slope to the top. The run is the horizontal distance between the top and the bottom. This will allow you to determine how many terraces you will need. For example, if your run is 20 feet and the rise is 8 feet, and you want each bed to be 5 feet wide, you will need four beds. The rise of each bed will be 2 feet.

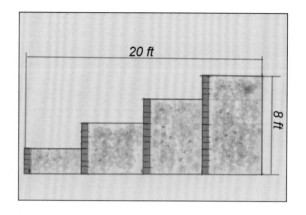

3. Start building the beds at the bottom of your slope. You will need to dig a trench in which to place your first tier. The depth and width of the trench will vary depending on how tall the terrace will be and the specific building materials you are using. Follow the manufacturer's instructions carefully when using masonry products, as many of these have limits on the number of tiers or the

height that can safely be built. If you are using landscape timbers and your terrace is low (less than 2 feet), you only need to bury the timber to about half its thickness or less. The width of the trench should be slightly wider than your timber. Make sure the bottom of the trench is firmly packed and completely level, and then place your timbers into the trench.

4. For the sides of your terrace, dig a trench into the slope. The bottom of this trench must be level with the bottom of the first trench. When the depth of the trench is one inch greater than the thickness of your timber, you have reached the back of the terrace and can stop digging.

5. Cut a piece of timber to the correct length and place it into the trench.

6. Drill holes through your timbers and pound long spikes, or pipes, through the holes and into the ground. A minimum of 18 inches of pipe length is recommended, and longer pipes may be needed in higher terraces for added stability.

7. Place the next tier of timbers on top of the first, overlapping the corners and joints. Pound a spike through both tiers to fuse them together.

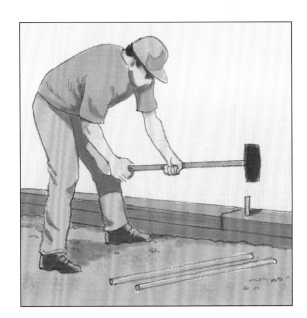

8. Move the soil from the back of the bed to the front of the bed until the surface is level. Add another tier as needed.

9. Repeat, starting with step 2, to create the remaining terraces. In continuously connected terrace systems, the first timber of the second tier will also be the back wall of your first terrace. The back wall of the last bed will be level with its front wall.

10. When finished, you can start to plant and mulch your terraced garden.

Other Ways to Make Use of Slopes in Your Yard

If terraces are beyond the limits of your time or money, you may want to consider other options for backyard slopes. If you have a slope that is hard to mow, consider using groundcovers on the slope rather than grass. There are many plants adapted to a wide range of light and moisture conditions that require little care (and do not need mowing) and provide soil erosion protection. These include:

- Juniper
- Wintercreeper
- Periwinkle
- Cotoneaster
- Potentilla
- Heathers and heaths

Strip-cropping is another way to deal with long slopes in your yard. Rather than terracing to make garden beds level, plant perennial beds and strips of grass across the slope. Once established, many perennials are effective in reducing erosion. Adding mulch also helps reduce erosion. If erosion does occur, it will be basically limited to the gardened area. The grass strips will act as filters to catch much of the soil that may run off the beds. Grass strips should be wide enough to mow easily, as well as wide enough to reduce erosion effectively.

Heather grows beautifully on hills and can help prevent erosion.

TREES FOR SHADE OR SHELTER

T rees in your yard can become home to many different types of wildlife. Trees also reduce your cooling costs by providing shade, help clean the air, add beauty and color, provide shelter from the wind and the sun, and add value to your home.

Choosing a Tree

Choosing a tree should be a well thought-out decision. Tree planting can be a significant investment, both in money and time. Selecting the proper tree for your yard can provide you with years of enjoyment, as well as significantly increasing the value of your property. However, a tree that is inappropriate for your property can be a constant maintenance problem, or even a danger to your and others' safety. Before you decide to purchase a tree, take advantage of the many references on gardening at local libraries, universities, arboretums, native plant and gardening clubs, and nurseries. Some questions to consider in selecting a tree include:

1. **What purpose will this tree serve?**

Trees can serve numerous landscape functions, including beautification, screening of sights and sounds, shade and energy conservation, and wildlife habitat.

2. **Is the species appropriate for your area?**

Reliable nurseries will not sell plants that are not suitable for your area. However, some mass marketers have trees and shrubs that are not fitted for the environment in which they are sold. Even if a tree is hardy, it may not flower consistently from year to year if the environmental factors are not conducive for it to do so. If you are buying a tree for its spring flowers and fall fruits, consider climate when deciding which species of tree to plant.

* Be aware of microclimates. Microclimates are localized areas where weather conditions may vary from the norm. A very sheltered yard may support vegetation not normally adapted to the region. On the other hand, a north-facing slope may be significantly cooler or windier than surround-

ing areas, and survival of normally adapted plants may be limited.

* Select trees native to your area. These trees will be more tolerant of local weather and soil conditions, will enhance natural biodiversity in your neighborhood, and be more beneficial to wildlife than many non-native trees. Avoid exotic trees that can invade other areas, crowd out native plants, and harm natural ecosystems.

3.   How big will it get?

When planting a small tree, it is often difficult to imagine that in 20 years it will most likely be shading your entire yard. Unfortunately, many trees are planted and later removed when the tree grows beyond the dimensions of the property.

4.   What is the average life expectancy of the tree?

Some trees can live for hundreds of years. Others are considered "short-lived" and may live for only 20 or 30 years. Many short-lived trees tend to be smaller, ornamental species. Short-lived species should not necessarily be ruled out when considering plantings, as they may have other desirable characteristics, such as size, shape, tolerance of shade, or fruit, that would be useful in the landscape. These species may also fill a void in a young landscape, and can be removed as other larger, longer-lived species mature.

5.   Does it have any particular ornamental value, such as leaf color or flowers and fruits?

Some species provide beautiful displays of color for short periods in the spring or fall. Other species may have foliage that is reddish or variegated and can add color in your yard year round. Trees bearing fruits or nuts can provide an excellent source of food for many species of wildlife.

6.   Does it have any particular insect, disease, or other problem that may reduce its usefulness in the future?

Certain insects and diseases can cause serious problems for some desirable species in certain regions. Depending on the pest, control of the problem may be difficult and the pest may significantly reduce the attractiveness, if not the life expectancy, of the tree. Other species, such as the silver maple,

are known to have weak wood that is susceptible to damage in ice storms or heavy winds. All these factors should be kept in mind, as controlling pests or dealing with tree limbs that have snapped in foul weather can be expensive and potentially damaging.

7.   How common is this species in your neighborhood or town?

Some species are over-planted. Increasing the natural diversity in your area will provide habitat for wildlife and help limit the opportunity for a single pest to destroy large numbers of trees.

8.   Is the tree evergreen or deciduous?

Evergreen trees will provide cover and shade year round. They may also be more effective as wind and noise barriers. On the other hand, deciduous trees will give you summer shade but allow the winter sun to shine in. If planting a deciduous tree, keep these heating and cooling factors in mind when placing the tree in your yard.

Placement of Trees

Proper placement of trees is critical for your enjoyment and for their long-term survival. Check with local authorities about regulations pertaining to placement of trees in your area. Some communities have ordinances restricting placement of trees within a specified distance from a street, sidewalk, streetlight, or other city utilities.

Before planting your tree, consider the tree's potential maximum size. Ask yourself these simple questions:

1.   When the tree nears maturity, will it be too close to your or a neighbor's house? An evergreen tree planted on your north side may block the winter sun from your next-door neighbor.
2.   Will it provide too much shade for your vegetable and flower gardens? Most vegetables and many flowers require considerable amounts of sun. If you intend to grow these plants in your yard, consider how the placement of trees will affect these gardens.

3. Will the tree obstruct any driveways or sidewalks?
4. Will it cause problems for buried or overhead power lines and utility pipes?

Once you have taken these questions into consideration and have bought the perfect tree for your yard, it is time to start digging!

Planting a Tree

Things You'll Need
- Tree
- Shovel
- Watering can or garden hose
- Measuring stick
- Mulch
- Optional: scissors or knife to cut the burlap or container, stakes, and supporting wires

A properly planted and maintained tree will grow faster and live longer than one that is incorrectly planted. Trees can be planted almost any time of the year, as long as the ground is not frozen. Late summer or early fall is the optimum time to plant trees in many areas. By planting during these times, the tree has a chance to establish new roots before winter arrives and the ground freezes. When spring comes, the tree is then ready to grow. Another feasible time for planting trees is late winter or early spring. Planting in hot summer weather should be avoided if possible as the heat may cause the young tree to wilt. Planting in frozen soil during the winter is very

difficult, and is tough on tree roots. When the tree is dormant and the ground is frozen, there is no opportunity for the new roots to begin growing.

Trees can be purchased as container-grown, balled and burlapped (B&B), or bare root. Generally, container-grown are the easiest to plant and to successfully establish in any season, including summer. With container-grown stock, the plant has been growing in a container for a period of time. When planting container-grown trees, little damage is done to the roots as the plant is transferred to the soil. Container-grown trees range in size from very small plants in gallon pots up to large trees in huge pots.

B&B trees are dug from a nursery, wrapped in burlap, and kept in the nursery for an additional period of time, giving the roots opportunity to regenerate. B&B plants can be quite large.

Bare root trees are usually extremely small plants. Because there is no soil around the roots, they must be planted when they are dormant to avoid drying out, and the roots must be kept moist until planted. Frequently, bare root trees are offered by seed and nursery mail order catalogs, or in the wholesale trade. Many state-operated nurseries and local conservation districts also sell bare root stock in bulk quantities for only a few cents per plant. Bare root plants are usually offered in the early spring and should be planted as soon as possible.

Be sure to carefully follow the planting instructions that come with your tree. If specific instructions are not available, here are some general tree-planting guidelines:

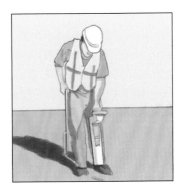

1. Before starting any digging, call your local utility companies to identify the location of any underground wires or lines. In the U.S., you can call 811 to have your utility lines marked for free.

2. Dig a hole twice as wide as, and slightly shallower than, the root ball. Roughen the sides and bottom of the hole with a pick or shovel so that the roots can easily penetrate the soil.

3. With a potted tree, gently remove the tree from the container. To do this, lay the tree on its side with the container end near the planting hole. Hit the bottom and sides of the container until the root ball is loosened. If roots are growing in a circular pattern around the root ball, slice through the roots on a couple of sides of the root ball. With trees wrapped in burlap, remove the string or wire that holds the burlap to the root crown; it is not necessary to remove the burlap completely. Plastic wraps must be completely removed. Gently separate circling roots on the root ball. Shorten exceptionally long roots and guide the shortened roots downward and outward. Root tips die quickly when exposed to light and air, so complete this step as quickly as possible.

4. Place the root ball in the hole. Leave the top of the root ball (where the roots end and the trunk begins) ½ to 1 inch above the surrounding soil, making sure not to cover it unless the roots are exposed. For bare root plants,

make a mound of soil in the middle of the hole and spread plant roots out evenly over the mound. Do not set the tree too deep into the hole.

5. As you add soil to fill in around the tree, lightly tap the soil to collapse air pockets, or add water to help settle the soil. Form a temporary water basin around the base of the tree to encourage water penetration, and be sure to water the tree thoroughly after planting. A tree with a dry root ball cannot absorb water; if the root ball is extremely dry, allow water to trickle into the soil by placing the hose at the trunk of the tree.

6. Place mulch around the tree. A circle of mulch, 3 feet in diameter, is common.

7. Depending on the size of the tree and the site conditions, staking the tree in place may be beneficial. Staking supports the tree until the roots are well established to properly anchor it. Staking should allow for some movement of the tree on windy days. After trees are established, remove all supporting wires. If these are not removed, they can girdle the tree, cut into the trunk, and eventually kill the tree.

Maintenance

For the first year or two, especially after a week or so of especially hot or dry weather, watch your tree closely for signs of moisture stress. If you see any leaf wilting or hard, caked soil, water the tree well and slowly enough to allow the water to soak in. This will encourage deep root growth. Keep the area under the tree mulched.

Some species of evergreen trees may need protection against winter sun and wind. A thorough watering in the fall before the ground freezes is recommended.

Fertilization is usually not needed for newly planted trees. Depending on the soil and growing conditions, fertilizer may be beneficial at a later time.

Young trees need protection against rodents, frost cracks, sunscald, lawn mowers, and weed whackers. In the winter months, mice and rabbits frequently girdle small trees by chewing away the bark at the snow level. Since the tissues that transport nutrients in the tree are located just under the bark, a girdled tree often dies in the spring when growth resumes. Weed whackers are also a common cause of girdling. In order to prevent girdling from occurring, use plastic guards, which are inexpensive and easy to control.

Frost cracking is caused by the sunny side of the tree expanding at a different rate than the colder, shaded side. This can cause large splits in the trunk. To prevent this, wrap young trees with paper tree wrap, starting from the base and wrapping up to the bottom branches. Sunscald can occur when a young tree is suddenly moved from a shady spot into direct sunlight. Light-colored tree wraps can be used to protect the trunk from sunscald.

Pruning

Usually, pruning is not needed on newly planted trees. As the tree grows, lower branches may be pruned to provide clearance above the ground, or to remove dead or damaged limbs or suckers that sprout from the trunk. Sometimes larger trees need pruning to allow more light to enter the canopy. Small branches can be removed easily with pruners. Large branches should be removed with a pruning saw. All cuts should be vertical. This will allow the tree to heal quickly without the use of any artificial sealants. Major pruning should be done in late winter or early spring. At this time, the tree is more likely to "bleed," as sap is rising through the plant. This is actually healthy and will help prevent invasion by many disease-carrying organisms.

Handy Household Hints

How to Prune Flowering Shrubs—Flowering shrubs may be pruned when their leaves fall off. Cut off all irregular and superfluous branches, and head down those that require it, forming them into handsome bushes, not permitting them to interfere with, nor overgrow other shrubs, nor injure lower growing plants near them. Put stakes to any of them that want support, and let the stakes be so covered with the shrug, that the stake may appear as little as possible.

Under no circumstance should trees be topped (topping is chopping off large top tree branches). Not only does this practice ruin the natural shape of the tree, but it increases its susceptibility to diseases and results in very narrow crotch angles (the angle between the trunk and the side branch). Narrow crotch angles are weaker than wide ones and more susceptible to damage from wind and ice. If a large tree requires major reduction in height or size, contact a professionally trained arborist.

Tree Grafting

Tree grafting is a method by which one plant is encouraged to fuse with another. It is most commonly used with trees and shrubs, to combine the trunk and roots of a hardier tree with branches that yield desirable fruit, flowers, or leaves. Grafting is far simpler than you think and requires two main things; some basic gardening skills and good, healthy tree parts.

When choosing a "stock"—the term used for the growing trunk part of the tree—pick a healthy specimen. Make sure the tree is not diseased. The trunk should be a near-wild variety, with little selective/cross breading.

The "scion," the part being grafted onto the stock, should also be from a healthy tree. Scions can be made with either summer or winter cuttings. For winter cuttings, simply bury them in a cool place after cutting and leave them until the spring.

There are several types of grafting: a whip graft involves a stock and a scion of about the same size, while a budding graft involves a single bud with its surrounding branch. No matter what method you are using, the goal of grafting is to bring together as closely as possible the "vascular cambium"- the inner bark. This is the part that will hopefully grow together to create a new tree.

Awl—This type of graft takes the least resources and the least time, however it should be done by a more experienced grafter. Use a flat-headed screwdriver to make a slit into the bark, making sure not to completely penetrate the vascular cambium. Then inset the wedged scion into the cut. Use only a bit of gardening tape or raffia to tie the graft on.

Whip grafting—Make a diagonal cut on the scion, sloping down from just below the base of a bud to the end of the stalk. The cut should be around two inches long. Near the top of the cut make another small one going upward toward the bud. The goal is to make a small tongue of wood. Prepare the stock in the same way, except making sure the cuts correspond to the scion, so that they will fit together.

Fit the scion onto the stock by slipping one tongue behind the other. Make sure that the two vascular cambium layers touch one another. Tie the two pieces together with gardening tape or raffia and cover it with grafting wax.

Cleft—This is best done in the spring and is useful for joining a thin scion of less than one inch in diameter to a thicker stock. It is best if the stock is one to three inches in diameter and has 3-5 buds. The stock should be split carefully down the middle to form a cleft about 1 inch deep. If it is a branch that is not vertical then the cleft should be cut horizontally. The end of the scion should be cut cleanly to a long shallow wedge, preferably with a single cut for each wedge surface, and not whittled. Attach the two and secure with gardening tape or raffia.

Budding—For this method use a healthy scion about one foot long. Put it in water and briefly soak. While the scion is soaking, prepare the stock. Cut a T-shaped slit on the stock and peel back the two flaps of bark from the long part of the T split. Remove your scion from the water and slice out an oval shaped piece of bark that contains a bud. Insert the oval into the T slice, under the bark flaps you loosened earlier. Make sure that the oval is secure inside the T slice, cutting off any excess bark from the oval to make it fit within the T slice. Put the bark flaps back on either side of the oval and bind everything together with gardening tape or raffia.

Cut off any stock that grows above the graft, to give the graft a chance to grab the nutrients it needs.

Stub—Stub grafting is a technique that requires less stock and retains the shape of a tree. Choose a scion with about 6-8 buds. Make a cut into the branch that is the stock of about one inch long, then wedge the scion into the branch. The scion should be angled no more than 35° to the parent tree so that the crotch remains strong. Cover the graft in grafting compound. After the graft has taken, the original stock branch should be removed and treated an inch above the graft, and be fully removed when the graft is strong.

Final Thoughts

Trees are natural windbreaks, slowing the wind and providing shelter and food for wildlife. Trees can help protect livestock, gardens, and larger crops. They also help prevent dust particles from adding to smog over urban areas. Tree plantings are key components of an effective conservation system and can provide your yard with beauty, shade, and rich natural resources.

VEGETABLE GARDENS

I f you want to start your own vegetable garden, just follow these simple steps and you'll be on your way to growing your own yummy vegetables—right in your own backyard!

How to Start a Vegetable Garden

1. Select a site for your garden and sketch a plan.

*   Vegetables grow best in well-drained, fertile soil (loamy soils are the best).
*   Some vegetables can cope with shady conditions, but most prefer a site with a good amount of sunshine—at least six hours a day of direct sunlight.
*   Refer to page 705 for tips on planning your garden layout.

2. Remove all weeds in your selected spot and dispose of them. If you are using compost to supplement your garden soil, do not put the weeds on the compost heap, as they may germinate once again and cause more weed growth among your vegetable plants.

3. Prepare the soil by tilling it. This will break up large soil clumps and allow you to see and remove pesky weed roots. This would also

be the appropriate time to add organic materials (such as compost) to the existing soil to help make it more fertile. The tools used for tilling will depend on the size of your garden. Some examples are:

- Shovel and turning fork—using these tools is hard work, requiring strong upper body strength.
- Rotary tiller—this will help cut up weed roots and mix the soil.

4. After the soil has been tilled you are ready to begin planting. If you would like straight rows in your garden, a guide can be made from two wooden stakes and a bit of rope.

5. Vegetables can be grown from seeds or transplanted:

- If your garden has problems with pests such as slugs, it's best to transplant older plants, as they are more likely to survive attacks from these organisms.
- Transplanting works well for vegetables like tomatoes and onions, which usually need a head start to mature within a shorter growing season. These can be germinated indoors on seed trays on a windowsill before the growing season begins.

6. Follow these basic steps to grow vegetables from seeds:

- Information on when and how deep to plant vegetable seeds is usually printed on seed packages

Label your garden rows as soon as you plant the seeds or seedlings so you'll remember what you planted where.

or on various websites. You can also contact your local nursery or garden center to inquire after this information.

- Measure the width of the seed to determine how deep it should be planted. Take the width and multiply by 2. That is how deep the seed should be placed in the hole. As a general rule, the larger the seed, the deeper it should be planted.

7. Water the plants and seeds well to insure a good start. Make sure they receive water at least every other day, especially if there is no rain in the forecast.

Things to Consider

- In the early days of a vegetable garden, all your plants are vulnerable to attack by insects and animals. It is best to plan multiples of the same plant in order to ensure that some survive. Placing netting and fences around your garden can help keep out certain animal pests. Coffee grains or slug traps filled with beer will also help protect your plants against insect pests.

- If sowing seed straight onto your bed, be sure to obtain a photograph of what your seedlings will look like so you don't mistake the growing plant for a weed.

- Weeding early on is very important to the overall success of your garden. Weeds steal water, nutrients, and light from your vegetables, which will stunt their growth and make it more difficult for them to thrive.

The Junior Homesteader

Starting a Mini Garden

You don't need a big field or even a backyard to grow some of your own food. You can grow some on a windowsill, balcony, porch, deck, or doorstep! Follow these steps to create your own mini garden:

Things You'll Need

- Container, such as milk carton, bleach jug, coffee can, ice cream tub, or ceramic pot
- Seeds
- Soil
- Plant fertilizer
- Tray or plate
- Water

1. Select seeds to plant. See "Seeds that grow well in containers" on page 748 for ideas of seeds to select.
2. Select a container. Match the container to the size of the plant. For example, tomatoes require a much bigger container than herbs. Rinse the container. Punch holes in the bottom, if there are none.
3. In a bucket, combine soil with water until the soil is damp. Fill your container with the damp soil to 1/2 inch from the top.
4. Read the seed packet to see how far apart and how deep to plant seeds. Cover seeds gently with soil.
5. Keep the seed bed watered well. The seeds need a lot of water, but don't add it all at once. Pour some on, let it sink in, and pour more on. Stop pouring when you see water coming out the bottom of the container. Keep a plate or tray under the plant container so the container will not leak. Keep the soil moist, but not sopping wet.
6. Place container(s) in a sunny location.
7. Once a week, add fertilizer following directions on the label.
8. Turn the containers often, so that sunlight reaches all sides of the growing plants.
9. As the plants grow larger, use scissors to trim the leaves of side-by-side plants, so they do not touch each other.

The Junior Homesteader

Root Vegetable "Magic"

Cut off the top 1 inch of a carrot, turnip, or beet. Put the top on a saucer, cut side down. Add just enough water to make the bottom of the vegetable top wet. Keep the saucer in a sunny window, add water every day so the bottom of the vegetable stays wet. Watch new leaves and roots grow!

Seeds that grow well in containers:

Tomatoes	Leaf lettuce
Peppers	Cucumbers
Radishes	Herbs

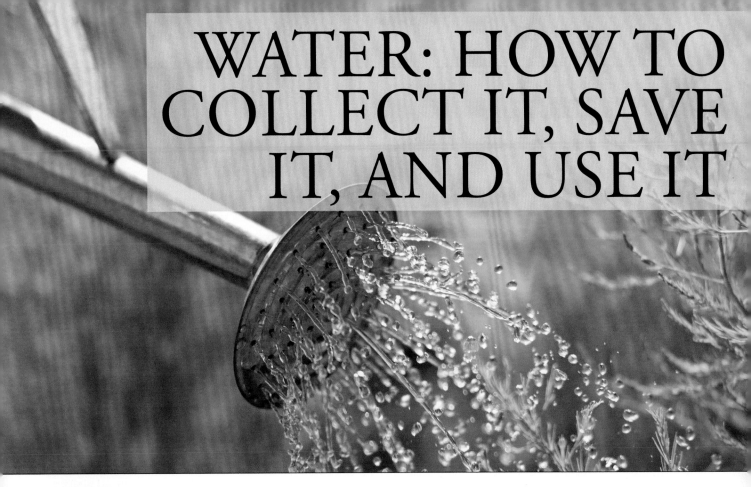

WATER: HOW TO COLLECT IT, SAVE IT, AND USE IT

Wise use of water for hydrating your garden and lawn not only helps protect the environment, but saves money and also provides optimum growing conditions for your plants. There are simple ways of reducing the amount of water used for irrigation, such as growing xeriphytic species (plants that are adapted to dry conditions), mulching, adding water-retaining organic matter to the soil, and installing windbreaks and fences to slow winds and reduce evapotranspiration.

You can conserve water by watering your plants and lawn in the early morning, before the sun is too intense. This helps reduce the amount of water lost due to evaporation. Furthermore, installing rain gutters and collecting water from downspouts—in collection bins such as rain barrels—also helps reduce water use.

How Plants Use Water

Water is a critical component of photosynthesis, the process by which plants manufacture their own food from carbon dioxide and water in the presence of light. Water is one of the many factors that can limit plant growth. Other important factors include nutrients, temperature, and amount and duration of sunlight.

Plants take in carbon dioxide through their stomata—microscopic openings on the undersides of the leaves. The stomata are also the place where water is lost, in a process called transpiration. Transpiration, along with evaporation from the soil's surface, accounts for most of the moisture lost from the soil and subsequently from the plants.

When there is a lack of water in the plant tissue, the stomata close to try to limit excessive water loss. If the tissues lose too much water, the plant will wilt. Plants adapted to dry conditions have developed numerous mechanisms for reducing water loss—they typically have narrow, hairy leaves and thick, fleshy stems and leaves. Pines, hemlocks, and junipers are also well adapted to survive extended periods of dry conditions—an environmental factor they encounter each winter when the frozen soil prevents the uptake of water. Cacti, which have thick stems and leaves reduced to spines, are the best example of plants well adapted to extremely dry environments.

The Junior Homesteader

See what happens when a plant (or part of a plant) doesn't get any light!

1. Cut 3 paper shapes about 2 inches by 2 inches. Circles and triangles work well, but you can experiment with other shapes, too. Clip them to the leaves of a plant, preferably one with large leaves. Either an indoor or an outdoor plant will do. Be very careful not to damage the plant.
2. Leave one paper cutout on for 1 day, a second on for 2 days, and a third on for a week. How long does it take for the plant to react? How long does it take for the plant to return to normal?

Photosynthesis means to "put together using light." Plants use sunlight to turn carbon dioxide and water into food. Plants need all of these to remain healthy. When the plant gets enough of these things, it produces a simple sugar, which it uses immediately or stores in a converted form of starch. We don't know exactly how this happens. But we do know that chlorophyll, the green substance in plants, helps it to occur.

Choosing Plants for Low Water Use

You are not limited to cacti, succulents, or narrow-leafed evergreens when selecting plants adapted to low water requirements. Many plants growing in humid environments are well adapted to low levels of soil moisture. Numerous plants found growing in coastal or mountainous regions have developed mechanisms for dealing with extremely sandy, excessively well-drained soils or rocky, cold soils in which moisture is limited for months at a time. Try alfalfa, aloe, artichokes, asparagus, blue hibiscus, chives, columbine, eucalyptus, garlic, germander, lamb's ear, lavender, ornamental grasses, prairie turnip, rosemary, sage, sedum, shrub roses, thyme, yarrow, yucca, and verbena.

Irrigation: Trickle Systems

Trickle irrigation and drip irrigation systems help reduce water use and successfully meet the needs of most plants. With these systems, very small amounts of water are supplied to the bases of the plants. Since the water is applied directly to the soil—rather than onto the plant—evaporation from the leaf surfaces is reduced, thus allowing more water to effectively reach the roots. In these types of systems, the water is not wasted by being spread all over the garden; rather, it is applied directly to the appropriate source.

Installing Irrigation Systems

An irrigation system can be easy to install, and there are many different products available for home irrigation systems. The simplest system consists of a soaker hose that is laid out around the plants and connected to an outdoor spigot. No installation is required, and the hose can be moved as needed to water the entire garden.

A slightly more sophisticated system is a slotted pipe system. Here are the steps needed in order to install this type of irrigation system in your garden:

1. Sketch the layout of your garden so you know what materials you will need. If you intend to water a vegetable garden, you may want one pipe next to every row or one pipe between every two rows.

2. Depending on the layout and type of garden, purchase the required lengths of pipe. You will need a length of solid pipe for the width of your garden, and perforated pipes that are the length of your lateral rows (and remember to buy one pipe for each row or two).

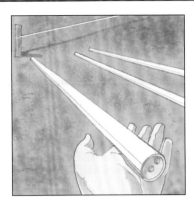

6. Cut the perforated pipe to the length of the rows.

3. Measure the distances between rows and cut the solid pipe to the proper lengths.

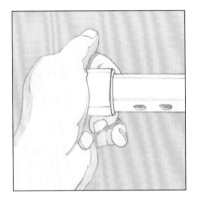

7. Attach the perforated pipes to the T-connectors so that the perforations are facing downward. Cap the ends of the pipes.

4. Place T-connectors between the pieces of solid pipe.

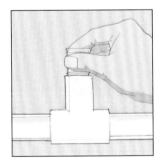

5. In the approximate center of the solid pipe, place a T-connector to which a hose connector will be fitted.

8. Connect a garden hose to the hose connector on the solid pipe. Adjust the pressure of the water flowing from the spigot until the water slowly emerges from each of the perforated pipes. And now you have a slotted pipe irrigation system for your garden.

The Junior Homesteader

Did you ever wonder how water gets from a plant's roots to its leaves? The name for this is "capillary action."

Things You'll Need

4 same-size stalks of fresh celery with leaves

4 cups or glasses

Red and blue food coloring

A measuring cup

4 paper towels

A vegetable peeler

A ruler

Some old newspapers

What to Do:

1. Lay the 4 pieces of celery in a row on a cutting board or counter so that the place where the stalks and the leaves meet matches up.

2. Cut all 4 stalks of celery 4 inches (about 10 centimeters) below where the stalks and leaves meet.

3. Put the 4 stalks in 4 separate cups of purple water (use 10 drops of red and 10 drops of blue food coloring for each half cup of water).

4. Label 4 paper towels in the following way: "2 hours," "4 hours," "6 hours," and "8 hours." (You may need newspapers under the towels).

5. Every 2 hours from the time you put the celery into the cups, remove 1 of the stalks and put it onto the correct towel. (Notice how long it takes for the leaves to start to change.)

6. Each time you remove a stalk from the water, carefully peel the rounded part with a vegetable peeler to see how far up the stalk the purple water has traveled.

7. What do you observe? Notice how fast the water climbs the celery. Does this change as time goes by? In what way?

8. Measure the distance the water has traveled and record this amount.

9. Make a list of other objects around your house or in nature that enable liquids to climb by capillary action. Look for paper towels, sponges, old sweat socks, brown paper bags, and flowers. What other items can you find?

Capillary action happens when water molecules are more attracted to the surface they travel along than to each other. In paper towels, the molecules move along tiny fibers. In plants, they move through narrow tubes that are actually called capillaries. Plants couldn't survive without capillaries because they use the water to make their food.

Rain Barrels

Another very efficient and easy way to conserve water—and save money—is to buy or make your own rain barrel. A rain barrel is a large bin that is placed beneath a downspout and that collects rainwater runoff from a roof. The water collected in the rain barrel can then be routed through a garden hose and used to water your garden and lawn.

Rain barrels can be purchased from specialty home and garden stores, but a simple rain barrel is also quite easy to make. Below are simple instructions on how to make your own rain barrel.

How to Make a Simple Rain Barrel

Instructions:

1. Obtain a suitable plastic barrel, a large plastic trash can with a lid, or a wooden barrel (e.g., a wine barrel) that has not been stored dry for too many seasons, since it can start to leak. Good places to find plastic barrels include suppliers of

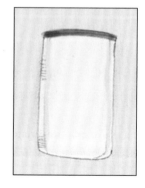

Things You'll Need

- A clean plastic barrel, tall trash can with lid, or a wooden barrel that does not leak—a 55-gallon plastic drum or barrel does a very good job at holding rainwater
- Two hose bibs (a valve with a fitting for a garden hose on one end and a flange with a short pipe sticking out of it at the other end)
- Garden hose
- Plywood and paint (if your barrel doesn't already have a top)
- Window screen
- Wood screws
- A drill
- A hacksaw
- A screwdriver

dairy products, metal plating companies, and bulk food suppliers. Just be sure that nothing toxic or harmful to plants and animals (including you!) was stored in the barrel. A wine barrel can be obtained through a winery. Barrels that allow less light to penetrate through will minimize the risk of algae growth and the establishment of other microorganisms.

2. Once you have your barrel, find a location for it under or near one of your home's downspouts. In order for the barrel to fit, you will probably need to shorten the downspout by a few feet. You can do this by removing the screws or rivets located at a joint of the downspout,

or by simply cutting off the last few feet with a hacksaw or other cutter. If your barrel will not be able to fit underneath the downspout, you can purchase a flexible downspout at your local home improvement store. These flexible tubes will direct the water from the downspout into the barrel. An alternative, and aesthetically appealing, option is to use a rain chain—a large, metal chain that water can run down.

3. Create a level, stable platform for your rain barrel to sit on by raking the dirt under the spout, adding gravel to smooth out lawn bumps, or using bricks or concrete blocks to make a low platform. Keep in mind that a barrel full of water is very heavy, so if you decide to build a platform, make sure it is sturdy enough to hold such heavy weight.

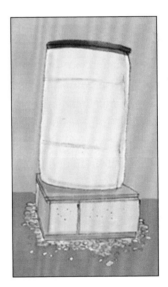

4. If your barrel has a solid top, you'll need to make a good-sized hole in it for the downspout to pour into. You can do this using a hole-cutting attachment on a power drill or by drilling a series of smaller holes close together and

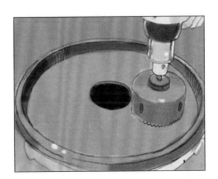

then cutting out the remaining material with a hacksaw blade or a scroll saw.

5. Mosquitoes are drawn to standing water, so to reduce the risk of breeding these insects, and to also keep debris from entering the barrel, fasten a piece of window screen to the underside of the top so it covers the entire hole.

6. Next, drill a hole so the hose bib you'll attach to the side of the barrel fits snugly. Place the hose bib as close to the bottom of the barrel as possible, so you'll be able to gain access to the maximum amount of water in the barrel. Attach the hose bib using screws driven into the barrel. You'll probably need to apply some caulking, plumber's putty, or silicon sealant around the joint between the barrel and the hose bib to prevent leaks, depending on the type of hardware you're using and how snugly it fits in the hole you drilled.

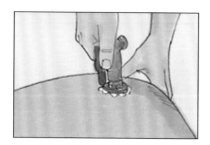

7. Attach a second hose bib to the side of the barrel near the top, to act as an overflow drain. Attach a short piece of garden hose to this hose bib and route it to a flowerbed, lawn, or another nearby area that won't be damaged by some running water if your barrel gets too full. (Or, if you want to have a second rain barrel for excess water, you can attach it to another hose bib on a second barrel.

If you are chaining multiple barrels together, one of them should have a hose attached to drain off the overflow).

8. Attach a garden hose to the lower hose bib and open the valve to allow collected rain water to flow to your plants. The lower bib can also be used to connect multiple rain barrels together for a larger water reservoir.

9. Consider using a drip irrigation system in conjunction with the rain barrels. Rain barrels don't achieve anything near the pressure of city water supplies, so you won't be able to use microsprinkler attachments, and you will need to use button attachments that are intended to deliver four times the amount of city-supplied water as you need.

10. Now wait for a heavy downpour and start enjoying your rain barrel!

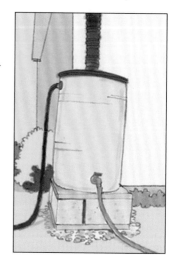

Things to Consider

- Spray some water in the barrel from a garden hose once everything is in place and any sealants have had time to thoroughly dry. The first good downpour is *not* the time to find out there's a leak in your barrel.

- If you don't own the property on which you are thinking of installing a rain barrel, be sure to get permission before altering the downspouts.

- If your barrel doesn't already have a solid top, cover it securely with a circle of painted plywood, an old trash-can lid screwed to the walls of the barrel, or a heavy tarp secured over the top of the barrel with bungee cords. This will protect children and small animals from falling into the barrel and drowning.

- As stated before, stagnant water is an excellent breeding ground for mosquitoes, so it would be a good idea to take additional steps to keep them out of your barrel by sealing all the openings into the barrel with caulk or putty. You might also consider adding enough non-toxic oil (such as vegetable cooking oil) to the barrel to form a film on top of the water that will prevent mosquito larvae from hatching.

Always double check to make sure the barrel you're using (particularly if it is from a food distribution center or other recycled source) did not contain pesticides, industrial chemicals, weed killers, or other toxins or biological materials that could be harmful to you, your plants, or the environment. If you are concerned about this, it is best to purchase a new barrel or trash can so there is no doubt about its safety.

A wooden barrel can also be used for a rain barrel.

PART EIGHT

WELL-BEING FOR BODY, MIND, AND SPIRIT

COMMUNITY

The challenges of a self-sufficient lifestyle can become overwhelming, especially if you are trying to do it alone, or even with just your family. Gardening, raising animals, preparing and preserving food, maintaining your energy-producing systems, and perhaps selling your produce or crafts all take a great deal of time and energy. Some people find that living together with a group of other like-minded individuals or families can ease these burdens and offer many additional benefits.

Intentional communities are groups of people who choose to live together because of common ideals, spirituality, or political views, or simply to accomplish a common purpose. Intentional communities differ in purpose, legal structure, location, financial resources, and demographics. Ecovillages in particular focus on creating self-sufficient, sustainable lifestyles, while other communities may exist to support families with mentally disabled children, or to intensify a particular spiritual walk. In some cases all the members live and work within the community, possibly even pooling all financial resources. In other cases, people may hold regular outside jobs and contribute to the community life in the evenings or on weekends. Remember, when considering location, to consider the zoning laws in that area and how they relate to your vision. Some places may have restrictions on the number of people who can live in a particular area, whether you can subdivide a property, or whether you can have a business there. Whether the intentional community shares a city apartment building or an organic farm, whether they're a group of artists or yogis, the goal is to create an environment where the common goal can more likely be achieved.

Many intentional communities fail, mainly due to financial constraints or interpersonal conflict. If you are interested in starting an intentional community, there are several things you should do to set a firm foundation for your endeavors.

1. **Create a clear vision.** Know specifically why you want to form a community and put it in writing. First, dream of what your "utopia" would be, and then outline clear, realistic goals to move toward that environment. No community is perfect, but you should at least know what perfection would look like for you, and have ideas

of how you can come closer to that vision. Giving your community a name can help to create a clearer identity and make it easier to explain to others.

2. **Determine how people can become members and what members' roles will be.** At first, this may be as simple as getting a few friends excited about the community and asking them to join. However, the most successful communities are generally those that take the time to consider what kind of people should become members, and how they'll be involved. You may want someone to keep records, someone to run meetings, and someone to oversee the chores. Or maybe everyone will share all the responsibilities. Be very clear about what sort of time, financial, or even emotional commitment is expected of members. An application process can be helpful in determining who becomes a member.

3. **Figure out finances.** You may need a substantial amount of money to buy land, rent an apartment, or otherwise prepare a location for your community. Determine how much your startup costs will be and where that money will come from. Then decide how members will contribute financially to the community (whether they'll pay monthly dues, or share a percentage of their income, etc). Some communities are almost entirely self-sustaining, but you'll likely at least need money for taxes or rent.

4. **Lay down the law.** It's important to determine early on how decisions within the community will be made—who can vote on group decisions, whether decisions must be unanimous or whether the majority rules, and so on. Generally, the more that people are involved in decisions that will affect the community, the less conflict and resentment there will be. Be sure to write these bylaws down.

5. **Be honest and open.** Don't try to hide anything from members, potential members, or those outside your intentional community. Doing so will only create problems, build tension, and threaten your vision. If there's a financial problem, share it with the other members. If your vision for the community begins to shift, talk about it. Being genuine and communicating freely with those within and around your community may be the biggest factor in your success.

- Even with the best planning and preparation, once your community is up and running, you are bound to run into interpersonal conflict. Personalities may clash, or practical problems such as a noisy dog, an unruly child, or a lazy worker will pop up. Expect problems to arise, and you won't feel so overwhelmed when they actually do. Here are a few tips for dealing with interpersonal conflict:

- When a disagreement between community members arises, determine whether the conflict is about the facts pertaining to a situation, or how they feel about the facts. If they are disagreeing about the facts, some research or questioning of other members may be helpful. If, however, they agree on the facts but are reacting differently to them, simply helping them to communicate their feelings adequately may be enough to settle the dispute.

- When a member feels very strongly about an issue, ask that person why she feels the way she does. Understanding each member's underlying values, beliefs, and assumptions will help to create harmony within the group.

- Play devil's advocate. Agree to argue the opposite side of a dispute for 15 minutes, and see if you discover some validity in the other person's argument.

- Stop discussions before they become explosions. If an argument is getting too heated, call a time out. Return to the discussion after 15 minutes, or once emotions have settled.

- Practice active listening. Listen to what someone has said, paraphrase it back to them, and ask relevant questions. This will facilitate clearer and deeper communication.

- Understand that everyone has a different way of perceiving and navigating the world. Another's worldview may seem threatening at first, but if you take the time to understand it, you may find your own worldview sharpened.

- Know when to let go. Some conflicts need immediate resolution, and some do not. In some cases, it's best to give the issue a rest and revisit it in a week or a month. By then circumstances or perceptions may have shifted.

COSMETICS

Homemade Lip Gloss

You only need a few ingredients to make your own lip gloss, though once you understand the basic recipe you can begin to experiment by adding different essential oils, aloes, and food products to create your own, unique type of gloss.

Homemade lip gloss containers can be any small glass jar or tin, or you can reuse an old lip gloss container (just make sure all the old gloss is out of the container). To sterilize the container, wash with soap and hot water, dunk the container in a jar of rubbing alcohol, rinse clean, and then allow the container to completely dry before pouring in your melted gloss. Allow the gloss mixture to cool completely before using (you can speed up this process by placing the container of gloss into the refrigerator for a few hours).

Honey Lip Gloss

Ingredients

1 tsp beeswax (you can find this at a craft store or at your local farmers' market)
½ tsp honey
2 tsp almond oil (optional)
Put ingredie Vitamin E oil from a capsule (optional)

Directions

1. Melt the beeswax and honey in a heat-proof jar in the microwave or use a double boiler method.
2. When the wax and honey are just melted, remove from the heat source and whisk in the almond oil and vitamin E oil, if you so desire. To remove the vitamin E oil from the capsule, simply prick the end of the capsule with a safety pin and squeeze it out.

Ingredients

  Petroleum jelly
  Blush, mineral eye shadow with shimmer, lipstick (only use one or two of these for your balm)
  Essential oil for flavoring (optional)

Directions

1. Mix together the petroleum jelly and either the blush (add a little at a time until the desired color is attained), eye shadow, or the last remnants of any lipstick. Add essential oil and mix thoroughly.
2. Scoop the mixture into containers and put in the refrigerator to harden.

Note: You can also experiment by melting the jelly with some beeswax and then adding in the leftover makeup. The possibilities are endless.

Homemade Bath Products

Lavender Bath Salt

Pour several tablespoons of this into your bath as it fills for an extra-soothing, relaxing, and cleansing experience. You can also add powdered milk or finely ground old-fashioned oatmeal to make your skin especially soft. Toss in a few lavender buds if you have them.

Ingredients

  2 cups coarse sea salt
  ½ cup Epsom salts
  ½ cup baking soda
  4 to 6 drops lavender essential oil
  Red and blue food coloring, if desired (use more red than blue to achieve a lavender color)

Mix all ingredients thoroughly and store in a glass jar or other airtight container.

3. Pour the mixture into the containers and allow to cool fully before using.

Note: If you want to add a citrus flavoring to this lip gloss, you can add a few drops of lemon or lime essential oil during the whisking stage.

"Make-up" Lip Balm

If you have leftover make-up (such as blush, lipstick, or shimmering eye shadow), don't let it go to waste. You can use it in this "recycled" lip balm.

Citrus Scrub

Use this invigorating scrub to wake up your senses in the morning. The vitamin C in oranges serves as an astringent, making it especially good for oily skin.

Ingredients

½ orange or grapefruit
3 tbsp cornmeal
2 tbsp Epsom salts or coarse sea salt

Squeeze citrus juice and pulp into a bowl and add cornmeal and salts to form a paste. Rub gently over entire body and then rinse.

Healing Bath Soak

This bath soak will relax tired muscles, help to calm nerves, and leave skin soft and fragrant. You may also wish to add blackberry, raspberry, or violet leaves. Dried or fresh herbs can be used.

2 tbsp comfrey leaves
1 tbsp lavender
1 tbsp evening primrose flowers
1 tsp orange peel, thinly sliced or grated
2 tbsp oatmeal

Combine herbs and tie up in a small muslin or cheesecloth sack. Leave under faucet as the tub fills with hot water. If desired, empty herbs into the bath water once the tub is full.

Rosemary Peppermint Foot Scrub

Use this foot rub to remove calluses, soften skin, and leave your feet feeling and smelling wonderful.

Ingredients

1 cup coarse sea salt
¼ cup sweet almond or olive oil
2 to 3 drops peppermint essential oil
1 to 2 drops rosemary essential oil
2 sprigs fresh rosemary, crushed, or ½ tsp dried rosemary

Combine all ingredients and massage into feet and ankles. Rinse with warm water and follow with a moisturizer.

Minty Cucumber Facial Mask

1 tbsp powdered milk

1 tsp plain yogurt (whole milk yogurt is best)

1 tsp honey

1 tsp fresh mint leaves

½ cucumber, peeled

Blend ingredients thoroughly, using a food processor or blender if available. Apply to face, avoiding eyes. Leave on for 10 to 15 minutes, then rinse.

After-Sun Comfrey Lotion

Comfrey root soothes skin and minimizes inflammation. Apply this lotion to sunburned skin for immediate relief and faster healing.

Ingredients

3 tbsp fresh comfrey root

1 cup water

1 tbsp beeswax, unrefined

¾ cup sweet almond oil or light cooking oil

¼ cup cocoa butter

4 vitamin E capsules

¼ cup aloe vera gel

1 tsp borax powder

12 to 16 drops essential oil (peppermint, lavender, or sandalwood are all good choices)

Directions

1. Place the comfrey root and water in a small pot and bring to a boil, simmering for about 30 minutes. Strain, retaining the water. Discard the root.

2. In a double boiler, combine beeswax, oil, and cocoa butter, stirring over low heat until melted. Remove from heat. Pierce the vitamin E capsules and add the oil from inside, stirring to combine.

3. In a separate saucepan, combine the comfrey water, aloe vera gel, and borax powder, stirring over low heat until the borax is fully dissolved. Allow to cool.

4. Once both mixtures are cooled to room temperature, pour the beeswax and oil mixture in a thin stream into the comfrey water mixture, whisking vigorously to combine (or use a food processor). Add the essential oils and continue mixing until thoroughly combined.

5. Cover and store in a cool, dark place.

Shampoo

Cleaning your hair can be as simple as making a baking soda and water paste, scrubbing it into your hair, and rinsing well. However, if you enjoy the feel of a sudsy, soapy, scented shampoo, try this recipe. You can substitute homemade soap flakes (see soap recipe on page 501) for the castile soap, if desired.

Ingredients

4 ounces liquid castile soap

3 tbsp fresh or dried herbs of your choice, boiled for 30 minutes in 2 cups water and strained

Pour the soap and herbal water into a jar, cover, and shake until well combined.

Handy Household Hints

Growth of Hair Increased and Baldness Prevented—Take four ounces of castor oil, eight ounces good Jamaica rum, thirty drops oil of lavender, or ten drops oil of rose, anoint occasionally the head, shaking well the bottle previously.

Hair Conditioner

This conditioner will add softness and volume to your hair. Avocado, bananas, and egg yolks are also great hair conditioners. Apply conditioner, allow to sit in hair a minimum of five minutes (longer for a deeper conditioning), and then rinse well. You may wish to shampoo a second time after using this conditioner.

Ingredients

1 cup olive oil

1 tsp lemon juice

1 tsp cider vinegar

2 tsp honey

6 to 10 drops essential oils, if desired

Whisk all ingredients together or blend in a food processor. Store in an airtight container.

Herbs for Your Hair

Herbs for dry hair	Burdock root, comfrey, elderflowers, lavender, marshmallow, parsley, sage, stinging nettle
Herbs for oily hair	Calendula, horsetail, lemon juice, lemon balm, mints, rosemary, witch hazel, yarrow
Herbs to combat dandruff	Burdock root, garlic, onion, parsley, rosemary, stinging nettle, thyme
Herbs for body and luster	Calendula, catnip, horsetail, licorice, lime flowers, nasturtium, parsley, rosemary, sage, stinging nettles, watercress
Herbs for shine	Horsetail, parsley, nettles, rosemary, sage, calendula
Herbs for hair growth	Aloe, arnica, birch, burdock, catmint, chamomile, horsetail, licorice, marigold, nettles, parsley, rosemary, sage, stinging nettle
Herbs for coloring	Brown: henna (reddish brown), walnut hulls, sage Blonde: calendula, chamomile, lemon, saffron, turmeric, rhubarb root

Fruits and Vegetables for Your Skin

These fruits and vegetables can be applied directly to your face or blended together to make a mask. Leave on skin for 20 to 30 minutes and then rinse thoroughly with clean water.

Handy Household Hints

To Make the Hair Grow Rich, Soft, Glossy, Etc—Take a half pint of alcohol, and castor oil quarter of a gill; mix, and flavor with bergamot, or whatever else may be agreeable. Apply it with the hand.

Beneficial for Oily Skin	Beneficial for Normal Skin	Beneficial for Dry Skin
Lemons, grapes, limes, strawberries, grapefruits, apples	Peaches, papayas, tomatoes, apricots, bananas, persimmons, bell peppers, cucumbers, kiwi, pumpkin, watermelons	Carrots, iceberg lettuce, honeydew melons, avocados, cantaloupes

Tropical Face Cleanser

The vitamin C in kiwi has enzymatic and cleansing properties, and the apricot oil serves as a moisturizer. The ground almonds act as an exfoliant to remove dead skin cells. Yogurt has cleansing and moisturizing properties.

1 kiwi

¾ cup avocado, banana, apricot, peach, strawberries, or papaya (or some of each)

2 tbsp plain yogurt (whole milk is best)

1 tbsp apricot oil (almond oil also works well)

1 tbsp honey

1 tsp finely ground almonds

Purée all ingredients together. Massage into face and neck and rinse thoroughly with cool water. Store excess in refrigerator for one to two days.

HERBAL REMEDIES

An herb is a plant or plant part used for its scent, flavor, or therapeutic properties. Herbs have been used in various forms for centuries for their health benefits. Many are now sold as tablets, capsules, powders, teas, extracts, and fresh or dried plants. However, some have side effects and may interact with other drugs you are taking.

To use an herbal product as safely as possible:

- Consult your doctor first.
- Do not take a bigger dose than the label recommends.
- Take it under the guidance of a trained medical professional.
- Be especially cautious if you are pregnant or nursing.

Herbal supplements are sold in many forms: as fresh or dried products; liquid or solid extracts; and tablets, capsules, powders, and tea bags. For example, fresh ginger root is often found in the produce section of food stores; dried ginger root is sold packaged in tea bags, capsules, or tablets; and liquid preparations made from ginger root are also sold. A particular group of chemicals or a single chemical may be isolated from a botanical and sold as a dietary supplement, usually in tablet or capsule form. Common preparations include teas, decoctions, tinctures, and extracts:

A *tea*, also known as an *infusion*, is made by adding boiling water to fresh or dried botanicals and steeping them. The tea may be drunk either hot or cold.

Some roots, bark, and berries require more forceful treatment to extract their desired ingredients. They are simmered in boiling water for longer periods than teas, making a *decoction*, which also may be drunk hot or cold.

A *tincture* is made by soaking a botanical in a solution of alcohol and water. Tinctures are sold as liquids and are used for concentrating and preserving a botanical. They are made in different strengths that are expressed as botanical-to-extract ratios (i.e., ratios of the weight of the dried botanical to the volume or weight of the finished product).

An *extract* is made by soaking the botanical in a liquid that removes specific types of chemicals. The liquid can be used as is or evaporated to make a dry extract for use in capsules or tablets.

Make Your Own Herbal Tincture

Tinctures help to concentrate and preserve the health benefits of your herbs. To use, mix 1 teaspoon of tincture with juice, tea, or water and drink no more than three times a day.

1. Pick the fresh herbs, removing any dirty, wilted, or damaged parts. Do not wash. Be sure you know whether it is the stems, leaves, roots, or flowers that have the health benefits, and use only those parts.
2. Coarsely chop the plant parts. Flowers can be left whole.
3. Clean and dry a small glass jar with an airtight lid and put the herbs inside. Fill the jar with 100-proof vodka or warm cider vinegar until plant parts are fully immersed. Screw the lid on securely and label the jar.
4. Store for 6 to 8 weeks, gently shaking a few times a week.
5. Strain out the herbs and store the liquid tincture in a clean, dry bottle. Be sure to label the jar with the ingredients and date and store it in a safe place away from children's reach.

Herbs for Your Ailments

Here is a list of common herbs that can be used to cure or alleviate the symptoms of conditions ranging from cancer to acne to the common cold. If you are taking any other medications or supplements, check with your doctor before trying any herbs. As with any medication, every body is unique and certain herbs can have adverse side effects for certain people, so pay attention to your body and cease taking any herbs that make you feel worse in any way. It's a good idea to try one herb at a time per condition and to keep a journal documenting what you take when and how you feel. This way you'll be able to tell more easily what effects the herbs are having.

Aloe Vera

Uses: The clear gel in aloe is used topically to treat osteoarthritis, burns, and sunburn. The green part can be made into a juice or dried and taken orally to treat a variety of conditions, such as diabetes, asthma, epilepsy, and osteoarthritis.

Aloe vera can be used topically to treat and soothe a variety of skin irritations.

Cautions: Using aloe vera on surgical wounds may inhibit their healing. If taken orally, aloe vera can produce abdominal cramps and diarrhea, which can decrease the absorption of many drugs.

If you have diabetes and take glucose-lowering medication, you should be careful of taking aloe orally, as studies suggest that aloe may decrease blood glucose levels.

Astragalus

Uses: Astragalus was traditionally used in Chinese medicine in combination with other herbs to help boost the immune system. It is still used widely in China for chronic hepatitis and as an additional cancer therapy. Astragalus is commonly used to boost the immune system to help colds and upper respiratory infections, and has also been used to fight heart disease. The astragalus plant root is used in soups, teas, extracts, and capsules and is generally used with other herbs, like ginseng, angelica, and licorice.

Cautions: Astragalus may interact with medications that suppress the immune system (such as those taken by cancer patients or organ transplant recipients).

Bilberry

Uses: Bilberry fruit is used to treat diarrhea, menstrual cramps, eye problems, varicose veins, and circulatory problems. The leaf of a bilberry is used to treat diabetes. It's claimed that bilberry fruit also helps improve night vision, but this is not clinically proven. The bilberry fruit can be eaten or made into an extract. Likewise, its leaves can be used in tea or made into an extract.

Cautions: Though bilberry fruit is considered safe, high doses of the leaf or leaf extract may have possible toxic side effects.

Chamomile

Uses: Chamomile has a calming effect and is often used to counteract sleeplessness and anxiety, as well as diarrhea and gastrointestinal conditions. Topically, chamomile is used in the treatment of skin conditions and for mouth ulcers (particularly due to cancer treatment). The chamomile plant has flowering tops, which are used to make tea, extracts, capsules, and tablets. It can also be applied as a skin cream or ointment or even be used as a mouth rinse.

Bilberries are a close relative of blueberries and can be eaten whole or made into an extract.

Chamomile flowers can be
used to make a relaxing tea.

Cautions: Some people have developed rare allergic reactions from eating or coming into contact with chamomile. These reactions include skin rashes, swelling of the throat, shortness of breath, and anaphylaxis. People allergic to related plants, such as daisies, ragweed, or marigolds, should be careful when coming into contact with chamomile.

Cranberry

Uses: Cranberry fruit and leaves are used in healing many conditions, including wounds, urinary disorders, diarrhea, diabetes, and stomach and liver problems. Cranberries are often utilized in treating urinary tract infections and stomach ulcers. They may also be useful in preventing dental plaque and in preventing *E. coli* bacteria from adhering to cells along the urinary tract wall. Cranberry fruit can be eaten straight, made into juice, or used in the form of extracts, tea, or tablets and taken as a dietary supplement.

Cautions: Drinking copious amounts of cranberry juice can cause an upset stomach and diarrhea.

Dandelion

Uses: Dandelions, throughout history, have been most commonly used to treat liver and kidney diseases and spleen problems. Dandelions are sometimes used in liver and kidney tonics, as a diuretic, and for simple digestive issues. The dandelion's leaves and roots (and sometimes the entire plant) are used in teas, capsules, and extracts. The leaves are used in salads or are cooked, and the flowers are used to make wine.

Cautions: While using dandelions is typically safe, there are a few instances of upset stomach and diarrhea caused by the plant, as well as allergic reactions. If your gallbladder is inflamed or infected, you should avoid using dandelion products.

Echinacea

Uses: Traditionally, echinacea has been used to boost the immune system to help prevent colds, flu, and various infections. Echinacea can also be used for wounds, acne, and boils.

The roots and exposed plant are used, either fresh or dried, for teas, juice, extracts, or in preparations for external use.

Cautions: Echinacea, taken orally, generally does not cause any problems. Some people do have allergic reactions (rashes, increased asthma, anaphylaxis), but typically only gastrointestinal problems were experienced. If you are allergic to any plants in the daisy family, it may be best to steer clear of echinacea.

Evening Primrose Oil

Uses: Since the 1930s, evening primrose oil has been used to fight eczema and recently, it has been used for other inflammatory conditions. Evening primrose oil is also used in the treatment of breast pain during the menstrual cycle, symptoms of menopause, and premenstrual issues. It may also relieve pain associated with rheumatoid arthritis.

The oil is extracted from the evening primrose seeds. You'll find it in capsule form at many health food stores.

Cautions: There may be some mild side effects, such as gastrointestinal upset or headache.

Flaxseed and Flaxseed Oil

Uses: Flaxseed is typically used as a laxative and to alleviate hot flashes. Flaxseed oil is used for treating arthritis pain. Both herbs are used to fight high cholesterol and can be beneficial for those with heart disease. Flaxseed, in either its whole or crushed form, may be mixed with water or juice and ingested. It is also available as a powder. Flaxseed oil can be taken in either a liquid or capsule form.

Cautions: It is essential to take flaxseed with lots of water, or constipation could worsen. Further, flaxseed fiber may decrease the body's ability to absorb other oral medications and so should not be taken together.

Garlic

Garlic grows best in soil that is pH 6.5 to 7.0.

Uses: Garlic is typically used as a dietary supplement for those with high cholesterol, heart disease, and high blood pressure. It may help decrease the hardening of the arteries

Echinacea is beautiful as well as useful medicinally. It grows well in moderately dry soil.

and is also used in the prevention of stomach and colon cancer. It is also used topically or orally to heal some infections, including ear infections. Garlic cloves may be eaten either raw or cooked or they may be dried or powdered and used in capsules. Oil and other extracts can be obtained from garlic cloves.

Cautions: Some common side effects of garlic are breath and body odor, heartburn, upset stomach, and allergic reactions. Garlic can also thin blood and so should not be used before surgeries or dental work, especially if you have a bleeding disorder. It also has an adverse effect on drugs used to fight HIV.

Ginkgo leaves

Ginger

Uses: Ginger is commonly used in Asian medicines to treat stomachaches, nausea, and diarrhea. Many U.S. dietary supplements containing ginger are used to help fight cold and flu and can be used to relieve post-surgery nausea or nausea related to pregnancy. It has also been used for arthritis and other joint and muscle pain. Ginger root can be found fresh or dried, in tablets, capsules, extracts, and teas.

Cautions: Side effects are rare but can include gas, bloating, heartburn, and, for some people, nausea.

Ginkgo

Uses: Traditionally, extract from ginkgo leaves has been used in the treatment of illnesses such as asthma, bronchitis, fatigue, and tinnitus. People use gingko leaf extract in the hopes that it will help improve their memory (especially in the treatment of Alzheimer's disease and dementia). It is also taken to treat sexual dysfunction, multiple sclerosis, and other health issues. Ginkgo leaf extracts are made into tablets, capsules, or teas. Sometimes the extracts can also be found in skin care products.

Cautions: Some common side effects are headache, nausea, gastrointestinal upset, diarrhea, dizziness, or skin irritations. Ginkgo may also increase bleeding risks, so those having surgery or with bleeding disorders should consult a doctor before using any ginkgo products. Uncooked ginkgo seeds are toxic and can cause seizures.

Ginseng (Asian)

Uses: Ginseng is used to help boost the immune system and contribute to the overall health of an individual. It has been used traditionally and currently for improving those who are recovering from illnesses, increasing stamina and mental and physical performance, treating erectile dysfunction and symptoms of menopause, and lowering blood glucose levels and blood pressure. In some studies, ginseng has been proven to lower blood glucose levels and boost immune systems. The ginseng root is dried and made into tablets, capsules, extracts, and teas. It can also be made into creams for external use.

Cautions: Limiting ginseng intake to three months at a time will most likely reduce any potential side effects. The most common side effects are headaches and sleep issues, along with some allergic reactions. If you have diabetes and are taking blood-sugar lowering medications, it is advisable not to use ginseng, as it too lowers blood sugar.

Grape Seed Extract

Uses: Grape seed extract is used for treating heart and blood vessel conditions, such as high blood pressure, high cholesterol, and low circulation. It is also used for those struggling with complications from diabetes, such as nerve and eye damage. Grape seed extract is also used in treating vision problems, reducing swelling after surgery, and cancer prevention. Extracted from grape seeds, it is readily available in tablets and capsules.

Cautions: Common side effects of prolonged grape seed oil use are headaches; dry, itchy scalp; dizziness; and nausea.

Green Tea

Uses: Green tea and its extracts have been used in preventing and treating breast, stomach, and skin cancers, as well as improving mental alertness, aiding weight loss, lowering cholesterol, and preventing the sun from damaging the skin. Green tea is typically brewed and drunk. Extracts can be taken in capsule form and sometimes green tea can be found in skin care products.

Cautions: While green tea is generally safe for most adults, there have been a few reports of liver problems occurring in those who take green tea extracts. Thus, these extracts should always be taken with food and should not be taken at all by those with liver disorders. Green tea also contains caffeine and can cause insomnia, anxiety, irritability, nausea, diarrhea, or frequent urination.

Lavender

Uses: Lavender, in the past, has been used as an antiseptic and to help with mental health issues. Now it is more commonly taken for anxiety, restlessness, insomnia, and

Lavender has a soothing, relaxing aroma. It can also be ingested in the form of tea or extracts, or even in baked goods.

depression, and can also be used to fight headaches, upset stomach, and hair loss.

Most commonly used in aromatherapy, lavender essential oil can also be diluted with other oils and rubbed on the skin. When dried, lavender flowers can be made into teas or liquid extracts and ingested.

Cautions: Lavender oil applied to the skin may cause some irritation and is poisonous if ingested. Lavender tea may cause headache, appetite change, and constipation. If used with sedatives, it may increase drowsiness.

Licorice Root

Uses: Traditionally, licorice root is used as a dietary supplement for the treatment of stomach ulcers, bronchitis, and sore throat. It is also used to help cure infections caused by viruses. When licorice root is peeled, it can be dried and made into powder. It is available in capsules, tablets, and extracts.

Cautions: If taken in large doses, licorice root can cause high blood pressure, water retention, and low potassium levels, leading to heart conditions. Taken with diuretics, it could cause the body's potassium levels to fall to dangerously low levels. If you have heart disease or high blood pressure, you should practice caution when taking licorice root. Large doses of licorice root may cause preterm labor in pregnant women.

Milk Thistle

Uses: Milk thistle is used as a protective measure for liver problems and in the treatment of liver cirrhosis, chronic hepatitis, and gallbladder diseases. It is also used to lower cholesterol, reduce insulin resistance in those with type 2 diabetes, and reduce the growth of cancerous cells in the breast, cervix, or prostate. Milk thistle seeds are used to make capsules, extracts, and strong teas.

Cautions: Occasionally, milk thistle may cause diarrhea, upset stomach, or bloating. It may also cause allergic reactions, especially in those with allergies to the daisy family.

Mistletoe

Uses: For hundreds of years, mistletoe has been used to treat seizures and headaches. In Europe, mistletoe is used to treat cancer and to boost the immune system. The shoots and berries of mistletoe are used in oral extracts. In Europe, these extracts are prescription drugs, available only by injection.

Milk thistle grows in a wide range of soil types and will thrive in sunny or partly shady areas.

Cautions: Eating raw and unprocessed mistletoe may cause vomiting, seizures, a slowing of the heart rate, and even death. American mistletoe cannot be used for medical purposes. Injected mistletoe extract can irritate the skin and produce low-grade fevers or flu-like symptoms. There is also a slight risk for severe allergic reactions that could cause difficulty breathing.

Peppermint Oil

Uses: Usually, peppermint oil is used to treat nausea, indigestion, and cold symptoms and it can also be used to allay headaches, muscle and nerve pain, and irritable bowel syndrome. Peppermint essential oil can be taken orally in small doses. It can also be diluted with other oils and applied to the skin.

Cautions: Common side effects include allergic reactions and heartburn, though peppermint oil is relatively safe in small doses.

Red Clover

Uses: Red clover has been used for treating cancer, whooping cough, asthma, and indigestion. It is also used to allay menopausal symptoms, breast pain, high cholesterol, osteoporosis, and enlarged prostate. The red clover flower is used in preparing extracts in tablets and capsules as well as teas.

Cautions: No serious side effects have been reported, though it is unclear if it is safe for use by pregnant women, women who are breastfeeding, or women with breast or other hormonal cancer. The estrogen in red clover may also increase a woman's chance of contracting cancer in the uterus.

Soy

Uses: Soy products are typically used for treating high cholesterol, menopausal symptoms, osteoporosis, problems with memory, breast and prostate cancer, and high blood pressure. Available in dietary supplements, soy can be found

St. John's Wort has been used for centuries as a mood enhancer and to ease pain.

in tablet or capsule form. Soybeans may be cooked and eaten, or made into tofu, soy milk, and other foods.

Cautions: Using soy supplements or eating soy products can create minor stomach and bowel problems, and in rare cases, allergic reactions causing breathing difficulties and rashes. While there is no conclusive evidence linking soy with increased risk of breast cancer, women who have or are at risk of getting breast cancer should consult a doctor about using soy products.

St. John's Wort

Uses: St. John's wort has been used for hundreds of years to treat mental illness and nerve pain. It has also been used as a sedative, in malaria treatment, and as a balm for wounds, burns, and insect bites. It is commonly used to treat depression, anxiety, and sleep disorders. The flowers are used, in extract form, for tea and capsules.

Cautions: A possible side effect of using St. John's wort is increased light sensitivity. Other common side effects are anxiety, dry mouth, dizziness, gastrointestinal symptoms, fatigue, headache, and sexual dysfunction. St. John's wort also interacts with drugs and may interfere with the way in which our body breaks down those drugs. It may affect antidepressants, birth control pills, cyclosporine, digoxin, indinavir and other HIV drugs, irinotecan and other cancer drugs, and anticoagulants.

If you are taking antidepressants, be careful if also taking St. John's wort, as it may increase the likelihood of nausea, anxiety, headache, and confusion.

Turmeric

Uses: Traditionally used in Chinese medicine, turmeric was supposed to aid digestion and liver function and to relieve arthritis pain. It was also taken to regulate the menstrual cycle. Applied directly to the skin, it was used to treat eczema and wounds. Now, turmeric is used in the treatment of heartburn, stomach ulcers, and gallstones. Turmeric is also used to reduce inflammation and in the prevention and treatment of certain cancers.

The underground stems of the turmeric plant are dried and taken orally in capsules, teas, or liquid extracts. It can also be made into a paste to be used on the skin.

Cautions: Considered safe for most adults, long-term use of turmeric may cause indigestion. Those with gallbladder problems should avoid turmeric, however, as it may worsen the condition.

Valerian

Uses: For many years, valerian has been used for sleep disorders and to treat anxiety. Further, valerian has been used to alleviate headaches, depression, irregular heartbeat, and trembling. The roots and underground stems of the valerian plant are usually made into supplements in capsule, tablet, or liquid extract form. It can also sometime be made into teas.

Cautions: Valerian is typically safe to use for short periods of time (no more than six weeks) but there is no proof about its long-term effectiveness. Some common side effects of valerian use are headaches, dizziness, upset stomach, and grogginess the morning after use.

Common valerian flowers

HERBAL TEAS

Herbal teas can be very tasty and deliver between 50 and 90 percent of the medicinal qualities of the herbs used. Teas you make yourself will be more potent and flavorful than those you can buy at the store, and much less expensive. Try experimenting with different herbal combinations, but be careful to avoid any plants you cannot confidently identify as edible, or any plants sprayed with pesticides. If using dried herbs, you can store your tea mixes in sealed containers for months. Be sure to label each container with the name of the tea.

Use 1 to 2 teaspoons of dried herbs per cup of hot water or 3 teaspoons of fresh herbs per pint of water. Steep the herbs for about 10 minutes and then strain. The following plants can all be safely used in teas:

Flowers

Alliums, bee balm, carnations, echinacea (roots and flowers), hibiscus, hollyhocks, honeysuckle (avoid the poisonous berries), lavender, marshmallow (use the roots), red clover, nasturtiums, roses (flowers or hips), violets.

Herbs

Basil, chamomile flowers, chives, dill, eucalyptus, ginger root, lemon balm, lemongrass, marjoram, mint, oregano, parsley, peppermint, linden leaves, mint, rosemary, sage, thyme, valerian root, verbena.

Bushes and Trees

Birch leaves, blackberry leaves, citrus blossoms, elderberry flowers, gardenia, pine needles, raspberry leaves.

Weeds

Chickweed, chicory, dandelions, goldenrod, stinging nettle.

Tea for the Common Cold

Combine the following herbs in any proportion you like. Boil for 10 minutes, strain, and add honey to taste.

Marshmallow root (eases body aches, reduces inflammation)

Peppermint (reduces congestion, eases headaches, soothes stomach)

Echinacea roots and flowers (boosts the immune system)

Thyme (reduces chest and nasal congestion, increases circulation)

Cinnamon (reduces inflammation and fights infection)

Rosehips, finely chopped (full of vitamin C, which boosts the immune system and energizes)

Ginger root, peeled and finely chopped (warms from the inside out)

Lavender, crushed (eases migraines)

Lemon peel, finely grated (full of vitamin C)

Calming Tea

Combine the following calming herbs, using about ¼ as much valerian as the other herbs (valerian can be very potent). Boil for 10 minutes, strain, and add honey to taste.

Lemon balm leaves
Chamomile flowers
Valerian root, crushed
Ginger root, peeled and finely chopped

Fertility Tea

Drink one cup of fertility tea a day to help balance your hormones and to get nutrients that can aid in becoming pregnant. Combine the herbs in equal proportion, boil for 10 minutes, strain, and add honey to taste.

Red clover blossoms (nourishes the uterus, promotes estrogen, rich in magnesium and calcium)

Nettle leaves (rich in calcium, potassium, phosphorous, iron, and sulfur)

Red raspberry leaves (aids the fertilized egg in attaching to the uterine lining, rich in minerals, helps to tone muscles in the pelvic region)

Peppermint (aids in absorption of red raspberry leaf nutrients)

Handy Household Hints

Sage is very useful both as a medicine, for the head-ache—when made into tea—and for all kinds of stuffing, when dried and rubbed into powder. It should be kept tight from the air.

Motherwort tea is very quieting to the nerves. Students, and people troubled with wakefulness, find it useful.

Thoroughwort is excellent for dyspepsy, and every disorder occasioned by indigestion. If the stomach be foul, it operates like a gentle emetic.

Sweet-balm tea is cooling when one is in a feverish state.

Catnip, particularly the blossoms, made into tea, is good to prevent a threatened fever. It produces a fine perspiration. It should be taken in bed, and the patient kept warm.

Lungwort, maiden-hair, hyssop, elecampane and hoarhound steeped together, is an almost certain cure for a cough. A wine-glass full to be taken when going to bed.

Cleansing Tea

The herbs in this tea will improve your digestion, help your body in its natural detoxification process, and give you more energy. Combine the herbs in any proportion (go easy on the cayenne), boil for 10 minutes, strain, and add honey if desired.

Peppermint leaves
Dandelion root
Whole allspice berries
Ginger root, peeled and finely chopped
Licorice root, crushed
Cayenne pepper

HOME SCHOOLING

"Home schooling" is the education of a child/ children at home, typically by parents but sometimes by tutors. It provides an alternative to the formal settings of a public or private school. Prior to the introduction of compulsory school attendance laws, most childhood education occurred within the family or community.

Home schooling is a legal option for parents to provide their children with an alternative learning environment. Some of the reasons parents may choose home schooling for their children are; better academic test results, the undesirable public school environment, religious reasons, the expense of private education, and objections over what their children could be taught in public school. Home schooling is also an alternative for families living in isolated rural locations or living temporarily abroad. Home schooling can be supplemented by using various community resources, such as public libraries, museums, parks, wildlife preserves, and athletic clubs.

Every state within the United States has laws regarding home schooling. Some states are more restrictive over conditions leading to home schooling and when parents are allowed to keep their children from formal schools. Some states require parents to teach approved curriculum or require the children to take standardized tests to prove they are receiving an adequate education. It is best to check state laws before making any decisions about home schooling.

Types of Home schooling

There are many different types of home schooling methods that parents can utilize in order to meet their children's educational needs. Many parents try several different methods and experiment before finding the right fit for their child/children.

"Unschooling" and "Radical Unschooling" are based on a curriculum-free philosophy of home schooling where the child is largely self-directed in choosing what and when to study. Parents facilitate the children's learning in subjects that interest them, but do not try to direct study. The expression was coined in 1977 by American educator and author John Holt. This type of schooling acknowledges that children can learn in many different ways, and one type of schooling might not work well for everyone.

The "Unit Study" method is an entire curriculum based around themes. Parents will choose a theme that needs to be studied and incorporate many different subjects, such as art, science, and history within the context of the theme. This method works well with home schooling multiple children of different grade levels, as different activities for each subject can still fit into the theme.

"All-in-One" is a method where parents get packets full of material, usually enough for an entire school year, and use them in a particular order. The packets include textbooks and writing materials. This is the method that most closely imitates the feel of a formal school environment at home and often closely follows the curriculum taught in public schools.

Charlotte Mason (1842-1923) was an English teacher who believed in a rounded liberal education for every child. Her method places a heavy focus on literature, building and reinforcing good habits, and instilling a love of learning. According to Mason, textbooks should be written on a single subject by someone who is both learned in, and passionate about their particular subject.

The "Classical" method is based on the teaching methods popular in the Western World throughout the early modern era. Classical home schooling teaches children in three stages; grammar, logic, and rhetoric. This style often includes a heavy focus on memorization and emphasizes logic and eloquent expression as a child matures.

The "Montessori" method is based on the works of Maria Montessori (1870-1952), an Italian doctor. She believed children have a natural desire to learn and are capable of teaching themselves. Children are typically given independence, freedom, and respect. The focus is on the process of learning, not the outcome; there are no exams and no grades are given.

A "Waldorf Education" is based on the educational philosophy of Rudolf Steiner and was first implemented at the school provided for the workers of the Waldorf-Astoria cigarette company in Germany. Steiner's philosophy states that there are three different stages of childhood development, and that each stage requires a different teaching approach. Early childhood learning is largely experiential, imitative, and sensory-based. The education emphasizes learning through activity. From age seven to fourteen, learning is artistic and imaginative and in these years, the approach emphasizes developing children's emotional life and artistic expression. During adolescence, to meet the developing capacity for abstract thought and conceptual judgment, the emphasis is on developing intellectual understanding and ethical ideals.

Like the "Waldorf" method the "Thomas Jefferson Education" method emphasizes three separate phases of childhood development and different teaching methods for each. The founders of the philosophy, Oliver and Rachael DeMille, created the "Seven Keys of Great Teaching" to direct educators; Classics, Mentors, Inspire, Structure Time, Quality, Simplicity, and You.

Home schooling can also refer to instruction in the home under the supervision of correspondence schools or umbrella schools. Online schools are becoming increasingly popular, as they can provide lesson plans, materials, and even online administered tests for students. There are many different institutions that provide material and assistance to parents for home schooling.

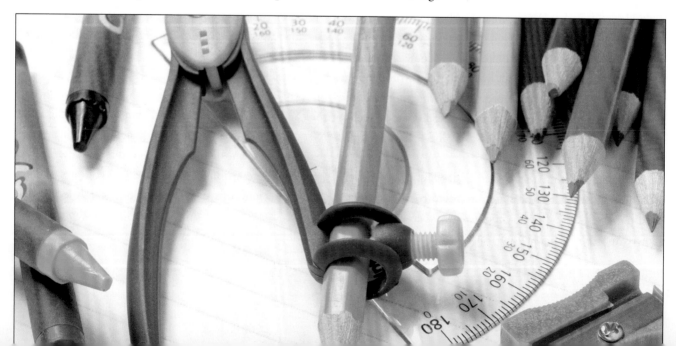

MASSAGE

Massage therapy (and, in general, the laying on of hands for health purposes) dates back thousands of years. References to massage have been found in ancient writings from many cultures, including those of Ancient Greece, Ancient Rome, Japan, China, Egypt, and the Indian subcontinent.

In the United States, massage therapy first became popular and was promoted for a variety of health purposes starting in the mid-1800s. In the 1930s and 1940s, however, massage fell out of favor, mostly because of scientific and technological advances in medical treatments. Massage has been gaining in popularity since the '70s, as more and more people recognize its ability to rehabilitate sports injuries, reduce stress, increase relaxation, address feelings of anxiety and depression, and aid general wellness.

There are more than 80 types of massage therapy. In all of them, therapists press, rub, and otherwise manipulate the muscles and other soft tissues of the body, often varying pressure and movement. They most often use their hands and fingers, but may use their forearms, elbows, or feet. Typi-cally, the intent is to relax the soft tissues, increase delivery of blood and oxygen to the massaged areas, warm them, and decrease pain.

A few popular forms of massage are:

Aromatherapy massage: This is similar to Swedish massage but incorporates strong-scented plant oils that contribute to a sense of relaxation and well-being. Aromatherapy has helped to cure a range of conditions, even including acne and whooping cough.

Deep tissue massage: The therapist uses patterns of strokes and deep finger pressure on parts of the body where muscles are tight or knotted, focusing on layers of muscle deep under the skin.

Shiatsu massage: The therapist applies varying, rhythmic pressure from the fingers on parts of the body that are believed to be important for the flow of a vital energy called *qi* (pronounced "chee"). In traditional Chinese medicine, *qi* is the vital energy or life force proposed to regulate a person's spiritual, emotional, mental, and physical health and to be influenced by the opposing forces of yin and yang.

Swedish massage: The therapist uses long strokes, kneading, and friction on the muscles and moves the joints to aid flexibility.

Trigger point massage: (also called pressure point massage): The therapist uses a variety of strokes but applies deeper, more focused pressure on myofascial trigger points. These "knots" that can form in the muscles are painful when pressed, and cause symptoms elsewhere in the body as well.

Massage is particularly powerful because it works on the physical, psychological, and often emotional and spiritual levels. When done with love, care, and skill, it soothes and stimulates the skin, muscles, organs, mind, emotions, and spirit, all of which are connected more than we often consider. We

can observe this in some obvious ways; if you are stressed, for example, your shoulders and neck tense up and the muscles can constrict and form knots. Likewise, when you have a bad headache, you're more likely to be irritable or emotional, which can then lead to stomach upset. Massage, when it is most effective, helps to heal the body, emotions, and spirit.

Preparation

A massage can be as simple as squeezing the top of your own shoulder while you stand in line at the grocery store or gently rubbing your child's back in a circular motion to soothe her to sleep. However, if you wish to give a friend or loved one a more complete massage, you may want to do a little setup.

Choose a draft-free, quiet room where there will be few distractions. Be sure it is warm enough that the person receiving the massage will not get cold while lying still for a length of time. Lighting should be soft and you may wish to have soothing music playing in the background. The surface where the recipient will be lying should be comfortable but firm; a futon covered with a clean sheet or several blankets layered on the floor will work.

If you wish to use massage oils or lotion to ease the friction on the skin, have these close at hand. You may also wish to light scented candles or burn incense, depending on the recipient's needs and desires. Finally, be sure your hands are clean before beginning a massage, and tie back your hair if it's likely to get in your way.

Basic Strokes and Principles

Always make the massage recipient's needs your priority. Pay attention to the recipient's reactions to your strokes and ask occasionally whether he or she would prefer a lighter or firmer touch, or if what you're doing feels good. It is normal to experience slight pain or discomfort when pressure is applied to tense muscles, but if the recipient experiences sharp jolts of pain, nausea, or ripping or tearing sensations, ease off the pressure or move your touch to a different location. The

recipient should always have the final say as to what is most beneficial in his or her massage.

For a basic full-body massage, the recipient should begin lying on her stomach with elbows at a right angle and hands on either side of the head. A small pillow can be placed under the chest or head if desired.

Do not massage areas where there is bruised or broken skin, infected areas, unusual swelling or inflammation, or where there are varicose veins present. Do not massage anyone who has a fever or high temperature.

Do massage toward the heart to improve blood circulation and lymph return, stay focused throughout the massage, respect the recipient's needs and desires, and start with a light touch, gradually using the leverage of your body weight to increase pressure as the recipient desires.

Long, gentle, flowing strokes using consistent light pressure on the full surface of the hands should be used to begin and end a massage. Use both hands, moving up and down the length of the back and then in sweeping motions across the width of the back. This stimulates the skin, gives

Use gentle pressure in small circles on the temples and along the eyebrows and jaw line for a relaxing face massage.

the recipient a chance to adjust to your touch, and allows you a few moments to sense the recipient's needs. At the end of the massage, it will leave the recipient with a sense of being calmed and nurtured.

Muscle kneading involves picking up the muscles with one hand and squeezing with the other, from the spinal column outward, following the course of the trapezius and latissimus muscles, first on one side of the back, then on the other. This is also effective moving from the shoulders down to the elbow and from the tops of the thighs to just above the knees. This increases circulation and the action of the nerves and helps to relax tired muscles.

Circular kneading involves using the middle three fingers or thumb of each hand, beginning at the neck on each side of the spine and applying the circular kneading outward, then beginning a little lower and working outward, one hand on each side, and so on to the end of the spine. This technique can also be used along the arms and legs. It increases blood flow, which will loosen and soothe muscles.

Muscle rolling involves putting both hands side by side on one of the recipient's shoulders, and making an alternating, very rapid pushing and pulling motion with the hands, gradually moving downward to the buttocks. The hands must be firmly on the patient so as to move his muscles from side to side, thereby causing a quick stretching and a vibration of them. First roll the muscles of one side of the back three to five times, from the neck to the end of the spine. The pressure should be made a little inward and upward and firmly, without jerking, in a slow and quiet way. Repeat two to four times to relieve backache and to stimulate nerve centers.

Percussion is applied with the edges of both hands and fingers alternately and very quickly from the neck downward on both sides of the spinal column. On the upper part of the back, from the shoulders to the lower end of the lungs, the

When massaging feet, start at the center just below the ball of the foot and work down toward the heel. Then gently squeeze each toe from the base to the tip. Finally, use your thumbs to press along the arch, sliding them from just above the heel up toward the base of the big toe.

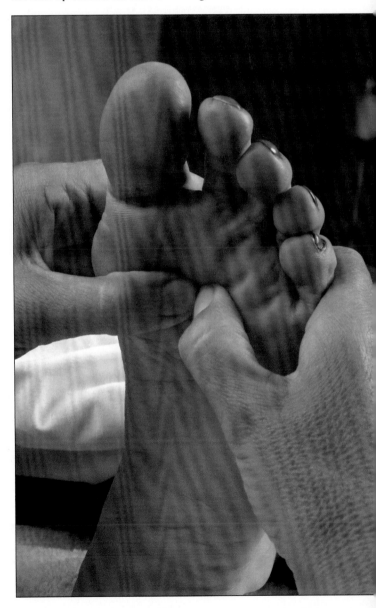

When using the circular kneading technique, pause and do a few extra circles in any area where you can feel knots or extra tension.

then the other side. This helps to increase circulation, which will help muscles to loosen and improve the skin.

Vibration involves putting both hands, with the fingers spread out, one on each side of the patient's back at the shoulders. Pull downward with a firm pressure and a rapid vibration of hands and fingers. Repeat three to five times to stimulate the nerves.

Spinal nerve compression involves pressing with the middle and index fingers on each side of the spinal column

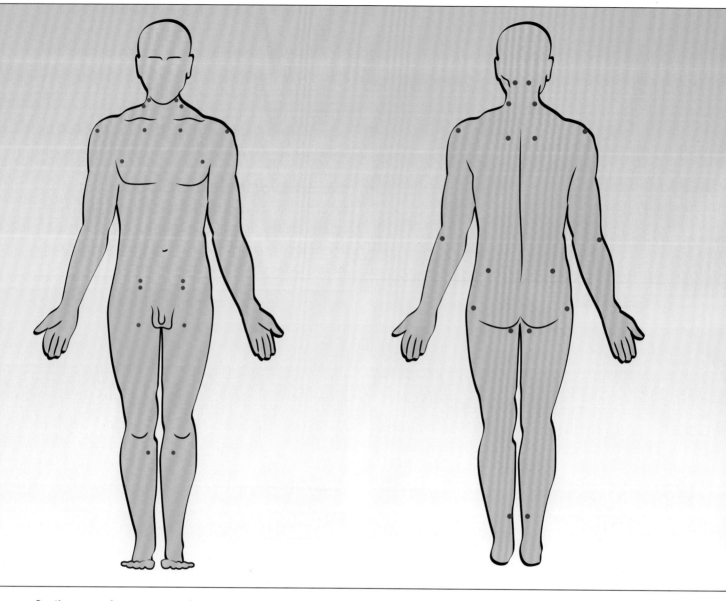

Gently massaging pressure points can help to relieve pain and promote healing.

percussion may also be applied outward to the sides. This technique can also be applied to the fleshy parts of the arms and legs, but avoid elbows, knees, the underside of the arms, and any other sensitive areas. This has a very stimulating and strengthening effect on the nerve centers.

Homemade Massage Oil

1 cup jojoba, grape seed, or almond oil (olive oil can also be used, but can cause breakouts on sensitive skin)

12 to 15 drops essential oil of your choice (lavender, jasmine, myrrh, orange, or mint are all good choices)

APPENDIX I

CO-OPS

FOOD CO-OPS: GROCERY SHOPPING IN COMMUNITY

What Is a Food Co-op?

Food co-ops are non-profit, democratic, and member-owned businesses that provide low-cost organic or natural foods to members and, in some cases, non-members.

Since food co-ops are established and operated by members, each one has a voice regarding what types of foods will be sold, maintenance issues, and management of the stores. Food co-ops are democratically run, so each member has one vote in any type of election. Members generally elect a board of directors to oversee the everyday running of the co-op and to hire staff.

What Are the Different Types of Food Co-ops?

Typically, there are two types of food co-ops: the co-op grocery store and the buying club. Both are owned and run by members but they vary in structure and number of members.

Co-op Grocery Store

Co-op grocery stores are basically regular grocery stores that are member-owned and -operated and provide low cost, healthy foods to members and often to the public as well. There are around 500 co-op grocery stores in the United States alone.

Buying Club

A buying club consists of a small group of people (friends, neighbors, families, or colleagues) who get together and buy food in bulk from a co-op distributor (a co-op warehouse or natural foods distributor) or from local farms. By ordering in bulk, the members are able to save money on grocery items. The members of the buying club share the responsibilities of collecting money from the other members, placing orders with the distributor, picking up the orders from the drop-off site, and distributing the food to the individual members or families.

How Do You Become a Food Co-op Member?

In order to become a member of a food co-op, you must pay a small initial fee and then typically invest a certain amount of money into the co-op to purchase a share. Sometimes members can accumulate more shares (by paying an annual fee, for example). Members can also help run the co-op by volunteering their time. Members reap the benefits of their membership by having access to discounted prices on food products. However, if you decide not to become a member, some food co-ops still allow non-members to shop at their stores without the membership discount.

How Do You Become a Buying Club Member?

If you are looking to join a buying club in your area, it is best to contact your regional co-op distributor and ask them for information on local buying clubs. Check out the websites of local distributors to see if they have links to buying clubs near you. Or ask friends and neighbors if they're aware of any buying clubs that are active in your area.

How Do You Start Your Own Food Co-op?

Here are some steps you need to follow in order to establish your own food co-op:

1. Invite potential members to meet and discuss the start-up of a food co-op. Identify how a food co-op may help the finances of the members.
2. Hold a meeting in which potential members vote to continue the process of forming a co-op and then select a committee for this purpose.
3. Determine how often the co-op will be used by the potential members.
4. Discuss the results of any surveys at another meeting and then vote to see if the plans should proceed.
5. Do a needs analysis (determine what the members will need in order to establish a food co-op).
6. Hold a meeting to discuss the outcome of the needs analysis and vote (anonymously) on whether or not to proceed with the co-op.
7. Develop a business plan for the co-op and decide the financial contribution needed to start the co-op.
8. At another meeting, have members vote on the business plan and if members want to continue, decide on whether or not to keep the committee members.
9. Prepare all legal documents and incorporate.
10. Hold a meeting for all potential members to review and accept the bylaws (terms of operation, responsibilities of members and board of directors). Hold an election for the board of directors.
11. At the first board of directors meeting, elect officers and assign them certain responsibilities in carrying out the business plan.
12. Hold a membership drive—try to recruit new members to the food co-op.
13. Pool monetary resources and create a loan application package.
14. Employ a manager for the co-op store.
15. Find a building or storefront to house the co-op.
16. Start your business!
    Example of bylaws for a food co-op:
- Establish membership requirements.

- Formulate the rights and responsibilities of all members and the board of directors.
- Stipulate the grounds for member expulsion.
- Establish rules for calling and implementing membership meetings.
- Establish how members will vote.
- Provide election procedures for board members and officers.
- Specify the number of board members and officers and how long their terms in office will be and what sort of compensation they will be awarded.
- Establish what time and where meetings will be held.
- Specify the co-op's fiscal year dates.
- Provide information on the distribution of net earnings.
- Include any other rules of management for the co-op.

How Do You Start Your Own Buying Club?

To start your own buying club, you will need to collect a group of people (preferably more than five households). If no one in your new group has any experience with organizing a buying club, it may be beneficial for you to temporarily join a buying club to see how it works. Once you are confident in your understanding of a buying club, it's time to begin!

1. Find a co-op distributor's (wholesaler's) pricing guide to share with the others in your buying club, so you all have an understanding of the products available and the savings from which you'll benefit. If you'll be buying

from local farms, discuss pricing and bulk discounts with the farmers.

2. Have a meeting and invite all those who are interested in joining your buying club. Emphasize that a buying club requires its members to share in all responsibilities—from placing orders to picking up deliveries to collecting the money—and that they will all reap the benefits of obtaining great organic and natural foods at wholesale prices.

3. Establish an organizational committee. Discuss areas such as coordination, price guide distribution, orderings, potential delivery location, what supplies will be needed, bookkeeping, and how to orient new members.

4. Draw up any membership requirements you think necessary.

5. Brainstorm possible delivery sites, such as churches, firehouses, or other public buildings. Your optimal site should be able to accommodate a large truck and have long hours of operation. Make sure you will have enough space at the site to go through the products and distribute them accordingly.

6. Develop a name for your buying club and fill out a membership application with the co-op distributor of your choice. You should receive some sort of confirmation, complete with order deadlines, date of delivery, and a simple orientation to the buying club.

7. Start enjoying your healthy foods for lower prices!

Food Co-ops and Distributors by State

Alabama

Grow Alabama Distribution Center
2301-B Finley Boulevard
Birmingham, Alabama 35234
(205) 991-0042
www.growalabama.com

Alaska

Organic Alaska
3404 Willow Street
Anchorage, Alaska 99517
(907) 306-3931
organic@alaska.com
www.organicalaska.com

Arizona

Food Conspiracy Co-op
412 North 4th Avenue
Tucson, Arizona 85705
(520) 624-4821
natural@foodconspiracy.org
www.foodconspiracy.org

Shop Natural Market
350 South Toole
Tucson, Arizona 85701
(520) 622-3911

Arkansas

Ozark Natural Foods Co-op
1554 North College Avenue
Fayetteville, Arkansas 72703
(479) 521-7558
jerry@ozarknaturalfoods.com
www.ozarknaturalfoods.com

Summercorn Foods
1410 West Cato Springs Road

Fayetteville, Arkansas 72701
(479) 695-1099

California

BriarPatch Co-op
290 Sierra College Drive, Suite A
Grass Valley, California 95945
(530) 272-5333
info@briarpatch.coop
www.briarpatch.coop

Co-opportunity Consumers Co-op
1525 Broadway
Santa Monica, California 90404
(310) 451-8902
service@coopportunity.com
www.coopportunity.com/

Davis Food Co-op
620 G Street
Davis, California 95616
(530) 758-2667
msimposon@davisfood.coop
www.daviscoop.com

Isla Vista Food Co-op
6575 Seville Road
Isla Vista, California 93117
(805) 968-1401
gm@islavistafood.coop
www.islavistafoodcoop.blogspot.com

Kresge Food Co-op
600 Kresge Court
Kresge College UCSC
Santa Cruz, California 95064
(831) 426-1506

North Coast Co-op, Arcata
811 I Street

Arcata, California 95521
(707) 822-5947
co-oparc@northcoastco-op.com
www.northcoastco-op.com

North Coast Co-op, Eureka
25 4th Street
Eureka, California 95501
(707) 443-6027
co-opeka@northcoastco-op.com
www.northcoastco-op.com

Ocean Beach People's Organic Food
  Co-op
4765 Voltaire Street
San Diego, California 92107
(619) 224-1387
editor@oceanbeachpeoples.com
www.obpeoplesfood.coop

Other Avenues Food Store
  Cooperative
3930 Judah Street
San Francisco, California 94122
(415) 661-7475
info@otheravenues.coop
www.otheravenues.coop

Quincy Natural Foods Co-op
269 Main Street
Quincy, California 95971
(530) 283-3528
qnf@snowcrest.net
www.qnf.coop

Rainbow Grocery Co-op
1745 Folsom Street @ 13th Street
San Francisco, California 94103
(415) 863-0620
comments@rainbowgrocery.coop

www.rainbow.coop
Sacramento Natural Foods Co-op
1900 Alhambra Boulevard
Sacramento, California 95816
(916) 455-2667
comments@sacfoodcoop.com
www.sacfoodcoop.com

Santa Rosa Community Market &
Café
1899 Mendocino Avenue
Santa Rosa, California 95401
(707) 546-1806
info@srcommunitymarket.com
www.srcommunitymarket.com

Ukiah Natural Foods
721 South State Street
Ukiah, California 95482
(707) 462-4778
unf@ukiahcoop.com
www.ukiahcoop.com

Colorado

Fort Collins Food Co-op
250 East Mountain Avenue
Fort Collins, Colorado 80524
(970) 484-7448
info@fcfood.coop
www.ftcfoodcoop.com

High Plains Food Cooperative
7900 East Union Avenue. Suite 200
Denver, Colorado 8023
(785) 626-6082
info@highplainsfood.org
www.highplainsfood.org

Connecticut

Willimantic Food Co-op
91 Valley Street

Willimantic, Connecticut 06226
(860) 456-3611
willifoodcoop@snet.net
www.willimanticfood.coop

Delaware

Newark Natural Foods Cooperative
280 East Main Street, East Market
   Plaza
Newark, Delaware 19711
(302) 368-5894
info@newarknaturalfoods.com
www.newarknaturalfoods.com

Florida

Ever'man Natural Foods
315 West Garden Street
Pensacola, Florida 32502
(850) 438-0402
info@everman.org
www.everman.org

Fresh Organics
2106 Flora Avenue
Ft. Myers, Florida 33907
(239) 939-2529
freshorganics@embarqmail.com
www.freshorganicsswfl.com

Homegrown Co-op
2310 North Orange Avenue
Orlando, Florida 32804
(407) 895-5559
www.homegrowncoop.org

Kissimmee's Green Place for Organic
   Vegetables
4130 Cardinal Lane
Kissimmee, Florida 347464
(407) 384-4699
www.kissimmeesgreenplacefororgan-
   icvegetables.com

Seminole Natural Foods
1550 Warwick Place
Longwood, Florida 32750
(407) 699-1883
www.seminolenaturalfoods.com

Sunseed Food Co-op, Inc.
6615 North Atlantic Avenue
Cape Canaveral, Florida 32920
(321) 784-0930
www.sunseedfoodcoop.com

Georgia

Daily Groceries Co-op
523 Prince Avenue
Athens, Georgia 30601
(706) 548-1732
dailygroceriescoop@gmail.com
www.dailygroceries.org

Life Grocery & Café Natural Foods
1453 Roswell Road
Marietta, Georgia 30062
(770) 977-9583
customerservice@lifegrocery.com
www.lifegrocery.com

Sevananda Natural Foods Market
467 Moreland Avenue NE
Atlanta, Georgia 30307
(404) 681-2831
info@sevananda.com
www.sevananda.coop

Hawaii

Kokua Market Natural Foods
   Co-op
2643 South King Street
Honolulu, Hawaii 96826
(808) 941-1922
info@kokua.coop
www.kokuamarket.ning.com

Idaho

Boise Consumer Co-op
888 West Fort Street
Boise, Idaho 83702
(208) 472-4500
info@boisecoop.com
www.boisecoop.com

Moscow Food Co-op
121 East 5th Street
Moscow, Idaho 83843
(208) 882-8537
outreach@moscowfood.coop
www.moscowfood.coop

Pocatello Co-op
515 North Main Street
Pocatello, Idaho 83204
(208) 232-2181
www.pocatellocoop.com

Illinois

Common Ground Food Co-op
300 South Broadway Suite 166
Urbana, Illinois 61801
(217) 352-3347
info@commonground.coop
www.commonground.coop

Duck Soup Co-op
129 East Hillcrest Drive
DeKalb, Illinois 60115
(815) 756-7044
ducksoupcoop@tbc.net
www.ducksoupcoop.com

Neighborhood Co-op Grocery
1815 West Main Street
Carbondale, Illinois 62901
(618) 529-3533
info@neighborhood.coop
www.neighborhood.coop

South Suburban Food Co-op
208 Forest Boulevard
Park Forest, Illinois 60466
(708) 747-2256
info@southsuburbanfoodcoop.com
www.southsuburbanfoodcoop.com

Stone Soup Cooperative
4637 North Ashland
Chicago, Illinois 60614
(773) 669-7687
www.stonesoupcoop.org

West Central Illinois Food
   Cooperative
P.O. Box 677
Galesburg, Illinois 61402
www.wcifoodcoop.com
www.unfi.com/default.aspx

Indiana

Bloomingfoods Downtown
419 East Kirkwood Avenue
Bloomington, Indiana 47408-4023
(812) 336-5300
kirkwood@bloomingfoods.coop
www.bloomingfoods.coop

Bloomingfoods Market & Deli
3220 East Third Street
Bloomington, Indiana 47401-5427
(812) 336-5400
east@bloomingfoods.coop
www.bloomingfoods.coop

Bloomingfoods Near West Side
316 West Sixth Street
Bloomington, Indiana
   47404-3912
(812) 333-7312
west@bloomingfoods.coop
www.bloomingfoods.coop

Clear Creek Food Co-op
701 East Main Street
Richmond, Indiana 47374
(765) 939-4390
coop@clearcreekcoop.org
www.clearcreekcoop.org

Lost River Market & Deli
26 Library Street
Paoli, Indiana 47454
(812) 723-3735
www.lostrivercoop.com

Maple City Market
314 South Main Street
Goshen, Indiana 46526
(574) 534-2355
marketing@maplecitymarket.com
www.maplecitymarket.com

River City Food Co-op
116 Washington Avenue
Evansville, Indiana 47713
(812) 401-7301
Co-op@rivercityfoodcoop.org
www.rivercityfoodcoop.org

Three Rivers Food Co-op's Natural
   Grocery
1612 Sherman Street
Fort Wayne, Indiana 46808
(260) 424-8812
gm@3riversfood.coop
www.3riversfood.coop

Iowa

New Pioneer Food Co-op &
   Bakehouse
1101 2nd Street
Coralville, Iowa 52241
(319) 358-5513
jangerer@newpi.com
www.newpi.com

New Pioneer Food Co-op
22 South Van Buren Street
Iowa City, Iowa 52240
(319) 338-9441
jangerer@newpi.com
www.newpi.com

Oneota Community Food Co-op
312 West Water Street
Decorah, Iowa 52101
(563) 382-4666
customerservice@oneotacoop.com
www.oneotacoop.com

Wheatsfield Cooperative Grocery
413 Northwestern Avenue
Ames, Iowa 50010
(515) 232-4094
shop@wheatsfield.coop
www.wheatsfield.coop

Kansas

The Merc Community Market & Deli
901 Iowa Street
Lawrence, Kansas 66044
(785) 843-8544
themerc@themerc.coop
www.communitymercantile.com

Prairieland Market
138 South 4th Street
Salina, Kansas 67401-2802
(785) 827-5877
prairielandmarket@sbcglobal.net
www.prairielandmarket.com

SE KS Buying Club
11th & Magnolia
Independence, Kansas 67301
(620) 205-7095
www.seksbuyingclub.wordpress.com

Topeka Natural Food Coop
503 Southwest Washburn Avenue

Topeka, Kansas 66606
(785) 235-2309
www.topekafoodcoop.wordpress.com

Kentucky

Good Foods Market & Café
455 Southland Drive
Lexington, Kentucky 40503
(859) 278-1813
goodfoods@goodfoods.coop
www.goodfoods.coop

Otherworld Food Co-op, Inc.
1865 Celina Road
Burkesville, Kentucky 42717
(270) 433-7400
otherworld@duo-county.com
www.duo-county.com/~otherworld/

Louisiana

New Orleans Food Co-op
621B North Rendon Street
New Orleans, Louisiana 70119
(504) 656-6632
info@nolafoodcoop.org
www.nolafoodcoop.org

Maine

Belfast Co-op
123 High Street
Belfast, Maine 04915
(207) 338-2532
info@belfast.coop
www.belfast.coop/

Blue Hill Co-op Community Market
  & Cafe
4 Ellsworth Road, P.O. Box 1133
Blue Hill, Maine 04614
(207) 374-2165
info@bluehill.coop
www.bluehill.coop

Fare Share Market
443 Main Street
Norway, Maine 04268
(207) 743-9044
www.faresharecoop.org

Oxford Hills Food Cooperative
35 Cottage Street
Norway, Maine 04268
(207) 743-8049
coopcoord@megalink.net

Rising Tide Co-op
323 Main Street

Damariscotta, Maine 04543
(207) 563-5556
customercare@risingtide.coop
www.risingtide.coop

Maryland

Common Market Co-op
5728 Buckeystown Pike, Unit 1-B
Frederick, Maryland 21704
(301) 663-3416
cm@commonmarket.coop
www.commonmarket.coop

Glut Food Co-op
4005 34th Street
Mt. Rainier, Maryland 20722
(301) 779-1978
www.glutfood.org

Maryland Food Co-op
University of Maryland B-0203 Adele
   H. Stamp Student Union
College Park, Maryland 20742
(301) 314-8089

Takoma Park Food Co-op
201 Ethan Allen Avenue
Takoma Park, Maryland 20912
(301) 891-2667
board@tpss.coop
www.tpss.coop

Takoma Park Silver Spring Coop
8309 Grubb Road
Silver Spring, Maryland 20910
(240) 247-2667
board@tpss.coop
www.tpss.coop

Massachusetts

Artichoke Food Co-op
800 Main Street
Worcester, Massachusetts 01610
(508) 752-3533
artichokefoodcoop@gmail.com
www.artichokecoop.org

Berkshire Co-op Market
42 Bridge Street
Great Barrington, Massachusetts
    01230
(413) 528-9697
community@berkshire.coop
www.berkshire.coop

Green Fields Market
144 Main Street
Greenfield, Massachusetts 01301
(413) 773-9567
email@greenfieldsmarket.coop
www.greenfieldsmarket.coop

Harvest Co-op Markets
581 Massachusetts Avenue
Cambridge, Massachusetts 02139
(617) 661-1580
cdurkin@harvest.coop
www.harvest.coop

Harvest Co-op Markets
57 South Street
Jamaica Plain, Massachusetts 02130
(617) 524-1667
cdurkin@harvest.coop
www.harvest.coop

River Valley Market
330 North King Street
Northampton, Massachusetts
    01061
(413) 584-2665
info@rivervalleymarket.coop
www.rivervalleymarket.coop

Michigan

Brighton Food Cooperative
2715 West Coon Lake Road
Howell, Michigan 48843
(517) 546-4190
bfcoop@brightonfoodcoop.org
www.brightonfoodcoop.org

Cass Corridor Food Co-op
456 Charlotte Street
Detroit, Michigan 48201
(313) 831-7452
casscorridorfoodcoop@yahoo.com

Dibbleville Food Cooperative
106 East Elizabeth Street
Fenton, Michigan 48430
(810) 629-1175
contact@dibbleville.com
www.dibbleville.com

East Lansing Food Co-op
4960 Northwind Drive
East Lansing, Michigan 48823
(517) 337-1266
www.elfco.coop

Grain Train Natural Food Market
220 Mitchell Street
Petoskey, Michigan 49770
(231) 347-2381
info@graintrain.coop
www.graintrain.coop

Hartland's Nutrition Connection
    Cooperative
1898 Korte Street
P.O. Box 530
Hartland, Michigan 48353
(810) 632-7952
hnccinc@ismi.net
www.hnccinc.com

Ionia Natural Food Co-op
Healthy Basics 2
576 State Street
Ionia, Michigan 48846
Infc_mc@hotmail.com
http://health.groups.yahoo.com/
    group/ionianaturalfoodscoop

Keweenaw Food Co-op
1035 Ethel Avenue
Hancock, Michigan 49930
(906) 482-2030
info@keweenaw.coop
www.keweenaw.coop

Marquette Food Co-op
109 West Baraga Avenue
Marquette, Michigan 49855
(906) 225-0671
www.marquettefood.coop

Oryana Food Cooperative
260 East 10th Street
Traverse City, Michigan 49684
(231) 947-0191
info@oryana.coop
www.oryana.coop

People's Food Co-op & Café Verde
214–216 North 4th Avenue
Ann Arbor, Michigan 48104
(734) 994-9174
info@peoplesfood.coop
www.peoplesfood.coop

People's Food Co-op of Kalamazoo
436 Burdick Street South
Kalamazoo, Michigan 49007
(269) 342-5686
market@peoplesfoodco-op.org
www.peoplesfoodco-op.org

Simple Times Farm Market & Buying
    Club
6081 East Baldwin Road
Grand Blanc, Michigan 48439
(810) 874-6463
info@simpletimesfarm.com
www.simpletimesfarm.com

West Michigan Co-op
1111 Godfrey Southwest, Suite S250
Grand Rapids, Michigan 49503
(616) 951-3287
membership@westmichigan-coop.com
www.westmichigancoop.com

Ypsilanti Food Co-op and River Street
    Bakery
312 North River Street
Ypsilanti, Michigan 48198
(734) 483-1520
www.ypsifoodcoop.org

Minnesota

Bluff Country Co-op
121 West 2nd Street
Winona, Minnesota 55987
(507) 452-1815
liz@bluff.coop
www.bluff.coop

Cook County Whole Foods Co-op
20 East 1st Street
Grand Marais, Minnesota 55604
(218) 387-2503
info@cookcounty.coop
www.cookcounty.coop

Crow Wing Food Co-op
720 Washington Street
Brainerd, Minnesota 56401
(218) 828-4600
Cwfoodco-op@clearwire.net
www.crowwingfoodco-op.com

Eastside Food Co-op
2551 Central Avenue Northeast
Minneapolis, Minnesota 55418
(612) 788-0950
amy@eastsidefood.coop
www.eastsidefood.coop

Hampden Park Food Co-op
928 Raymond Avenue
St. Paul, Minnesota 55114
(651) 646-6686
www.hampdenparkcoop.com

Harmony Natural Food Co-op
117 3rd Street Northwest
Bemidji, Minnesota 56601
(218) 751-2009
harmonyg@paulbunyan.net
www.harmonycoop.com

Just Food Co-op
516 Water Street South
Northfield, Minnesota 55057
(507) 650-0106
www.justfood.coop

Lakewinds Natural Foods
1917 2nd Avenue South
Anoka, Minnesota 55303
(763) 427-4340
lakewinds@lakewinds.coop
www.lakewinds.coop

Lakewinds Natural Foods
435 Pond Promenade
Chanhassen, Minnesota 55317
(952) 697-3366
lakewinds@lakewinds.coop
www.lakewinds.coop

Lakewinds Natural Foods
17523 Minnetonka Boulevard
Minnetonka, Minnesota 55345
(952) 473-0292

lakewinds@lakewinds.coop
www.lakewinds.coop

Linden Hills Food Co-op
2813 West 43rd Street
Minneapolis, Minnesota 55410
(612) 922-1159
info@lindenhills.coop
www.lindenhills.coop

Linden Hills Natural Home
3815 Sunnyside Avenue
Minneapolis, Minnesota 55410
(612) 922-1159
info@lindenhills.coop
www.lindenhills.coop

Mississippi Market Natural Food Co-op
622 Selby Avenue
St. Paul, Minnesota 55104
(651) 310-9499
info@msmarket.coop
www.msmarket.coop

Mississippi Market Natural Foods
    Co-op
1500 West 7th Street
St. Paul, Minnesota 55105
(651) 690-0507
info@msmarket.coop
www.msmarket.coop

Natural Harvest Whole Food Co-op
505 3rd Street, on Bailey's Lake
Virginia, Minnesota 55792
(218) 741-4663
www.naturalharvestcoop.com

Pomme De Terre Food Co-op
613 Atlantic Avenue
Morris, Minnesota 56267
(320) 589-4332
manager@pdtfoods.org
www.pdtfoods.org

River Market Community Co-op
221 North Main Street, Suite 1
Stillwater, Minnesota 55082
(651) 439-0366
info@rivermarket.coop
www.rivermarket.coop

Rochester Good Food Co-op
1001 6th Street Northwest
Rochester, Minnesota 55901
(507) 289-9061
gmphil@rochestergoodfood.coop
www.rochestergoodfood.com

Seward Co-op Grocery & Deli
2823 East Franklin Avenue
Minneapolis, Minnesota 55406
(612) 338-2465
www.seward.coop

St. Peter Food Co-op & Deli
228 Mulberry Street

St. Peter, Minnesota 56082
(507) 934-4880
info@stpeterfood.coop
www.stpeterfood.coop
Valley Natural Foods Co-op
13750 County Road 11
Burnsville, Minnesota 55337
(952) 891-1212
info@valleynaturalfoods.com
www.valleynaturalfoods.com

Wedge Community Co-op
2105 Lyndale Avenue South
Minneapolis, Minnesota 55405
(612) 871-3993
wedge@wedge.coop
www.wedge.coop

Whole Foods Co-op
610 East 4th Street
Duluth, Minnesota 55805
(218) 728-0884

info@wholefoods.coop
www.wholefoods.coop

Wild Rice Buying Club
912 20th Avenue South
Moorhead, Minnesota 56560
(218) 236-1181
m-s.amick@juno.com

Mississippi

Rainbow Whole Foods Co-op
    Grocery, Deli and Café
2807 Old Canton Road
Jackson, Mississippi 39216
(601) 366-1602
info@rainbowcoop.org
www.rainbowcoop.org

Missouri

City Food Co-op (Buying Group of
    St. Louis)

639 Cherokee Street
St. Louis, Missouri 63118-3129
(314) 865-0567
jpciv@mindspring.com

Montana

Community Food Co-op
908 West Main Street
Bozeman, Montana 59715
(406) 587-4039
info@bozo.coop
www.bozo.coop

Nebraska

Open Harvest Natural Foods Co-op
1618 South Street
Lincoln, Nebraska 68502
(402) 475-9069
harvest@openharvest.com
www.openharvest.com

Nevada

Great Basin Community Food Coop
542 ½ Plumas Street
Reno, Nevada 89509
(775) 324-6133
info@greatbasinfood.coop
www.greatbasinfood.coop

New Hampshire

Concord Cooperative Market
24 South Main Street
Concord, New Hampshire 03301
(603) 225-6840
infof@concordfoodcoop.coop
www.concordfoodcoop.coop

Co-op Community Food Market
43 Lyme Road
Hanover, New Hampshire 03755
(603) 643-2725

comment@coopfoodstore.coop
www.coopfoodstore.coop

Hanover Co-op Food Store
45 South Park Street
Hanover, New Hampshire 03755
(603) 643-2667
comment@coopfoodstore.coop
www.coopfoodstore.coop

Hanover Co-op: Lebanon
12 Centerra Park Street
Lebanon, New Hampshire 03766
(603) 643-4889
comment@coopfoodstore.coop
www.coopfoodstore.coop

New Jersey

George Street Co-op
89 Morris Street
New Brunswick, New Jersey 08901
(732) 247-8280
gscoop@georgestreetcoop.com
www.georgestreetcoop.com

Purple Dragon Co-op
289 Washington Street
Glen Ridge, New Jersey 07028
(973) 429-0391
info@purpledragon.com
www.purpledragon.com

Sussex County Food Co-op
30 Moran Street
Newton, New Jersey 07860
(973) 579-1882
scfc@beithe.com
www.sussexcountyfoods.org

New Mexico

Silver City Food Co-op
520 North Bullard Street

Silver City, New Mexico 88061
(575) 388-2343
doug@silvercityfoodcoop.com
www.silvercityfoodcoop.com

Wild Sage Natural Foods Cooperative
226 West Coal Avenue
Gallup, New Mexico 87301
(505) 863-5383
sage@patmail.com

New York

Abundance Cooperative Market
62 Marshall Street
Rochester, New York 14607
(585) 454-2667
info@abundance.coop
www.abundance.coop

Flatbush Food Cooperative
1415 Cortelyou Road
Brooklyn, New York 11226-5604
(718) 284-9717
info@flatbushfoodcoop.com
www.flatbushfoodcoop.com

4th Street Food Co-op
58 East 4th Street
New York, New York 10003-8914
(212) 674-3623
www.4thstreetfoodcoop.org

GreenStar Cooperative Market
701 West Buffalo Street
Ithaca, New York 14850
(607) 273-9392
www.greenstar.coop

High Falls Food Coop
1398 State Road 213
High Falls, New York 12440-5721
(845) 687-7262
www.highfallscoop.com

Honest Weight Food Coop
484 Central Avenue
Albany, New York 12206
(518) 482-2667
coop@hwfc.com
www.hwfc.com

Lexington Co-operative Market
807 Elmwood Avenue
Buffalo, New York 14222
(716) 886-2667
tim@lexington.coop
http:/www.lexington.coop

Park Slope Food Co-op
782 Union Street
Brooklyn, New York 11215
(718) 622-0560
www.foodcoop.com

Potsdam Consumer Co-op
24 Elm Street
Potsdam, New York 13676
(315) 265-4630
mail@potsdamcoop.com
www.potsdamcoop.com

South Bronx Food Cooperative
3103 Third Avenue
Bronx, New York 10451
(718) 401-3500
info@sbxfc.org
www.sbxfc.blogspot.com

Syracuse Real Food Co-op
618 Kensington Road
Syracuse, New York 13210
(315) 472-1385
contactus@syracuserealfood.coop
www.syracuserealfood.coop

The Village Store Co-op
1 West Main Street

Cambridge, New York 12816
(518) 677-5731
goodfood@cambridgefoodcoop.com
www.villagestorecoop.com

North Carolina

Chatham Marketplace
480 Hillsboro Street, Suite 320
Pittsboro, North Carolina 27312
(919) 542-2643
mary@chathammarketplace.coop
www.chathammarketplace.coop

Company Shops Market
P.O. Box 848
Burlington, North Carolina 27216
(336) 223-0390
info@companyshopsmarket.coop
www.companyshopsmarket.coop

Deep Roots Market
3728 Spring Garden Street
Greensboro, North Carolina 27407
(336) 292-9216
www.deeprootsmarket.coop

Durham Co-op Grocery
1101 West Chapel Hill Street
Durham, North Carolina 27701
(919) 490-0929
Michael@durhamcentralmarket.org
www.durhamfoodcoop.org

French Broad Food Co-op
90 Biltmore Avenue
Asheville, North Carolina 28801
(828) 255-7650
mlc@frenchbroadfood.coop
www.frenchbroadfood.coop

Hendersonville Community Co-op
715 South Grove Street

Hendersonville, North
   Carolina 28792
(828) 693-0505
mail@hendersonville.coop
www.hendersonville.coop

Tidal Creek Cooperative Food
   Market
5329 Oleander Drive, Suite 100
Wilmington, North Carolina 28403-
   5841
(910) 799-2667
mail@tidalcreek.coop
www.tidalcreek.coop

North Dakota

Amazing Grains Natural Food Market
214 Demers Avenue
Grand Forks, North Dakota 58201
(701) 775-4542
info@amazinggrains.org
www.amazinggrains.org

Ohio

Clintonville Community Market
200 Crestview Road
Columbus, Ohio 43202
(614) 261-3663
info@communitymarket.org
www.communitymarket.org

Good Food Co-op
113 West College Street
Oberlin, Ohio 44074
(440) 775-6533
Victoria.werner@oberlin.net
http://www.oberlin.edu/stuorg/gfc/
   index.html

Great Food Co-op at Trinity United
   Church of Christ
915 North Main Street

Akron, Ohio 44310

(330) 376-7186

www.trinucc.org

Kent Natural Foods Co-op

151 East Main Street

Kent, Ohio 44240

(330) 673-2878

www.kentnaturalfoods.org

Oregon

Ashland Food Co-op

237 North 1st Street

Ashland, Oregon 97520

(541) 482-2237

outreach@ashlandfood.coop

www.ashlandfood.coop

Astoria Cooperative

1355 Exchange Street, Suite 1

Astoria, Oregon 97103

(503) 325-0027

www.astoriacoop.org

Brooking Natural Food Co-op

630 Fleet Street

P.O. Box 8051

Brookings, Oregon 97415-0344

(541) 469-9551

www.bnf-foodcoop.org

First Alternative Natural Foods Co-op

2855 Northwest Grant Avenue

Corvallis, Oregon 97330

(541) 452-3115

fanorth@firstalt.coop

www.firstalt.coop

First Alternative Natural Food Co-op

1007 Southeast 3rd Street

Corvallis, Oregon 97333

(541) 753-3115

firstalt@firstalt.coop

www.firstalt.coop

Food Front Cooperative Grocery

2375 Northwest Thurman Street

Portland, Oregon 97210

(503) 222-5658

info@foodfront.coop

Lents Food Buying Club

9635 Southeast Boise Street

Portland, Oregon 97266

(503) 282-5819

Oceana Natural Foods Co-op

159 Southeast 2nd Street

Newport, Oregon 97365

(541) 265-8285

Rhonda@oceanfoods.org

www.oceanafoods.org

People's Food Co-op

3029 Southeast 21st Street

Portland, Oregon 97202

(503) 674-2642

info@peoples.coop

www.peoples.coop

Pennsylvania

East End Food Co-op

7516 Meade Street

Pittsburgh, Pennsylvania 15208

(412) 242-3598

memberservices@eastendfood.coop

www.eastendfood.coop

Selene Whole Foods Co-op

305 West State Street

Media, Pennsylvania 19063

(610) 566-1137

manager@selenecoop.org

www.selenecoop.org

Swarthmore Co-op

341 Dartmouth Avenue

Swarthmore, Pennsylvania 19081

(610) 543-9805

swacoop@comcast.net

www.swarthmore.coop

Weavers Way Cooperative

  Association

559 Carpenter Lane

Philadelphia, Pennsylvania 19119

(215) 843-2350

editor@weaversway.coop

www.weaversway.coop

Whole Foods Co-op

1341 West 26th & Brown Avenue

Erie, Pennsylvania 16508

(814) 456-0282

wfcboard@gmail.com

www.wholefoodscoop.org

Rhode Island

Alternative Food Co-op

357 Main Street

Wakefield, Rhode Island 02879

(401) 789-2240

www.alternativefoodcoop.com

South Carolina

Upstate Food Co-op

404 John Holiday Road

Six Mile, South Carolina 29682

(864) 868-3105

upstatecoop@bellsouth.net

www.upstatefoodcoop.com

South Dakota

The Co-op Natural Foods

2504 South Duluth Avenue

Sioux Falls, South Dakota 57105

(605) 339-9506

www.coopnaturalfoods.com

Tennessee

Marketplace Buying Club Co-op
3511 Belmont Boulevard
Nashville, Tennessee 37215
(615) 899-4938
jackiw@marketplaceco-op.org
www.marketplaceco-op.org

Morningside Buying Club
215 Morningside Lane
Liberty, Tennessee 37095
(615) 563-2353
www.morningsidefarm.com

Naturally Good Food
4242 Port Royal Road
Spring Hill, Tennessee 37174
(931) 486-3192

Three Rivers Market
937 North Broadway
Knoxville, Tennessee 37917
(865) 525-2069
gm@threeriversmarket.coop
www.threeriversmarket.coop

Texas

Central City Co-op
2515 Waugh Drive
Houston, Texas 77006
(713) 524-9408
info@centralcityco-op.com
www.centralcityco-op.org

Wheatsville Food Co-op
3101 Guadalupe
Austin, Texas 78705
(512) 478-2667
gm@wheatsville.coop
www.wheatsville.coop

Utah

The Community Food Co-op of Utah
1726 South 700 West
Salt Lake City, Utah 84104
(801) 746-7878 or (866) 959-COOP
    (2667)
community@crossroads-u-c.org
www.foodco-op.net/

Vermont

Brattleboro Food Co-op
2 Main Street/Co-op Plaza
Brattleboro, Vermont 05301
(802) 257-0236
adminbfc@sover.net
www.brattleborofoodcoop.com

City Market–Onion River Co-op
82 South Winooski Avenue
Burlington, Vermont 05401
(802) 861-9700
info@citymarket.coop
www.citymarket.coop

Hunger Mountain Co-op
623 Stone Cutters Way
Montpelier, Vermont 05602
(802) 223-8000
info@hungermountain.com
www.hungermountain.com

Putney Food Co-op
8 Carol Brown Way
P.O. Box 730
Putney, Vermont 05346
(802) 387-5866
ptnycoop@sover.net
www.putneycoop.com

St. J. Food Co-op
490 Portland Street #101
St. Johnsbury, Vermont 05819
(802) 748-9498
info@stjfoodcoop.com
www.stjfoodcoop.com

Virginia

Eats Natural Foods Co-op
708 North Main Street
Blacksburg, Virginia 24060
(540) 552-2279
eatsnaturalfoods@gmail.com
www.eatsnaturalfoods.com

Healthy Foods Co-op
110 West Washington Street
Lexington, Virginia 24450
(540) 463-6954
healthyfoods@embarqmail.com
www.healthyfoodscoop.org

Roanoke Natural Foods Co-op
1319 Grandin Road Southwest
Roanoke, Virginia 24015
(540) 343-5652
info@roanokenaturalfoods.coop
www.roanokenaturalfoods.com

Valley Market
P.O. Box 23
Staunton, Virginia 24402-0023
www.valleymarket.org

Washington

Community Food Co-op
1220 North Forest Street
Bellingham, Washington 98225
(360) 734-8158
diana@communityfood.coop
www.communityfood.coop

The Food Co-op
414 Kearney Street

Port Townsend, Washington 98368
(360) 385-2883
membersservicesr@ptfoodcoop.coop
www.foodcoop.coop
Madison Market/Central Co-op
1600 East Madison
Seattle, Washington 98122
(206) 329-1545
info@madisonmarket.com
www.madisonmarket.com

Okanogan River Co-op
21 West 4th Street
P.O. Box 591
Tonasket, Washington 98855
(509) 486-4188
okrivercoop@planettonasket.com

Puget Consumers' Co-op–Fremont
600 North 34th Street
Seattle, Washington 98103
(206) 632-6811
fremont.storems@pccsea.com
www.pccnaturalmarkets.com

Puget Consumers' Co-op–Greenlake
7504 Aurora Avenue North
Seattle, Washington 98103
(206) 525-3586
green.storems@pccsea.com
www.pccnaturalmarkets.com

Puget Consumers' Co-op–Issaquah
1810 12th Avenue Northwest
Issaquah, Washington 98027
(425) 369-1222
issaquah.storems@pccsea.com
www.pccnaturalmarkets.com

Puget Consumers' Co-op–Kirkland
10718 Northeast 68th Street
Kirkland, Washington 98033
(425) 828-4622

kirk.storems@pccsea.com
www.pccnaturalmarkets.com

Puget Consumers' Co-op–Offices
4201 Roosevelt Way Northeast
Seattle, Washington 98105
(206) 547-1222
www.pccnaturalmarkets.com

Puget Consumers' Co-op–Seward Park
5041 Wilson Avenue South
Seattle, Washington 98118
(206) 723-2720
park.storems@pccsea.com
www.pccnaturalmarkets.com

Puget Consumers' Co-op–View Ridge
6514 40th Street Northeast
Seattle, Washington 98115
(206) 526-7661
view.storems@pccsea.com
www.pccnaturalmarkets.com

Puget Consumers' Co-op–West Seattle
2749 California Avenue Southwest
Seattle, Washington 98116
(206) 937-8481
west.storems@pccsea.com
www.pccnaturalmarkets.com

Sno-Isle Natural Foods Co-op
2804 Grand Avenue
Everett, Washington 98201
(425) 259-3798
snoisle.coop@verizon.net
www.snoislefoods.coop

Yelm Food Co-op
404 First Street Southeast
Yelm, Washington 98597

(360) 400-2210
www.yelmfood.coop

West Virginia

Mountain People's Market
1400 University Avenue
Morgantown, West Virginia
    26505-5520
(304) 291-6131
coop@labs.net
www.mountaincoop.com

Wisconsin

Basic Cooperative
1711 Lodge Drive
Janesville, Wisconsin 53545
(608) 754-3925
gm@basicshealth.com
www.basicshealth.com

Kickapoo Exchange Food Co-op
Box 276, 209 Main Street
Gays Mills, Wisconsin 54631
(608) 735-4544
kickapooexchange@yahoo.com
www.geocities.com/kickapooexchange

Mega Pik N Save
1201 South Hastings Way
Eau Claire, Wisconsin 54701
(715) 836-8700
mathew@megafoods.com
www.megafoods.com

Menomonie Market Food Co-op
521 2nd Street East
Menomonie, Wisconsin 54751-1864
(715) 235-6533
info@menomoniemarket.org
wwwmmfc.org

Nature's Bakery Co-op
1019 Williams Street
Madison, Wisconsin 53703
(608) 257-3649
mail@naturesbakery.coop
www.naturesbakery.coop

Outpost Natural Foods Co-op
100 East Capital Drive
Milwaukee, Wisconsin 53212
(414) 961-2597
www.outpost.coop

Outpost Natural Foods Co-op:
  Wauwatosa
7000 West State Street
Wauwatosa, Wisconsin 53213
(414) 778-2012
www.outpost.coop

People's Food Co-op
315 South 5th Avenue
La Crosse, Wisconsin 54601
(608) 784-5798
www.peoplesfoodcoop.com

Riverwest Co-op Grocery & Café
733 North Clark Street
Milwaukee, Wisconsin 53212
(414) 264-7933
vinceb@riverwestcoop.org
www.riverwestcoop.org

Viroqua Food Cooperative
609 North Main Street
Viroqua, Wisconsin 54665
(608) 637-7511
marketing@viroquafood.ccp
viroquafood.coop

Willy Street Co-op
1221 Williamson Street
Madison, Wisconsin 53703
(608) 251-6776

l.olson@willystreet.coop
www.willystreet.coop

Yahara River Grocery Cooperative
229 East Main Street
Stoughton, Wisconsin 53589
(608) 877-0947
info@yaharagrocery.coop
www.yaharagrocery.coop

Wyoming

Sweet Grass Food Co-op
169 Esterbrook Road
Douglas, Wyoming 82633
(307) 358-0582
sweetgrassclub@yahoo.com
http://sweetgrassfood.org/index.html

APPENDIX 2

ANIMAL TRACKS

ARMADILLO

The armadillo, well-known for its tough, armor-like covering, is commonly found from Texas to Argentina. Being around the size of a normal cat, its tracks are small. With 4 toes on the front foot and 5 on the hind, its tracks are distinguishable by powerful, long claws on each toe—claws that the armadillo uses to dig. Those clawed toes almost resemble hooves in mud or dust. Armadillos frequent water holes and enjoy mud baths, so look for impressions of their noses and armor plates in the mud.

BADGER

The west and Midwest, ranging from Canada to Mexico, is badger territory. When looking for badgers, keep an eye out not only for its tracks, but the holes that the animal digs along its trail in search for rodents. The badger's front feet are endowed with lengthy claws for that specific purpose, and those gigantic claws help indicate that it's a badger to which the trail belongs. Compared to the front ones, the badger's hind claws barely express themselves. Each foot has five toes that register in each track. The trail of the badger implies that it's a clumsy animal because of the double row, close together footprints. The hind feet toe-in a little and tend to touch the earlier tracks made by the forefeet.

BEAVER

Found around water in much of the northern hemisphere, these animals leave obvious traces through their lumbering activities. Beavers are exceptional engineers, renowned for their dams, which are constructed with branches, sticks, mud, and roots, and also their lodges, which are mounds of mud and sticks, often reaching 20 feet in diameter and 4-6 feet in height. Other signs of beavers are in the water, too, like floating plant roots with teeth marks or even a straight, slow trail in water—a black arrowhead—which indicates that a beaver is swimming by. Alternatively, look on land for mud-pies, the romantic calling cards of male beavers, scented with drops of beaver castoreum. Beavers have 5 toes on both their front and hind feet, although the latter set is webbed for swimming purposes, and this shows itself in their trails. Each toe has a claw, a characteristic that appears in the track. However, the claw on the second toe of the hind foot is split, which seldom registers. The beaver can use this odd toe as a comb. The beaver's tail is broad and flat and often drags in the trail, weaving from side to side.

BLACK BEAR

Black bears are the only bears that climb trees into adulthood, so they're commonly found in wooded sections of the arctic all the way down to Mexico. Their tracks are reminiscent of grizzly and giant brown bears' except for the much smaller size and claw marks. A black bear's hind foot measures about 6 inches and its claws are not nearly as large as a grizzly bear's. Often, the claws don't show up in the track at all. Black bears have five toes on both front and hind feet, and though the toe is well-developed in both places, it doesn't register. Black bears typically hibernate during winter, so their tracks are seldom seen in snow.

BOBCAT

The bobcat, found in almost all the wooded sections of the United States, has tracks similar to a domestic cat, but larger. Bobcats also have a more complex palm and heel pad outline than other cats. It walks correctly, meaning the hind feet step into the marks made by the front feet, which helps the animal stalk its prey.

BROWN BEAR

The brown bear is the most massive meat eater living in the world today, weighing in at around 1200-1400 pounds. Tracks of brown bear hind feet measure an impressive 14-16 inches long. These animals make it a habit to step in tracks they have already made, meaning that the hind foot and front foot register, and so claw marks are generally not visible in their tracks. The fifth toe in both front and hind feet is well developed, but it also does not show up in the trail. Other signs of a brown bear, or any bear, for that matter, are gnawed and gouged trees acting as signposts or billboards, droppings, and logs that have been torn open. Brown bears are found along the Pacific coast, ranging from Alaska to British Columbia.

CAT

Domestic cats are found all over America, and other than the size, their tracks could easily be mistaken for any of their more wild relatives. Cats walk on their toes, in a nearly straight line, and they are great predators because they stealthily place their hind feet in tracks made by their forefeet. Only four toes are visible in the hind and forefeet tracks, and claws do not register thanks to cats having retractable claws. If their claws were always out, they would deteriorate, and they must be sharp for hunting.

COTTONTAIL RABBIT

This small, widespread animal is famous for its hopping, bounding movements. Rabbits place their hind feet in front of the tracks made by their forefeet; they are able to travel 10-15 ft in one bound. Cottontail rabbits are fast, but often resort to tricks to avoid their pursuers, like doubling back or leaping to the side. Often, they travel the same route every day, and they also do not stray too far from their home, not beyond an acre. They have 5 toes on their forefeet and 4 on their hind ones, although only 4 of the toes on the front foot make any impact in their track.

COYOTE

Like cats, coyotes walk on their toes, and their tracks are similar to a wolf's, except smaller, although their prints are not as small as a fox's. A coyote's claws are always apparent in the trail, and the outer toes of the hind foot are larger than the inner toes. These animals are found in a variety of places, but especially in the prairies and open woodlands of North America. Their reach even stretches into Alaska.

CROW

Crows can be found almost anywhere except for deserts and pine forests. Because they are birds, the signs that identify them are sometimes more difficult to find. Their characteristic "caw-caw" bird call identifies them immediately, but if they are not in the vicinity, look for regurgitated pellets at the base of their roosting places. Composed of things that can't be digested, crow pellets will reflect their varied diet. If you happen to spot a nest, crows make 4-6 green-colored eggs that have brown spots.

DEER

The white-tailed deer is regularly found in the east, living in fields and woodlands that are close to communities. They have nimble cloven hoofs and their hind feet always step into the tracks made by the forefeet. The tracks of the doe are more petite and narrow than those of the buck, and her tracks will often face straight forward or inward, while his toes point out. A buck also spreads his feet more widely and drags them during rutting season. Other buck signs are pawed-up ground and trees and bushes scraped by antlers. A whitetail's track measures about 3 inches for a buck, 2½ inches for a doe, and ¾ inches for a fawn. A mule deer, native of the west and heaviest type of deer at 200 pounds, leaves tracks that are rounder in appearance than a whitetail, and when it bounds, the deer will produce a peculiar, close group of tracks that is much different from their eastern counterparts. A mule deer's track measures 3¼ inches long. The coast blacktail deer lives on the Pacific coast, ranging from Alaska to California. Their tracks resemble those of a mule deer, especially when bounding, but their walk is like any other. They weigh around 150 pounds.

DOG

Domestic dogs can be found anywhere, and their tracks, especially those of a large dog, can give a very good idea of what a wolf's footprint may look like; they are very similar to their wild relatives. Dogs walk on their toes and leave prominent claw marks. They have 5 toes, although only 4 register in both the fore and hind feet. The former feet are wider than the latter. Dogs are not as elegant as cats in their gaits, as their footprints overlap often. It is not uncommon for a dog to drag its toes. Some dogs actually have deformed feet stemming from intensive breeding. It's very difficult to distinguish dog tracks from a wolf, but dogs will often lack the suspicious, hunting traits of a wolf, and tend to approach objects directly.

FOX

A red fox can be found all over North America, but will more likely be in northern areas. They frequent mixed woodlands, open country, and especially farming areas. Like other members of the dog family, a fox has prominent claws, but their track itself is more delicate than the others. They leave little, widely spaced, pad marks, and that distinguishes them. They are also good trackers because their hind feet step into the tracks made by the forefeet, their trail often leading in a straight line. Another sign of a fox is digging; their holes are similar to the ones dug by skunks, although they are deeper and narrower. A gray fox is likely to be found in the south and resembles a dog more than a red fox; their tracks are somewhat smaller because the toes of a gray fox make a larger print. A desert fox can be found in Southwest and lower California, while a kit fox is only found in the prairies and grows to be about the size of a housecat—their tracks resemble a cat's, too, other than the claws. An arctic fox lives between the timberline and then northward as far as the land reaches. Their tracks are smaller than the red fox, and in the winter resemble little oval pads because they grow extra fur as protection.

GOAT

Found in America from Alaska to Washington and Montana, the goat is a very steady animal on its feet. It has 4 hoofs on each foot, but only 2 make any impression in their tracks. Goats are said to have rubber heels because their heels are so large and rubberlike; the prints they leave behind are rarely as sharp as other hoofed animals. During mating season, they leave a scent trail by rubbing their horns on underbrush as they wander, and so their smell may give them away as well. Also look for various mineral and salt licks or hollows that have been scraped out by goats in which they roll and sprawl.

GRIZZLY BEAR

Inhabiting the west, the grizzly bear is known as the giant of the western mountains. Grizzly bears have excessively long claws that help them dig for burrowing animals. All 5 of their toes leave marks when they walk, and a grizzly's hind foot often measures about 12 inches. They are known to bury their kills with fresh dirt, and they also make use of bear trees as signposts and billboards to communicate with other bears.

HORSE

Domestic horses are found all over the world, although some of their wild relatives live in the western United States. Horses are toenail walkers, named because their hoof is actually the nail at the end of its middle toe. The hoofs of the front feet are rounder and winder than those of its hind feet, and the larger size in general distinguishes a horse's hoof from a donkey or mule. Another sign of a horse having been in your proximity is the pungent aroma of its waste.

MARTEN

These shy, weasel-like animals live in the forest areas from the northern limit of the trees to southern United States. A marten is much longer than a weasel—at least 3 feet in length—and its front paw is about 1½ inches.

 # MOOSE

Moose reside in the north: Canada and some of the northern areas in the United States. These animals have hoofs, although theirs are long and pointed because they are used as weapons. Measuring in at around 7 inches long, they may be the largest of the native ungulates.

Their hoofs are easily distinguishable from those of a domestic cow. Other signs of moose, particularly bull moose, are twigs and branches broken down by antlers, pawed ground, and acrid smelling wallows.

OPOSSUM

Found in the eastern United States, the Midwest, and Pacific coast, opossums live in any woodland habitats, farming areas, and even some urban areas. They climb trees, and their feet have evolved into resembling hands with thumbs rather than having claws; their hind feet is one of their distinguishing qualities, as is that each toe, other than the thumb, has claws. Opossums waddle when they walk, moving the feet of one side at the same time. Their tracks are normally not larger than 1¾ inches in width, and they are often spotted along streams.

OTTER

Otters live from the northern timberline to southern South America always near water. There are many different species, but the Canadian otter is most common throughout Canada and the United States. Otters measure around 4-5 feet long and can usually be found in the water, searching for fish. Their water trail is quite distinctive because they swim quickly, leaving a zig-zag trail in the water. In winter, otters have to travel on land in order to find open water, and they leave a very characteristic trail: they bound, each one leaving a full-length, well defined imprint in the snow. Their padded footprints that can often be seen. Otters have 5 curled toes on both their fore and hind feet, but the 5th toe does not make an impression. Both feet leave a rounded track, but the hind foot is slightly longer. Otters can also be identified by their slides as they enjoy sliding down slippery mud banks which cuts down travel time.

PORCUPINE

Famous for its pricks, a porcupine is the size of a small dog and lives in forests or similarly brushy areas in the western and northern United States, spreading their influence as far as Canada and Alaska. Because they spend much of their time in trees, they have strong climbing claws on both their front and hind feet. The forefoot has 4 toes while the hind one has 5, and because porcupine do not hibernate, their tracks are visible in the snow. Look for teeth marks on evergreen trees or on axe handles or canoe paddles; porcupines love salt and will gnaw on salty leftovers.

RACCOON

Another tree climber, the raccoon's claws are very important in distinguishing its track. Raccoons have bare soles and toes that can be plainly seen from the impression they leave, 5 well developed toes that have claws on each foot, and hind feet that are longer than the forefeet. The forefoot measures about 3¼ inches while the hind is about an inch longer. When it walks, it steps upon the track of its forefoot with its hind foot. Other signs of a raccoon are its den which is usually found in a hollow tree and hairs that catch in wood and bark. They are found all over the United States.

SHEEP

Domesticated sheep can be found on any form, but mountain sheep normally live in between British Columbia and northern New Mexico. All sheep tend to congregate in family groups or flocks. Their hoofs spread, but their toes are not as widely spread as mountain goats. They have dewclaws very rarely leave a mark in their track. Their hoofs are more blunt than the ones of deer.

SKUNK

The easiest way to spot a skunk is by its smell thanks to their poison gas scent glands. They're located anywhere from the Hudson Bay in Canada south to Guatemala. Skunks are flat-footed with 5 toes on each foot, although their claws usually do not show on their hind foot.

Their prints look somewhat similar to those of a small bear. When walking, skunks toe-in and place the fore and hind foot of the same side close together. Like foxes, skunks dig holes, although theirs are more shallow and round than the holes dug by foxes.

SQUIRREL

Gray squirrels populate the East: southern Canada to southern United States. These are animals that bound with hind legs that are longer than the fore. When running, squirrels pair their forefeet behind their hind feet, a nod to their tree climbing capabilities. Squirrels have 5 toes on their hind feet that clearly show up and 4 in the front; the thumb of the front has disappeared from their anatomy over the centuries. They have flat feet and walk on the palms and soles of their aforementioned feet. Fox squirrels are bigger and stockier than gray ones, and they live in eastern hardwood forests from the southern tip of the Great Lakes to southern United States. Red squirrels can be found in the northern United States and Canada. Traces of those squirrels can be seen even in winter; they only sleep during the darkest, most frigid days and dig tunnels in the snow otherwise.

TURKEY

Turkeys are distinctly American, found in many open woodlands. They are the largest game bird, therefore their tracks are also the largest. Their middle toe curves slightly inward, which adds to their already toe-in imprints. Their stiff feathers on their wings can leave marks on the trail when they strut around.

WOLF

Wolves are found in the wilderness of the northern forests and tundra. They also appear in most other habitats, except deserts and high mountains. The tracks of a dog and wolf are so similar that many concede that there is not one reliable feature that will distinguish between the two. A wolf's tracks tend to be larger and also their suspicious, shy nature leads wolf to use every bit of cover as they approach an object, while dogs approach things openly. The inner toes of the hind foot of a wolf leave a larger track than do the outer toes.

ACKNOWLEDGMENTS

This was a big project, and I most certainly didn't do it alone.

This book may never have been completed without the assistance of Melanie Trice, whose research, writing, and positive attitude made this project manageable. I'm also grateful to Katherine Jansen, who jumped right in to help toward the end as if she'd been writing about tanning leather and butchering pigs her whole life.

Illustrator James Balkovek has spoiled me forever—he interpreted my words with expert precision and with very little guidance on my part, for which I'm very grateful.

Thanks, always, to the rest of the Skyhorse team, who make writing and publishing such a joy. An especially warm thanks to Tony Lyons who has given me countless opportunities to tackle "impossible" projects and been wonderfully supportive through each one. He had the idea for this book years ago and I'm grateful that he entrusted it to me. Thanks also to Bill Wolfsthal and Katherine Mennone, who do the side of publishing I know least about and do it very well, to Julie Matysik and Yvette Grant for their careful eyes, and to Brian Peterson for the great cover.

Finally, thanks to my husband Tim, who can chop wood, haul logs, tap maple trees, build anything, and cook some of the best meals I've ever tasted. He's the real deal.

SOURCES

Adams, Joseph H. *Harper's Outdoor Book for Boys*. New York: Harper & Brothers, 1907.

American Heart Association. *How Can I Manage Stress?* http://americanheart.org/downloadable/heart/1196286112399ManageStress.pdf (accessed June 24, 2009).

American Wind Energy Association. *Wind Energy Fact Sheet*. http://www.awea.org/pubs/factsheets/HowWind-Works2003.pdf (accessed June 22, 2009).

Andersen, Bruce and Malcolm Wells. *Passive Solar Energy Book*. Build it Solar (2005). www.builditsolar.com/Projects/SolarHomes/PasSolEnergyBk/PSEBook.htm (accessed June 23, 2009).

Anderson, Ruben. "Easy homemade soap." *Treehugger: A Discovery Company*. http://treehugger.com/files/2005/12/easy_homemade_s.php (accessed June 24, 2009).

Anderson, Tiffany. "All About Rototillers." *Home & Garden Ideas*. March, 03 20011. http://www.homeandgardenideas.com/outdoor-living/lawn-care/lawn-equipment/all-about-rototillers.

Andress, Elizabeth L. and Judy A. Harrison, ed. *So Easy to Preserve, 5th ed.* Athens: The University of Georgia Cooperative Extension, 2006.

Ashbrook, Frank Getz, Georg Andress Anthony, and Frants Peter Lund. "Pork on the Farm: Killing, Curing, and Canning." *Farmers' Bulletin*. 1186 (1921): Print.

Autumn Hill Llamas & Fiber. "Llama Fiber Article." *Autumn Hill Llamas & Fiber*. http://autumnhillllamas.com/llama_fiber_article.htm (accessed June 24, 2009).

Bailey, Henry Turner, ed. *School Arts Book*, vol. 5. Worcester, MA: The Davis Press, 1906.

"Basics of Choosing a Tractor." *Buyer Zone*. http://www.buyerzone.com/industrial/tractors/bg-choosing-tractor-basics/.

Beard, D.C. *The American Boy's Handy Book*. With Foreword by Noel Perrin. Jaffrey, NH: David R. Godine, Publisher, Inc., 1983.

Beard, Linda and Adelia Belle Beard. *The Original Girl's Handy Book*. New York: Black Dog & Leventhal Publishers Inc., 2007.

Bell, Mary T. *Food Drying with an Attitude*. New York: Skyhorse Publishing, Inc., 2008.

Bellows, Barbara. "Solar Greenhouse Resources." *ATTRA: National Sustainable Agriculture Information Service* (2009). attra.ncat.org/attra-pub/solar-gh.html (accessed June 24, 2009).

Ben. "My Inexpensive 'Do It Yourself' Geothermal Cooling System." *Trees Full of Money.* www.treesfullofmoney. com/?p=131 (accessed June 29, 2009).

Benton, Frank. U.S. Department of Agriculture. *The Honey Bee: A Manual of Instruction in Apiculture.* Washington: Government Printing Office, 1899.

Brooks, William P. *Agriculture vol. III: Animal Husbandry, including The Breeds of Live Stock, The General Principles of Breeding, Feeding Animals; including Discussion of Ensilage, Dairy Management on the Farm, and Poultry Farming.* Springfield, MA: The Home Correspondence School, 1901.

Bower, Mark. "Building an Inexpensive Solar Heating Panel." *Mobile Home Repair* (Aberdeen Home Repair, 2007). www.mobilehomerepair.com/article17c.htm (accessed June 22, 2009).

Boy Scouts of America. *Handbook for Boys.* New York: The Boy Scouts of America, 1916.

"Build a Solar Cooker." *The Solar Cooking Archive.* www. solarcooking.org/plans/default.htm (accessed June 22, 2009).

Byron, A. Hugh and William F. Hubbard. U.S. Department of Agriculture. "The Production of Maple Sirup and Sugar." *Farmers' Bulletin.* 516. (1917): Print. http://ddr.nal.usda. gov/bitstream/10113/32894/1/CAT87201975.pdf.

California Integrated Waste Management Board. "Compost—What Is It?" http://ciwmb.ca.gov/organics/ CompostMulch/CompostIs.htm (accessed June 24, 2009).

California Integrated Waste Management Board. "Home Composting." http://ciwmb.ca.gov/Organics/ HomeCompost (accessed June 24, 2009).

Call Ducks: Call Duck Association UK. http://callducks.net (accessed June 24, 2009).

"Candle making." *Lizzie Candles and Soap.* http://lizziecandle. com/index.cfm/fa/ home.page/pageid/12.htm (accessed June 24, 2009).

"Caring and Cleaning you Equine Tack." *Shane's Tack.* http:// www.shanestack.com/blog/2009/02/caring-and-cleaning- your-equine-tack.html (accessed Junr 11, 2011).

"Cleaning and Care of a Leather Saddle." *State Line Tack.* http://www.statelinetack.com/statelinetack-articles/ cleaning-and-care-of-a-leather-saddle/9583/ (accessed June 11, 2011).

Colnar, Rebecca. "How to save your tack; Tack lasts longer when it's clean." *Horses and Horse Information.* www. horses-and-horse-information.com/articles/0397tack. shtml (accessed June 11, 2011).

Comstock, Anna Botsford. *How to Keep Bees; A Handbook for the Use of Beginners.* Doubleday, Page & Co., 1905.

Cook, E.T., ed. *Garden: An Illustrated Weekly Journal of Horticulture in all its Branches,* vol. 64 (London: Hudson & Kearns, 1903).

Corie, Laren. "Building a Very Simple Solar Water Heater." *Energy Self Sufficiency Newsletter* (Rebel Wolf Energy Systems, September 2005). www.rebelwolf.com/essn/ ESSN-Sep2005.pdf (accessed June 22, 2009).

"Craft instructions: how to make hemp jewelry." *Essortment.* http://essortment.com/hobbies/makehempjewelr_sjbg. htm (accessed June 24, 2009).

Dahl-Bredine, Kathy. "Windshield Shade Solar Cooker." *Slow Cookers World Network.* solarcooking.wikia.com/wiki/ Windshield_shade_solar_funnel_cooker (accessed June 22, 2009).

Dairy Connection Inc. http://dairyconnection.com (accessed June 24, 2009).

Danlac Canada Inc. http://danlac.com (accessed June 24, 2009).

d'Argent, Renee. "Common Problems with Garden Tiller Motors." *Home & Garden Ideas.* http://www. homeandgardenideas.com/outdoor-living/lawn-care/lawn- equipment/common-problems-garden-tiller-motors.

Davis, Michael. "How I Built an Electricity Producing Solar Panel." *Welcome to Mike's World.* www.mdpub.com/ SolarPanel/index.html (accessed June 22, 2009).

Department of Energy. "Energy Kid's Page." *Energy Information Administration.* November 2007. www.eia. doe.gov/kids/energyfacts/sources/renewable/solar.html (accessed June 26, 2009).

Dharma Trading Co., San Rafael, CA 94901 www.dharmatrading.com

Dickens, Charles, ed. *Household Worlds*, vol. 1. London: Charles Dickens & Evans, 1881.

"DIY Home Solar PV Panels." *Green TerraFirma*. greenterrafirma.com/home-solar-panels.html (accessed June 23, 2009).

"Do-It-Yourself Wind Turbine Project." *Green TerraFirma* (2007). greenterrafirma.com/DIY_Wind_Turbine.html (accessed June 23, 2009).

Druchunas, Donna. "Pattern: Fingerless Gloves for Hand Health." *Subversive Knitting*. http://sheeptoshawl.com (accessed June 24, 2009).

"Dry Stone." *Wikipedia*. Web. http://en.wikipedia.org/wiki/Dry_stone (accessed June 6, 2011).

Earle, Alice M. *Home Life in Colonial Days*. New York: Macmillan Company, 1899.

Earthsong Fibers, Osceola, WI 54020. www.earthsongfibers.com.

"Easy Cold Process Soap Recipes for Beginners." *TeachSoap.com: Cold Process Soap Recipes*. http://teachsoap.com/easycpsoap.html (accessed June 24, 2009).

Farmer, Fannie Merritt. *The Boston Cooking-School Cookbook*. Boston: Little, Brown and Company, 1896. Print.

"Features." *Buyer Zone*. http://www.buyerzone.com/industrial/tractors/bg-tractor-features/.

Flach, F., ed. *Stress and Its Management*. New York: W. W. Norton & Co. 1989.

"Fun-Panel." *Solar Cook World Netrwork*. solarcooking.wikia.com/wiki/Fun-Panel (accessed June 22, 2009).

Gegner, Lance. "Llama and Alpaca Farming." *Appropriate Technology Transfer for Rural Areas (ATTRA)*, December 2000. http://attra.ncat.org/attra-pub/llamaalpaca.html (accessed June 24, 2009).

Gehring, Abigail. *Back to Basics*. Third edition. New York: Skyhorse Publishing, 2008. Print.

Georgia Dept. of Agriculture. *Publications*. 6. (1880): Print.

Glengarry Cheesemaking and Dairy Supply Ltd. http://glengarrycheesemaking.on.ca (accessed June 24, 2009).

"Guide to Herbal Remedies." *Natural Health and Longevity Resource Center*. http://all-natural.com/herbguid.html (accessed June 24, 2009).

Hall, A. Neely and Dorothy Perkins. *Handicraft for Handy Girls: Practical Plans for Work and Play*. Boston: Lothrop, Lee & Shepard Company, 1916.

Hill, Thomas E. *The Open Door to Independence: Making Money From the Soil*. Chicago: Hill Standard Book Company, 1915.

"Homemade Solar Panel." pyronet.50megs.com/RePower/Homemade%20Solar%%20Panels.htm (accessed June 24, 2009).

"Homemade Teat Dip & Udder Wash Recipe." *Fias Co Farm*. http://fiascofarm.com/ goats/teatdip-udderwash.html (accessed June 24, 2009).

"Horse Bridles Care and Cleaning of a Horse Bridle." *The Equestrian and Horse*. www.equestrianandhorse.com/tack/bridles/cleaning_a_bridle.html (accessed June 11, 2011).

"How to Build a Mortared Stone Wall." *HGTV*. http://www.hgtv.com/landscaping/mortared-stone-wall/index.html.

"How to Build a Stone Wall." *DIY Network*. http://www.diynetwork.com/how-to/how-to-build-a-stone-wall/index.html.

"How to Knit a Hat." *Knitting for Charity: Easy, Fun and Gratifying*. http://knittingforcharity.org/how_to_knit_a_hat.html (accessed June 24, 2009).

"How to Knit a Scarf for Beginners." *AOK Coral Craft and Gift Bazaar*. http://aokcorral.com/how2oct2003.htm (accessed June 24, 2009).

"How to Make Hemp Jewelry." *Beadage: All About Beading!* http://beadage.net/ hemp/index.shtml (accessed June 24, 2009).

"How to make Taper candles" *How To Make Candles.info*. http://howtomakecandles.info/cm_article.asp?ID=CANDL0603 (accessed June 24, 2009).

"How to Milk a Goat." *Fias Co Farm*. http://fiascofarm.com/goats/ how_to_milk_a_goat.htm (accessed June 24, 2009).

"How to Sell Your Crafts on eBay." *Craft Marketer: DIY Home Business Ideas*. http://craftmarketer.com/sell-your-crafts-on-ebay-article.htm (accessed June 24, 2009).

"How to Use a Rototiller." *Ehow*. http://www.ehow.com/how_4235_rototiller.html.

J.G. "The Fragrance of Potpourri." *Good Housekeeping*, January 1917. New York: Hearst Corp., 1916.

Junket: Making Fine Desserts Since 1874. http://junketdesserts.com (accessed June 24, 2009).

Kellogg, Scott and Stacy Pettigrew. *Toolbox for Sustainable City Living: A Do-It-Ourselves Guide*. Cambridge, MA: South End Press, 2008.

Kendall, P. and J. Sofos. "Drying Fruits." *Nutrition, Health & Food Safety*. Colorado State University Cooperative Extension: No. 9.309 (2003). http://uga.edu/ nchfp/how/dry/csu_dry_fruits.pdf (accessed June 24, 2009).

Kleen, Emil, and Edward Mussey Hartwell. *Handbook of Massage*. Philadelphia: P. Blakiston Son & Co.,1892.

Kleinheinz, Frank. *Sheep Management: A Handbook for the Shepherd and Student, 2nd ed.* Madison, WI: Cantwell Printing Company, 1912.

Ladies' Work-Table Book, The: Containing Clear and Practical Instructions in Plain and Fancy Needlework, Embroidery, Knitting, Netting and Crochet. Philadelphia: G.B. Zeiber & Co., 1845.

Lambert, A. *My Knitting Book*. London: John Murray, 1843.

Lamon, Harry M. and Rob R. Slocum. *Turkey Raising*. New York: Orange Judd Publishing Company, 1922.

"Lawn Mower Buying Guide." *How Stuff Works*. http://products.howstuffworks.com/lawn-mowers-buying-guide.htm.

"Learn to Make Beeswax Candles." *MyCraftBook*. http://mycraftbook.com/ Make_Beeswax_Candles.asp (accessed June 24, 2009).

Lindstrom, Carl. *Greywater*. www.greywater.com (accessed June 25, 2009).

Llucky Chucky Llamas. http://llamafarm.com/welcome.html (accessed June 24, 2009).

Lynch, Charles. *American Red Cross Abridged Text-book on First Aid: General Edition, A Manual of Instruction*. Philadelphia: P. Blakiston's Son & Co., 1910.

"Make Your Own Maple Syrup." *Massachusetts Maple Producers Association*. http://www.massmaple.org/make.php.

"Make Your Own Paper." *Environmental Education for Kids!* http://dnr.wi.gov/org/caer/ ce/eek/cool/paper.htm (accessed June 24, 2009).

Marino, Kristina. "It's Easy Being Green." *All About Lawns*. August 22, 2006. http://www.allaboutlawns.com/lawn-mowing-mowers/its-easy-being-green.php.

"Marketing your homemade crafts." *Essortment*. http://essortment.com/all/ craftsmarketing_mfm.htm (accessed June 24, 2009).

McGee-Cooper, Ann. *You Don't Have to Go Home From Work Exhausted!: The energy engineering approach*. Dallas, Texas: Bowen & Rogers, 1990.

Moore, Donna. "Shear Beauty." *International Lama Registry*. http://lamaregistry.com/ilreport/2005May/shear_beauty_may.html (accessed June 24, 2009).

Moorlands Cheesemakers: Suppliers of Farm and Household Dairy Equipment. http://cheesemaking.co.uk (accessed June 24, 2009).

Morais, Joan. "Beeswax Candles." *Natural Skin and Body Care Products*. http://naturalskinandbodycare.com/2008/12/beeswax-candles.html.

Morris, Gloria. "Buying a Tractor." *Floyd County in View*. CountryView Studios Publishing. http://www.floydcountyinview.com/buyingatractor.html.

Murphy, Karen. "How to make beeswax candles." *SuperEco*, February 14, 2009. http://supereco.com/how-to/how-to-make-beeswax-candles/ (accessed June 24, 2009).

N., Beth. "How to Make Taper Candles." *Associated Content*, September 3, 2007. http://associatedcontent.com/article/360786/how_to_make_taper_candles.html?cat=24 (accessed June 24, 2009).

National Ag Safety Database. "Basic First Aid: Script." *Agsafe*. http://nasdonline.org/docs/d000101-d000200/d000105/d000105.html (accessed June 24, 2009).

National Center for Complementary and Alternative Medicine. "Herbal Medicine." *MedlinePlus: Trusted Health Information for You*. http://nlm.nih.gov/medlineplus/herbalmedicine.html (accessed June 24, 2009).

National Center for Complementary and Alternative Medicine. *Herbs at a Glance*. http://nccam.nih.gov/health/herbsataglance.htm (accessed June 24, 2009).

National Center for Complementary and Alternative Medicine. *Massage Therapy: An Introduction*. http://nccam.nih.gov/health/massage/#1 (accessed June 24, 2009).

National Center for Home Food Preservation. "Drying: Herbs." http://uga.edu/ nchfp/how/dry/herbs.html (accessed June 24, 2009).

National Center for Home Food Preservation. "General Freezing Information." http://uga.edu/nchfp/how/freeze/dont_freeze_foods.html (accessed June 24, 2009).

National Center for Home Food Preservation. "USDA Publications: USDA Complete Guide to Home Canning, 2006." http://uga.edu/nchfp/publications/ publications_usda.html (accessed June 24, 2009).

National Institutes of Health: Office of Dietary Supplements. "Botanical Dietary Supplements: Background Information." *Office of Dietary Supplements*. http://ods.od.nih.gov/factsheets/BotanicalBackground.asp (accessed June 24, 2009).

National Renewable Energy Laboratory. "Wind Energy Basics." *Learning about Renewable Energy*. www.nrel.gov/learning/re_wind.html (accessed June 24, 2009).

New England Cheesemaking Supply Company. http://cheesemaking.com (accessed June 24, 2009).

Nissen, Hartvig. *Practical Massage in Twenty Lessons*. Philadelphia: F.A. Davis Company, 1905.

Nucho, A. O. *Stress Management: The Quest for Zest*. Illinois: Charles C. Thomas, 1988.

Nummer, Brian A. "Fermenting Yogurt at Home." National Center for Home Food Preservation: 2002. http://uga.edu/nchfp/publications/nchfp/factsheets/ yogurt.html (accessed June 24,2009).

Ostrom, Kurre Wilhelm. *Massage and the Original Swedish Movements: their application to various diseases of the body*. 6th ed. Philadelphia: P. Blakiston's Son & Co., 1905.

Ponder, T. *How to Avoid Burnout*. California: Pacific Press Publishing Association, 1983.

Powell, Albrecht. "All About Maple Syrup." *About*. http://pittsburgh.about.com/cs/pennsylvania/a/maple_syrup.htm.

Reyhle, Nicole. "Selling Your Homemade Goods." *Retail Minded*, January 23, 2009. http://retailminded.com/blog/2009/01/selling-your-homemade-goods (accessed June 24, 2009).

Retail Minded. http://retailminded.com/blog (accessed June 24, 2009).

"Rotary Tiller or Rototillers Which One in Your Garden's Future?." *Plant-Care*. http://www.plant-care.com/rototillers-garden-tiller.html.

"Rototiller Parts." *Rototiller Store*. http://www.rototillerstore.com/rototiller-parts/.

Sanford, Frank G. *The Art Crafts for Beginners*. New York: The Century Co., 1906.

Sell, Randy. "Llama" *Alternative Agriculture* Series, no. 12, August 1993. http://ag.ndsu.edu/pubs/alt-ag/llama.htm (accessed June 24, 2009).

Singleton, Esther. *The Shakespeare Garden*. New York: The Century Co., 1922.

Smith, Kimberly. "Where to Sell Your Homemade Crafts Offline." *Associated Content*, April 29, 2009. http://associatedcontent.com/article/1678550/where_to_sell_your_homemade_crafts.html (accessed June 24, 2009).

"Soap making – General Instructions." *Walton Feed, Inc.* http://waltonfeed.com/old/old/ soap/soap.html (accessed June 24, 2009).

"Soy candle making." *Soya – Information about Soy and Soya Products*. http://soya.be/soy-candle-making.php (accessed June 24, 2009).

Swenson, Allan A. *Foods Jesus Ate and How to Grow Them*. New York: Skyhorse Publishing, Inc., 2008.

Szykitka, Walter. *The Big Book of Self-Reliant Living: Advice and Information on Just About Everything You Need to Know to Live on Planet Earth*. Second edition. Guilford, CT: The Lyons Press, 2004.

Table Rock Llamas Fiber Art Studio, Black Forest, CO 80908. www.tablerockllamas.com.

Taylor, George Herbert. *Massage: Principles and Practice of Remedial Treatment by Imparted Motion*. New York: John B. Alden, 1887.

Thompson, Nita Norphlet and Sue McKinney-Cull. "Soothing Those Jangled Nerves: Stress Management." *ARCH Factsheet*, no. 41, September 1995, revised February 2002. http://archrespite.org/archfs41.htm (accessed June 24, 2009).

"Tips and Techniques." *Hub UK*. http://www.hub-uk.com/ cooking/tipsmaplesyrup.htm.

"Tractor Attachments." *Buyer Zone*. http://www.buyerzone. com/industrial/tractors/bg-tractor-attachments/.

U.S. Department of Agriculture. *Farmers' Bulletins, Nos. 1176-1200*. Washington D.C.: Government Printing Office, 1922. http://books.google.com/ books?id=1H8EAAAAYAAJ&dq=baking%20 bread&lr&as_drrb_is=b&as_minm_is=0&as_miny_ is=1750&as_maxm_is=0&as_maxy_is=1923&as_ brr=0&pg=PP1#v=onepage&q=baking%20 bread&f=false.

U.S. Department of Agriculture: Food Safety and Inspection Service. *Fact Sheets: Egg Products Preparation*. http://fsis. usda.gov/Factsheets/ Focus_On_Shell_Eggs/index.asp (accessed June 24, 2009).

U.S. Department of Agriculture: Food Safety and Inspection Service. *Fact Sheets: Poultry Preparation*. http://fsis.usda. gov/Fact_Sheets/ Chicken_Food_Safety_Focus/index.asp (accessed June 24, 2009).

U.S. Department of Agriculture: Natural Resources Conservation Service. "Backyard Conservation: Composting." http://nrcs.usda.gov/feature/ backyard/ compost.html (accessed June 24, 2009).

U.S. Department of Agriculture: Natural Resources Conservation Service. "Backyard Conservation: Nutrient Management." http://nrcs.usda.gov/ feature/backyard/ nutmgt.html (accessed June 24, 2009).

U.S. Department of Agriculture: Natural Resources Conservation Service. "Composting in the Yard." http:// nrcs.usda.gov/feature/backyard/ compyrd.html (accessed June 24, 2009).

U.S. Department of Agriculture: Natural Resources Conservation Service. "Home and Garden Tips: Composting." http://nrcs.usda.gov/feature/ highlights/ homegarden/compost.html (accessed June 24, 2009).

U.S. Department of Agriculture: Natural Resources Conservation Service. "Home and Garden Tips: Lawn and Garden Care." http://nrcs.usda.gov/ feature/highlights/ homegarden/lawn.html (accessed June 24, 2009).

U.S. Department of Agriculture: National Agricultural Library. "Organic Production." http://afsic.nal.usda.gov/ nal_display/index.php?info_center=2&tax_level=1&tax_ subject=296 (accessed June 24, 2009).

U.S. Department of Energy. "Active Solar Heating." *Energy Efficiency and Renewable Energy: Energy Savers*. www. energysavers.gov/your_home/space_heating_cooling/ index.cfm/mytopic=12490 (accessed June 26, 2009).

U.S. Department of Energy. "Benefits of Geothermal Heat Pump Systems." *Energy Efficiency and Renewable Energy: Energy Savers*. www.energysavers.gov/your_home/space_ heating_cooling/index.cfm/mytopic=12660 (accessed June 25, 2009).

U.S. Department of Energy. "Energy Efficiency and Renewable Energy." *Wind and Hydropower Technologies Program*. www1.eere.energy.gov/windandhydro/ (accessed June 24, 2009).

U.S. Department of Energy. "Energy Technologies." *Efficiency and Renewable Energy: Solar Energy Technologies Program*. www1.eere.energy.gov/solar/want_pv.html (accessed June 26, 2009).

U.S. Department of Energy. "Geothermal Heat Pumps." *Energy Efficiency and Renewable Energy: Energy Savers*. www. energysavers.gov/your_home/space_heating_cooling/ index.cfm/mytopic=12650 (accessed June 26, 2009).

U.S. Department of Energy. "Heat Pump Water Heaters." *Energy Efficiency and Renewable Energy: Energy Savers*. www.energysavers.gov/your_home/water_heating/index. cfm/mytopic=12840 (accessed June 26, 2009).

U.S. Department of Energy. "Hydropower Basics." *Energy Efficiency and Renewable Energy: Wind and Hydropower Technologies Program*. www1.eere.energy.gov/windandhydro/ hydro_basics.html (accessed June 26, 2009).

U.S. Department of Energy. "Renewable Energy." *Energy Efficiency and Renewable Energy: Energy Savers*. www. energysavers.gov/renewable_energy/solar/index.cfm/ mytopic=50011 (accessed June 26, 2009).

U.S. Department of Energy. "Selecting and Installing a Geothermal Heat Pump System." *Energy Efficiency and Renewable Energy: Energy Savers*. www.energysavers. gov/your_home/space_heating_cooling/index.cfm/ mytopic=12670 (accessed June 25, 2009).

U.S. Department of Energy. "Technologies." *Energy Efficiency and Renewable Energy: Geothermal Technologies Program*. www1.eere.energy.gov/geothermal/faqs.html (accessed June 25, 2009).

U.S. Department of Energy. "Your Home." *Energy Efficiency and Renewable Energy: Energy Savers*. www.energysavers.

gov/your_home/space_heating_cooling/index.cfm/ mytopic=12300 (accessed June 26, 2009).

U.S. Department of Energy: National Renewable Energy Laboratory. "Direct Use of Geothermal Energy." *Office of Geothermal Technologies.* www1.eere.energy.gov/ geothermal/pdfs/directuse.pdf (accessed June 26, 2009).

U.S. Department of Energy: National Renewable Energy Laboratory. "Wind Energy Myths." *Wind Powering American Fact Sheet Series.* www.nrel.gov/docs/ fy05osti/37657.pdf (accessed June 26, 2009).

U.S. Department of Energy. "Solar." *Energy Sources.* www. energy.gov/energysources/solar.htm (accessed June 26, 2009).

U.S. Department of Energy. "Toilets and Urinals." *Greening Federal Facilities.* Second edition. www.eere.energy.gov/ femp/pdfs/29267-6.2.pdf (accessed June 29, 2009).

U.S. Environmental Protection Agency. "Composting Toilets." *Water Efficiency Technology Fact Sheet.* www.epa.gov/owm/ mtb/comp.pdf (accessed June 29, 2009).

U.S. House of Representatives. United States Department of Agriculture. *Report of the Commissioner of Patents for the Year 1831: Agriculture.* 37th congress, 2nd sess., 1861.

University of Maryland. *National Goat Handbook.* http:// uwex.edu/ces/ cty/richland/ag/documents/national_goat_ handbook.pdf (accessed June 24, 2009).

Volk, Bill. "Building a Natural Stone Wall." DIY Life. AOL, January 27, 2008. Web. http://www.diylife. com/2008/01/27/building-a-natural-stone-wall/ (accessed June 7, 2011).

West, Dawn. "What Mower Should I Use?." *All About Lawns.* August 22, 2006. http://www.allaboutlawns.com/lawn-mowing-mowers/what-mower-should-i-use.php

"Where to sell crafts? Consider these often overlooked alternative markets…" *Craft Marketer: DIY Home Business Ideas.* http://craftmarketer.com/ where_to_sell_crafts.htm (accessed June 24, 2009).

Whipple, J. R. "Solar Heater." *J. R. Whipple & Associates.* www. jrwhipple.com/sr/solheater.html (accessed June 23, 2009).

Wickell, Janet. *Quilting.* Teach Yourself Books. Chicago: NTC/Contemporary Publishing, 2000.

Williams, Archibald. *Things Worth Making.* New York: Thomas Nelson and Sons, Ltd., 1920.

"Wind Energy Basics." *Wind Energy Development Programmatic EIS.* windeis.anl.gov/guide/basics/index.cfm (accessed June 25, 2009).

Wolok, Rina. "How to Build a Composting Toilet." *Greeniacs,* June 15, 2009. greeniacs.com/GreeniacsGuides/How-to-Build-a-Composting-Toilet.html (accessed June 29, 2009).

Woods, Tom. "Homemade Solar Panels." *Forcefield,* 2003. www.fieldlines.com/story/2005/1/5/51211/79555 (accessed June 24, 2009).

Woolman, Mary S. and Ellen B. McGowan. *Textiles: A Handbook for the Student and the Consumer.* New York: The Macmillan Company, 1921.

Worcester Polytechnic Institute. "A Passive Solar Space Heater for Home Use." *Solar Components Corporation.* www. solar-components.com/SOLARKAL.HTM#doityourself (accessed June 22, 2009).

Young Ladies' Journal, The: Complete Guide to the Work-Table. London: E. Harrison, 1885.

RESOURCES

Animals

Bees

American Beekeeper Federation
Helps promote beekeeping and the existence of honeybees. Check out the section on kids and bees.
www.abfnet.org

American Beekeeping Federation
3525 Piedmont Road, Building 5, Suite 300
Atlanta, Georgia 30305
info@afbnet.org
Phone: (404) 760-2875

American Honey Producers Association
Learn more about beekeeping.
www.americanhoneyproducers.org

Backyard Beekeepers Association
A club dedicated to sharing knowledge about beekeeping and furthering general education.
www.backyardbeekeepers.org

Bee Culture
A magazine for all beekeeping needs.
www.beeculture.com
Phone: (800) 289-7668

Beekeeping: Self-Sufficiency, by Joanna Ryde
A cute little book with a lot of good information.

Heartland Apicultural Society
An association dedicated to raising and keeping bees.
www.heartlandbees.com

International Bee Research Association
The world's largest bee research center.
www.ibra.org.uk
I.B.R.A.
16 North Road
Cardiff, CF10 3DY
U.K.
info@ibra.org.uk
Phone: 00 44 (0)29 2037 2409

Western Apicultural Society
An educational organization dedicated to beekeeping.
http://groups.ucanr.org/WAS

Chickens

American Pastured Poultry Producers Association
A nonprofit organization that encourages raising poultry on pasture.
http://apppa.org
PO Box 87
Boyd, WI 54726

Backyard Chickens
A site dedicated to spreading information about raising chickens in urban, suburban, or rural areas.
www.backyardchickens.com

Beginning Farming
Great site for new farmers; has a section on raising chickens.
www.beginningfarmers.org/information-about-raising-chickens

International Poultry Breeders Association
Good information about breeds of chickens and how to raise them.
www.featheredfamilies.com

Joy of Keeping Chickens by Jennifer Megyesi
A compact but comprehensive guide.

Poultry Youth Association
A site with vast resources for raising chickens.
www.poultryyouth.com

US Poultry and Egg Association
The world's largest poultry association.
www.poultryegg.org
1530 Cooledge Road
Tucker, Georgia 30084-7303
Phone: (770) 493-9401

Cows

American Dairy Association and Dairy Council, Inc.
An organization that spreads awareness of dairy farming in the Tri-State area.
www.adadc.com

American Milking Shorthorn Society
Good information about milking, breeding, and raising cows. Great classifieds listing.
www.milkingshorthorn.com

Farm Sanctuary
An organization that promotes raising cows at home, and highlights the negative effects of factory farming.
www.farmsanctuary.org/issues/factoryfarming/dairy
P.O. Box 150
Watkins Glen, NY 14891
    Phone: (607) 583-2225

National Sustainability Agriculture Information Service
A wealth of information on raising cows at home.
www.cias.wisc.edu/dairysch.html

School for Beginning Dairy and Livestock Farmers
An organization that provides training on how to become a dairy farmer.
www.cias.wisc.edu/dairysch.html

Small Dairy
A site providing information to small dairy farmers.
www.smalldairy.com

Small Farms
A publication of Oregon State University that provides information about all aspects of dairy farming.
http://smallfarms.oregonstate.edu/dairy

Dogs

American Kennel Club
A national organization that provides information about rescuing and adopting dogs. You can search by breed.
www.akc.org/breeds/rescue.cfm

Great Dog Site
The name says it all.
www.greatdogsite.com

Next Day Pets
A site for buying dogs which also offers information on dog care.
www.nextdaypets.com

Pet Education.com
This site is dedicated to educating people about proper pet care.
www.peteducation.com

Pet Owners Association
An organization that educates dog and other pet owners.
www.pet-owners.co.uk
Crosslands House Nynehead, Wellington Somerset, TA21 0BS

Terrific Pets.com
This site has an extensive amount of information on dogs.
www.terrificpets.com

Ducks

Duck Hobby
Complete online backyard guide to raising ducks.
http://duckhobby.com

Game Bird
A site designed to help beginners learn about raising game birds, including ducks.
www.gamebird.com/ducksgeeseswans.html

The New Agrarian
A site by a suburban hobby farmer. Read the "Duckling Diaries."
www.newagrarian.com/category/ducks

Poultry One
Good, straightforward instructions.
http://poultryone.com/raisingducks.php

Poultry Pages
All different components of raising ducks and poultry.
www.poultrypages.com/raising-ducks.html

Raising Ducks
A publication by the University of California's Division of Agriculture and Natural Resources that provides information on raising ducks.
http://animalscience.ucdavis.edu/avian/ducks.pdf

Raising Ducks
A publication put out by Holderread farms, a waterfowl farm and preservation center.

www.metzerfarms.com/Articles/RaisingDucksHolderread.pdf
Dave Holderread
Holderread Waterfowl Farm & Preservation Center
PO Box 492
Corvallis, OR 97330

Fish

American Tilapia Association
A national organization for tilapia fish farmers.
http://ag.arizona.edu/azaqua/tilapia.html

Fish Farming Business
Lots of information for beginning fish farmers.
www.fishfarmingbusiness.com/

Fish Farming News
A national aqua culture news publication that provides up-to-date information on the fish farming industry.
www.fish-news.com/ffn.htm/
PO Box 600
Deer Isle, ME 04627
Phone: (207) 348-1057

National Cooperative Grocers Association
An organization that promotes growing and raising produce and fish in your backyard.
www.northernaquafarms.com/hobbyfarms.html
14 S Linn Street,
Iowa City, Iowa 52240
Phone: (319) 466-9029

Northern Aqua Farms
This site details how to start your own fish farm.

www.northernaquafarms.com/hobbyfarms.html

United States Trout Farmers Association
National association of trout farmers.
www.ustfa.org
PO Box 1647
Pine Bluff, AR 71613 USA
Phone: (870) 850-7900
ustfa@thenaa.net

Goats

Dairy Goat Factbook
Solid information on raising dairy goats.
http://members.toast.net/dawog/Goats/DairyGoatFactBook.htm

Dairy Goat Journal
A publication that provides up-to-date information on raising milking goats.
www.dairygoatjournal.com/
45 Industrial Drive
Medford, WI 54451
Phone: (715) 785-7979

Fias Co Farm
Good listing of all goat breeds.
http://fiascofarm.com/goats/breeds.htm

Joy of Keeping Goats by Laura Childs.
This whole "Joy" series is rich with useful advice. This one's no exception.

Maryland Small Ruminant Page
Lots of articles on raising goats.
www.sheepandgoat.com

The Meat Goat Breeds
A 4-H fair factsheet of all the meat goat breeds.
www.ansci.cornell.edu/4H/meatgoats/meatgoatfs2.htm

Raising Goats
Clear, concise inforamtion.
www.raisinggoats.org

Raising Goats Guide.com
Complete guide to goat care.
www.raisinggoatsguide.com
PO Box 14079
Spokane Valley, WA 99214

Horses

Acreage Equine
Basic information for the novice horseperson.
www.acreageequines.com/horsecare

Alpha Horse
A site providing horse care advice.
www.alphahorse.com/horse-care.html

American Horse Council
An association of horse owners. The Unwanted Horse Coalition is particularly worthwhile.
www.horsecouncil.org/
1616 H Street NW, 7th floor
Washington, DC 20006
Phone: (202) 296-4031

ASPCA
An animal protection agency with a complete guide to proper horse care on their site.
www.aspca.org/pet-care/horse-care/

Equerry.com
A site specifically for first-time horse owners.
http://equerry.com/html/ftho/eq_ftho-feeding-care.htm

Horse Channel.com
Has a complete list of horse breeds.
www.horsechannel.com.

Horses Never Lie, 2nd Ed by Mark Rashid
All of Rashid's books give incredible insight into horse behavior and horsemanship.

Wow Horses
A site with a full list of horse diseases and how to raise a healthy horse.
www.wowhorses.com/horse-care.html

Llamas

Alternative Farming Systems Information Center
A part of the United States Division of Agriculture's National Agricultural Library that provides information on raising llamas. Type "llamas" into the search box.
http://afsic.nal.usda.gov

Hurricane Creek Llamas
A llama farm with a site detailing how to raise your pet llama.
www.hurricanecreekllamas.com/care.html

Kent Rock Meadows Llamas
A llama farm with excellent online resources providing information on how to care for llamas.
www.krmllamas.com/KRMllamas.html

Llama
Fun videos and information.
www.llama.org

Llama Web
A site dedicated to llamas and llama care.
www.llamaweb.com

Philo Llamas
An online guide to llama care.
www.philollamas.com/basics.htm

UMass Extension
The University of Massachusetts Amherst puts out an online guide to llama shearing.
www.umass.edu/cdl/BMPs/Llama%20Shearing%2008-10.pdf

Pigs

Country Living, Country Skills
Good information for small-scale pig raising.
www.kountrylife.com/articles/pigs.htm

Countryside and Small Stock Journal
All-around interesting publication that often has info on raising pigs.
www.countrysidemag.com

National Swine Registry
A national organization that registers purebred pigs.
www.nationalswine.com/
2639 Yeager Road
West Lafayette, IN 47906
Phone: (765) 463-3594

Pig Health
Factsheets on pig health.
www.pighealth.com

The Pig Site
Good information the different types
of breeds.
www.thepigsite.com

Rabbits

American Rabbit Breeders Association
An organization that promotes the
development of rabbits and gives
information to rabbit owners.
www.arba.net
PO Box 5667
Bloomington, IL 61702
Phone: (309) 664-7500
info@arba.net

House Rabbit Society
An organization that rescues rabbits
and educates the public about
rabbit care.
www.rabbit.org/

My House Rabbit
An organization dedicated to
spreading information about
rabbits as pets.
www.myhouserabbit.com/

Raising Rabbits
Comprehensive and straightforward
information on rabbit care.
www.raising-rabbits.com

Survival Homestead
A site for homesteaders that includes
some useful articles on rabbit
breeds.
www.survival-homestead.com

Sheep

American Sheep Industry Association
A national association of sheep
farmers.
www.sheepusa.org

National Sheep Organization
The United Kingdom's national
organization for sheep farmers.
www.nationalsheep.org.uk

Ontario Ministry of Agriculture,
Food, and Rural Affairs
Ontario's agriculture site with links to
sheep care information.
www.omafra.gov.on.ca/english/
livestock/sheep/links.html

Sheep 101
The name says it all.
www.sheep101.info

Sheep!
A magazine dedicated to sheep
farmers.
www.sheepmagazine.com
145 Industrial Drive
Medford, WI 54451
Phone: (715) 785-7979

U Mass Extension: Sheep Shearing
The University of Massachusetts
Amherst's publication on sheep
shearing.
www.umass.edu/cdl/BMPs/Sheep%20
Shearing%2008-08.pdf

Turkeys

Common Diseases and Ailments of
Turkeys
A useful article about the diseases
contracted by turkeys.
www.albc-usa.org/documents/
turkeymanual/ALBCturkey-5.pdf

Heritage Turkey Foundation
An organization that tries to preserve
heritage turkeys.
www.heritageturkeyfoundation.org
300 Plum Street #97
Capitola, CA 95010
Phone: (831) 477-1501

Minnesota Turkey Growers
Association
Good information on raising turkeys.
www.minnesotaturkey.com

Turkey Management Guide
A PDF guide to turkey farming.
www.cpdosrbng.kar.nic.in/
TURKEY%20FARMING%20
GUIDE.pdf

Baking, Preserving, and More

Beer

Beer Week
A site for America's beer drinkers.
www.beerweek.com

Brewers Association
An organization dedicated to brewers.
www.brewersassociation.org

Brewing Techniques
Great instructions and tips.
www.brewingtechniques.com/

Here's to Beer
A site for beer enthusiasts with tips on
how to brew your own beer.
www.htbeerconnoisseur.com

Homebrew Heaven
Online brewery shop.
http://store.homebrewheaven.com

Bread

Bread
This site is really for folks in the
bread industry, but check out the
glossary.
http://bread.com

Bread World
Lots of great recipes.
www.breadworld.com

History of Bread
A look into the timeline of bread
throughout history. (Great for a
homeschool study!)
www.breadinfo.com/history.shtml

Types of Bread in the World
A baker's online list of all the types of
bread in the world.
www.abigailsbakery.com/bread-
recipes/types-of-bread-in-the-
world.htm

Yeast Bread Recipes
A site with yeast bread recipes.
www.breadexperience.com/yeast-
bread-recipes.html

Butchering

Beginning Farming
This farming online resource has an
excellent guide to butchering for
beginners.
www.beginningfarmers.org/
butchering-for-beginners

The Modern Homestead
Has a wealth of articles, including
some on butchering poultry.
www.themodernhomestead.us

The Rabbit Revolution
Includes useful articles about
butchering rabbits.
http://therabbitrevolution.com

Butter

Butter through the Ages
Interesting article with a recipe for
butter.
www.webexhibits.org/butter/
doityourself.html

History of Butter
The Wisconsin Cheese Organization
explaining the history of butter.
www.eatwisconsincheese.com/
wisconsin/other_dairy/butter/
butter_basics/history_of_butter.
aspx
8418 Excelsior Dr.
Madison, WI 53717
Phone: (608) 836-8820
info@EatWisconsinCheese.com

Muller Lane's Farm
I love this site for all things
homesteading-related. Includes
instructions for making butter.
www.mullerslanefarm.com.

Canning

Backwoods Homes Magazine
A homesteading magazine that often
has articles on canning.
www.backwoodshome.com
Phone: (800) 835-2418

Canning Basics
A site that outlines all of the basic
canning methods.
www.canningbasics.com/guidelines-
for-food-safety.html

Canning Recipes for Preserving Food
Lots of yummy recipes.
www.canning-food-recipes.com/

National Center for Home Food
Preservation
A national food preservation
organization with full online
resources describing the process of
canning.
www.uga.edu/nchfp/how/can_home.
html
The University of Georgia
208 Hoke Smith Annex
Athens, GA 30602-4356

Pick Your Own
A site for finding pick-your-own
farms that also details how to can
your own food.
www.pickyourown.org/
allaboutcanning.htm

Simply Canning
A site dedicated to canning tricks and
tips.
www.simplycanning.com/canning-
safety.html

Cheese

American Cheese Society
An association for cheese lovers and
makers.
www.cheesemaking.com/store/pg/88-
American-Cheese-Society.html
54B Whately Rd
South Deerfield, MA 01373
Phone: (413) 397-2012
info@cheesemaking.com

Beginning Cheese Making
A site for beginner cheese makers.
http://biology.clc.uc.edu/fankhauser/
cheese/cheese_course/cheese_
course.htm

Cheese and Dairy-Related Industry
Resources
A compilation of cheese industry
societies, buyers, and consumers.
www.cheesemarketnews.com/
cdresources.html

Cheese Making Illustrated
A site detailing how to make cheese
step by step.
http://biology.clc.uc.edu/fankhauser/
cheese/cheese_5_gallons/
cheese_5gal_00.htm

Cheese Making Recipes
Lots of fun recipes.
www.cheesemakingrecipe.com

Real Milk
A rather opinioned site that promotes
raw milk sale and consumption.
Has a great listing of where you
can find raw milk.
www.realmilk.com

Specialist Cheese Makers
Organization
An organization dedicated to protect
cheese makers' economic interests
in the United Kingdom and to
encourage people to make their
own cheese.
www.specialistcheesemakers.co.uk/
17 Clerkenwell Green
London, EC1R ODP
Phone: 020 7253 2114

Cider

All About Apples
A site that describes the various
different types of apples.
www.allaboutapples.com/varieties

Apples—History and Legends of
Apples
A site dedicated to the history of
apples; also gives a cider recipe.
http://whatscookingamerica.net/Fruit/
Apples.htm

Cider Apple Varieties
An article about the best apples to use
to make the best cider.
http://homepage.ntlworld.com/
scrumpy/cider/ciderapp.htm

Making Hard Cider
An article published by the Cornell
science faculty detailing how to
make your own hard apple cider.
www.hort.cornell.edu/department/
faculty/merwin/hardcider2.htm

Drying

A Review of Solar Food Drying
An article on drying food through
solar processes.
http://solarcooking.org/dryingreview
.htm

Drying Food
A detailed article written by the
University of Illinois at Urbana-
Champaign's Department of
Agriculture explaining the process
of drying.
www.aces.uiuc.edu/vista/html_pubs/
DRYING/dryfood.html

Drying Food
Good info on the length of time
certain foods take to dry.
www.barlowscientific.com/technotes/
home/dry_food.htm

University of Minnesota Extension:
Drying Food
The agricultural department of the
University of Minnesota that
published an article on how to dry
food.
www.extension.umn.edu/foodsafety/
components/handouts/
dryingfoods.pdf

University of Missouri Extension:
Drying for Keeps
The agricultural department of
the University of Missouri that
published an article on drying
food.
http://extension.missouri.edu/
publications/DisplayPub.
aspx?P=GH1562

Edible Wild Plants

Science Reference Services: Edible
    Wild Plants
The Library of Congress
www.loc.gov/rr/scitech/tracer-bullets
    /edibleplantstb.html

Wild Food!
A survival man's site that describes
    edible wild plants.
www.wildmanstevebrill.com/

Wilderness Survival
A survival site with a section about the
    edibility of wild plants.
www.wilderness-survival.net/plants-1.
    php

Wildwood Survival
A survival skills site with a list of
    edible wild plants.
www.wildwoodsurvival.com/survival/
    food/edibleplants/index.html

Carpentry and Woodworking

Basic Tools

Basic Woodworking Tools
A site with projects and plans for
    woodworking at home.
http://woodworkingprojectsonline.
    com/basic-woodworking-tools/

Mike's Tools
A consumer site for woodworking
    tools and supplies.
www.mikestools.com/woodworking-
    tools.aspx

Woodworking and Carpentry
A site full of resources for specific tasks and projects.
http://alsnetbiz.com/homeimprovement/woodwork.html

Wood Projects Using Basic Carpentry
A how-to site on woodworking do-it-yourself projects.
www.mycarpentry.com

Bedroom Furniture

Woodworker's Workshop
Free online plans and guides.
www.woodworkersworkshop.com

Woodworking Plans
A site with wooden furniture plans available for download.
www.furnitureplans.com

Birdhouses

Building 50 Bird Nestboxes
A brief article on the construction of nesting boxes.
www.50birds.com/dbuildingnestboxes.htm

Building Bird Houses for Wrens, Bluebirds, and More
Charts for appropriate sizes for a variety of birds.
www.wild-bird-watching.com/Building_Bird_Houses.html

Free Bird House Plans
Includes plans for different varieties of birds.
www.freebirdhouseplans.net/freebuildingbirdhouseplans.html

Materials Used for Building Birdhouses
A brief article on what and what not to use.
www.birdhouses101.com/Materials-Used-Building-Birdhouses.asp

Bridges

Arched Bridge
A design plan for a wooden path and small arched bridge.
http://pages.areaguides.com/ubuild/landscapearchedbridge.htm

Building a Balsa Bridge for Competition
A site with techniques and plans for building your own balsa wood bridge.
www.science-projects.com/Bridges/BridgeBulsa.htm

Fun & Learning About Bridges
A resource site with links for all things bridges.
www.bridgesite.com/funand.htm

How to Build an Arched Footbridge
Plans and techniques for building bridges in your garden.
www.redwoodbridges.com/build_footbridge.html

Doghouses

Doghouse Instructions
A site that provides links and helpful tips on building a home for your dog.
www.unchainyourdog.org/Doghouse.htm

How to Build a Doghouse
A beginner's project available from Lowes.
www.lowes.com/cd_Build+a+Doghouse_939249501_

Only Dog Beds
A site with information on a variety of styles of custom doghouses.
www.onlydogbeds.net/build_a_dog_house

Gates and Fences

Building a Fence Gate
A brief tutorial on building a privacy fence.
www.instructables.com/id/Building-a-fence-gate/

Building and Hanging a Double Gate
A double gate project available from Home Depot.
www.homedepot.com/webapp/wcs/stores/servlet/ContentView?pn=KH_PG_FN_Build_and_Hang_Double_Gate&langId=-1&storeId=10051&catalogId=10053

Farm Fences
A comprehensive guide to building long-lasting farm fences.
www.gatewayalpacas.com/alpaca-farming/fence-building.htm

Free Woodworking Plans and Projects
A series of free plans to build a variety of fences and gates.
www.buildeazy.com/fences.html

Kitchen Furniture

How to Reface and Refinish Kitchen Cabinets
A site with a step-by-step tutorial with photos.
www.diynetwork.com/how-to/how-to-reface-and-refinish-kitchen-cabinets/index.html

Making Your Own Kitchen Cabinets
Tips and techniques for building your own cabinet project.
www.rockler.com/articles/build-your-own-kitchen-cabinets.cfm

Modern Kitchen Furniture and How to Build Cabinets
Tips and techniques on installing kitchen furniture with inspirational design photos.
http://luxury-furniture-design.net

Lumber: Converting Trees To Logs

Traditional Timber Farming
A site explaining how to build with timber. Included is a section on how to turn logs into lumber.
http://environment.uwe.ac.uk

Units of Measure and Conversion Factors for Forest Products
A scholarly bulletin explaining the scales of measure and conversion.
www.umext.maine.edu/onlinepubs/PDFpubs/7103.pdf

Poultry Houses

Housing Chickens in a Chicken Coop
Tips and techniques on how, when, and where to build or place your coop.
http://poultryone.com

Poultry Housing
A guide to animal husbandry with techniques on poultry farming.
www.smallstock.info/info/genhusb/poultry-house.htm

Poultry Housing Help
A site with various resources on building, maintaining, and repairing poultry houses.
www.poultryhelp.com/link-housing.html

Small Scale Poultry Housing
A guide to building and maintaining coops in your home.
http://pubs.ext.vt.edu/2902/2902-1092/2902-1092_pdf.pdf

Root Cellars

Build a Root Cellar
A blog with resources, links, and photos.
www.survival-spot.com/survival-blog/build-root-cellar/

Building a Root Cellar
A brief guide to building a root cellar.
www.organicgardening.com/learn-and-grow/building-root-cellar

Produce Bound Underground
An article summing up everything you need to know on root cellars.
www.hobbyfarms.com/food-and-kitchen/root-cellars-14908.aspx

Sheds, Toolhouses, and Workshops

Designer Shed Plans
Shed plans both free and for purchase with illustrations of finished products.
www.designer-shed-plans.com

My Shed Plans Center
A brief guide on where to choose the perfect location for your new shed.
www.myshedplanscenter.com/location-for-backyard-shed

Smokehouses

Building a Small Smokehouse
A blueprint and overview of building your own smoker.
www.kountrylife.com/articles/smkhse.htm

Building a Smokehouse
A step-by-step do-it-yourself guide.
www.diy-guides.com/building-a-smoke-house/

How to Build a Smoke House
An article with tips, techniques, and photos.
www.squidoo.com/how-to-build-smoker

Simple Smoking by Paul Kirk
More than 80 great recipes.

Smokehouse Plans

A brief summary with photos on building your own smoke house.

www.smoking-meat.com/smokehouse-plans.html

Stables

Building a Horse Stable

A brief overview, tips, and techniques for building a stable.

www.colorado-horse.com/building-a-horse-stable

Horse Barn Plans

Blueprints and plans available for purchase with illustrations of the finished products.

www.applevalleybarns.com/

Horse Barns

A consumer site for custom-built stables.

www.horizonstructures.com/horse-barns.asp

Phone: (888) 447-4337

Horse Barns, Stables, and Fencing Information

A brief article with online resources about your horse's home.

www.horses-and-horse-information.com/horsebarns.shtml

Tree Houses

Build the Ultimate Tree House

A step-by-step tutorial provided by Disney.

http://familyfun.go.com/crafts/build-the-ultimate-tree-house-708814/

Build Your Own Tree House

A comprehensive site for everything tree house.

www.thetreehouseguide.com/

Workshop Furniture

Woodworker's Workshop

Free online plans and guides.

www.woodworkersworkshop.com

Crafts

Basketry

Basket Makers

Good site for discussion of various basket weaving techniques, including links to sites that focus on everything basket weaving, from retailers to tips and tutorials.

http://basketmakers.com

Basket Making

A site with illustrated step-by-step instructions on various forms of basket weaving.

www.basket-making.com

Basket Patterns

Your online source for basket patterns and materials

www.basketpatterns.com/mm5/merchant.mvc

3741 W Houghton Lake Dr.

Houghton Lake, MI 48629

info@basketpatterns.com

Phone: (800) 563-2356

The Country Seat, Inc.

A company that weaves baskets as well as lists weaving teachers nationwide, offers weaving classes and tips, and lists weaving conventions.

www.countryseat.com

1013 Old Philly Pike

Kempton, Pennsylvania 19529-9321

weaving@countryseat.com

Phone: (610) 756-6124

National Basketry Organization, Inc.

A non-profit organization that is dedicated to the art of basketry, holding biennial conferences, workshops, and seminars for those interested in the craft.

www.nationalbasketry.org/index.html

PO Box 277

Brasstown, NC 28902

info@nationalbasketry.org

Phone: (828) 837-1280

Native American Basketry

A comprehensive site of Native American basketry complete with history, slideshow tutorials, and games.

www.nativetech.org/basketry/index.html

Simply Baskets

A site for basket-related information including basket weaving patterns, a directory of basket weaving suppliers and articles about baskets.

www.simplybaskets.net

Blown Eggs

Adventures in Cooking: Herb Stenciled Easter Eggs

A home cooking blog that includes an article on how to use herbs and natural ingredients for stenciling and dying eggs.

http://adventurescooking.blogspot.com

The Artful Crafter: How to Dye
    Easter Eggs Naturally
A blog dedicated to crafts and
    how-tos. Includes instructions
    on how to dye eggs using natural
    ingredients.
http://the-artful-crafter.blogspot.
    com/2011/04/how-to-dye-easter-
    eggs-naturally.html

Baby Jayne's
A step-by-step tutorial on how to
    blow an egg so it can be decorated
    for Easter
www.babyjaynes.com/2011/04/hand-
    blown-egg-tutorial.html

South Texas Unit of the Herb Society
    of America
An organization dedicated to
    promoting the knowledge, use,
    and delight of herbs including
    instructions on how to naturally
    dye eggs.
www.herbsociety-stu.org/DyeingEggs.
    htm
South Texas Unit of the HSA
PO Box 6515
Houston, TX 77265-6515
Phone: (713) 513-7808

Simple Organic: Back to Nature, Back
    to Basics
How to Color Eggs with Natural Dyes
A site focusing on working with
    organic materials, including
    tutorials on how to blow eggs and
    dye them with natural plants.
http://simpleorganic.net/how-to-
    color-eggs-with-natural-dyes/

Book Binding

About Bookbinding
A site dedicated to the art of
    bookbinding, including
    information on book anatomy,
    links to free binding books online,
    and bookbinding companies.
www.aboutbookbinding.com/index.html
info@aboutbookbinding.com

Bookbinding
A site dedicated to bookbinding,
    including how-to links for every
    experience level, essays, and live
    feed for bookbinding news.
http://bookbinding.com/

Bookbinding for Beginners
An informational site about
    bookbinding for beginners and
    how to go about it, including a
    glossary of definitions, diagrams,
    and supplies store.
http://bookbindingfb.com/
    index.html
P.O. Box 1081
Summerfield, FL 24492
bookbindingfb@hotmail.com

The Care of Fine Books by Jane
    Greenfield
A classic that includes instructions for
    re-binding old books.

The Center for Book Arts
An organization that is dedicated to
    preserving book binding as a craft
    and offers classes and events on it.
www.centerforbookarts.org/
28 West 27th Street, 3rd Floor
    New York, New York 10001
Phone: (212) 481-0295
info@centerforbookarts.org

Dave's Book Tutorial
A site that gives an in-depth tutorial
    on how to bind your own book.
www.davethedesigner.net/booktut/
    index.html

DIY Bookbinding
A comprehensive site on the
    advantages of do-it-yourself
    bookbinding, including discussion
    groups and how-to articles on the
    craft.
www.diybookbinding.com

Candles

Cajun's Candle Making Supplies
A company that sells candle making
    supplies and provides step-by-step
    instructions on how to make
    them.
www.cajuncandles.com
Phone: (337) 643-8344

Candle and Soap Making Techniques
A site that gives instruction
    on different candle making
    techniques.
www.candletech.com

Candle Help
Dozens of candle making tutorials for
    all experience levels and project
    ideas.
www.candlehelp.com

How to Make Candles
An instructive site all about making
    candles and allowing you to
    contact other candle makers online
www.howtomakecandles.info

Let's Make Candles
A site with instructions on candle
making, as well as project ideas and
tips on how to sell candles.
www.letsmakecandles.com/index.asp
info@letsmakecandles.com

My Candle Connection
A site and blog dedicated to candles,
from home décor to candle making
as a hobby.
www.mycandleconnection.com/

Soy Wax Candles
A site focusing specifically on the
making of soy wax candles,
including instructions and
information on everything from
types of waxes to containers.
www.soywaxcandles.org/

Cornhusk Dolls

All Homemade
Learn how to make homemade crafts
from the past.
www.all-homemade.com

Claude Moore Colonial Farm
A farm that reenacts the life of people
from 1771 and teaches children
skills such as wool making, butter
churning, and making corn husk
dolls.
www.1771.org/cd_doll.htm
Claude Moore Colonial Farm
6310 Georgetown Pike
McLean, Virginia 22101
webmaster@1771.org
Phone: (703) 442-7557

Manataka American Indian Council
Learn about the Manataka Tribe and
other Native American customs
such as making corn husk dolls.
www.manataka.org/page67.html

Native Tech
A site focusing on the techniques used
by Native Americans, including a
history and instructional page on
making corn husk dolls.
www.nativetech.org/cornhusk/
corndoll.html

Snoww Owl
A site that focuses on Native
American traditions and
provides the history, legends,
and instructions on how to make
various native crafts.
www.snowwowl.com/index.html

Teachers First
An informational site for teachers
providing project ideas and lesson
plans for children, including a
diagramed tutorial on making corn
husk dolls.
www.teachersfirst.com/summer/
cornhusk.htm

Flower Arranging

The Charlottesville Garden Club
A garden club that provides general
rules for flower arranging along
with links to books and site on the
same subject.
http://cgc.avenue.org/
flowerarrangingpublic.html#sites

Classic Flower Arrangements
A site on how to create flower
arrangements.
www.classic-flower-arrangements.com

DIY Flower Craft
A site focusing on how to arrange
flowers, including videos, setting
up classes, and arrangement ideas.
www.diyflowercraft.com

Fearless Flowers
A site based on video tutorials of how
to arrange flowers.
www.fearlessflowers.com
fearlessflowers@gmail.com
Phone: (877) 331-8454

Flower Arrangement Advisor
A site about how to make flower
arrangements for all occasions.
www.flower-arrangement-
advisor.com

Home Made Simple
A site devoted to home decoration
and organization, including do-it-
yourself floral arrangements.
www.homemadesimple.com/
en-US/Garden/Pages/do-it-
yourself-floral-arrangements.
aspx?TID=78972bb2-b0f5-41e7-
a5cd-7a481558628d

Soapmaking

Natural Pure Soap
Lots of information, directions, and
recipes for homemade soaps.
www.naturalpuresoap.com

Soap Making Fun

Directions for both cold process and hot process soap making, as well as melt and pour.

www.soapmakingfun.com

Super Soap Making

Tips, techniques, recipes, and more for organic soapmaking

www.supersoapmaking.com

Teach Soap

Soapmaking tutorials, recipes, tips, and more.

www.teachsoap.com

Disaster-preparedness

The Disaster Preparedness Handbook, by Arthur T. Bradley, PhD

The best resource I've seen for protecting your home and family.

Earthquakes

The Disaster Center

A complete step-by-step guide on what to expect—and what to do—when an earthquake hits.

www.disastercenter.com/guide/earth.html

Equipped to Survive

A good resource indicating what to put in an earthquake preparedness kit.

www.equipped.org/earthqk.htm

FEMA: Earthquake

Index of earthquake related publications, preparedness tools and resources for families, building designers, businesses, and more.

www.fema.gov/hazard/earthquake/

Federal Emergency Management Agency

U.S. Department of Homeland Security

500 C Street SW, Washington, D.C. 20472

Phone: (202) 646-2500

QuakeKare

Provides ready-to-buy earthquake preparedness kits as well as news on disasters happening all over the world.

www.quakekare.com/

Ready America: Earthquakes

Immediate information about earthquake preparedness.

www.ready.gov/america/beinformed/earthquakes.html

Red Cross

Emergency contacts and disaster relief when earthquakes hit.

www.redcross.org/

USGS: Earthquake Hazards Program

Presents details of past earthquakes, monitors present earthquakes, and extends educated information about where earthquakes are likely to hit.

http://earthquake.usgs.gov/earthquakes/

Earthquake Hazards Program Headquarters Office

12201 Sunrise Valley Drive, MS 905

Reston, VA, 20192

First Aid

American Heart Association

CPR and other first aid classes available to prepare you for any emergency.

www.heart.org/

American Heart Association

7272 Greenville Ave.

Dallas, TX 75231

Phone: (800) 242-8721

Family Doctor

Assesses common household medical situations and how to administer immediate assistance.

http://familydoctor.org/online/famdocen/home/healthy/firstaid.html

The First Aid Kits

Allows you to buy first aid kits online, fully stocked with everything you could need to administer first aid.

www.thefirstaidkits.com/

Kids Health

Prepares parents and caretakers to make sure their children are safe and how to administer first aid if necessary.

http://kidshealth.org/parent/firstaid_safe/

Mayo Clinic: First Aid

Lists maladies that need immediate aid and how you should react to each one.

www.mayoclinic.com/health/FirstAidIndex/FirstAidIndex

Ready America: First Aid

Lists important items every first aid kit should contain, whether you plan on buying one or making a kit by yourself.

www.ready.gov/america/getakit/firstaidkit.html

Equipped to Survive: Home First Aid Kit

Invaluable source when creating a first aid kit—based on the recommendations from the American College of Emergency Physicians

www.equipped.org/home1staid.htm

Flood

Country Survival: Flood Preparedness

What to do when a flood occurs and how to be ready for it.

www.countrysurvival.com/flood-preparedness/

The Disaster Center

Complete guide to the causes of floods, how to be prepared, and how to recover.

www.disastercenter.com/guide/flood.html

FEMA: Flood

Complete guide to the significance of preparing for floods, the type of damage they cause, and how to be ready.

www.fema.gov/areyouready/flood.shtm

Federal Emergency Management Agency

U.S. Department of Homeland Security

500 C Street SW, Washington, D.C. 20472

Phone: (202) 646-2500

Flood Preparedness and Response

A complete, detailed guide on what to do when floods occur, from preparedness to how to react when the waters rise.

www.wvdhsem.gov/WV_Disaster_Library/Library/FLOODS/Flood%20Preparedness%20and%20Response.htm

New York State Department of Health

Quick guide to floods: what to expect and how to be prepared.

www.health.state.ny.us/environmental/emergency/flood/

Public Health Emergency

Information about current flood warnings and disasters, as well as emergency preparedness tips.

www.phe.gov/emergency/naturaldisasters/Pages/flood.aspx

Hurricanes

Be Prepared

Everything you need to pack to be prepared for the storm, from creating a disaster supply kit to planning evacuation ahead of time.

www.nhc.noaa.gov/HAW2/english/disaster_prevention.shtml

The Disaster Center

What causes disasters and how to protect yourself and your belongings from them.

www.disastercenter.com/guide/hurricane.html

FEMA: Hurricane

What to expect in a hurricane, how to prepare, and how to react.

www.fema.gov/hazard/hurricane/index.shtm

Federal Emergency Management Agency

U.S. Department of Homeland Security

500 C Street SW

Washington, D.C. 20472

Phone: (202) 646-2500

National Hurricane Center

Videos, audio files, and more information about hurricanes presented during National Hurricane awareness week.

www.nhc.noaa.gov/outreach/prepared_week.shtml

NOAA/National Weather Service

National Centers for Environmental Prediction

National Hurricane Center

11691 SW 17th Street

Miami, FL 33165-2149 USA

Ready America: Hurricanes

Everything you need to know about how to be informed and how to prepare for a hurricane emergency.

www.ready.gov/america/beinformed/hurricanes.html

Top Ten Hurricane Tips

If you're in a hurry, these ten essential tips will help you be prepared for evacuation, as well as preparing an emergency kit.

www.chiff.com/a/hurricane-tips.htm

Wildfires

CDC: Wildfires
Health concerns and safety issues
involved with wildfires.
www.bt.cdc.gov/disasters/
wildfires
Centers for Disease Control and
Prevention
1600 Clifton Rd.
Atlanta, GA 30333, USA
Phone: (800) CDC-INFO

Every Life Secure: Wildfire
How to prepare for a wildfire.
www.everylifesecure.com/wildfire/

FEMA: Wildfires
What to expect when a wildfire
occurs, how to be prepared, and
how to react.
www.fema.gov/hazard/wildfire/wf_
prepare.shtm
Federal Emergency Management
Agency
U.S. Department of Homeland
Security
500 C Street SW
Washington, D.C. 20472
Phone: (202) 646-2500
Firewise Communities
Encourages communities to take
initiative in the planning and
prevention of wildfires.
www.firewise.org/
Phone: (617) 984-7486

National Fire Protection Agency
The authority on fire, building and
electrical safety. Includes extensive
safety information for fires and fire
prevention.
www.nfpa.org
National Fire Protection Agency

1 Batterymarch Park
Quincy, MA 02169-7471
Phone: (617) 770-3000

Wildfire Preparedness
A complete to being prepared for a wildfire disaster.
http://disaster-emergency-preparedness.com/wildfire-preparedness-wildfire-readiness-steps/2011/05/
Emergency Packs: Wildfire Disaster Info
Tips for surviving a wildfire, and the resources to do so, complete with emergency kits available for sale.
www.essentialpacks.com/Wildfire-Disaster-Info

Energy

Composting Toilets

BioLet
Composting toilets and their waterless predecessors can be purchased. Also includes information about the importance of composting toilets.
www.biolet.com/

Clivusmultrum Inc.
Manufactures and sells both compost toilets and greywater systems.
www.clivusmultrum.com/

Composting Toilet World
A site dedicated to educating about composting toilets and promoting their use, including where to buy one and how to make your own.
http://compostingtoilet.org

The Humanure Handbook
Pictures and step-by-step instructions on how to create your own composting toilet.
http://humanurehandbook.com/humanure_toilet.html

Let's Go Green
Practical alternatives to sewer and septic systems, including composting toilets.
www.letsgogreen.com/

Sustainable Sources
An honest discussion about the use and cost of composting toilets, and the issues that arise.
http://composttoilet.sustainablesources.com/

Geothermal Energy

Clean Energy Ideas
Explanation of the use of geothermal energy.
www.clean-energy-ideas.com/geothermal_power.html

Energy Consumer's Edge: Geothermal Energy
A complete guide to the process of obtaining geothermal energy, as well as the pros and cons associated with it.
www.energy-consumers-edge.com/pros_and_cons_of_geothermal_energy.html

Geothermal Education Office
An introduction to geothermal energy and how it is used.
http://geothermal.marin.org/pwrheat.html

Geothermal Energy Association
An association with the intention of expanding the use of geothermal energy through public policy and activity, complete with guides to understanding geothermal energy and how it will help.
www.geo-energy.org/
Geothermal Energy Association
209 Pennsylvania Avenue SE, Washington, DC 20003

Union of Concerned Scientists: How Geothermal Energy Works
A thorough article outlining the uses of geothermal energy and how it is obtained.
www.ucsusa.org/clean_energy/technology_and_impacts/energy_technologies/how-geothermal-energy-works.html

Greywater

Clivusmultrum Inc.
Manufactures and sells both compost toilets and greywater systems.
www.clivusmultrum.com/

Greywater
Explanation of greywater and blackwater, how to tell the difference, and why it is important to make that distinction.
www.greywater.com/

Oasis Designs
Links for information on using greywater
http://oasisdesign.net/greywater/references.htm

ReWater Systems
Creates a filtering system that will separate greywater from blackwater to route the greywater for reuse.
www.rewater.com

Sustainable Sources
Discusses the process, costs, and benefits to creating a greywater irrigation system.
http://greywater.sustainablesources.com

Hydropower

Alternative Energy News
Provides news and information about hydroelectric power technologies.
www.alternative-energy-news.info/technology/hydro/

Energy
Provides useful links and information about hydropower.
www.energy.gov/energysources/hydropower.htm

Interesting Energy Facts
Easy-to-digest facts about hydropower to help you understand its significance.
http://interestingenergyfacts.blogspot.com/2008/03/hydropower-facts.html

International Hydropower Association
Resources for moving forward hydropower initiatives. Includes publications, useful links, and the opportunity to make a difference.
www.hydropower.org/

National Hydropower Association
Explains uses and production of hydropower, as well as information on its implementation.
http://hydro.org/
National Hydropower Association
25 Massachusetts Ave., NW
Suite 450
Washington, DC 20001
Phone: (202) 682-1700
help@hydro.org

Renewable Energy World
Explains the process of using hydropower as a renewable energy source, as well as providing links to blogs and news sources involving renewable energy.
www.renewableenergyworld.com/rea/tech/hydropower

Solar Energy

Alternate Energy Sources
Thirty facts about solar energy, with useful links to other informational sites.
www.alternate-energy-sources.com/facts-about-solar-energy.html

The American Solar Energy Society
The nation's leading nonprofit association of solar professionals and grassroots advocates. Includes ways to get involved in promoting solar energy.
http://ases.org
2400 Central Ave, Suite A
Boulder, CO 80301
Phone: (303) 443-3130

Energy Efficiency and Renewable Energy
Works to develop cost-competitive solar energy systems for America and has information about the development of research and solar resources.
www1.eere.energy.gov/solar/about.html

National Renewable Energy Laboratory
Educational resources about solar energy
www.nrel.gov/learning/re_solar.html

Solar Energy
Nonprofit organization whose mission is to help others use renewable energy and environmental building technologies through education.
www.solarenergy.org

Solar Energy Industries Association
Works to expand the use of solar technologies, strengthen research and development, remove market outreach, and educate more people on solar.
www.seia.org/
Solar Energy Industries Association
575 7th Street, NW, Suite 400
Washington DC 20004

Science Daily
A regularly updated site, including information on the development of solar energy.
www.sciencedaily.com/news/matter_energy/solar_energy

Wind Energy

Alternate Energy Sources
Thorough facts about wind energy
and the process of obtaining it.
www.alternate-energy-sources.com/
wind-energy-facts.html

American Wind Energy Association
Complete facts and explanations
of wind energy, as well as
opportunities to get involved.
www.awea.org/
American Wind Energy Association
1501 M Street, NW, Suite 1000
Washington, DC 20005
Phone: (202) 383-2500

Clean Energy Ideas
The advantages as well as the
disadvantages of wind energy.
www.clean-energy-ideas.com/articles/
advantages_and_disadvantages_of_
wind_energy.html

Energy Refuge
Informational resource on wind energy.
www.energyrefuge.com/archives/
wind-energy-facts.htm

Interesting Energy Facts
Easy-to-digest one-line quick facts
about wind energy.
http://interestingenergyfacts.blogspot.
com/2008/03/wind-energy-facts.
html

Wind Energy Development
Programmatic EIS
A public information and
involvement resource
http://windeis.anl.gov/guide/basics/
index.cfm

Gardening

Bees, Birds, and Butterflies: How To Attract Them To Your Garden

Backyard Birding by Julie Zickefoose
and the Editors and Writers of
Bird Watcher's Digest.
Lovely illustrations and lots of useful
tips.

Gardening to Attract Wild Birds,
Butterflies & Hummingbirds
This is a consumer site with products
and extensive descriptions to help
you decide which butterflies and
birds you wish to attract.
www.birds-n-garden.com/birdgarden.
html

Here's How to Befriend the Birds, Bees,
and Butterflies on Your Balcony
An article geared to those in more
urban areas.
http://lifeonthebalcony.com/heres-
how-to-befriend-the-birds-bees-
and-butterflies-on-your-balcony/

How to Attract Butterflies,
Hummingbirds, and Bees
A site with articles and lists about
attracting pollinators to your
garden and how pollination works.
www.dutchgardens.com/butterflies.asp
Plant a Garden that Attracts Birds and
Butterflies
Brief listings of which plants attract
certain wildlife.
www.acmehowto.com/garden/flowers/
garden-birds-bees.php

Community Gardens

American Community Gardening
Association
Start your garden in an urban area.
http://communitygarden.org/
American Community Gardening
Association
1777 East Broad Street
Columbus, OH 43203
info@communitygarden.org
Phone: (877) ASK-ACGA

The Food Project
Building sustainable local food
systems.
http://thefoodproject.org/
555 Dudley Street
Dorchester, MA 02125

GardenWeb
Good site for discussion of various
garden types and gardening
techniques, including directories
of garden retailers and local garden
resources.
http://gardenweb.com/

Local Harvest
Find a farmer's market nearest to
you!
http://localharvest.org/

Oasis NYC
A site covering urban gardening
organizations in the New York City
area.
http://oasisnyc.net/gardens/resources.
htm

Composting in Your Backyard

CalRecycle
Tips, techniques, and troubleshooting
for home composting.
www.calrecycle.ca.gov/Organics/
HomeCompost/

DIYlife
An article on creating your own
compost pile. The site provides
links and how-to videos.
www.diylife.com/2008/04/14/create-
a-compost-pile-in-your-backyard/

How to Compost
A comprehensive site for all things
compost.
www.howtocompost.org/

Master Gardening
A consumer site that provides
materials for home gardening,
including compost bins.
www.mastergardening.com/
composters

Natural Resources Conservation
Service
A comprehensive site administered by
the USDA for conservation of land
and animals.
www.nrcs.usda.gov/FEATURE/
backyard/compost.html
National Headquarters
Mailing Address:
Natural Resources Conservation
Service
Attn: Public Affairs Division
PO Box 2890
Washington, DC 20013
Street Address:

Natural Resources Conservation
Service
14th and Independence Avenue, SW
Washington, DC 20250

U.S. Environmental Protection
Agency
A short guide on both indoor and
outdoor composting.
www.epa.gov/osw/conserve/rrr/
composting/by_compost.htm
Mailing Address:
US EPA
Office of Resource Conservation and
Recovery (5305P)
1200 Pennsylvania Avenue, NW
Washington, DC 20460

Container Gardening

Container Gardening Guide
A site with articles, forums, and FAQ
for all your container gardening
needs.
www.containergardeningtips.com/

Container Gardening Ideas
Photos and tips from the experts at
HGTV.com
www.hgtv.com/topics/container-
gardening/index.html

Container Gardens
A comprehensive site dedicated to
container gardening.
www.container-gardens.com/

Gardener's Supply Company
A consumer site for indoor and
outdoor planters.
www.gardeners.com/Container-
Gardening

Guide to Container Gardening
Articles and community forums for
gardeners.
www.gardenguides.com/685-guide-
container-gardening.html

iVillage Garden Web
A community forum and discussion
group all about container and
other types of gardening.
http://forums2.gardenweb.com/
forums/contain

Farmers' Markets

Farmers Market Coalition
A site dedicated to education and
resources for farmers' markets
managers and members.
http://farmersmarketcoalition.org/
Mailing Address:
Farmers Market Coalition
PO Box 331
Cockeysville, MD 21030

Farmers Markets Search
A search engine for USDA approved
farmers' markets located
throughout the US
http://apps.ams.usda.gov/
FarmersMarkets

Local Harvest
A comprehensive site with a shop,
community forum, blogs, photos,
and resources to locate your local
farmers' market.
www.localharvest.org/farmers-markets

Flower Gardens

Beginner's Guide to Perennial
Gardening
Everything you need to know to begin
planting your own perennial flower
garden.
www.lewisgardens.com/

Flower Gardening Made Easy
A comprehensive guide to all things
flower garden.
www.flower-gardening-made-easy.com

Garden Ideas
A site dedicated to inspiring successful
flower garden designs.
www.gardenideas.com/garden_
designs/

Fruit Bushes and Trees

How to Grow Fruit Trees and Fruit
Bushes
A guide and forum for gardening
made easy.
www.gardening-advice.net/growing-
fruit-trees.html

Where to Buy Fruit Trees, Fruit
Bushes and Fruit Plants
An article detailing how, when, and
where to buy fruit trees, bushes,
and plants.
www.thegardeningbible.com/where-
to-buy-fruit-trees-fruit-bushes-and-
fruit-plants

Spraying and Pruning Fruit Trees and
Fruit Bushes
An article on how to care for your
fruit trees, bushes, and plants.
www.gardeninginfozone.com/
spraying-and-pruning-fruit-trees-
and-fruit-bushes

Indoor Fruit Trees
A community and forum on indoor
fruit trees, bushes, and plants.
http://forums.treehugger.com/
viewtopic.php?f=1&t=14881

Nature Hills Nursery
A consumer site for online shopping
and shipping.
www.naturehills.com/
Phone: (888) 864-7663

Grains: Growing and Threshing

Homegrown Grains: The Key to Food
Security—How to Grow and Make
Your Own Wheat Flour
An article on how to grow grain and
make it into an edible product.
www.organicconsumers.org/articles/
article_18082.cfm

How to Grow Grains
A comprehensive site to all things
grains. This article includes an
FAQ section.
www.islandgrains.com/how-to-grow-
grains

iVillage Garden Web
A community forum and discussion
group all about growing grains and
other types of gardening.
http://forums.gardenweb.
com/forums/load/legumes/
msg0515542921424.html

Greenhouses and Hoophouses

Hoop House Greenhouse
An article describing the pros and
cons of backyard hoophouses.
www.123-greenhouse-gardening.com/
hoop-house.html

Hoop House Greenhouse Kits
A consumer site with products and tips.
www.hoophouse.com/

How to build a PVC Hoop House for
Your Garden
A how-to guide with photos, charts,
and step-by-step instructions to
do-it-yourself.
http://westsidegardener.com/howto/
hoophouse.html

How to Build My 50 Dollar
Greenhouse
A how-to guide with photos, charts,
and step-by-step instructions to
do-it-yourself.
http://doorgarden.com/10/50-dollar-
hoop-house-green-house

Harvesting Your Garden

Harvesting Herbs from Your Garden
Five tips and tricks on harvesting from
your herb garden.
www.bhg.com/gardening/vegetable/
herbs/harvesting-herbs-from-your-
garden

How to Harvest Seeds from Your
Garden
An article on how and when to
harvest your fruit, veggie, and
flower seeds.
www.howtogardenadvice.com/seeds/
harvesting_seeds.html

Preserving Your Garden Harvest
A site with recipes and how-tos in order to preserve and eat your produce in the months to come.
www.bettycrocker.com/tips/tipslibrary/misc/preserving-your-garden-harvest

When to Harvest Your Garden Vegetables
A chart with brief descriptions on when and how to harvest your veggies.
www.savvygardener.com/Features/harvesting_vegetables.html

Herb Gardens

Creating Herb Gardens
An article geared towards getting children involved with herb and other forms of gardening.
www.kidsgardening.com/growingideas/projects/may04/pg1.html

Herbal Gardens
A site dedicated to providing information on how to grow, harvest, and utilize herbs for various activities.
www.herbalgardens.com

Herb Gardening
A comprehensive guide to growing and maintaining your own herbs.
http://herbgardening.com

Herb Kits
A consumer site with products for indoor herb and other gardening kits.
www.herbkits.com

Hydroponics and Aggregate Culture

A Guide to the Secret World of Hydroponic Cultivation
A beginner's guide to hydroponic gardening.
http://hydroponiclight.com/a205643-a-guide-to-the-secret-world.cfm

Hydroasis
A consumer site with tips, tricks, and FAQ.
www.hydroasis.com/growingtips/growingveggieswithoutsoil.shtml
Phone: (888) 355-4769

Hydroponics Live
A series of articles dedicated to hydroponics and aggregate culture and equipment.
www.hydroponicslive.com/aggregate-culture.html

The Hydroponic System
A guide to building and maintaining your own hydroponic garden.
www.myhydroponicgarden.net/

Machinery For Your Garden and Farm

Farm and Garden
A buyer's guide and comprehensive site dedicated to farming and gardening.
www.ag1.biz

Tiny Farm Gear
A site dedicated to small farm and gardening tools, tips, and techniques.
http://gear.tinyfarmblog.com

Tractors and Farm Equipment
An article pertaining to the top three pieces of equipment you will need for small farming in the spring.
www.motherearthnews.com/blogs/blog.aspx?blogid=2147484070&tag=spring%20chores

Mulching: Why and How To Do It

All About Mulch
A brief guide on everything you need to know on mulch.
www.savvygardener.com/Features/mulch.html

Compost Guide
An article explaining just how much mulch you really need.
http://compostguide.com/how-much-mulch-do-you-need/

Making Mulch from Leaves
A step-by-step guide to making your own mulch, with photos.
www.hereandthere.org/making-mulch/making-mulch-from-leaves.html

Tree Care Information
An article dedicated to proper mulching techniques.
www.treesaregood.com/treecare/mulching.aspx

Organic Gardening

iVillage Garden Web
A community forum and discussion group all about organic and other types of gardening.
http://forums2.gardenweb.com/forums/organic/

Organic Gardening
A site dedicated entirely to organic gardening.
www.organicgardening.com/

Organic Gardening Guru
A site with tips, tricks, articles, and glossaries.
www.organicgardeningguru.com/

Organic Gardening Information
A comprehensive site with unique takes on organic home gardening, including pest control tips and guides on lunar gardening.
www.organicgardeninfo.com/

Pest and Disease Management

How to Go Organic
A site with brief articles on pest, insect, weed, and disease management with many other resources provided.
www.howtogoorganic.com/index.php?page=pest-disease-management

Pest and Disease Management Handbook
An older free online guide to pest and disease management.
http://onlinelibrary.wiley.com/book/10.1002/9780470690475

Resource Guide for Organic Insect and Disease Management
An excellent guide geared specifically towards the organic farmer.
http://web.pppmb.cals.cornell.edu/resourceguide/

Planning a Garden

Kitchen Garden Planner
Virtual and pre-planned kitchen gardens.
www.gardeners.com/Kitchen-Garden-Planner/kgp_home,default,pg.html

Plan a Garden
A virtual garden planner available from Better Homes and Gardens.
www.bhg.com/gardening/design/nature-lovers/welcome-to-plan-a-garden

Royal Horticultural Society
A comprehensive site for all things plants and gardening.
http://apps.rhs.org.uk/advicesearch/Search.aspx?oa=true
Mailing Address:
80 Vincent Square
London
SW1P 2PE
Phone: 0845 260 5000

Vegetable Garden Guide
Photos, tips, and articles on planning and cultivating your veggie garden.
www.marthastewart.com/photogallery/planning-your-vegetable-garden

Planting and Tending Your Garden

Garden Tools You Will Need
A brief article on some materials to get you started.
www.urbanorganicgardening.org/planting-tending-tools.html

On Mother Plants and Tending Your Organic Garden
An article on tending rather than planting, leaning towards hydroponics.
www.plantpropagation.homehydroponics.info/mother-plants/how-to-tend-to-your-organic-garden

Tending Garden
An article for beginner organic farmers.
www.organic-gardening-tips.co.uk/html/tending_garden.html

Raised Beds

Cedar Raised Bed Gardens
A site dedicated to landscape ideas, kits, and designs.
www.raised-garden-beds.com/

Grow a Vegetable Garden in Raised Beds
A slideshow with beautiful photos and techniques for raised bed landscaping.
www.bhg.com/gardening/vegetable/vegetables/grow-a-vegetable-garden-in-raised-beds/

How to Build the Perfect Raised Bed
A step-by-step guide with photos and video.
www.sunset.com/garden/backyard-projects/ultimate-raised-bed-how-to-00400000011938/

Raised Beds
A consumer site with products for raised beds and other gardening supplies.

www.gardeners.com/Raised-
Beds/VegetableGardening_
RaisedBeds,default,sc.html

Rooftop Gardens

Green Roof
A site with blogs, news, video
dedicated to bringing awareness to
rooftop gardening.
www.greenroofs.com/

Rooftop Garden
A community site dedicated to rooftop
gardening. Includes a variety of
resources, video, and links.
http://rooftopgarden.com/

Green Roof
A site with blogs, news, video
dedicated to bringing awareness to
rooftop gardening.
www.greenroofs.com/

Manhattan Rooftop Garden Project
A blog dedicated to rooftop gardening
in a densely packed city.
http://nycroofgardenproject.blogspot.
com/

Urban Roof Gardens
Roof Gardening in the UK
www.urbanroofgardens.com/

Saving Seeds

Basic Seed Saving
A comprehensive site on saving and
planting seeds.
www.seedsave.org/issi/issi_904.html

Saving Seed
A quick fact sheet devoted to
gardeners who wish to harvest their
own seeds.
www.ext.colostate.edu/pubs/
garden/07602.html

Save Your Seeds!
A how-to guide.
www.care2.com/greenliving/save-your-
seeds-how-to-directions.html#
Vegetable Seed Saving Handbook
An online handbook with specific
instructions for a variety of veggies.
http://howtosaveseeds.com/

Soil and Fertilizer

Soil Fertilizer and Soil Additives

A brief article accompanied by
additional resources.
www.improve-your-garden-soil.com/
soil-fertilizer-soil-additives.html

Home and Garden Information
Center
A month-to-month guide to care for
your garden soil.
www.hgic.umd.edu/content/soil.cfm

Organic Fertilizer
A site dedicated to maintaining your
organic farm with the correct form
of fertilizers.
www.the-organic-gardener.com/
organic-fertilizer.html

Vegetable Garden Soil
An informative how-to guide on
garden soil.
http://home.howstuffworks.com/
vegetable-garden-soil.htm

Terracing

Building a Terraced Hillside Garden
A brief article and how-to guide with
photos.
www.learn2grow.com/projects/
landscaping/construction/
BuildingTerracedGarden.
aspx?page=1

iVillage Garden Web
A community forum and discussion
group all about terraced and other
types of gardening.
http://forums.gardenweb.
com/forums/load/calif/
msg052100451852.html

Natural Resources Conservation
Service
A comprehensive site administered by
the USDA for conservation of land
and animals.
www.nrcs.usda.gov/feature/backyard/
terrac.html
National Headquarters
Mailing Address:
Natural Resources Conservation
Service
Attn: Public Affairs Division
PO Box 2890
Washington, DC 20013
Street Address:
Natural Resources Conservation
Service
14th and Independence Avenue, SW
Washington, DC 20250

Terraced Gardening
A brief guide to backyard terracing.
www.mastergardenproducts.com/
gardenerscorner/terrac.htm

Trees For Shade or Shelter

Arbor Day Foundation
A site dedicated to educating and
inspiring people about trees. Trees
are also available for purchase.
www.arborday.org/index.cfm

Plants for Shade Gardening
A comprehensive list of trees and
plants which provide shade, for
how long, and where they grow.
www.mortonarb.org/images/stories/
pdf/our_work/Plants_Shade_
Gardening.PDF

Pruning Field Grown Shade and
Flowering Trees
An article with techniques and timing
for pruning trees around your
home.
www.ces.ncsu.edu/depts/hort/hil/hil-
406.html

Vegetable Gardens

Home Farming
A community site with tips, tricks,
plenty of information, fun guides,
and recipes.
www.homefarming.com/
Cornell Gardening Resources
A site with links and online resources
for everything you need to know
about how to grow and maintain
your own veggie garden.
www.gardening.cornell.edu/
vegetables/index.html

Vegetable Garden Ideas
A series of articles explaining
everything you need to know about
growing and harvesting your own
food.
www.vegetablegardeningideas.com/

The Vegetable Garden
A comprehensive site all about veggie
gardens.
www.thevegetablegarden.info/

Water: How To Collect It, Save It, And Use It

Build it Solar
A site with do-it-yourself plans for
conserving and harvesting water
around your home. Complete with
online resources and links.
www.builditsolar.com/Projects/Water/
Water.htm
American Water and Energy Savers
A site with forty-nine ways to
conserve water both indoors and
outdoors.
www.americanwater.com/49ways.php

Save Water—Collect It
A site dedicated to "being green." This
particular article is accompanied by
an instructional video.
www.dothegreenthing.com/blog/
save_water_collect_it

Water: Use It Wisely
A comprehensive site designed to
educate the public on water
conservation.
www.wateruseitwisely.com/

Well-being

Community

Societal Cultural Competence and
Cultural Community Well-Being.
A scholarly article on community and
cultural awareness.
www.forumonpublicpolicy.com/
archive07/woodroffe.pdf

Cosmetics

Aromantic
Instructions and recipes for
homemade skincare products.
www.aromantic.com

Cranberry Lane
Tips, supplies, and recipes for
make-it-yourself body care.
www.cranberrylane.com

Make Your Cosmetics
Lots of recipes for homemade
natural cosmetics.
www.makeyourcosmetics.com

Herbal Remedies

Herbal Remedies
A comprehensive chart with ailments
and their respective herbal cures.
www.gardensablaze.com/
HerbRemedies.htm

Herbal Remedies Info
A comprehensive site dedicated to
alternative medicine and therapies.
www.herbalremediesinfo.com

Natural Herbs Guide
A comprehensive site and buyers' guide complete with recipes, dietary information, and herb search engine.
www.naturalherbsguide.com/herbal-remedies.html

Strictly Health
A consumer site with alternative medicinal products.
www.strictlyhealth.com/

Herbal Teas

How to Make Herbal Tea
A chart and guide on how to brew herbal teas.
www.gardensablaze.com/HerbTea.htm

Natural Herbs Guide
Information and descriptions on different herbal teas.

www.naturalherbsguide.com/herbal-teas.html

Homeschooling

HomeSchool
A comprehensive community site with resources, links, and forums.
www.homeschool.com/
Homeschooling
A site available from Home Education Magazine.
www.homeedmag.com/

Homeschool World
A site available from Practical Homeschooling Magazine.
www.home-school.com/

HSLDA
A site dedicated to homeschooling laws and practices across the US
www.hslda.org/hs/default.asp

Peace Hill Press
Excellent resource for curriculum and books on education and learning.
www.welltrainedmind.com/store

Massage

Taking Charge of Your Health
A comprehensive site dedicated to the awareness of the benefits of massage therapy, how it works, and how to choose the right therapist and treatment.
www.takingcharge.csh.umn.edu/explore-healing-practices/massage-therapy/how-can-massage-help-my-health-and-well-being

INDEX